D1757088

Ε/ιc

PRINCIPLES AND PRACTICES
OF RICE PRODUCTION

PRINCIPLES AND PRACTICES
OF RICE PRODUCTION

Surajit K. De Datta

Head, Department of Agronomy
The International Rice Research Institute
Los Baños, The Philippines

A WILEY-INTERSCIENCE PUBLICATION

JOHN WILEY & SONS,   NEW YORK • CHICHESTER • BRISBANE • TORONTO

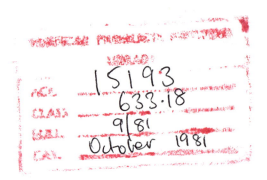
Library of Congress Cataloging in Publication Data:

De Datta, Surajit K   1933-
  Principles and practices of rice production.

  "A Wiley-Interscience publication."
  Includes index.
  1. Rice. I. Title.

SB191.R5D38   633.1'8   80-28941
ISBN 0-471-08074-8

Printed in the United States of America

10 9 8 7 6 5 4 3 2 1

To my late parents, Dinanath and Birahini,
with gratitude

and to my family,
Vijji and Raj, for their understanding
and inspiration

Foreword

The threat of widespread hunger, especially in Asia where a burgeoning population depends mainly on rice, looms as ominously now as it did at the start of the so-called green revolution in food grains in the 1960s. Growth in rice production resulting from the modern technology flowing to farmers from developing national and international research programs has been matched year by year and country by country by population growth. As we enter the 1980s, with two decades of rice research experience to build on, it is increasingly clear that the need for an accelerated effort to increase rice production is acute, real, and—in truth—critical to world stability.

The growth in rice production that has kept pace with the demands for food in the past 20 years has come largely from success in research and in application of that research to grow rice in farmers' fields. Most of that research has been done in relatively young national research programs. The current challenge is for those programs to grow—in size, in scope, and in ability to serve the pressing need for more rice.

I believe this book will help meet that challenge. Surajit K. De Datta's 17 years of agronomic research in rice and his service to the world's rice farmers make him well acquainted with the broad range of Asian rice farm conditions and with the research needs to improve those conditions. Dr. De Datta has been a collaborator and friend to rice scientists who now carry the responsibility in national rice research programs. He has been teacher and counselor to hundreds of them—in both research and extension—as they have been associated with training at the International Rice Research Institute.

Based on his extensive work with the problems of the rice crop in the areas where the crop is grown, on his close association with the world's key rice scientists, on his vast experience as a key IRRI scientist, and especially on his perception of the needs to assure increased rice production, Dr. De Datta was well qualified to draw from experience, and to collect and summarize the information in this book.

This book, as the author points out, should serve scientist and student alike. It provides a digest of information about rice research and production. It also provides the reader with references on where to seek added information. In the labor of creating this work, Dr. De Datta has established a permanent link with his colleagues of today and the rice scientists of the future and has provided a valuable addition to the world rice literature.

N. C. BRADY
Director-General
International Rice Research Institute

Preface

For the last two decades a significant portion of the world's agricultural research has focused on rice. That research resulted in the development of modern rice varieties and a modern rice technology that has greatly increased rice production, especially in Asia.

A great body of knowledge emanating from rice research has remained scattered. My purpose in writing this treatise, which I direct primarily toward students and agricultural scientists interested in rice production, was to assemble and summarize pertinent available information from rice research in the world's major rice-growing areas. The significant developments in recent years make a complete review almost impossible in a single book. However, drawing on my 17 years of rice research at the International Rice Research Institute (IRRI), and on the knowledge gained through close personal and professional working relationships with the world's leading rice scientists during that period, I believe the most significant developments are covered. Particular attention is given to inclusion of the latest available information, a factor I deem important in view of the tremendous mass of research done in the 1970s.

I intend this book for worldwide usage. The scope of the coverage by the 14 chapters is:

Chapter 1 establishes the importance of rice as an important food crop and discusses the yield and production of the crop in the rice-growing countries.

Chapters 2 and 3 cover the environments in which rice grows, specifically the types of landscape and soil, as well as climatic conditions.

Chapter 4 deals with the chemistry of submerged rice soils, a subject which no book on rice has dealt with in the last two decades.

Chapter 5 is devoted to the growth and development stages of the plant.

Chapter 6 deals with varietal development and seed production. Most of the significant developments in tropical and temperate rice varieties are summarized in this chapter. Also covered in this chapter are breeding methods followed for developing rice varieties in various rice-growing countries.

Chapters 7 to 10 are concerned with the systems of cultivation of rice and the management of land, water, soil fertility, and fertilizer in relation to growth, nutrition, and production of rice.

Chapters 11 and 12 cover the various agents, such as insects, diseases, weeds, rodents, and others, that have deleterious effects on the growth of rice, and methods for their control. Because this book is intended for worldwide usage it cannot deal in detail with various cultural practices and conditions encountered in different countries. Therefore, nothing in this book should be considered as a recommendation for the use of any particular pest control practice or chemical in any country. Local authorities should be consulted for information on local regulations and recommendations.

Chapter 13 is directed to the technology of harvesting and postharvesting treatments.

The final chapter considers the modern rice technology in relation to the world's food supply and to the biological and socioeconomic barriers to high yields that prevent farmers from getting rice yields as high as those obtained by rice researchers. I discuss also a number of problems that remain unresolved.

For all of the material covered in various chapters, principles are enumerated before practices are described. The list of references at the end of each chapter provides an added source of knowledge, not only for the research worker but for the advanced student as well.

Although I accept personal responsibility for the information presented, many authorities on different subjects covered in this book have reviewed the manuscript to make it technically correct. Most of these reviewers are my colleagues at IRRI, and the others are from various universities and institutions in the United States, the Netherlands, India, Sri Lanka, Thailand, and the Philippines.

I hope this book will serve as a textbook for courses on cereal crops, especially rice, and as a reference for agricultural scientists and development workers interested in the science of rice.

SURAJIT K. DE DATTA

Los Baños, Philippines
January 1981

Acknowledgments

In writing this book I have consulted many of my colleagues at IRRI and elsewhere. It is impossible for me to acknowledge all of them individually here, but my sincere appreciation is noted for all who greatly contributed to improvement of the manuscript. I acknowledge the help of the individuals at IRRI who reviewed the manuscript extensively: Keith Moody (Agronomist) and Benito S. Vergara (Plant Physiologist) for reviewing the entire manuscript, and Shoichi Yoshida (Plant Physiologist) and Te Tzu Chang (Geneticist) for reviewing several chapters.

I wish to thank Lina M. Vergara, IRRI Librarian, and her staff and also the staff members of the Office of Information Services for providing valuable assistance. Our Agronomy secretarial staff and others in the Department spent long hours typing and proofreading the manuscript and drawing the illustrations, for which I am most appreciative.

My sincere appreciation to Nyle C. Brady, IRRI Director General, for granting me the study leave to write major parts of the manuscript at the University of California, Davis.

Finally, I express my profound gratitude to Walter G. Rockwood, IRRI Editor, who did so much to make this book readable.

I appreciate permission from authors and publishers to reproduce figures and tables from the following sources: P. A. Sanchez. Puddling tropical rice soils: effects of water losses. *Soil Science* 115. Williams and Wilkins Co., © 1973 (Fig. 8.1); R. E. Shapiro. Effect of organic matter and flooding on availability of soil and synthetic phosphates. *Soil Science* 85 (Table 4.5). Williams and Wilkins Co., © 1958. The Williams and Wilkins Co., Baltimore; K. Kawaguchi and K. Kyuma. 1977. *Paddy soils in tropical Asia. Their material nature and fertility*. The Center of Southeast Asian studies, Kyoto (Fig. 3.2, Table 3.5); Reprinted from "Nitrogen Loss from Flooded Soils" by W. H. Patrick, Jr. and M. E. Tusneem (in *Ecology* 53:735–737) by permission of the Ecological Society of America. Copyright © 1972 by the Ecological Society of America (Fig. 4.6); G. S. Khush and W. R. Coffman. 1977. Genetic Evaluation and Utilization (GEU) The Rice Improvement Program of the International Rice Research Institute.

Theoretical and Applied Genetics 51. Springer-Verlag New York Inc. (Fig. 6.7); From *Tropical Meteorology* by H. Rhiel. Copyright © 1954. McGraw-Hill International Book Co. Used with the permission of McGraw-Hill Book Company, New York (Fig. 2.9); Reprinted by permission from *Rice in Asia,* pp. 325, 389 (Fig. 11.9, Table 11.7). University of Tokyo Press, 1975. M. Matsubayashi et al. (ed. comte.). 1963. *Theory and practice of growing rice.* Fuji Publishing Co. Ltd., Tokyo (Fig. 9.10); Reprinted from *Rice postproduction technology in the tropics* by Merle Esmay, Soemangat, Eriyatno, and Allan Philipps; an East-West Center Book published by the University Press of Hawaii. Copyright © 1979 by The East-West Center (Fig. 13.10, Table 13.2); W. H. Patrick, Jr. and I. C. Mahapatra. 1968. Transformation and availability to rice of nitrogen and phosphorus in waterlogged soils. *Advances in Agronomy* 20. Academic Press, Inc., New York (Fig. 4.14); N. V. Nguu and S. K. De Datta. 1979. (Fig. 10.14, Table 7.2). L. T. Evans and S. K. De Datta. 1979. (Fig. 2.20). *Field Crops Research* 2. Elsevier Scientific Publishing Co., Amsterdam; D. S. Mikkelsen and S. Kuo. 1977. *Zinc fertilization and behavior in flooded soils.* Special Publication 5. Commonwealth Bureau of Soils. Commonwealth Agricultural Bureau, England (Figs. 4.18, 4.19). Reprinted with permission from *Soil Biology and Biochemistry* 7, K. R. Reddy and W. H. Patrick, Jr., Effect of alternate aerobic and anaerobic conditions on redox potential, organic matter decomposition and nitrogen loss in a flooded soil, copyright © 1975, Pergamon Press, Ltd. (Table 4.2); P. A. Sanchez. 1976. *Properties and management of soils in the tropics,* John Wiley & Sons, Inc., New York (Table 3.3); K. Maramorosch and K. F. Harris (eds.). 1979. *Leafhopper vectors and plant disease agents.* Academic Press, Inc., New York (Table 11.3); S. K. De Datta, F. R. Bolton and W. L. Lin. 1979. Prospects for using minimum and zero tillage in tropical lowland rice. *Weed Research* 19. By permission of Blackwell Scientific Publications (Table 8.7).

S.K.D.

Contents

PRINCIPLES AND PRACTICES
OF RICE PRODUCTION

Rice in Perspective

In a Balinese legend, the lord *Vishnu,* male god of fertility and water, came to earth to provide better food for the people who had only sugarcane juice as food. *Vishnu* made Mother Earth give birth to rice and then fought *Indra,* lord of the heavens, to force him to teach men to grow rice. Thus rice, as a source of life and wealth and as a gift from the gods, was born from a union of the divine creative forces represented in earth and water. Rice, therefore, was treated with reverence and respect and its culture developed into an elaborate ritual. Even today the Balinese are considered efficient rice-growers in the Indonesian archipelago.

The importance of rice as a daily food is expressed differently in different countries. For example, some people in Indonesia believe that the rice grain has a soul like a human being. In Sri Lanka, astrologers are often consulted and prayers offered before rice is planted. In southern China and parts of India, people greet each other by saying, "Have you eaten your rice?"

In Japan, rice was considered second only to the emperor in sacredness. In fact, the emperor, in annual solemn ceremonies, planted a few grains of rice in the palace grounds (Hammond 1961). Money can be squandered by the prodigal and he will be forgiven, but there is no forgiveness for a person who willfully throws away a handful of rice. This veneration for rice among the Japanese people has its roots in mythology and the dawn of history, and it is not unlike the significance attached to the food gods by all ancient people. The difference in Japan, however, is that rice was in ancient days considered to be the food of the gods and the samurais (Rabbitt 1940).

Most of these beliefs and practices have changed over the years but rice has remained truly "life itself" for most of the world's densely populated regions.

The ancient home of rice is monsoonal Asia. And it remains the area of the world where rice is practically the whole of the people's diet, nearly all of their agriculture, and much of their hopes. Rice is clearly the most important food crop of the world if one considers the area under rice cultivation and the number of people depending on the crop.

In 1976–1978 rice occupied about 143.5 million hectares—more than 90% of which was in Asia. India has the world's largest rice-growing area, with 39.6 million hectares, followed by the People's Republic of China, with 36.0 million hectares (Fig. 1.1). Globally, rice ranks second to wheat in terms of area

1

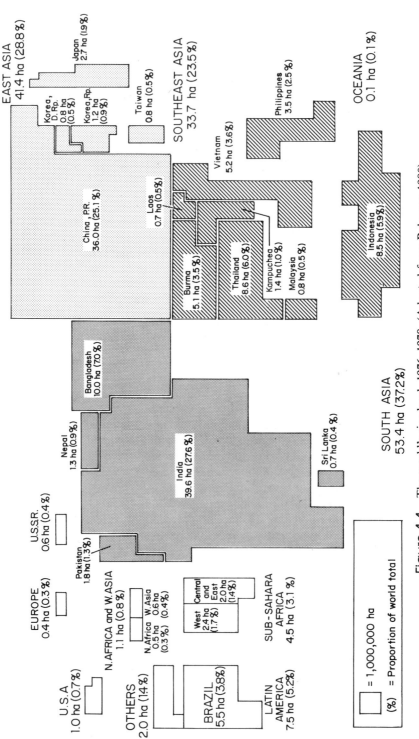

Figure 1.1 The world's rice land, 1976-1978. (Adapted from Palacpac 1980)

harvested; but in terms of importance as a food crop, rice provides more calories per hectare than any other cereal crop. For example, at average world yields, a hectare of rice could sustain 5.7 persons for a year compared to 5.3 for maize and 4.1 for wheat. The total caloric output of all world food is equal to 3119 kcal/person per day at the farm gate, with rice accounting for 552 kcal/person per day, or 18% of the total.

It is estimated that 40% of the world's population use rice as a major source of calories. For 1.3 billion people rice provides more than half of their food; for another 400 million people rice provides from 25 to 50% of their total food. Table 1.1 shows estimates of persons in some Asian countries whose major food is rice. Note that 90% of the population in Bangladesh, Burma, Sri Lanka, Vietnam, and Kampuchea depend on rice for their major food intake.

East Asian countries such as Japan, Republic of Korea, and Taiwan have higher apparent per capita caloric consumption than countries in South and Southeast Asia (Fig. 1.2). With the exception of India, Pakistan, and Sri Lanka, in most Asian countries annual rice consumption per capita exceeds 100 kg of milled rice. Outside Asia, only Liberia, Malagasy, Mauritius, Sierra Leone, Guyana, and Surinam have levels of rice consumption comparable with those of Asia.

Besides its importance as food, rice provides employment to the largest sector of the rural population in most of Asia.

Table 1.1 Estimate of Persons in Selected Asian Nations Whose Major Food is Rice (R. E. Huke, International Rice Research Institute, personal communication, 1980)

Country	Population (× 1,000,000)	% Who Are Rice Eaters	No. of Rice Eaters (× 1,000,000)
People's Republic of China	956	63	602
India	660	65	429
Indonesia	147	80	118
Japan	116	70	81
Bangladesh	90	90	81
Pakistan	80	30	24
Vietnam	50	90	45
Philippines	49	75	37
Thailand	48	80	38
Republic of Korea	38	75	29
Burma	35	90	32
Taiwan	17	70	12
Sri Lanka	15	90	14
Nepal	15	60	9
Kampuchea	9	90	8
Total			1559

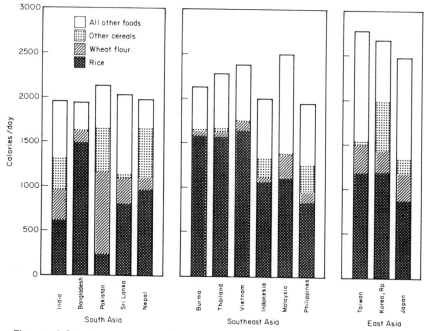

Figure 1.2   Apparent per capita daily calorie consumption, 1971–1975. (Adapted from Palacpac 1980)

RICE-GROWING COUNTRIES

There are 111 rice-growing countries in the world. They include all Asian countries, most countries of West and North Africa, some countries in Central and East Africa, most of the South and Central American countries, Australia, and at least four states in the United States. Figure 1.1 shows the estimated area of rice lands for major rice-growing countries for the years 1976–1978. Although the bulk of rice production is centered in wet tropical climates, the crop flourishes in humid regions of the subtropics and in temperate climates such as Japan, Korea, China, Spain, Portugal, Italy, France, Romania, Czechoslovakia, USSR, and the United States.

Japan and Spain have historically produced the highest average rice yield per hectare (6.0 t/ha). In 1977, however, the Republic of Korea took the lead with an average 6.8 t/ha.

Among the 111 rice-producing countries, 3 countries produce an average of 6 t/ha or more, 17 countries produce 4 t/ha or more, 78 countries produce 3 t/ha or less, of which 57 produce 2 t/ha or less, and 13 produce less than 1 t/ha.

Figure 1.3 shows the average yield per hectare by the important rice-growing countries of the world. When rice yield is compared with average yields for all cereal crops, it ranks second in terms of national average and world average yields by crop (Table 1.2).

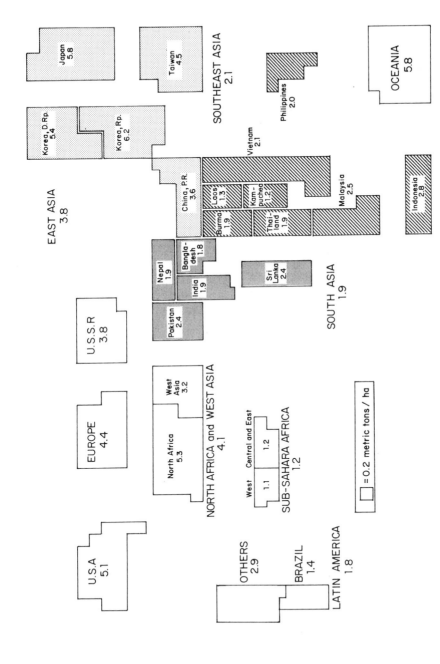

Figure 1.3 The world's rice yields, 1976–1978. (Adapted from Palacpac 1980)

Table 1.2   National and World Average Yields of Cereals (Average 1974–1976) (Adapted from IADS 1979)

Cereals	Number of Producing Countries	Yield (t/ha)		
		Highest National Average	World Average	Lowest National Average
Wheat	92	5.4	1.6	0.3
Rice	111	6.8	2.4	0.4
Maize	131	8.0	2.8	0.4
Sorghum	72	4.3	1.2	0.4
Millet	65	3.9	0.7	0.3
Barley	74	4.6	1.9	0.3
Rye	39	4.2	1.7	0.2
Oat	51	4.6	1.6	0.2

The importance of the comparison between the highest national yields and record yields, usually achieved under experimental conditions, is to determine the potential productivity levels. The world average yield for each crop, including rice, is generally one-third, or less, of that achieved by the country with the highest national average (IADS 1979). However, whereas most of the rice crop is used for human food, a substantial portion of the maize crop is used as livestock feed.

South and Southeast Asia, which have the largest rice-growing area and one of the highest concentrations of people, produce only 2 t/ha or less. Although most of the rice in South and Southeast Asia is grown as a lowland crop, yield per hectare is not much higher than in Latin America where rice is mostly grown as an upland crop. Both regions grow primarily rainfed rice. However, South and Southeast Asia have considerably higher hectarage compared with Latin America. A monsoonal climate causes more variability in water control in South and Southeast Asia than in Latin America, and although drought is a common problem in both regions, flooding is not a problem in Latin America as in South and Southeast Asia.

Total annual world rice production averaged for 1976–1978 was 363,940,000 tons. About 330 million tons were produced in Asia. Within Asia, China and India produce more than half of the world's crop. Figure 1.4 shows the production of rice by country and percentage production in a given country in relation to total rice production.

RICE SCIENCE AND TECHNOLOGY

Rice is the only major food crop that can be grown under various degrees of flooding. It is primarily grown on the vast areas of flat, low-lying river basins and

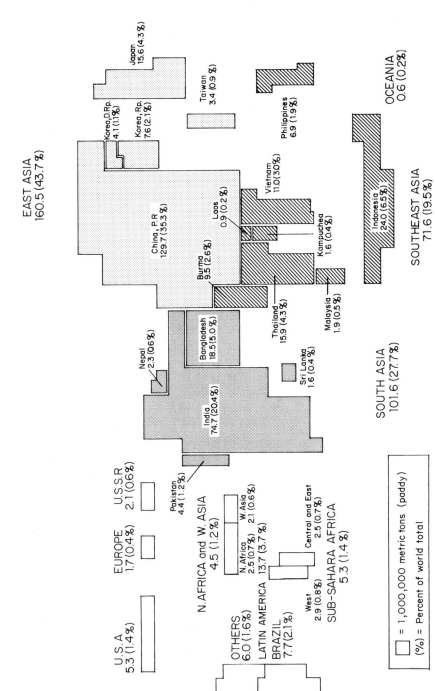

Figure 1.4   The world's annual rice production, 1976–1978. (Adapted from Palacpac 1980)

delta areas in Asia that are flooded to various depths during the monsoonal season. If it were not for rice, the unprecedented population growth in the vast wetlands in Asia would have never taken place.

In the succeeding chapters, the principles and practices of rice production that relate to today's needs are discussed. But rice science is highly complex and dynamic. To put it in perspective I quote Bradfield (1964), "The nations that are most advanced agriculturally are, in general, those that have made a substantial investment in science and education during the past century." Understanding the science and technology of rice will greatly facilitate efforts on research and production of rice in the coming decades.

REFERENCES

Bradfield, R. 1964. The role of educated people in agricultural development. Pages 95–114 *in* American Association for the Advancement of Science. *Agricultural sciences for the developing nations*. Publ. 76. Washington, D.C.

Hammond, W. 1961. *Rice, food for a hungry world.* (Student ed.). Fawcett Publications, Inc., New York. 143 pp.

IADS (International Agricultural Development Service). 1979. *Report for 1978.* 95 pp.

Palacpac, A. C. 1980. World rice statistics. Department of Agricultural Economics, International Rice Research Institute, Los Baños, Philippines. 130 pp. (unpubl. mimeo.)

Rabbitt, J. A. 1940. Rice in the cultural life of the Japanese people. Pages 189–257 + 12 pl. *in Transactions of the Asiatic Society of Japan.* Second ser., Vol. 19.

2

The Climatic Environment and its Effects on Rice Production

For a farmer to get maximum input efficiency and optimum grain yields from a crop he must have a thorough knowledge of the crop's environment. That environment is affected by two components:

- Nature, which includes weather, soil, and biotic surroundings.
- The farmer, who attempts to maximize his grain yields by the use of cultural practices such as plowing and harrowing and inputs such as fertilizers, insecticides, and herbicides.

The soil and biotic environment is discussed in Chapter 3, and the influence of man on the environment is covered in subsequent chapters.

To delineate the effects of climate on rice production, understanding of both weather and climate is essential. Weather, which depends upon the heating and cooling of the earth's atmosphere, is a condition of the atmosphere at a given moment; climate is the condition of atmosphere over a period of time. Climatic differences must be carefully considered when comparing the performance of a rice crop or a rice variety grown at different sites.

RICE-GROWING REGIONS

The extreme latitudes in which rice is grown are in temperate regions. A report by Moomaw and Vergara (1965) suggested that rice cultivation is limited to as far north as 49° in Czechoslovakia and as far south as 35° in New South Wales, Australia. There is evidence that rice once grew at 53°N in Moho, northeastern China. In fact, rice is still grown in China's Aihwei county at about 50°N, which is considered the northernmost rice-growing area in the world (IRRI 1978). However, most of the world's rice is grown in the tropics and the critical determining factor for growing rice appears to be temperature.

9

The tropical region comprises the area between the Tropic of Cancer (23°27'N latitude) and the Tropic of Capricorn (23°27'S latitude). The tropics include most of the rice-growing regions in India and all of the other South and Southeast Asian countries—Bangladesh, Burma, Kampuchea, Indonesia, Laos, Malaysia, Pakistan, Philippines, and Thailand. It also includes all of the West African rice-growing areas and most of the rice-growing areas of Central and South America.

Although rice is primarily a tropical and subtropical crop, the best grain yields are obtained in temperate regions such as the Po Valley, Italy (45°45'N); northern Honshu, Japan (38°N); Korea (37°N); and New South Wales, Australia (35°S) (Moomaw and Vergara 1965).

CLIMATIC ENVIRONMENT

The effect of climate on the environment for the rice crop is major. Basic to understanding climate is knowledge of its elements, especially rainfall, solar radiation, temperature, and relative humidity.

Rainfall

The evolution of rice as a food crop was influenced primarily by amount and distribution of rainfall. Huke (1976), for example, classified the rice-growing world into four distinct climatic environments (Fig. 2.1) based primarily on rainfall.

1 A home area of rice where climate is not optimum but allows one crop at low risk.
2 A second area where the mean annual rainfall totals are marginal for rice. The onset, amount, and distribution of rainfall are highly variable. Here, an excellent rice crop can be produced wherever water is provided by irrigation.
3 A third area, which represents widely scattered rice-growing areas, where water supply must be entirely from irrigation for stable rice production.
4 Areas with no important production of rice.

For upland rice production, the influence of amount and variability of rainfall is far different than for lowland rice. For example, upland rice is grown in areas of heavy rainfall in Assam and West Bengal in India and in Bangladesh but it is grown in areas with low rainfall in Madhya Pradesh in India. The growing season is extremely short and rainfall is highly variable in eastern Uttar Pradesh in India. In upland areas in Burma, rainfall from May to November can be as low as 500 mm or as high as 2000 mm (De Datta and Vergara 1975).

In West Africa, where rice is grown mostly as an upland crop, the amount and the distribution of rainfall are of paramount importance. The rain may begin

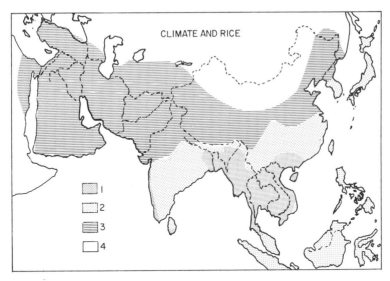

Figure 2.1   Climate and rice. (1) Home area of rice where the crop may be raised year after year with no climate modification by man. (2) Important rice-producing areas where at least one parameter of climate is frequently less than the ideal for successful crops. (3) Areas where rice production is widely scattered and where climate must be modified to produce a crop. (4) Areas with no important production of rice. (From Huke 1976)

anytime from March to July (it begins later at higher latitudes). Rainfall is unimodal (having one peak) in areas with a short rainy season, but it is bimodal (having two peaks), with a 1- to 2-month break from July to August, in areas with long rainy seasons. The region of bimodal rainfall includes southeastern Ivory Coast, southern Ghana, southern Togo and Benin, and southern Nigeria. Figure 2.2 shows the bimodal pattern for southern Nigeria. Less important areas of bimodal distribution include southeastern Guinea and northeastern Liberia. In West African areas of less than 1000 mm of annual rainfall, the rainy season is from June to October. In areas of more than 1800 mm of annual rainfall, the season may begin as early as late March (Food and Agriculture Organization Inventory Mission 1970).

For Latin America, Brown (1969) reported that 1000 mm of annual rainfall, with 200 mm of monthly rainfall during the growing season, is adequate for growing upland rice.

Brazil, which has by far the largest upland rice-growing area in Latin America, has a distinct rainy season that begins in October and ends in April (Fig. 2.2). Annual rainfall varies from 1300 to 1800 mm and 70–80% of the rain falls during the upland rice-growing season. The rainfall diminishes in February, leading to an ideal harvesting period (De Datta and Vergara 1975).

Rainfall in most of Peru's Amazon Basin ranges from 2000 to 4000 mm annually. In many Central American countries more than 2000 mm of annual

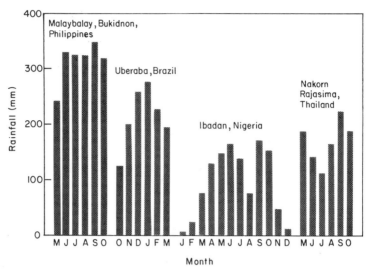

Figure 2.2   Rainfall patterns of selected upland rice regions. (From De Datta and Vergara 1975)

rainfall is fairly common. This amount of rainfall is more than enough to grow one upland rice crop.

In most rice-growing areas the year is divided into fairly distinct wet and dry seasons and most of the rice is grown during the wet season. The amount of rainfall received during the dry season is usually insufficient to grow a crop of rice without irrigation. Because of lack of irrigation facilities, the bulk of the rice grown in the tropics in the immediate future will depend on monsoonal rains. Rainfall data indicate the need to develop rice varieties with wide ranges of maturity and cultural practices that conform with rainfall distribution patterns.

Solar Radiation

Solar radiation is radiant energy from the sun, measured as a total amount (direct beam solar radiation plus sky radiation) expressed in calories per cm^2 per hour (cal/cm^2 per hour). Classification of solar radiation according to wave length is shown in Table 2.1. Only the "visible" part (380–720 nm) of the total solar energy is important for photosynthesis.

Three instruments are used for measuring solar radiation:

1 PYRHELIOMETERS   The pyrheliometer is designed to measure the direct beam solar radiation at normal incidence, usually called solar intensity. Pyrheliometers are the most accurate of all radiation measurement instruments and are commonly used as calibration standards. They are expensive and usually

Table 2.1 Classification of Solar Radiation According to Wavelength[a]

Micron (μ)	Angstrom (Å)	Nanometer (nm)	
$< 10^{-3}$	< 10	< 1	X-rays and γ-rays
10^{-3}–0.2	10–2,000	1–200	Far ultraviolet rays (UV)
0.2–0.315	2,000–3,150	200–315	Middle ultraviolet (UV)
0.315–0.38	3,150–3,800	315–380	Near ultraviolet (UV)
0.38–0.72	3,800–7,200	380–720	Visible
0.72–1.5	7,200–15,000	720–1,500	Near infrared (IR)
1.5–5.6	15,000–56,000	1,500–5,600	Middle infrared (IR)
5.6–10^3	56,000–10^7	5,600–10^6	Far infrared (IR)
$> 10^3$	$> 10^7$	$> 10^6$	Micro- and radiowaves

[a]1 cm = 10^8 Å = $10^4 \mu$ = 10^7 nm

found only at special research calibration and observing stations (De Datta 1970).

2 ACTINOGRAPH OR CASELLA PYRANOMETER   Like the pyrheliometer, an actinograph measures the total solar radiation.

3 NET RADIOMETER   The net radiometer measures total solar radiation plus infrared radiation from the atmosphere minus the outgoing reflected and infrared radiation from the crop and the underlying surface. The net radiometer consists of two freely exposed blackened elements placed back to back—one to face upward, the other to face downward—with a thin layer of insulation material between the two elements. The elements are assumed to absorb all radiation incident upon them.

During the ripening period of the crop in the monsoonal tropics, the intensity of solar radiation during an average day is about 350 cal/cm^2 per day, which is similar to the values reported during the rice-growing season in temperate Asia such as in Japan (Munakata et al. 1967) and Korea (IRRI 1972). Fukui (1971) reported that in the temperate Asian countries, solar radiation during the rice-growing season is nearly the same as that for the rainy season in humid tropical regions, that is, 400 cal/cm^2 per day. Because the day is longer during the main cropping season in Sapporo (43°N) or Konosu (36°N) in Japan, or Suweon (37°N) in Korea (IRRI 1972), than in Los Baños in the Philippines (14°N), the intensity of solar energy per hour is higher in Los Baños (De Datta 1973). These solar radiation data suggest that if rice is grown under irrigation during the dry season, the amount of solar radiation available per unit of time is greater in the tropics than in Asian temperate rice-growing regions.

In the rice-growing areas in the Mediterranean countries, the United States, and southern Australia, average solar energy for the rice-growing season is about 100 cal/cm^2 per day greater than in either Asian temperate or monsoon tropical countries. Solar energy during the ripening period of rice in southern Australia is

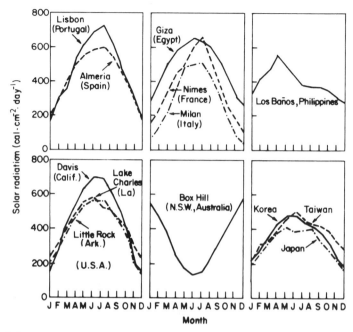

Figure 2.3   Average daily solar radiation values by month in some rice-growing areas in temperate countries and in Los Baños, Philippines, in the tropics. (Adapted from Fukui 1971, De Datta and Malabuyoc 1976)

100 cal/cm^2 per day more than in Asian temperate or monsoon tropical conditions (Fukui 1971). At least 700 cal/cm^2 per day is recorded in some rice-growing regions in Portugal and the United States. Some examples of range of solar radiation values recorded in rice-growing countries are cited in Fig. 2.3.

Day Length

The natural day length, or photoperiod, which affects growth of the rice plants, consists of the length of the period of daylight and the duration of the civil twilight. Day length is the interval between sunrise and sunset. Civil twilight is the interval between sunrise or sunset and the time when the position of the center of the sun is 6° below the horizon. At that time stars and planets of the first magnitude are just visible and darkness forces the suspension of normal outdoor activities. Data on day length and twilight can be obtained from standard meteorological tables or books. The day length (including civil twilight) pattern during the main rice-cropping season varies most in high latitudes such as in Sapporo, Japan, and least near the equator such as in Bogor, Indonesia, or Bukit Merah, Malaysia (Fig. 2.4). Similarly, the day length patterns of selected upland rice-growing areas in some countries show that greater changes take place in Uberaba, Brazil, than in Ibadan, Nigeria (Fig. 2.5).

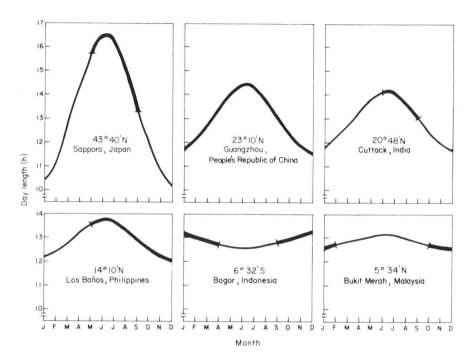

Figure 2.4 Day length (including civil twilight) patterns during the main crop season in various rice-growing countries. Heavy lines indicate the main cropping season for lowland rice.

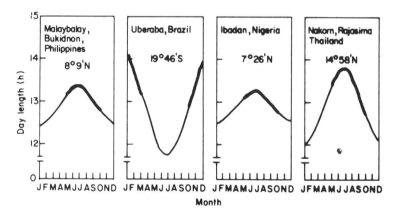

Figure 2.5 Day length of selected upland rice areas. Heavy lines indicate the cropping season. (From De Datta and Vergara 1975)

15

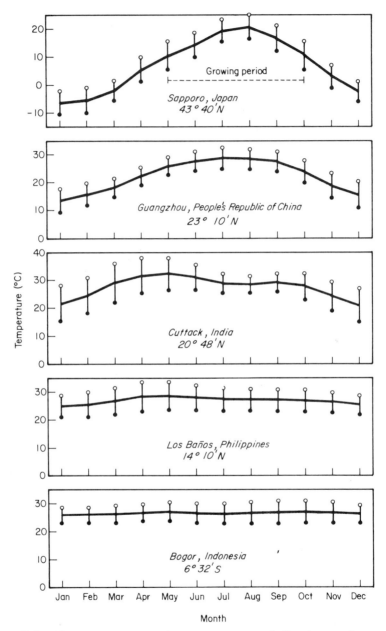

Figure 2.6 Mean temperatures and temperature ranges in five rice-growing countries in Asia.

16

Temperature

Temperature is a relative term, which indicates capacity to transfer heat by conduction. Figure 2.6 shows the temperature range during the rice-growing months at five sites in Asia, where primarily lowland rice is grown. Examples of temperature ranges in upland rice-growing areas in selected countries are in Fig. 2.7. Temperature regimes are discussed later in this chapter.

Relative Humidity

Relative humidity refers to water vapor, exclusive of condensed water, in the atmosphere. It is the ratio, expressed as a percentage, of vapor pressure (e) to saturation vapor pressure (em) at the existing temperature. Relative humidity is measured by a hygrothermograph (hair hygrometer) or a psychrometer (wet bulb and dry bulb).

Wind

Wind refers to air in motion and is best expressed in kilometers per hour. Other usage includes miles per hour and knots per hour. An anemometer is used to measure wind speed.

Cyclone is a general term for the storm created as polar easterlies (cold air mass) and prevailing westerlies (warm air mass) move in opposite directions parallel to each other. The drag along the interface between the two, plus the

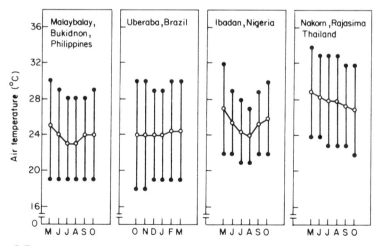

Figure 2.7   Mean temperature and temperature ranges during the cropping seasons of selected upland rice areas. (From De Datta and Vergara 1975)

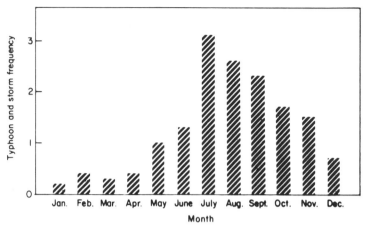

Figure 2.8   Mean monthly typhoon and storm frequencies within 250 km of Los Baños, Laguna (1959–1972). (From De Datta 1973)

differences in air density, sets up a wavelike motion. The cold front emerges into the warm front and low atmospheric pressure develops, thus creating a "low" or "cyclonic depression" or cyclone. Tropical cyclones are also called hurricanes or typhoons. The mean monthly typhoon frequency within 250 km of Los Baños, Laguna, Philippines, is seen in Fig. 2.8.

CLIMATIC EFFECTS ON RICE PRODUCTION

In many rice-growing areas, the year is divided into fairly distinct wet and dry seasons. In most areas, most of the rice produced comes from wet-season harvests. In rainfed areas, the rice-cropping season is determined by the rainfall pattern. In most of the temperate rice-growing countries in Asia (Japan, Korea, and China) and in other regions such as North America, Australia, and Europe, rice cropping is determined primarily by the temperature pattern; rice is almost entirely an irrigated crop. With irrigation, planting can be adjusted to take advantage of favorable climatic conditions such as optimum temperature and high solar radiation.

Effects of Variability of Rainfall

Variability in the amount and distribution of rainfall is the most important factor limiting yields of rainfed rice, which constitutes about 80% of the rice grown in South and Southeast Asia. For the same amount of rainfall, the coefficient of variability of the rainfall is higher in the tropics than in the temperate areas. In low-rainfall areas, variability is high regardless of latitude (De Datta 1970).

Most of the tropical Southeast Asian countries, such as major parts of Burma, Kampuchea, Indonesia, Philippines, Thailand, and Vietnam, receive about 2000 mm of rainfall annually. This should be adequate for one rice crop provided rainfall distribution is reasonably uniform. Even in areas where the annual rainfall is 1200–1500 mm, if rainfall is concentrated in the monsoonal season (as is usual), it is adequate for a single rice crop.

Unfortunately, the world's two largest rice-growing countries, India and China, have many areas that receive less than 1200–1500 mm of rainfall. India, with the largest rice-growing area in the world, often has inadequate or excessive rainfall during the rainy season. As a result, drought or flood, and sometimes both, cause substantial damage to rainfed rice production. China now grows rice almost entirely under irrigation. Pakistan also grows rice primarily with irrigation.

Riehl (1954) provides a global view of annual rainfall variability (Fig. 2.9). In many regions, the total rainfall is apparently adequate but strong tendencies to depart from the norm add to the risk of rainfed agriculture.

The tropics can be divided into two climatic types (Money 1972) according to rainfall distribution. One is the tropical rainy or perennially wet climate areas: west coast of Indian subcontinent, coastal Bangladesh, Burma, Sumatra and Kalimantan in Indonesia, Philippines, equatorial West Africa, east coast of Central America, highlands of Peru, northwestern and northern Brazil, central Venezuela, Surinam, and so forth. Those areas generally receive adequate rainfall and are not considered drought-prone.

The other type consists of tropical wet-dry climates, which are characterized by a distinct dry season: major inland rice-growing areas of India, Burma, eastern Indonesia, and some other Southeast Asian countries; also, West Africa, west coast of Central America, inland areas of Colombia and Venezuela, central Brazil, Bolivia, and Paraguay. In those areas, the total rainfall is generally adequate but variability in its distribution causes drought damage to rainfed crops, including rice.

Rainfed Lowland Rice

Variability in the onset of the monsoon season is a factor that determines the beginning of planting season for transplanted rainfed rice. This is because of the water required for land preparation and soil puddling (see Chapter 8).

Because of the uncertainty in the amount and distribution of rainfall, millions of rice farmers hold as much water as they can on the field by bunds. The resulting water depth induces a high percentage of seedling mortality and at times causes lodging at later stages of crop growth.

Variability of rainfall affects the rice crop at different times. If the variability is associated with the onset of the rain, stand establishment and the growth duration of rice are affected. If variability is associated with an untimely cessation at the reproductive or ripening stage of the rice crop, yield reduction is severe.

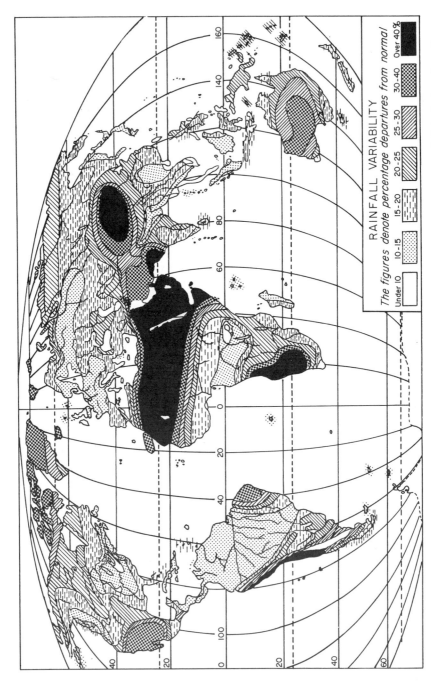

Figure 2.9 Annual rainfall variability of the globe. Percent relative variability denotes the ratio of the sum of all deviations from the mean, to the mean. (From Riehl 1954)

Table 2.2   Maximum Rainfall Intensity in 24 Hours
Reported from Several Tropical Stations (After Watts
1955, Moomaw and Vergara 1965)

Station	Rainfall (mm)
Baguio, Philippines	1168
Cherrapunji, India	1036
Funkiko, Taiwan	1034
Honomu, Hawaii	810
Los Baños, Philippines	305

Table 2.2 shows some of the record high-intensity rainfall occurrences at tropical weather stations. When 30–50% of the annual rainfall occurs in a 24-hour period, control and use of this water is usually impossible.

Upland Rice

Rainfall variability is more critical for upland rice than for lowland rice. Moisture stress can damage, or even kill, plants in an area that receives as much as 200 mm of precipitation in a day and then receives no rainfall for the next 20 days. An evenly distributed precipitation of 100 mm/month is preferable to 200 mm/month that falls in 2 or 3 days.

In examining the rainfall patterns of some upland rice-growing areas (Fig. 2.2), using 200 mm/month as the base line, it appears that varieties that mature in less than 100 days would be desirable for the unimodal pattern in Uberaba, Brazil. In Bukidnon, Philippines, the rainfall pattern is also unimodal but duration is longer and varieties that mature in 100–150 days should perform well. Alternatively, a short-duration upland rice at Bukidnon can be followed by another short-duration upland crop, particularly cash crops like onion, garlic, or even mung bean.

The rainfall pattern in Ibadan, Nigeria, is bimodal and monthly rainfall is lower than the minimum requirement suggested by Brown (1969). If, as Brown reported, upland rice cannot be grown with monthly rainfall less than 200 mm, then many areas of the world, including West Africa and Northeastern Thailand, are not suited for upland rice production.

In Brazil, droughts of 5–20 days duration occur in the Cerrado region, with a high frequency of drought periods (Table 2.3) during the period locally known as *veranico* (Fig. 2.10). Upland rice, which occupies 3.7 million hectares in the Cerrado, is adversely affected by the *veranico*. It often causes severe damage to the rice crop at the reproductive and ripening stages (Moraes 1978).

Table 2.3 Frequency of Short Drought Periods
(*Veranicos*) in Brazil, Based on 42 Years of Records
(From Moraes 1978)

Period Without Rainfall (days)	Frequency
8	3 per year
10	2 per year
13	1 per year
18	2 years in 7
22	1 year in 7

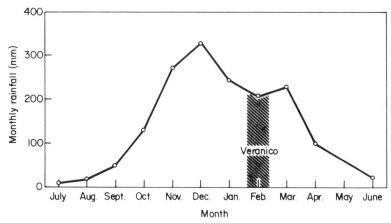

Figure 2.10 Rainfall distribution and identification of the chronically drought-prone period (*veranico*) in the Cerrado of central Brazil. (From Moraes 1978)

Effects of Solar Radiation

The importance of solar energy in tropical agriculture was recognized only after World War II (Best 1962). Calculated by De Wit's (1958) method, the average daily solar radiation available during the monsoon season in the tropics is one-and-a-half times lower than that available in the temperate rice-growing regions such as the Po Valley, Italy; Suicca, Spain; New South Wales, Australia; or Davis, California. But, because of his dependence on rainfall, the farmer of rainfed rice in the tropics must grow his crops when there is low sunlight intensity (Fig. 2.11). On the other hand, where irrigation water is available, rice can be grown in the dry season and the grain yield will be higher than in the wet season because of the higher intensity of solar radiation.

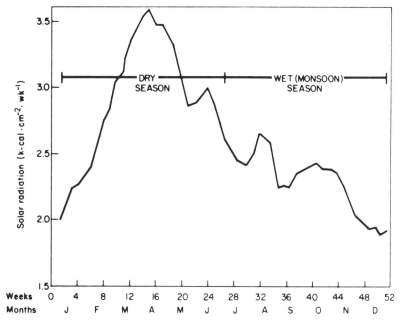

Figure 2.11 In the monsoonal tropics, the intensity of solar energy is considerably higher during the dry season compared with the wet season. Average (1959–1966) annual solar radiation curve (3-point moving average) for Los Baños, Philippines.

Solar Energy and Grain Yield

Based on experiments in Texas, Stansel et al. (1965) and Stansel (1975) suggest that the rice plant's most critical period of solar energy requirement is from panicle initiation until about 10 days before maturity (Fig. 2.12).

In the tropics, the correlation between solar radiation for the 45-day period before harvest (from panicle initiation to crop maturity) and grain yield, plotted by harvest month (Fig. 2.13), was highly significant (De Datta and Zarate 1970). Earlier experiments indicated a strong correlation between grain yield and solar radiation during the last 30 days of crop growth (Moomaw et al. 1967). Subsequent IRRI research indicated that the increase in dry matter between panicle initiation and harvest was highly correlated with grain yield (De Datta et al. 1968).

These results indicate that the amount of solar energy received from as early as panicle initiation until crop maturation is important for the accumulation of dry matter during that period. This may be explained from the results obtained by Murata (1966), which showed that the accumulation of starch in the leaves and culms begins about 10 days before heading. Starch accumulates markedly in the grain during the 30-day period following heading (Murata 1966, Yoshida and

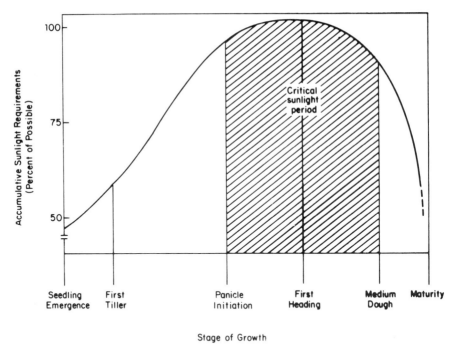

Figure 2.12 Solar energy requirements of rice at different stages of growth and development. (Adapted from Stansel 1975)

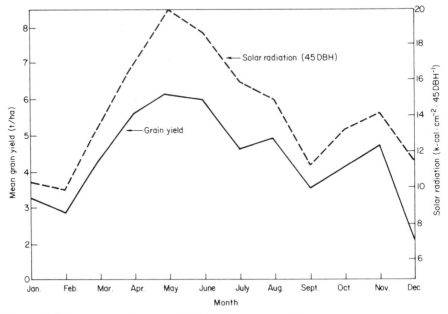

Figure 2.13 Mean grain yield of IR8 for three levels of nitrogen (0, 30, and 90 kg/ha) and three spacings (15 × 15, 25 × 25, and 35 × 35 cm) plotted against solar radiation totals during 45 days before harvest (DBH). (Adapted from De Datta and Zarate 1970)

Ahn 1968) and the total period of 40 days before maturity may be considered as the period of grain production.

With irrigation, the dry-season rice yields in the tropics (11 t/ha reported at IRRI) should be similar to those reported by Best (1962) for the temperate region (12.5 t/ha). However, the grain yield obtained during the wet season is lower than those in the dry season because of the lower level of solar radiation received during the crops' grain-filling and ripening stages (De Datta and Zarate 1970).

Effects of Day Length

Rice is generally a short-day plant and sensitive to photoperiod. Thus, long days can prevent or considerably delay flowering (Vergara and Chang 1976). In lowland rainfed rice culture, delay in transplanting photoperiod-sensitive varieties because of delayed rains does not usually affect the grain yields. It is because of that flexibility at the seedling stage that photoperiod-sensitive varieties have traditionally been grown in monsoonal Asia (Vergara 1976). The photoperiod-sensitive traditional varieties have provided stability in rice production even though their yield levels have been low. Those varieties produced some yields regardless of lodging, typhoon damage, or inadequate management practices such as no fertilizer or no weeding (Vergara et al. 1966). In some areas, the need to delay harvesting until monsoon-season floodwater has receded makes it essential to grow varieties with long growth duration. This is possible only by using photoperiod-sensitive varieties.

But day length-, or photoperiod-insensitive rice varieties enable the farmer in the tropics and subtropics to plant rice at any time of the year without great changes in growth duration (Chang and Vergara 1972). In irrigated areas, and in rainfed areas where flooding is limited to a maximum water depth of 15–20 cm, improved photoperiod-insensitive varieties have partially replaced photoperiod-sensitive varieties. Using these short-duration varieties, such as IR8, rice can be planted in any month in the tropics (Fig. 2.14) and will mature in a fixed number of days.

Thus, it is obvious that insensitivity to day length is essential in one situation and a liability in another.

Effects of Temperature

Temperature regime greatly influences not only the growth duration but also the growth pattern of the rice plant. During the growing season, the mean temperature, and the temperature sum, range, distribution pattern, and diurnal changes, or a combination of these, may be highly correlated with grain yields (Moomaw and Vergara 1965). Critical temperatures for germination, tillering, inflorescence initiation and development, dehiscence, and ripening of rice have been identified (Table 2.4).

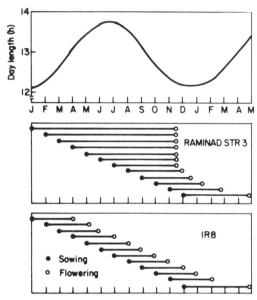

Figure 2.14 Date of sowing and flowering of Raminad Str. 3, a traditional photoperiod-sensitive variety, and IR8, an improved rice variety; sown at Los Baños, Philippines (14° N). (From Vergara 1976)

Table 2.4 Response of the Rice Plant to Varying Temperature at Different Growth Stages (Adapted from Yoshida 1978)

Growth Stage	Critical Temperature (° C)		
	Low	High	Optimum
Germination	16–19	45	18–40
Seedling emergence and establishment	12–35	35	25–30
Rooting	16	35	25–28
Leaf elongation	7–12	45	31
Tillering	9–16	33	25–31
Initiation of panicle primordia	15	—	—
Panicle differentiation	15–20	30	—
Anthesis	22	35–36	30–33
Ripening	12–18	>30	20–29

26

In northern latitudes, rice plants are sown when temperature is low, complete the early growth stages in a rising temperature cycle, and after flowering, complete their growth in a regime of declining mean temperature (see Sapporo, Japan, Fig. 2.6). For tillering, the optimum temperature reported in Japan is 32–34°C (Matsuo 1959). This optimum temperature is never attained in the northern areas.

Areas in the lower latitudes have highest temperature at sowing time and a slowly declining temperature until maturity. Near the equator little change in temperature occurs (Moomaw and Vergara 1965). The range of diurnal temperature change for any site will depend on elevation and proximity to a large body of water. During the crop season, areas in the northern latitudes and at high elevation have greater diurnal change than low-latitude areas (Fig. 2.6).

In northern Japan, a rather low night temperature (16–21°C) except during tillering and late ripening favors grain production (Matsushima and Tsunoda 1958). Higher grain yields in temperate countries than in tropical countries have generally been attributed to the lower temperatures during ripening. This is because the ripening period is extended due to low temperature, giving more time for grain filling. Long day length and a high level of solar energy during the ripening period also contribute to high grain yield in temperate rice-growing regions—the United States, southern Australia, and parts of Europe.

Low Temperature Effects

Injury of rice plants by low temperature occurs in temperate and tropical regions. Kaneda (1972) reported 20 countries, mainly in lower-latitude areas, where cold injury in rice was confirmed. Those countries included Australia, Bangladesh, Colombia, Indonesia, India (Kashmir), Nepal, Peru, Sri Lanka, China, and the United States.

Injury due to low temperature is a major constraint to rice production in the hill areas in the tropics and subtropics. For example, the hill zones of northern India have about 1.8 million hectares of rice land spread over Kashmir (Jammu), Himachal Pradesh, Manipur, Meghalaya, Assam, Arunachal Pradesh, Uttar Pradesh, and West Bengal (Hamdani 1979). Short cropping period and temperature fluctuations adversely affect rice production in those areas. The temperatures in the Kashmir rice-growing areas are almost the same as in Hokkaido, Japan. For example, the temperatures during early crop growth (April–May) in Kashmir are 3.9–7.5°C minimum and 14.0–22.3°C maximum. During ripening (September–October), minimum temperature is 8.8–14.7°C and maximum is 25.6–26.9°C.

In Nepal, 15–20% of 1.3 million hectares of rice land are in a temperate region. Large areas of that land are at altitudes of 1000–2000 m and cold damage to rice is common. The highest altitude at which rice is grown is in Nepal's Jumla Valley (2621 m) in the far western Himalayas (Shahi and Heu 1979).

In temperate regions, cold injury is the main constraint limiting the rice-growing area and length of growing season.

In Korea, low temperature often causes low rice yields (Chung 1979). In the Beijing area of China, where the temperature can go as low as 5°C, rice seedlings have to be protected from cold injury. In California, two major types of cold problems have been cited since rice became a commercial crop about 1912:

·   Seedling vigor and establishment in cool water (18°C or below).
·   Sterility caused by cool night temperatures (below 15°C) 10–14 days before heading (Rutger and Peterson 1979).

The most comprehensive work on the effect of cold injury to the rice crop has been done in Japan where low temperature is the main limiting factor in rice production. In 22 of 90 years, the island of Hokkaido had low yield due to cool temperature in the rice-growing season (Fig. 2.15).

Severe cold injury no longer causes near-total crop failure as it did in 1902 and 1913 because of progress made in alleviating cold injury in rice (Satake 1976). Some highlights of that research are:

·   STAGE SUSCEPTIBLE TO COOLNESS   For many years it was generally assumed that sterility resulted from cool summer temperature at anthesis. Satake and Hayase (1970) found that the stage most sensitive to coolness is the young microspore stage after meiotic division.
·   CRITICAL LOW TEMPERATURE   Nishiyama et al. (1969) showed that the critical low temperature for inducing sterility is 15–17°C in the highly cold-tolerant

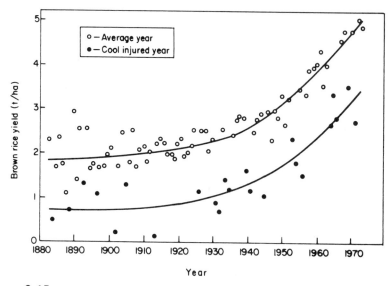

Figure 2.15   Effect of technical improvement on average brown rice yields (t/ha) in Hokkaido during the past 90 years. (From Satake 1976)

varieties and 17–19°C in the cold-sensitive varieties at the meiotic stage of crop growth. Their studies suggest that the critical temperature for sterility is about 15–20°C. This sterility is primarily due to injuries occurring in anthesis (Satake 1976).

Two factors cause cold injury to rice—cool weather and cold irrigation water. The common types of symptoms caused by low temperature are (Kaneda and Beachell 1974):

- Poor germination
- Slow growth and discoloration of seedlings
- Stunted vegetative growth characterized by reduced height and tillering
- Delayed heading
- Incomplete panicle exsertion
- Prolonged flowering period because of irregular heading
- Degeneration of spikelets
- Irregular maturity
- Sterility
- Formation of abnormal grains

Cold injury to the crop is ultimately reflected in reduced yields (Fig. 2.16). Delay in flowering and increase in growth duration are especially marked in modern rice varieties grown in low-temperature (32–25°C) zones. In some areas, two rice crops a year will be possible if cold-tolerant varieties are developed.

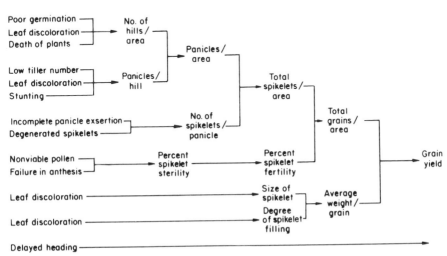

Figure 2.16 Modified path diagram of different types of cold damage affecting rice fields. (Adapted from Kaneda 1972)

High Temperature Effects

High temperature is a critical factor in rice grain production in Pakistan, the Middle East, and tropical Africa (Satake and Yoshida 1977a). Sato (1967) reported a high percentage of sterility and empty spikelets in rice crops in oasis areas of Egypt. In Thailand, Osada et al. (1973) reported empty spikelets in an indica rice crop in the dry, hot season. Recently, Satake and Yoshida (1978) reported high-temperature-induced sterility in rice in areas with high temperature and where the rice-growing season had changed since the introduction of modern varieties. Similar high-temperature-induced sterility is reported from Punjab, India. If Iran's rice-growing area is to expand, it will be in the southern area where high temperature and salinity may be serious constraints to successful rice cultivation. The same authors (Satake and Yoshida 1978) reported high temperature as an important constraint for rice cultivation in Senegal and other tropical African countries.

Several reports of high-temperature-induced sterility in experiments in a controlled environment are cited by Satake and Yoshida (1978). The same authors (Satake and Yoshida 1977a, b, 1978) reported heading to be the stage at which the rice plant is most sensitive to high temperature. Figure 2.17 shows the relationships between spikelet fertility and number of days before or after flowering day when the rice plants were exposed to high temperature for 5

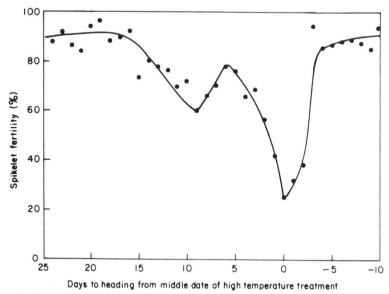

Figure 2.17   Fertility of the spikelets in rice plants (variety BKN 6624-46-2) subjected to high temperature (35°C for 5 days) at different stages of panicle growth. (Adapted from Satake and Yoshida 1978)

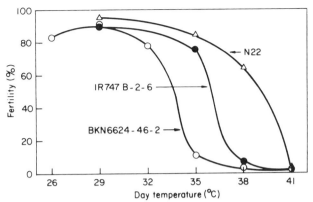

Figure 2.18   Relation between temperature at flowering and fertility of spikelets. (Day temperature was kept for 8 hours, and night temperature was fixed at 21°C in each treatment.) (From Satake and Yoshida 1977b)

consecutive days. Fertility of spikelets was 75% for plants held at 35°C for 4 hours, about 55% at 38°C for 4 hours, and about 15% at 41°C for 2 hours.

There are distinct varietal differences in response to critical high temperature. For convenience, Satake and Yoshida (1977b) designated the temperature at which percent fertility begins to decrease below 80% as CT (80) and the temperature at which percent fertility becomes 50% as CT (50). The difference in CT (80) between the heat-tolerant variety N22 and a heat-susceptible line BKN 6624-46-2 is more than 4°C (Fig. 2.18).

It is common to have maximum daily temperatures from 35–41°C or higher in semiarid regions and during hot months in tropical Asia. In these areas, a heat-susceptible variety may suffer from a high percentage of sterility induced by high temperature (Satake and Yoshida 1977b). According to Satake and Yoshida (1978), spikelet sterility from high temperature is induced largely on the day of flowering. Within the flowering day, high temperature during anthesis was the most detrimental to spikelet fertility, high temperature just before anthesis was the second most detrimental, and high temperature after anthesis had little effect on spikelet fertility. Two important characteristics for heat tolerance of rice varieties at flowering are good pollen shedding, and early morning anthesis.

Effects of Relative Humidity

The effects of relative humidity in the tropics are generally confused with the effects of solar energy and temperature. The average relative humidity before harvest follows a trend opposite that of the solar radiation values for the same period. Therefore, no importance is attributed to the high negative correlation between relative humidity and grain yield. However, a long dew period often causes increased incidence of blast disease in rice. In such cases, the effects of high relative humidity are often confounded by the night temperature regime, which causes a long dew period.

Effects of Wind

A gentle wind during the growing period of the rice plant is known to improve grain yields because it increases turbulence in the canopy. The air blown around the plants replenishes the carbon dioxide supply of the plant (Matsubayashi et al. 1963).

Photosynthesis of the plant community increases with the wind speed but wind speeds greater than 0.75–2.25 cm/sec have no further effect on increasing photosynthesis in different plants (Wadsworth 1959). Strong winds such as those that accompany cyclones, if they occur after heading, cause severe lodging and shattering in some rice varieties (De Datta and Zarate 1970). Strong winds often desiccate the panicles of a rice crop, increasing floret sterility and sometimes increasing the number of abortive endosperms. Strong winds also enhance the spread of bacterial leaf diseases of rice (Matsubayashi et al. 1963).

Dry winds have been known to cause desiccation of rice leaves. Wind can also cause mechanical damage of leaves. Such damage is more severe for upland than for lowland rice crops (Vergara 1976).

Climatic Effects on Yield and Yield Components

Climate directly influences the physiological processes that affect the rice plant's growth, development, and grain formation. Indirectly, climate influences the incidence of crop insects and diseases and hence, grain yield (Yoshida and Parao 1976).

To study the effects of climate on yield differences, all other factors, such as varieties and soil and crop management practices, must be optimal. Several studies in Japan (Murata 1964, Hanyu et al. 1966, Murakami 1973), reported close correlation between climatic parameters and rice yield. These and other studies led to the conclusion by Murata (1972) that local differences in rice productivity in Japan were largely accounted for by differences in solar radiation and temperature during the ripening period.

Harvest Index

To obtain high grain yields, balanced growth at different growth stages must be achieved. Balanced growth is reflected in the high ratio of the weight of the panicles to that of the total dry matter produced. This ratio is known as the harvest index (HI). Tall plants and excessive vegetative dry matter production reduce HI. Low solar energy and high temperature are detrimental to high HI.

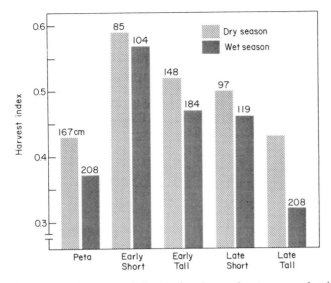

Figure 2.19 Changes in harvest index during dry- and wet-season plantings of Peta and Peta isogenic lines. (From Vergara and Visperas 1977)

This explains why the tall tropical variety, Peta, and its isogenic lines showed a marked decrease in HI during the wet season compared with the dry season (Fig. 2.19).

For the same reason, grain production in temperate areas will be more efficient for a given rice variety because the HI is higher. The physiological causes for variation in HI within a variety and among varieties of rice are not well understood (Yoshida 1972).

Climatic Influence on Rice Yield

Rice yields are influenced by many interrelated and often diverse environmental and biological factors, with the result that it is difficult to separate their effects. The relationship between grain yield and yield components can be expressed as:

Grain yield = number of panicles × number of spikelets per panicle
× percentage of filled spikelets × weight of 1000 grains

The main stage at which the number of panicles is determined is the tillering stage; the number of panicles is affected most strongly by climatic conditions during that period. High day temperature and solar radiation and low night temperatures are apparently conducive to producing more panicles without much reduction in spikelet number.

Murata (1969) in a discussion of the relative importance of yield capacity and assimilate supply for grain yield in the temperate regions cited the situations wherein:

- Yield capacity is limiting.
- Assimilate supply is limiting.
- Yield capacity and assimilate supply are well balanced.

An optimum number of grains appears to exist for a maximum grain yield under a given condition. For the tropics, Yoshida et al. (1972) showed that the grain yield is closely correlated with spikelet number per unit land area. The filled spikelet percentage is about the same for the dry and wet seasons. Thus, grain yield can be integrated as:

$$\text{Yield (t/ha)} = \text{spikelets/m}^2 \times \text{grain weight (g/1000 grains)} \times \text{filled spikelets \%} \times 10^{-5}$$

All of these determinants are developed during the reproductive and ripening stages of rice crop. Because a major portion of grain carbohydrate comes from photosynthesis during the ripening period (Yoshida 1972), it is obvious that photosynthesis, and hence, climatic factors (Stansel et al. 1965) during that period are important. De Datta and Zarate (1970) correlated solar radiation during the 45 days before maturity, which included about 15 days before flowering, with grain yield and obtained coefficients ranging from .50** to .77**.

Thus, although postflowering solar radiation is clearly an important determinant of grain yield in rice, it is possible that radiation at earlier periods in the development of the crop may be as important, or even more so. In fact, in an experiment by Yoshida and Parao (1976), in which IR747-B2-6 rice was shaded to various degrees for periods of 25 days during the vegetative, reproductive, and ripening stages, shading during the vegetative stage had little effect on grain yield whereas shading during the reproductive stage had even greater effect than that during grain ripening (Table 2.5).

During the reproductive stage, solar radiation affects spikelet number per square meter, and during ripening it affects filled spikelet (grain) percentage (Yoshida and Parao 1976). Earlier work in Texas had similar results (Stansel et al. 1965).

Besides correlating yield with solar radiation during periods of equal duration at various stages of crop growth, Evans and De Datta (1979) correlated yield with cumulative radiation for various periods working either forward from the date of planting or backward from the date of maturity. Figure 2.20 shows the results for IR8. The slight effect on yield of radiation in the vegetative period, and the importance of radiation both before and just after flowering, are evident.

Regardless of whether solar radiation was rising or falling progressively, high radiation at any stage after panicle initiation was associated with higher yields in both traditional and modern varieties. Plant response to radiation was greater at higher levels of nitrogen fertilizer application (Evans and De Datta 1979).

Temperature during the ripening period is another important factor for determination of rice yield. Murata (1964) and Hanyu et al. (1966) used the term climatic productivity index to express the effects of radiation and temperature on

Table 2.5 Effect of Shading at Different Growth Stages on Yield and Yield Components of IR747-B2-6; IR RI, 1974 (Adapted from Yoshida and Parao 1976)

Percent Sunlight	Grain Yield (t/ha)	Harvest Index	Yield Components		
			Spikelets (no./m^2)	Filled Spikelets (%)	1000-grain wt (g)
Vegetative stage					
100	7.11	0.49	41.6	88.9	20.0
75	6.94	0.48	40.6	88.9	19.9
50	6.36	0.51	38.3	89.5	19.9
25	6.33	0.51	38.1	84.3	19.8
Reproductive stage					
100	7.11	0.49	41.6	88.9	20.0
75	5.71	0.47	30.3	87.8	20.3
50	4.45	0.40	24.4	89.4	19.5
25	3.21	0.36	16.5	89.4	19.1
Ripening stage					
100	7.11	0.49	41.6	88.9	20.0
75	6.53	0.49	41.1	81.1	20.0
50	5.16	0.44	40.6	64.5	19.5
25	3.93	0.38	41.7	54.9	19.1

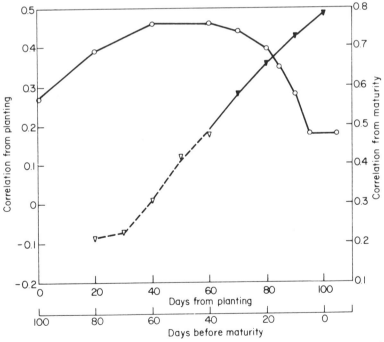

Figure 2.20 Correlation coefficients between the yield of IR8 and cumulative irradiance either forward to planting (∇) or backward from maturity ($\bigcirc$). Solid symbols indicate (probability) $p < .05$. (From Evans and De Datta 1979)

35

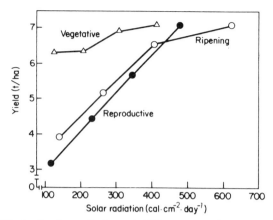

Figure 2.21   Effect of solar radiation at different growth stages on grain yield of IR747-B2-6. (From Yoshida and Parao 1976)

grain yield of rice. Yoshida and Parao (1976) found high correlation coefficients of that index with actually measured yield. From their studies, it was suggested that spikelet number per square meter was the most important factor limiting yield of IR747-B2-6 rice at Los Baños, Philippines, and spikelet number was highly correlated with solar radiation and temperature during the reproductive stage– the 25-day period before flowering.

From the same series of experiments, Yoshida and Parao (1976) contended that a yield of 4 t/ha can be obtained with 200 cal/cm^2 per day during the reproductive stage (Fig. 2.21). Thus, it is unlikely that incident solar energy limits rice yields in farmers' fields in most tropical countries. But the national average yield for most countries is about 2 t/ha and management practices, particularly water control, must be substantially improved to realize 4 t/ha rice yields with 200 cal/cm^2 per day. This observation is substantiated by the results summarized by De Datta and Malabuyoc (1976).

REFERENCES

Best, R. 1962. Production factors in the tropics. *Neth. J. Agric. Sci.* 10 (Spec. issue):347–353.

Brown, F. B. 1969. Upland rice in Latin America. *Int. Rice Comm. Newsl.* 18:1–5.

Chang, T. T., and B. S. Vergara. 1972. Ecological and genetic information on adaptability and yielding ability in tropical rice varieties. Pages 431–453 *in* International Rice Research Institute. *Rice breeding.* Los Baños, Philippines.

Chung, G. S. 1979. The rice cold tolerance program in Korea. Pages 7–19 *in* International Rice Research Institute. *Report of a rice cold tolerance workshop.* Los Baños, Philippines.

De Datta, S. K. 1970. The environment of rice production in tropical Asia. Pages 53–67 *in* University of the Philippines College of Agriculture in cooperation with the International Rice Research Institute. *Rice production manual.* Los Baños, Philippines.

De Datta, S. K. 1973. Principles and practices of rice cultivation under tropical conditions. *ASPAC Food Fert. Technol. Cent. Ext. Bull.* 33. 28 pp.

De Datta, S. K., and J. Malabuyoc. 1976. Nitrogen response of lowland and upland rice in relation to tropical environmental conditions. Pages 509–539 *in* International Rice Research Institute. *Climate and rice.* Los Baños, Philippines.

De Datta, S. K., and B. S. Vergara. 1975. Climates of upland rice regions. Pages 14–26 *in* International Rice Research Institute. *Major research in upland rice.* Los Baños, Philippines.

De Datta, S. K., and P. M. Zarate. 1970. Environmental conditions affecting growth characteristics, nitrogen response and grain yield of tropical rice. *Biometeorology* 4:71–89.

De Datta, S. K., A. C. Tauro, and S. N. Balaoing. 1968. Effect of plant type and nitrogen level on the growth characteristics and grain yield of indica rice in the tropics. *Agron. J.* 60:643–647.

De Wit, C. T. 1958. Transpiration and crop yields. *Versl. Landbouwkd. Onderz.* 64(6):88.

Evans, L. T., and S. K. De Datta. 1979. The relation between irradiance and grain yield of irrigated rice in the tropics, as influenced by cultivar, nitrogen fertilizer application and month of planting. *Field Crops Res.* 2:1–17.

FAO (Food and Agriculture Organization of the United Nations) Inventory Mission. 1970. Development of rice cultivation in West Africa. Preliminary report of the FAO Inventory Mission, July, 1970. Conference of Plenipotentiaries for the establishment of a West Africa Rice Development Association, 1–4 September 1970. Dakar, Senegal. 40 pp. (mimeo.)

Fukui, H. 1971. Environmental determinants affecting the potential dissemination of high yielding varieties of rice—a case study of the Chao Phrraya river basin. Conference on agriculture and economic development, Tokyo and Hakone, 1971. 37 pp. (mimeo.)

Hamdani, A. R. 1979. Low-temperature problems and cold tolerance research activities for rice in India. Pages 39–52 *in* International Rice Research Institute. *Report of a rice cold tolerance workshop.* Los Baños, Philippines.

Hanyu, J., T. Uchijima, and S. Sugawara. 1966. Studies on the agro-climatological method for expressing the paddy rice products. I. An agro-climatic index for expressing the quantity of ripening of the paddy rice. *Bull. Tohoku Natl. Agric. Exp. Stn.* 34:27–36.

Huke, R. 1976. Geography and climate of rice. Pages 31–50 *in* International Rice Research Institute. *Climate and rice.* Los Baños, Philippines.

IRRI (International Rice Research Institute). 1972. *Annual report for 1971.* Los Baños, Philippines. 238 pp.

IRRI (International Rice Research Institute). 1978. *Rice research and production in China: an IRRI team's view.* Los Baños, Philippines. 119 pp.

Kaneda, C. 1972. Terminal report on studies on the breeding for cold resistance. International Rice Research Institute, Los Baños, Philippines 80 pp. (unpubl. mimeo.)

Kaneda, C., and H. M. Beachell. 1974. Response of indica-japonica rice hybrids to low temperatures. *SABRAO J.* 6:17-32.

Matsubayashi, M., R. Ito, T. Takase, T. Nomoto, and N. Yamada. 1963. *Theory and practice of growing rice.* Fuji Publishing Co., Ltd., Tokyo. 502 pp.

Matsuo, T. 1959. *Rice culture in Japan.* Ministry of Agriculture and Forestry, Japan, 128 pp.

Matsushima, S., and K. Tsunoda. 1958. Analysis of developmental factors determining yield and application of yield prediction and culture improvement of lowland rice. XLV. Effects of temperature and its daily range in different growth stages upon the growth, grain yield, and constitutional factors in rice plants. *Proc. Crop Sci. Soc. Jpn.* 26:243-244.

Money, D. C. 1972. *Climate, soils and vegetation.* University Tutorial Press, London. 272 pp.

Moomaw, J. C., and B. S. Vergara. 1965. The environment of tropical rice production. Pages 3-13 *in* International Rice Research Institute. *The mineral nutrition of the rice plant.* Proceedings of a Symposium at the International Rice Research Institute, February, 1964. The Johns Hopkins Press, Baltimore, Maryland.

Moomaw, J. C., P. G. Baldazo, and L. Lucas. 1967. Effects of ripening period environment on yields of tropical rice. Pages 18-25 *in International Rice Commission Newsletter* (Spec. issue). Symposium on problems in development and ripening of rice grain. 11th Pacific Science Congress, Tokyo (1966).

Moraes, J. F. V. 1978. Arroz de sequeiro. Paper presented at the 11 Congresso nacional sobre economia orizicola, Cuiaba, Mató Grosso. Feb. 1978. (unpubl. mimeo.)

Munakata, K., I. Kawasaki, and K. Kariya. 1967. Quantitative studies on the effects of the climatic factors on the productivity of rice. *Bull. Chugoku Agric. Exp. Stn.*, A,14:59-96.

Murakami, T. 1973. Paddy rice ripening and temperature. *JARQ* 7:1-5.

Murata, Y. 1964. On the influence of solar radiation and air temperature upon the local differences in the productivity of paddy rice in Japan [in Japanese; English summary]. *Proc. Crop Sci. Soc. Jpn.* 33:59-63.

Murata, Y. 1966. On the influence of solar radiation and air temperature upon the local differences in the productivity of paddy rice in Japan. *Int. Rice Comm. Newsl.* 15:20-30.

Murata, Y. 1969. Physiological responses to nitrogen in plants. Pages 235-259 *in* J. D. Eastin, F. A. Haskins, C. Y. Sullivan, and C. M. H. van Bavel, eds. *Physiological aspects of crop yield.* American Society of Agronomy and Crop Science Society of America. Madison, Wisconsin.

Murata, Y. 1972. Local productivities [in Japanese]. Pages 315-318 *in* Y. Togari, ed. *Photosynthesis and matter production of crop plants.* Yokendo, Tokyo.

Nishiyama, I., N. Ito, H. Hayase, and T. Satake. 1969. Protecting effect of temperature and depth of irrigation water from sterility caused by cooling treatment at the meiotic stage of rice plants [in Japanese]. *Proc. Crop Sci. Soc. Jpn.* 38:554-555.

Osada, A., V. Sasiprapa, M. Rahong, S. Dhammanuvong, and H. Chakrabandhu. 1973. Abnormal occurrence of empty grains of indica rice plants in the dry, hot season in Thailand. *Proc. Crop Sci. Soc. Jpn.* 42:103–109.

Riehl, H. 1954. *Tropical meteorology.* McGraw-Hill Book Co., Inc., New York. 392 pp.

Rutger, J. N., and M. L. Peterson. 1979. Cold tolerance of rice in California. Pages 101–104 *in* International Rice Research Institute. *Report of a rice cold tolerance workshop.* Los Baños, Philippines.

Satake, T. 1976. Sterile-type cool injury in paddy rice plants. Pages 281–300 *in* International Rice Research Institute. *Climate and rice.* Los Baños, Philippines.

Satake, T., and H. Hayase. 1970. Male sterility caused by cooling treatment at the young microspore stage in rice plants. V. Estimations of pollen developmental stage and the most sensitive stage to coolness. *Proc. Crop Sci. Soc. Jpn.* 39:468–473.

Satake, T., and S. Yoshida. 1977a. Mechanism of sterility caused by high temperature at flowering time in indica rice. *JARQ* 11:127–128.

Satake, T., and S. Yoshida. 1977b. Critical temperature and duration for high temperature-induced sterility in indica rice. *JARQ* 11:190–191.

Satake, T., and S. Yoshida. 1978. High temperature-induced sterility in indica rices at flowering. *Jpn. J. Crop Sci.* 47:6–17.

Sato, K. 1967. *The report on the technical support to develop the U.A.R. desert* [in Japanese]. Overseas Technical Cooperation Agency, Tokyo. 100 pp.

Shahi, B. B., and M. H. Heu. 1979. Low temperature problem and research activities in Nepal. Pages 61–68 *in* International Rice Research Institute. *Report of a rice cold tolerance workshop.* Los Baños, Philippines.

Stansel, J. W. 1975. Effective utilization of sunlight. Pages 43–50 *in* Texas Agricultural Experiment Station, in cooperation with the U.S. Department of Agriculture. *Six decades of rice research in Texas.* Res. Monogr. 4.

Stansel, J. W., C. N. Bollich, J. R. Thysell, and V. L. Hall. 1965. The influence of light intensity and nitrogen fertility on rice yields components. *Rice J.* 68(4):34–35, 49.

Vergara, B. S. 1976. Physiological and morphological adaptability of rice varieties to climate. Pages 67–86 *in* International Rice Research Institute. *Climate and rice.* Los Baños, Philippines.

Vergara, B. S., and T. T. Chang. 1976. *The flowering response of the rice plant to photoperiod: a review of literature.* Int. Rice Res. Inst. Tech. Bull. 8. 75 pp.

Vergara, B. S., and R. M. Visperas. 1977. Harvest index: criterion for selecting rice plants with high yielding ability. Paper presented at a Saturday seminar, 10 September 1977, International Rice Research Institute, Los Baños, Philippines. 24 pp. (unpubl. mimeo.)

Vergara, B. S., A. Tanaka, R. Lilis, and S. Puranabhavung. 1966. Relationship between growth duration and grain yield of rice plants. *Soil Sci. Plant Nutr.* (Tokyo) 12:31–39

Wadsworth, R. M. 1959. On optimum wind speed for plant growth. *Ann. Bot. N. S.* 23:195–199.

Watts, I. E. M. 1955. *Equatorial weather.* Pitman Publishing Co., New York. 223 pp.

Yoshida, S. 1972. Physiological aspects of grain yield. *Annu. Rev. Plant Physiol.* 23:437–464.

Yoshida, S. 1978. Tropical climate and its influence on rice. *IRRI Res. Pap. Ser.* 20. 25 pp.

Yoshida, S., and S. B. Ahn. 1968. The accumulation process of carbohydrate in rice varieties in relation to their response to nitrogen in the tropics. *Soil Sci. Plant Nutr.* (Tokyo) 14:153–161.

Yoshida, S., and F. T. Parao. 1976. Climatic influence on yield and yield components of lowland rice in the tropics. Pages 471–494 *in* International Rice Research Institute. *Climate and rice.* Los Baños, Philippines.

Yoshida, S., J. H. Cock, and F. T. Parao. 1972. Physiological aspects of high yields. Pages 455–469 *in* International Rice Research Institute. *Rice breeding.* Los Baños, Philippines.

3

Landscape and Soils on Which Rice Is Grown

Rice is grown from the equator to 50°N and from sea level to 2500 m. It is grown in the hot, wet valleys of Assam and in the irrigated deserts of Pakistan. The soils on which rice grows are as varied as the climatic regime to which the crop is exposed: texture ranges from sand to clay, pH from 3 to 10; organic matter content from 1 to 50%; salt content from almost 0 to 1%, and nutrient availability from acute deficiencies to surplus.

Productivity of land used for growing rice is to a large extent determined by soil and water conditions. Rice is the only major annual food crop (with the exception of aroids) that thrives on land that is water saturated, or even submerged, during part or all of its growth cycle.

LANDSCAPE OF RICE

Rice will grow, under appropriate temperature regimes, wherever there is enough water to sustain a crop. That includes low-lying areas in coastal plains, floodplains, and valleys, where there is often more than enough water to maintain lowland rice and where water control must be practiced. Also included are rice fields (paddies) on steep and mountainous areas and vast upland areas where rice is grown in unbunded fields.

Wherever conditions are favorable, lowland rice fields are formed into paddies that hold water during the land preparation and rice-growing periods. A number of land management systems for rice cultivation have evolved over the centuries to suit soil, climate, water supply, and socioeconomic conditions of any area. However, two broad systems of land management are used in all rice-growing areas:

- Lowland, or wetland.
- Upland, or dryland.

41

There are a number of other systems of rice culture where water regime, particularly flooding level, is variable (see Chapters 7 and 9). But by origin and preference, rice is primarily a lowland crop and its semiaquatic character was the key to the development of wet lowlands in Asia at an early stage in the history of rice culture. Rice grew and thrived in those lowlands without need for extensive drainage.

Based on physiography and hydrology, rice lands are classified by Moormann and van Breemen (1978) who proposed a new terminology (Table 3.1). Rice lands are categorized into irrigated (where water supply is assured) and rainfed (where water supply is uncontrolled).

Rainfed rice lands are grouped as:

- Pluvial
- Phreatic
- Fluxial

Table 3.1 Terminology of Rice Lands in the Function of Physiography and Hydrology (Moormann and van Breemen 1978)

Intensity of Irrigation	Physiographic Hydrologic Category	Bunding and Leveling	Flooding Regimes[a]	Terminology Proposed	Terminologies Replaced
Zero or low; availability of water depends on natural supply	Pluvial	Without	No flooding	Pluvial rice land	Upland, hill, dryland
		With	1, 2, (3)	Pluvial-anthraquic rice land	Upland, lowland, hill
	Phreatic	Without	No flooding	Phreatic rice land	Upland, lowland, hill
		With	1, 2, 3	Phreatic-anthraquic rice land	Lowland, hill
	Fluxial	With or without	2, 3, 5, 6, 7, 8	Fluxial rice land	Lowland swamp
High; water is available independent of natural supply in 4 out of 5 years	Irrigated	Rarely without	4	Irrigated rice land	Lowland

[a]Regime 1: shallow, irregular, brief flooding; Regime 2: shallow, irregular, prolonged flooding; Regime 3: shallow, continuous, uncontrolled flooding; Regime 4: shallow, continuous flooding controlled by irrigation; Regime 5: shallow to moderately deep seasonal flooding; Regime 6: deep seasonal flooding; Regime 7: moderately deep to shallow flooding after recession of deep floods; Regime 8: tidal flooding.

In pluvial rice lands, the field is generally well drained without free groundwater within the rooting zone of the rice plant. This system is currently known as upland (or dryland), and the fields are not bunded.

Phreatic rice lands, whether they are naturally sloping or flat, may or may not be bunded and free groundwater is present within the rooting zone of the rice plant during the growing season. This land form and land management system is currently known as upland or lowland depending on the kind of land preparation. To distinguish between pluvial and phreatic rice lands one generally needs to have information on water table depths.

Fluxial rice lands are located in lower aspects of the landscape or in flat areas and are flooded during the greater part of the growing season. These areas are currently known as lowland swamps in which the land could be either bunded or unbunded.

RICE SOILS

Rice is grown on a variety of soils ranging from waterlogged and poorly drained to well drained. It is also grown under many different climatic and hydrologic conditions. Consequently, there is a considerable range in pedogenetic and morphological characteristics of rice-growing soils.

Pedology of Rice Soils

The terms *rice soils* or *paddy soils* are not precise enough to be used as the names of a soil group. The terms are currently used for soils on which lowland rice is grown and they merely give an indication of the land use; they do not give any precise information about the soil (Dudal 1958). Therefore, the term *paddy soil* is not a pedological term, but signifies a type of land use. The word *paddy* comes from *padi,* a Malayan word meaning rice, suggesting it is synonymous to rice soil (Kawaguchi and Kyuma 1977).

The specific morphology and classification of paddy soils are unique. By and large, they occur in low-lying lands that are inundated naturally or into which water can be introduced by gravity. Almost all kinds of soils can be used to grow rice if water conditions are favorable.

Rice Soil Morphology

Wherever possible the soil is kept flooded, or at least saturated with fresh water, throughout the growing period of the rice crop. Flooded soils may undergo profound changes (see Chapter 4). For example:

- Iron and manganese compounds move from the upper soil layers and subsequently precipitate in lower layers. Because the process is governed by their respective oxidation potentials, manganic and ferric oxides tend to precipitate separately, forming an iron accumulation zone which is underlain by a manganese accumulation zone (Kawaguchi and Kyuma 1977).

- Land surface is modified by terracing in sloping landscape areas.
- Silt carried by irrigation water accumulates on top of the original soil (Dudal 1958).

These changes may occur over extensive areas, in a wide range of soils, and under various climatic regimes. Thus, the morphology of the altered soils shows definite characteristics resulting from human interference as well as those resulting from the original profile. Dudal (1958) contends that "paddy soils" or "lowland rice soils," although they have definite features in common, are very much diversified by the properties of the different soils from which they originated. Therefore, such rice soils should not be dealt with as one group.

Due to their widespread distribution, paddy soils possess large differences in soil properties expressed in terms of chemistry, physics, and biology. Figure 3.1 shows a sequence of paddy soils in the landscape in relation to groundwater table and formation of characteristic paddy horizons.

The dominant feature in all these types of paddy profile is the existence of the *gleyed* horizon, which is a layer where soil reducing conditions exist. This varies in degree in response to the position of the water table. Soils in paddies with water tables at or near the surface exhibit intense reduction characterized by the presence of ferrous iron and neutral or greenish or bluish gray soils. The color of the soil is one property that can be distinguished in morphological description of soil profiles. The color is determined by using a Munsell soil color chart with various notations and named colors.

In the paddy soils, gleying persists even though the water table is deep because of the practice of maintaining water on the soil surface during most of the rice growing season. The presence of floodwater, plus the practices applied to the soil, result in a soil system with diverse properties compared to its drained counterpart. Paddy soils are, in general, man-made (anthropic) because they have been puddled during land preparation and submerged for the major part of the growing season of rice.

Classification of paddy soils recognizes the occasions when they undergo partial drying even when rice is standing in the field and the fact that water is often withdrawn between cropping seasons. The drying and wetting give the paddy soil system a unique behavior, which is reflected in cyclical changes of the chemistry (see Chapter 4) and physics of the system (Briones 1979).

Over the years, Kawaguchi and Kyuma (1977) have extensively studied the paddy soils of tropical Asia. Their reports suggest that most soils studied had no clear iron and manganese illuvial horizons and therefore cannot be called Aquorizems. They remain as alluvial soils, low humic gley soils, and Grumusols even after long years of rice cultivation.

Because of the continuous inundation of most low-lying paddy soils, the topography does not favor the downward movement of water through the solum,

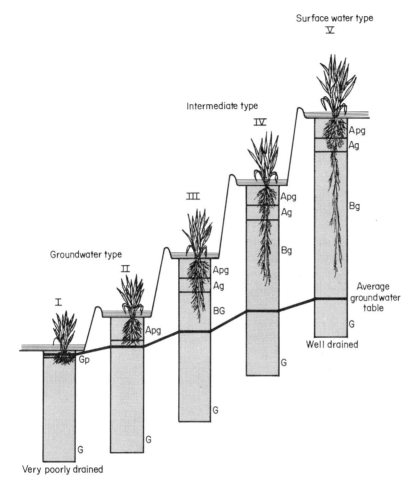

Figure 3.1 Sequence of paddy soil profiles as a function of groundwater depth. p = cultivated (plowed); g = periodically reduced (gleyed, mottled); G = periodically reduced horizon. (Adapted from Kanno 1956, Moormann and van Breemen 1978, R. Brinkman, IRRI, personal communication)

particularly in most deltaic soils. Because of the very heavy texture of these deltaic soils, many wide cracks occur during the dry season. The first few rains at the beginning of the rainy season may percolate fairly rapidly, but once the clay expands there is little or no possibility of water penetrating the solum.

Classification of Rice Soils

Various systems of classification of paddy soils, specific for conditions where rice is grown under temporary surface soil saturation, have been devised. However, Moormann (1978) contends that soils on which rice is grown can and should be fitted into a soil classification system, applicable to all soils irrespective of their land use, with adaptation if necessary where the specific land use has imposed major morphological changes.

Japan was the first country to attempt classification of rice soils as early as 1930. The most widely acclaimed system was developed by Kanno (1956, 1962) using a catenary sequence varying in drainage system from very poorly drained to well drained soils.

Each rice-growing country has in fact classified its rice soils using either a national or international system. In those systems, no special provision is made to include the classification of submerged rice-growing soils (Dudal and Moormann 1964).

A modern system of soil classification has evolved over the years and was published in 1975 by the United States Department of Agriculture (USDA 1975). In that system are 10 orders. Their key profile characteristics are summarized in Table 3.2.

The five categories in *Soil Taxonomy* (USDA 1975) are:

- Orders
- Suborders
- Great Groups

Table 3.2 Soil Orders: Comprehensive Classification System (USDA) (Adapted from USDA 1975)

Order	Key Profile Characteristics
Entisols	Recent soils; little or no change from parent material
Inceptisols	Light colored subsoils; weak soil development
Mollisols	Soft, deep, dark soils; high base status of surface horizon
Alfisols	Subsoil horizon of accumulated clay; high base saturation; high in weatherable minerals
Ultisols	Subsoil horizon of accumulated clay; low base saturation; few or no weatherable minerals
Oxisols	Uniform textured; friable profile high in oxides of iron and aluminum with kaolin clay; no weatherable minerals, low cation exchange capacity
Vertisols	Dark soils; high in montmorillonitic clay, prone to shrink and swell; high cation exchange capacity
Aridisols	Mineral soils of dry regions with either calcium carbonate or salt accumulation
Spodosols	Strong brown subsoil underlying a gray to brown surface horizon; strongly acid
Histosols	Soils with more than 30% organic matter to a depth of 40 cm

- Families
- Series

The lowest category, the soil Series, has thousands of taxa, making it impossible to describe all the taxa on which rice is grown. Therefore, the description of rice-growing soils is limited primarily to taxa of the highest category, the Order: Alfisols, Aridisols, Entisols, Histosols, Inceptisols, Mollisols, Oxisols, Spodosols, Ultisols, and Vertisols. Some examples of Suborders and Great Groups are cited, however. Details on the Orders and other categories are given elsewhere (USDA 1975, Sanchez 1976, Moormann and van Breemen 1978).

Most tropical rice soils in Asia are now classified using *Soil Taxonomy* (USDA 1975) and reported in a recent publication (IRRI 1978). Other areas, particularly the African countries, use the FAO-Unesco soil map of the world (Unesco 1974). Soil Orders in *Soil Taxonomy* and the former Great Soil Groups are compared in Table 3.3.

Table 3.3   Orders of the Soil Taxonomy System in Relation to Great Soil Groups of the Previous USDA Schemes and Other Classification Systems (From Sanchez 1976)

Order	Former Great Soil Groups Included
Entisols	Azonal soils, some Low Humic Gley, Lithosols, Regosols
Vertisols	Grumusols, Tropical Dark Clays, Regur, Black Cotton Soils, Dark Magnesium Clays
Inceptisols	Andosols, Hydrol Humic Latosols, Sol Brun Acide, some Brown Forest, Low Humic Gley, Humic Gley
Aridisols	Desert, Reddish Desert, Sirozem, Solonchak, some Brown and Reddish Brown soils, associated Solonetz
Mollisols	Chestnut, Chernozem, Brunizem, Rendzina, some Brown Forest, Brown, associated Humic Gley and Solonetz
Spodosols	Podzols, Brown Podzolic, Ground-Water Podzols
Alfisols	Gray-Brown Podzolic, Gray Wooded, Noncalcic Brown, Degraded Chernozem, associated Planosols and Half Bog, some Terra Roxa Estruturada and eutric Red-Yellow Podzolics, some Latosols and Lateritic soils
Ultisols	Red-Yellow Podzolic, Reddish Brown Lateritic, Humic Latosols, associated Planosols, and some Half Bogs, Latosols, Lateritic soils, Terra Roxa, and Ground-Water Laterites
Oxisols	Low Humic Latosols, Humic Ferruginous Latosols, Aluminous Ferruginous Latosols, some Latosols, Lateritic soils, Terra Roxa Legítima, Ground-Water Laterites
Histosols	Bog soils, Organic soils, Peat, Muck

Rice soils of a given country can be classified using the comprehensive USDA system (*Soil Taxonomy*), the FAO-Unesco soil map, or the early USDA system. An example for representative Philippine soil series is presented in Table 3.4.

LOWLAND RICE SOILS

Rice soils are more readily formed on acidic than on basic materials. Changes occurring in the latter during soil formation may be less noticeable, but are equally important. If conditions are favorable, lowland rice soils can develop in 50–100 years (Dudal 1958). They occur most extensively in alluvial lowlands, where inundation is natural or water is easily introduced by gravity. Therefore, most rice soils are either Entisols or Inceptisols that have undergone very little soil formation:

- Entisols are soils in which the effects of any major set of soil-forming processes are so minor that they do not have distinct pedogenic horizons.
- Inceptisols are soils that have available water for more than three consecutive months during a warm season and that have one or more pedogenic horizons other than those used to characterize other orders.

In most cases material deposited as alluvial sediments has been barely altered by the natural and artificial processes inherent in rice cultivation (Kyuma 1978). Vertisols, Alfisols, and Ultisols are other soils commonly used for growing lowland rice. Vertisols are clay soils with indications of regular mixing of the soil that prevents the development of diagnostic horizons, and with pronounced changes resulting from differences in moisture content (the soil expands and shrinks depending on moisture supply). Alfisols and Ultisols are described later in the section on upland rice soils of Southeast Asia.

Properties of Lowland Rice Soils

The productivity of lowland rice soils is heavily dependent on soil fertility or the chemical nature of the soils. Soil tilth, which is considered highly important for an upland crop is generally considered unimportant for lowland rice. However, there are a number of physical properties of soil that are of importance for growing rice or for rice-based cropping systems.

Physical Properties

Because rice is grown primarily in a lowland (paddy) soil, the physical properties of the soil are relatively unimportant as long as sufficient water is available (Kawaguchi and Kyuma 1977).

SOIL COMPONENTS   All soils, including those in partially drained paddies, are characterized as a three-phase system of solid, liquid, and gas. For mineral soils, the solid component constitutes at least 80% of its weight and is composed mostly of inorganic particles. Due to infinite variations in geometry (sizes and shapes), their arrangement results in formation of interstices and spaces between particles. These spaces are interconnected and are known as pore space.

Consideration of physical properties centers on the pore spaces where retention and movement of water and air occur. But the nature of the pore space is totally dependent on the solid component of soil volume and is influenced by its particle size distribution (soil texture) as well as the arrangement of soil particles (soil structure).

In submerged and completely saturated paddy profiles, all pores are filled with water so that aeration is drastically or totally curtailed. The water in the soil, also referred to as soil moisture, constitutes the soil solution. Considerable characterization of lowland rice soils is done by determining the kinetics of nutrient elements in the soil solution (Ponnamperuma 1978).

SOIL TEXTURE Texture is an expression of the distribution of the various particle sizes present in soils. A soil is described as having a coarse, medium, or fine texture depending on the predominant particle size. Soil texture determines whether an area is suitable for maintaining a paddy system because texture influences transmission and storage of water, flow of air in the soil, and the soil's capacity to supply nutrients.

In general, fine-textured soils are more fertile than coarse-textured soils primarily because of the influence of higher clay content (and organic matter), which relates to nutrient supply. Fine-textured soils are also called heavy soils because tillage is often difficult to perform at certain moisture conditions. In contrast, the term light soil refers to the coarse-textured soils, and implies a lesser degree of effort required for tillage (see Chapter 8). Soil texture plays an important role in the management of irrigation and fertilizers. Small pore size in fine-textured soils makes water movement slower than in coarse-textured soils. Excessive application of fertilizer or water to coarse-textured soils may increase losses due to leaching or deep percolation.

For rice cultivation, soils of fine to medium texture are most commonly used.

A study by Kawaguchi and Kyuma (1974) with 410 surface soil samples collected from nine tropical Asian countries suggests that the clayey texture of rice soils results either from the sedimentation process in wide floodplains and deltas or from the basic nature of parent materials in the mountain regions. Sandy texture is a result of either severe weathering of the materials of acidic rock origin, or of direct inheritance from sandy sedimentary materials. Soils from Sri Lanka are predominantly sandy and contain little silt, whereas surface soils of Bangladesh are generally silty (Fig. 3.2).

Kyuma (1978) explained that the sandy nature and low silt content of Sri Lankan soils are due to their development on residual or local alluvial materials, mostly poorly sorted materials derived from weathered gneissic rocks. The silty nature of Bangladesh soils is explained by the sedimentation process of the Ganges-Brahmaputra rivers. Soils from the other seven countries show similar textural patterns (Fig. 3.2).

Table 3.4 Comparison of Terms (Approximate) Used in Current Classification Systems, with Representative Philippine Soil Series, 1973 (Adapted from De Datta and Feuer 1975)

Great Soil Group (Early USDA System)	Soil Units, FAO-Unesco Soil Map of the World	Soil Taxonomy (USDA)		
		Order	Great Soil Group Suborder	Representative Philippine Soil Series
Alluvial	Eutric Fluvisols	Entisol	Tropofluvent	San Manuel Mandawe
Ando soil	Humic Andosol	Inceptisol	Eutrandepts	Tupi Mayon Taal
Rendzina and Brunizems	Rendzina and Phaeozems	Inceptisol and Mollisol	Eutropepts Xerochrepts and Lithic subgroups and Rendolls	Faraon Lugo Cataingan Sevilla Sibul
		Inceptisol	Tropepts of many kinds	Lithic, Vertic, Rendollic, Dystric, Andic, Oxic, Eutric subgroups of soils developed from basic igneous rocks, mostly hilly to mountainous
Either Latosolic Brown Forest or Gray Brown, Podzolic	Eutric Cambisol or Luvisol	Mollisol or Alfisol	Typic Rendoll; Haplustic Rendoll or Udalfs Tropudalfs	Lipa Maahas Ibaan Tigaon Magallanes

Either Gray Brown Podzolic or Non-Calcic Brown	Luvisol	Alfisol	Tropudalfs	Quinqua Umingan
Red-Yellow Podzolic and Reddish Brown Lateritic	Acrisols and Dystric Nitosols	Ultisol	Tropudult Ustult Xerult	Castilla Annam Alaminos Barotac San Rafael Bolinao Alimodian Carig Ilagan Sara Kidapawan
Latosol, Red Latosol, Earthy Latosol Laterite	Rhodic Ferralsols	Oxisol	Orthox Acrorthox Ustox Acrustox	Luisiana Antipolo Sigcay Adtuyon (also Andic phase) Gimbalaon Tugbok
Grumusol	Chromic Vertisols	Vertisol	Chromusterts	Prenza Rugao
Grumusol	Pellic Vertisols	Vertisol	Pellusterts	Bantog Buenavista Maligaya Bigaa San Fernando Toran Sta. Rita
Grumusol	Pellic Vertisols	Vertisol	Pelluderts	Pili Palo

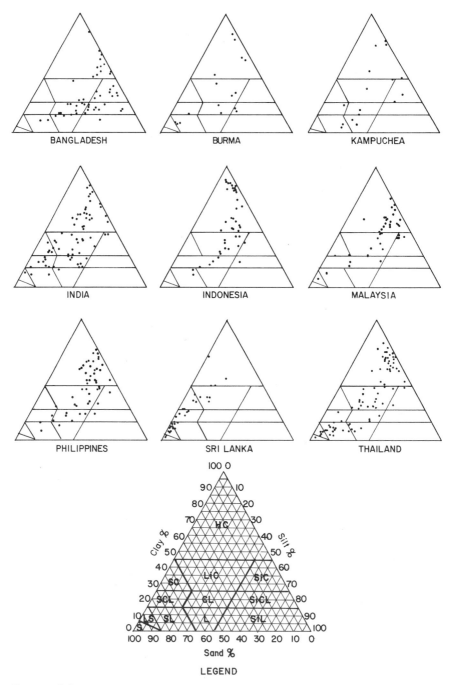

Figure 3.2 Distribution of sample soils in the triangular diagram for textural classification by countries. (Kawaguchi and Kyuma 1977)

SOIL STRUCTURE   Soil structure refers to the pattern of spatial arrangement of soil particles in a soil mass. This signifies size, shape, and durability of peds and their stability. For puddled lowland fields, soil structure is irrelevant because the puddling process destroys structure (see Chapter 8). However, soil particles, particularly clays and finer silts, seldom occur as individual units in paddies even when allowed to dry. Dried paddies often display a massive type of structure, particularly in the hard pan or traffic pan, that impedes water movement and helps keep the water ponded on the surface (Briones 1979).

Soil structure is usually evaluated by indirect means using a number of indices such as bulk density, porosity, pore-size distribution, and soil aggregation, which includes stability and size distribution of aggregates.

Bulk Density.   Bulk density is the ratio of the mass of a soil's solids (oven-dried soil) to the bulk volume of that soil. In a given soil mass, the total volume V_T is composed of the volume of solids V_s and the volume of pore space V_{ps} that contains the water and air. The total mass M_T is composed of the mass of soil solids M_s plus mass of whatever water M_w is present. The mass of air is negligible and is ignored in calculation. Bulk density D_b is represented by

$$D_b = \frac{M_s}{V_T}$$

Bulk density is expressed as mass per unit volume, normally as g/cm^3. The quantity D_b includes the overall volume of the soil and any change in the volume of pore space resulting from soil compaction or compression will be reflected in a corresponding change in bulk density. Bulk density values tend to be smaller as texture of the soil becomes finer. This is due to the tendency of finer particles to aggregate, which results in larger amounts of pore space than in a soil with coarse particles.

Several methods are used for determining bulk density of paddy soils. The clod method (Blake 1965a) is considered suitable for dry soil sampled before the start of the puddling process or after harvesting a lowland rice crop. For a moist soil, a core method is used (Blake 1965a, Foth and Turk 1972).

Particle Density.   Particle density D_p is defined as the ratio of mass of solids (oven-dried soil) to the volume of the solids alone and is expressed as g/cm^3. The volume of solids is part of the total soil volume and does not include the volume of water and air. Particle density is a stable property and depends on soil texture and the mineral composition of the particles. For calculation purposes, an average value of 2.65 g/cm^3 is used. The method for determining particle density is described by Blake (1965b) and ASTM (1958). Particle density measurement may be significant to pedogenesis and is often used in soils studies.

Porosity. The amount of pore space, known as total porosity, is defined as the fraction of total soil volume occupied by pores. Based on the volume of the pores V_{ps} and total bulk volume V_T,

$$\text{pore space} = \frac{V_{ps}}{V_T}$$

The air-filled porosity is the fraction of the pore space not occupied by water and is expressed in terms of total soil volume. In saturated paddies, all the pores are expected to be filled with water, thus, the porosity in this case is waterfilled porosity.

Total porosity is seldom determined directly because it can be estimated from bulk density and particle density relations of the soil. To calculate pore space the formula $1 - D_b/D_p$ is generally used. Because particle density is constant, the formula implies that any change in porosity is due to changes in bulk density. Porosity depends both on texture and structure because spaces will form between particles of an aggregate (intra-aggregate pores) as well as between aggregates (interaggregate pores).

Apparent Specific Volume of Soil. Change in the apparent specific volume of soil reflects the susceptibility of a soil to puddling. Puddlability is the change in apparent specific volume per unit of work expended. The change in the apparent specific volume of soil is the difference between apparent specific volume after and before puddling (Bodman and Rubin 1948):

$$\Delta ASV = ASV_{ap} - ASV_{bp}$$

where ap = after puddling, bp = before puddling. The data are expressed as cm^3/g.

If the density of water is considered equal to 1 g/cm^3, which is usual in engineering works (McCarthy 1977), the equation will be

$$ASV = m + \frac{1}{D_p}$$

where m = mass of water per unit mass of oven-dry soil, or gravimetric moisture content, and D_p = particle density (Bodman and Rubin 1948).

Soil Consistency. Soil consistency refers to the manifestations of the physical forces of cohesion and adhesion acting within the soil at various moisture contents. The forces involved are those bringing about compression, shear, or tension.

Consistency of soils has been described in terms of friability, plasticity, or stickiness. Consistency behavior of soils is affected by:

- Moisture content
- Forces applied or present
- Composition (clay mineralogy)
- Texture
- Structure of the material

Because most paddy soils are essentially puddled systems the existence of discrete particles of the finer sizes is enhanced. With drying, forces of cohesion and adhesion over the soil mass are accentuated and its consistency behavior approximates that exhibited by a soil paste upon drying (Briones 1979).

Within the range of normal shrinkage, decrease in volume of water is equal to the decrease in volume of soil. Continuous drying will reach a point where soil ceases to exhibit the properties of flow under the applied force; the soil changes from a liquid consistency to a plastic paste. The moisture content at this transition is identified as the liquid limit or upper plastic limit.

Within the plastic range, the soil mass can be molded to any shape. Water between soil particles provides sufficient cohesion but also permits sliding between particles. With continued drying, the mass loses its plasticity and a moisture content is reached where the consistency changes from plastic to semisolid. That moisture content is defined as the plastic limit or lower plastic limit.

Cohesion forces play an important role in the consistency transformation of a drying paddy soil mass. In the field, some aggregates are not destroyed and, because of alternate wetting and drying, some degree of aggregation is regenerated. In this case cohesion within the aggregate (intra-aggregate) and between aggregates (interaggregate) occurs, which influences the consistency of the soil mass. Figure 3.3 shows the consistency indices in relation to intra- and interaggregates cohesion.

IMPLICATIONS OF SOIL PHYSICAL PROPERTIES   Adequate water supply is a key to optimum rice growth and high grain yield. Because of the characteristic shrinkage and the accompanying changes in physical properties that occur, drying paddies are likely to pose conditions that will restrict root development for sensitive plants cultivated under upland conditions, which is also the case when an upland crop is grown in rotation with lowland rice. The nature of soil impedance to root growth takes cognizance not only of the role of water content as it affects mechanical resistance but also to the characteristics of the growing root tip.

Soil Mineralogy

Understanding of clay minerals is important for management of rice soils. Both clay content, as expressed by texture, and clay mineralogical characteristics have great bearing on the productivity of soils. In Japan, and in most tropical Asian

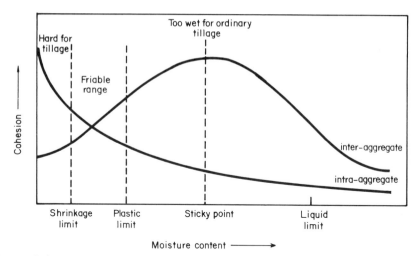

Figure 3.3   Relationship between moisture content and consistency indices to inter-
and intra-aggregate cohesion of soils. (From Briones 1979)

countries, soils with montmorillonitic clay have higher fertility and higher yield
potential than soils with kaolin materials or soils with allophane (amorphous
silicates) as the major clay constituents (Kawaguchi and Kyuma 1969). Clay
minerals play a significant role in physical and chemical properties of rice soils.
Clay minerals in soils are discussed extensively by Dixon and Weed (1977).
    Silicate clays are classified into four groups based on number and arrangement
of sheets of silica tetrahedra (one silicon atom surrounded by four oxygen atoms)
and alumina octahedra (one aluminum atom surrounded by six hydroxyl or
oxygen atoms). The four groups of clay minerals and the number of sheets within
the layers of the clay crystals are:

- 1 : 1 type minerals (one silica and one alumina).
- 2 : 1 expanding type minerals (two silica and one alumina), which can
  expand by water molecules moving between the layers between crystal
  units.
- 2 : 1 nonexpanding type minerals (two silica and one alumina) which do
  not expand between layers.
- 2 : 2 type (two silica and two magnesium).

The layer thickness of clay minerals is different for different groups and ranges
from 0.7 nm to 1.4 nm (1.8 nm for expanding clay minerals).

1 : 1 TYPE CLAY MINERALS The 1 : 1 type clay minerals have one silica
tetrahedral and one alumina octahedral sheet in the crystal structure, with layer

thickness of about 0.7 nm. Example: Kaolinite. In soils containing kaolinite, which is a nonexpanding clay mineral, cations and water do not enter between the layers. Therefore, rice soils with 1 : 1 type clay minerals have low cation exchange and water-holding capacities. Many of the Oxisols and Ultisols have kaolinite (1 : 1) type clay minerals. According to Kawaguchi and Kyuma (1969) the soils of Khorat in northeastern Thailand are generally kaolinitic.

2 : 1 EXPANDING TYPE   In the 2 : 1 expanding type, two tetrahedral silica sheets have one octahedral alumina sheet sandwiched between them. Example: Montmorillonite. Its layer thickness varies between 1.0 and 1.8 nm. It can expand because the layers are loosely held together by oxygen-to-oxygen linkages. As a result, swelling and shrinking capacity is high in soils rich in montmorillonite. It has high cation exchange and water-holding capacities. The clays of Vertisols are of the 2 : 1 expanding type.

2 : 1 NONEXPANDING TYPE   The 2 : 1 nonexpanding type structure is similar to the montmorillonite group but has much larger particles. Example: Illite (belonging to hydrous mica group). With regards to cation adsorption, hydration, swelling, and shrinkage, the 2 : 1 nonexpanding type is intermediate between montmorillonite and kaolinite. Soils in the northern and central valleys of Thailand have higher illite content than soils in Malaysia (Kawaguchi and Kyuma 1969).

2 : 2 TYPE CLAYS   The crystal unit in the 2 : 2 type clays contains two silica tetrahedral sheets and two magnesium octahedral sheets, with a fixed layer thickness of about 1.4 nm. Example: Chlorite. A typical unit has talc (similar to montmorillonite crystal unit) and brucite [$Mg(OH_2)_2$] layers. Particle size and cation exhange capacity are about the same as for kaolinite. There is little water adsorption between the layers.

Except for few soils in the northern valleys of Thailand, lowland rice soils in Thailand, Sri Lanka, and Malaysia do not generally contain chlorite (Kawaguchi and Kyuma 1969).

ALLOPHANES   Allophanes are X-ray amorphous, clay-sized materials. The terms Andosol and volcanic ash soil have been used as general expressions to designate soils that have been derived from volcanic ash and pumice (Kawaguchi 1973) which *Soil Taxonomy* (USDA 1975) classifies under the suborder Andepts. The dominant clay mineral is allophane. The A horizon of Andepts is dark in color and contains large quantities of organic matter. The soil has low bulk density, high porosity and water-holding capacity. Because soils containing allophanes are mostly acidic the effective cation exchange capacity is low; but cation exchange capacity is highly dependent on pH, and in soils with pH 7.0, the cation exchange capacity is high. Soils high in allophane fix large amounts of phosphorus in forms unavailable to plants.

Table 3.5 Distribution of Clay Mineral Groups in Surface Soils of South and Southeast Asia (Adapted from Kawaguchi and Kyuma 1977)

Distribution (%) Frequency of Clay Mineral Groups in Surface Soils

Country	Diameters (nm) of Clay Minerals[a]									
	0.7	0.7–1.0	0.7–1.4	1.0	1.0–0.7	1.0–1.4	1.4	1.4–0.7	1.4–1.0	0.7–1.0–1.4
Bangladesh	0	13	13	0	7	0	0	10	26	30
Burma	0	16	13	0	2	2	0	25	6	36
Kampuchea	31	0	34	0	0	0	0	28	0	6
India	3	6	14	0	10	4	9	23	9	23
Indonesia	20	0	18	0	0	0	28	31	0	0
Malaysia	34	2	24	0	0	0	0	39	0	0
Philippines	4	0	9	0	0	0	37	43	0	2
Sri Lanka	39	17	18	0	0	0	6	17	0	3
Thailand	19	21	46	0	1	0	3	7	0	4

[a]When more than one diameter heads a column the first number indicates the major component in a soil; following numbers indicate less dominant components within the percentage given.

CLAY MINERALS IN LOWLAND RICE SOILS   According to Kawaguchi and Kyuma (1974) the surface horizons of lowland rice soils have the following important clay mineral species:

- 0.7 nm minerals, mainly belonging to kaolin group.
- 1.0 nm minerals, mainly micaceous clays (illites).
- 1.4 nm minerals, mainly montmorillonite clay with varying amounts of vermiculite and aluminum interlayered vermiculite-chlorite intergrades.

Kawaguchi and Kyuma's data from nine countries on clay mineral species, based on their thickness, are summarized in Table 3.5.

Chemical Composition

Data on the chemical composition of tropical soils suggest that most are highly weathered,unlike many rice soils in Japan or the Philippines.

Soil pH, before and after flooding lowland fields, is an important determinant in evaluating fertility and management of rice soils. The pH values of lowland rice soils vary greatly from country to country. Examples of the range of pH values of paddy soils in some South and Southeast Asian countries are given in Table 3.6. With regard to mineral composition, many Indonesian soils contain low silica and high iron, aluminum, and manganese. Soils from Thailand and Kampuchea have extremely low phosphorus content. Soils from Sri Lanka are highly siliceous and low in iron oxide, reflecting their sandy texture. Soils from

Table 3.6   pH of Plow Layer Soils of the Paddy Fields in Selected Countries (In a Water Suspension of Air Dried Soil) (Adapted from Kawaguchi 1973)

Country	Mean pH	Standard Deviation	Number of Samples
Thailand	5.2	0.56	95
Malaysia (West)	4.7	0.57	41
Sri Lanka	5.9	0.84	33
Bangladesh	6.1	0.95	53
India	7.0	1.12	73
Kampuchea	5.2	0.76	16
Philippines	6.4	0.65	54
Indonesia (Java)	6.6	0.87	46
Australia	6.2	0.87	8
Republic of Korea	5.4	—	1038
Japan	5.5	—	1700

India, Bangladesh, and Burma have intermediate levels of minerals (Kyuma 1978). Details on chemical properties of lowland soils and the transformations of nutrients in them are in Chapter 4.

Microflora in Rice Soils

Flooding of rice soils provides a favorable environment for anaerobic microbes and the biochemical changes are varied and numerous. However, a thin surface layer of lowland soil generally remains oxidized and sustains aerobic microbes. The main biochemical processes in flooded soil, however, can be regarded as a series of successive oxidation-reduction reactions mediated by different types of bacteria. Yoshida (1975) reviewed the microbial metabolism of flooded soils.

In a complex ecosystem of lowland soils, the soil-water ratio varies and so does the microbial population. Three major types of microbes are present in lowland rice soils in variable proportion:

- Obligate aerobes that grow only in the presence of molecular oxygen.
- Obligate anaerobes that grow only in the absence of oxygen.
- Facultative anaerobes that can grow either with molecular oxygen or anaerobically when supplied with a suitable electron acceptor other than molecular oxygen.

Results of studies in Japan, India, Egypt, and the Philippines suggest that bacteria predominate in flooded soils, whereas fungi and actinomycetes are more abundant in upland soils. Bacteria such as *Mycobacteria, Bacillus,* and *Pseudomonas,* and other biologically active bacteria, are present in greater numbers in the rhizosphere than in the soil farther away (Miura and Yoshida 1972).

Some aerobic microbes, including fungi, nematodes, yeast, and protozoa, have occasionally been found inside the root tissue of the rice plant.

Flooding causes changes in the character of the microbial flora in soils. The sequence of changes is (Takai 1969)

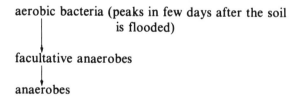

The microbial flora causes a large number of biochemical changes in the soil that largely determine the fertility of the soil. The major biochemical functions are solubilization, mineralization, immobilization, oxidation,and reduction.

For sustaining soil fertility and crop production, the important micro-biological functions are mineralization and immobilization of elements such as carbon, nitrogen, phosphorus, and sulfur; fixation of atmospheric nitrogen or carbon dioxide; and solubilization of phosphorus (Yoshida 1978).

There are many kinds of nitrogen-fixing microbes in lowland rice soils. A study in Thailand identified Azotobacter ($0-10^4$/g soil), Beijerinckia ($0-10$/g soil), Clostridia (10^4-10^6/g soil), nonsulfur purple bacteria (10^2-10^5 g/soil), and blue-green algae (10^2-10^3/g soil). The population density of these nitrogen fixers depends primarily on specific soil properties such as pH, organic matter content, and available phosphorus (Matsuguchi and Tangcham 1974).

The nonsymbiotic, heterotrophic nitrogen fixers (Azotobacters and Clostridia) are not very effective in lowland rice soils. On the other hand, photoautotrophs such as blue-green algae and photoautotrophic bacteria are important. Blue-green algae in the surface water and the surface soil layer can assimilate both carbon and nitrogen and therefore do not require any extraneous energy sources (Kawaguchi and Kyuma 1977).

Recently, Watanabe and Furusaka (1980) summarized the microbiology of flooded rice soils. It appears that the kinds and numbers of microorganisms identified in flooded rice soils vary greatly with the techniques used. Newer techniques result in discoveries of new types of microorganisms such as the identification of *Propionibacterium* (Hayashi and Furusaka 1979), N_2-fixing bacteria on rice roots (Watanabe and Barraquio 1979), and nitrate-dependent denitrifiers.

Among the modern techniques that can quantify and elucidate *in situ* physiological activities of the microorganisms are the immunofluorescent technique (Dommergues et al. 1978), and the microradiogram (Hoppe 1977). It is becoming increasingly apparent that the physiological activities of microorganisms in natural environments cannot be analyzed simply by the extrapolation of knowledge on the microbial behavior observed in laboratory-grown cultures.

Rice Soils in Different Countries

The soils on which rice is grown are extremely varied. Details on rice-growing soils of the world by country are given in a recent IRRI publication, *Soils and Rice* (IRRI 1978). Based on the importance of rice, and the description of rice soils using *Soil Taxonomy* (USDA 1975), the following are examples of tropical and temperate lowland rice soils.

Lowland Rice Soils in Tropical Asia

Most lowland rice soils in tropical Asia are considered either Entisols or Inceptisols (Kawaguchi and Kyuma 1977).

INDIA   India is the world's second largest rice producer. The rice-growing area for West Bengal and Bihar is the largest in India (more than 5 million hectares each), followed by Orissa (4.7 million hectares), Madhya Pradesh (4.5 million

Table 3.7 Distribution of Rice Soils in Different Agroclimatic Regions of India (From Murthy 1978)

Agroclimatic Region	States	Annual Rainfall	Temperature Range	Major Soil Groups
Humid western Himalayan region	Jammu & Kashmir, Himachal Pradesh & Kumaon, and Garhwal Divisions of Uttar Pradesh	<150 cm	Jan: 0–4°C July: 5–30°C	Submontane soils, hill soils, and Tarai soils
Humid Bengal-Assam Basin	West Bengal & Assam	<200 cm	Jan: 10–25°C July: 25–41°C	Riverine alluvium, Tarai soils, Laterite soils, red-yellow loams, red-sandy or gravelly soils
Humid eastern Himalayan region and Bay Islands	Arunachal Pradesh, Nagaland, Manipur, Mizoram, Tripura, Meghalaya and Andaman & Nicobar Islands	<200 cm	Jan: 11–24°C July: 25–33°C	Red loamy soils, laterite soils, red-yellow soils, and alluvial soils
Subhumid Sutlej-Ganga alluvial plains	Punjab, Uttar Pradesh, Bihar, and Delhi Territory	100–150 cm	Jan: 6–23°C July: 25–41°C	Calcareous alluvial soils, riverine alluvium, saline and alkaline soils, red-yellow loam, red-sandy or gravelly soils, and mixed red and black soils

Region	States / Union Territories	Rainfall	Temperature	Soils
Subhumid to humid eastern and southeastern uplands	Orissa, Andhra Pradesh, and Eastern Madhya Pradesh	75–150 cm	Jan: 16–28°C July: 27–36°C	Laterite soils, red-yellow loams, red-sandy or gravelly soils, mixed red and black soils, deltaic alluvium, deep and medium-deep black soils, red loamy soils, and coastal alluvium
Arid western plains	Haryana, Rajasthan, Gujarat, and Union Territory of Dadra Nagar Havelli	25–75 cm	Jan: 5–22°C July: 28–45°C	Alluvial soils, red-yellow soils, and medium-deep black soils
Semiarid Lava Plateaus and Central Highlands	Maharashtra, Western and Central Madhya Pradesh, and Union Territory of Goa, Daman and Diu	50–100 cm	Jan: 13–29°C July: 26–42°C	Riverine alluvium, coastal alluvium, mixed red and black soils, skeletal soils, deep-to-shallow black soils, and red-sandy or gravelly soils
Humid to semiarid Western Ghats and Karnataka Plateaus	Karnataka, Tamil Nadu, Kerala, Union Territory of Pondicherry and Lakshadweep Islands	75–200 cm	Jan: 20–29°C July: 28–38°C	Laterite soils, red-sandy or gravelly soils, deltaic alluvium, coastal alluvium, and red loamy soils

hectares), Andhra Pradesh (3.2 million hectares), Tamil Nadu (2.7 million hectares), Assam (2.0 million hectares), Maharashtra (1.4 million hectares) and Karnataka (1.1 million hectares). These areas cover 90% of India's total rice area of 39.6 million hectares. India's rice soils are grouped into eight agroclimatic regions (Table 3.7).

The soils of India for purposes of rice cultivation can generally be classified as:

- Alluvial-derived soils (Haplaquents, Ustifluvents, Udifluvents, Ustochrepts and Haplustalfs).
- Calcareous alluvial soils (Calciorthids).
- Coastal and deltaic alluvium (Tropaqualfs).
- Red Soils (Paleustalfs, Rhodustalfs, Haplustalfs).
- Red and yellow soils (Haplustults, Ochraquults, Rhodustults).
- Lateritic soils (Plinthaquults, Plinthustults, Plinthudults).
- Black soils (Ustochrepts, Ustropepts, Pellusterts, Chromusterts, Pelluderts).
- Mixed red and black soils (association of Alfisols and Vertisols).
- Gray-brown soils (Calciorthids).
- Brown-hill soils (Palehumults).
- Submontane soils (Hapludalfs).
- Tarai soils (Haplaquolls).
- Desert soils (Lithic Entisols, Psamments, Calciorthids).
- Saline-alkali soils (Salorthids and Natrargids).
- Peaty and saline-peaty soils (Histosols) (Murthy 1978).

Humid Western Himalayan Region.   The Western Himalayan region includes submountain areas that have:

- Karewa soils derived from deposits of lacustrine clays with moderate to high nitrogen and organic matter.
- Valley floor alluvium deposited by the Jhelum and Indus rivers. The alluvium has silt loam to clay loam texture and pH range of 5.4–8.5.

These soils cover sloping hill areas of Uttar Pradesh and Himachal Pradesh.

Humid Bengal-Assam Basin and Eastern Himalayan Region and Bay Islands. Rice is grown mostly on flat lands but is also grown on a wide range of slopes. Soils are mostly of alluvial silt deposited by numerous tributaries of the Ganges and Brahmaputra rivers. In lowlands, the water table is high and drainage is poor.

Soils in the Rarh region of West Bengal that cover the portion of Murshidabad, Bankura, the whole of Burdwan, the western half of Midnapore districts are classified as old alluvium.

The soils in Birbhum, Bankura, Burdwan and West Dinajpur districts are red, acidic, and deficient in phosphorus.

The soils in the Assam Valley are acidic but are high in available phosphorus and potassium with moderate organic matter and nitrogen contents.

Subhumid Sutlej-Ganga Alluvial Plain. In Uttar Pradesh, alluvial soils developed from the deposits from the Ganges and Yamuna rivers. The soils are sandy-textured in the west and more clayey in the east. In irrigated areas of Punjab, the soils are sandy-textured and alkaline in reaction. Organic matter and nitrogen contents are low.

Subhumid to Humid Eastern and Southeastern Uplands. The red and yellow soils in Madhya Pradesh are sandy in texture with pH ranging from 5.5 to 8.5.

Ultisols and Oxisols, which are deficient in phosphorus, are found in Orissa. In Andhra Pradesh, the alluvial soils of the delta are deep and well drained with clay loam to clay texture. The red soils occupy large areas. The laterite outwash materials are rich in iron and aluminum oxides with low silica and potassium. The pH range is 4.0–5.0. Poorly drained Vertisols occur in long stretches and in isolated pockets.

Arid Western Plains. In the states of Haryana, Rajasthan and Gujrat, alluvial red-yellow and moderately deep black soils constitute the major groups of the arid western plains.

In Rajasthan, rice is a minor crop. Except in Banaswara, where red-yellow soils are present, Rajasthan has alluvial soil. In Gujrat, the rice soils are medium-black riverine alluvium.

Semiarid Lava Plateau and Central Highlands. Maharashtra state rice-growing areas are divided into high rainfall zones with laterite soils (Ultisols and Oxisols) and nonlaterite soils. In western Madhya Pradesh, the soils are mostly alluvial, neutral to slightly alkaline in reaction, and sandy loam to clay loam in texture.

Humid to Semiarid Western Ghats and Karnataka Plateau. In coastal areas and in Malnad, the soils are Ultisols and Oxisols. The deep black Vertisols are found in all districts except Nilgiri.

The Tamil Nadu rice soils are in a relatively high-rainfall belt where irrigated rice is grown. In Thanjavur district, the soils are poor in nitrogen, phosphorus, and organic matter.

Acid Ultisols are found on the Malabar coast in Kerala. Soils that occur on the plains are deep and are deficient in phosphorus, potassium, and lime (Murthy 1978).

Classification of rice soils of India by rice-growing states are given in Table 3.8.

BANGLADESH In Bangladesh all major soils are used for rice, either solely or in rotation with upland crops. Eighty percent of the country is occupied by

Table 3.8 Classification (USDA 1975) of Rice Soils in the Rice-Growing States of India (Adapted from Murthy 1978)

State	Name of Soil Series Dominantly Growing Rice	Classification (Soil Taxonomy)
Bihar	Pagwara	Clayey mixed hyperthermic family of Ultic Paleustalfs
	Silphar	Fine clayey mixed hyperthermic family of Typic Haplustalfs
	Maheshjora	Fine loamy mixed hyperthermic family of Aquic Dystric Eutrochrepts
	Dudhan	Fine loamy mixed hyperthermic family of Lithic Haplustalfs
West Bengal	Banpara	Coarse silty mixed hyperthermic family of Typic Ochraqualfs
	Totpara	Fine silty mixed hyperthermic family of Vertic Ochraqualfs
	Hanar Gram	Fine loamy mixed hyperthermic family of Typic Ochraqualfs
	Sasanga	Fine loamy mixed hyperthermic family of Typic Haplaquepts
	Canning	Fine silty mixed hyperthermic family of Typic Haplaquents
Orissa	Pamra	Fine loamy mixed hyperthermic family of Aeric Ochraqualfs
	Bansimal	Fine loamy mixed hyperthermic family of Udic Plinthustalfs
	Sukhsoda	Fine mixed hyperthermic family of Aquic Haplustalfs
	Salod	Coarse loamy mixed hyperthermic family of Typic Ustifluvents
	Thorba	Fine mixed hyperthermic family of Aquic Paleustalfs
Maharashtra	Pen	Fine loamy mixed hyperthermic family of Lithic Ustochrepts
	Pale	Fine loamy mixed hyperthermic family of Paralithic Ustorthents
	Awasthi	Fine mixed hyperthermic family of Typic Haplustalfs

Region	Location	Soil classification
Goa Territory	Rivona	Fine mixed isohyperthermic family of Aquoxic Tropudalfs
	Calangute	Coarse loamy mixed isohyperthermic family of Orthic Hapludents
	Zuari	Fine mixed isohyperthermic family of Udic Haplaquents
Andhra Pradesh	Varigonda	Coarse loamy mixed isohyperthermic family of Udic Quartzipsamments
	Errakalva	Fine loamy mixed isohyperthermic family of Udic Haplustalfs
	Atmakur	Fine calcareous mixed isohyperthermic family of Typic Haplustalfs
	Dacha Palli	Fine calcareous montmorillonitic isohyperthermic family of Typic Pellusterts
Tamil Nadu	Adnur	Fine loamy over coarse loamy mixed isohyperthermic family of Ustifluvents
	Tiruvengadu	Fine loamy over coarse loamy mixed isohyperthermic family of Udipsamments
	Mudukkur	Coarse loamy over fine loamy mixed isohyperthermic family of Aquic Paleustalfs
	Alathur	Fine calcareous montmorillonitic hyperthermic family of Typic Chromusterts
	Cholachal	Fine loamy mixed isohyperthermic family of Tropaquepts
Kerala	Kunnathukal	Fine acid mixed isohyperthermic family of Aquic Dystrochrepts
	Marvkil	Coarse loamy mixed isohyperthermic family of Udifluvents
	Koipuram	Fine loamy mixed isohyperthermic family of Oxic Dystropepts
Karnataka	Arepalya	Coarse loamy mixed isohyperthermic family of Ustifluvents
	Chillahalli	Fine loamy mixed isohyperthermic family of Ustrochrepts
	Neginahalu	Coarse loamy mixed isohyperthermic family of Typic Haplaquents
North eastern Region	Sonparam	Fine loamy mixed thermic family of Aquic Udifluvents
	Lyampopki	Fine loamy mixed thermic family of Typic Haplaquepts

floodplain soils, mainly on the Ganges and Brahmaputra sediments. The most important characteristics of the floodplain soils are the complex patterns in which they occur, their rich mineralogical composition, the rapidity of profile development, and the dominant influence of hydrology on their agricultural use and potential (Brammer 1978).

Old terrace soils occupy 8% of the country. They occur in dissected landscapes on uplifted Madhupur clay low in weatherable sand minerals.

Brown acid hill soils occupy 12% of the country. Hill soils mainly comprise very strongly acid, dark brown, permeable, loamy soils overlying unconsolidated or fragmented sandstone or shale at 60–120 cm (except where eroded).

SRI LANKA   In Sri Lanka, 50% of the country's rice lands are in the dry zone. According to morphology of the landscape, the dry zone soils are grouped into three broad categories:

- Rice soils of the inland valleys (Aqualfs).
- Rice soils of the coastal plains (Aquepts).
- Rice soils of minor floodplains (Aquents and Fluvents) (Panabokke 1978).

In the intermediate zone, comprising 20% of the rice land in the country, soils are grouped into four categories based on climate, elevation, and the dominant suborder of the particular region:

- Rice soils of the semidry, low country (Aqualfs).
- Rice soils of the semiwet, low country (Aquults).
- Rice soils of the semiwet, middle country (Udalfs).
- Rice soils of the semidry, high country (Ustults).

In the wet zone, which has 30% of the country's rice lands, rice soils are considered under three broad categories based on elevation and the morphology of the landscape:

- Rice soils of the inland valleys, low country (Aquults).
- Rice soils of the west and southwest coastal plain and associated floodplains (Aquepts and Hemists).
- Rice soils of the narrow inland valleys and terraced slopes, in the middle country (Udults) (Panabokke 1978).

INDONESIA   In Indonesia, rice soils on alluvia and Regosols have the greatest variety in parent material and occur in humid and subarid regions. Lowland rice is grown on alluvial soils, gley soils, on Regosols, Grumusols, Podzolic, and Latosols, and partly on Andosols and Mediterranean soils (Soepraptohardjo and Suhardjo 1978).

Alluvial Soils (Entisols and Inceptisols). The alluvial soils occur in humid (Sumatra, West Java) and subarid regions (Sumbawa, Lombok). The total area of lowland rice on alluvial soils is about 2.9 million hectares.

Gley Soils (Aquepts). The alluvial soils, marked by gley phenomena in the subsurface layer, occur in West Java, Central Java, Lampung, North Sumatra, and South Sulawesi. The estimated area of gley soils growing rice is 200,000 ha.

Regosols (Entisols and Inceptisols). Lowland rice on Regosols is grown in the humid and subhumid regions that are irrigated. The total area is about 400,000 ha in Java, Bali, Lombok, Sumatra, and Sulawesi.

Grumusols (Vertisols). Rice is grown on Grumusols on 350,000 ha in three regions:

- Lowland plains and foot slopes adjacent to volcanoes (Central and East Java and South Sulawesi).
- Lowland plains and basins adjacent to, or surrounded by, limestone and marly hills of Central and East Java.
- Alluvial plains with subrecent deposits of admixtures of volcanic, marly materials, sometimes with other material from sandstone of marine origin.

Mediterranean soils (Alfisols, Luvisols). The Mediterranean soils occur in foot slopes and sloping plains of old volcanoes, and limestone and marly hills, usually in association with Grumusols, or as a transition from the Reddish Brown Latosols to the Grumusols. The area is about 200,000 ha.

Latosols (Inceptisols and Ultisols). Latosols are distributed in the eastern part of Java on the sides of old volcanoes; the foot slopes are usually occupied by Mediterranean soils. The area of lowland rice on Latosols is about 900,000 ha.

Andosols (Inceptisols and Tropohumults). A small area of lowland rice (50,000 ha) is grown on Andosols. The terrain of Andosols is usually sloping and rice is grown on terraced slopes, which are occasionally steep (West Java, Central Java, Bali, and North Sumatra).

Podzolic Soils (Ultisols and Acrisols). Lowland rice is grown on Podzolic soils on scattered plains in Banten (West Java), Lampung, East Sumatra, Aceh, and Southeast Sulawesi, comprising only about 300 ha.

THAILAND Rice occupies 25% of the land area of Thailand. On the Central Plain, 80% of the rice soils are Tropaquepts, of which at least 50% are acid sulfate soils. On the Northeast Plateau, Paleaquults, and Plinthaquults, together with

Table 3.9 Classification of Major Rice Soils in Thailand (From Rojanasoonthon 1978)

Great Soil Group (USDA 1975)	Distribution of Major Series			
	Central	Northeast	North	South
Alluvial				
Hydraquents				
Ustifluvents				
Sulfaquents				
Tropaquents	Tha Chin, Tha Muang, Sanphaya, Bang Pakong	Chiang Mai		Sai Kao
Hydromorphic Alluvial				
Tropaquents				
Sulfic-	Samutsongkham, Rungsit, Ayuthaya			
Vertic-				
Typic-	Bangkok, Bangkhen, Chachoengsao, Singburi, Ratchaburi, Chainat, Cha Am	Phimai		Tak Bai
Aeric-				
Sulfaquents			Chumsang	
Halaquepts				
Haplaquolls	Bang Len, Damnoen Saduak	Udon		
Grumusols				
Pelluderts				
Chromuderts	Lop Buri, Ban Mi, Chong Kae			
Low Humic Gley				
Tropaqualfs	Wichienburi	Phan	Hang Dong, Mae Sai	
Natraqualfs	Nong Kae	Kula Ronghai		
Paleaqualfs				
Paleaquults	Chonburi, Manorom, Lom Kao	Roi Et	Lampang, Chiang Rai	Chonburi, Bang Nara
Plinthaquults (Oxic)		Phen, On		Klaeng, Visai

Paleustults, are common associations. These are very infertile (Rojanasoonthon 1978). Table 3.9 indicates the classification of the major rice soils in Thailand.

The fertility level of the rice soils in Northeast Thailand is lower than that of the rice soils on the Central Plain (Kawaguchi and Kyuma 1969).

PHILIPPINES The lowland rice soils of the Philippines, especially those developed from mixed alluvia, often differ sharply from adjacent soils even with slight change in topographic conditions and position. Figures 3.4 and 3.5 illustrate the general grouping of Philippine rice soils on a coastal plain and a broad river valley. Included are:

- ENTISOLS Lowland rice soils on Fluvaquents, Tropaquents, and Tropofluvents are generally found lowest in the toposequence and normally at slopes of 0.5% or less.
- INCEPTISOLS Lowland rice fields on Tropaquepts, Fragiaquepts, and Eutropepts are normally on alluvial toeslopes or narrow coastal plains and small river valleys leading to the coastal areas and plains. The Inceptisols are higher in the toposequence than the associated paddies on the Entisols.
- VERTISOLS Pelluderts, Chromuderts, Pellusterts, and Chromusterts are extensively used for lowland rice production. They are found in all landscape positions generally used for lowland rice.
- ALFISOLS Tropaqualfs, Fragiudalfs, and Tropudalfs are found highest in the toposequence in the lowland rice system.

Lowland rice soils in the Philippines are generally high in natural fertility because of the geologically younger parent materials and because of periodic rejuvenation from volcanic ash, which readily weathers and releases nutrients in the humid tropical climate (Raymundo 1978).

Lowland Rice Soils in Temperate Countries

Rice soils of Japan and the United States are briefly described here as examples of soils on which flooded lowland rice is grown in temperate climates.

JAPAN Rice soils in Japan have been classified in a local system using descriptive terms such as *Brown lowland soils* and *Andosols*. More than 1 million hectares of *Gray lowland soils* constitute the largest soil group in which rice is grown. Gray lowland soils are widely distributed, mainly on the relatively well drained flat river and marine alluvial plains. Their productivity is usually medium to high. Sometimes physiological diseases like *Akiochi* occur in these soils. Intensive soil amendment and fertilization practices alleviate the physiological disease problems.

The soil group next in importance is the *Wet Andosols*, which are hydromorphic Andosols characterized by prominent iron mottles in their profiles.

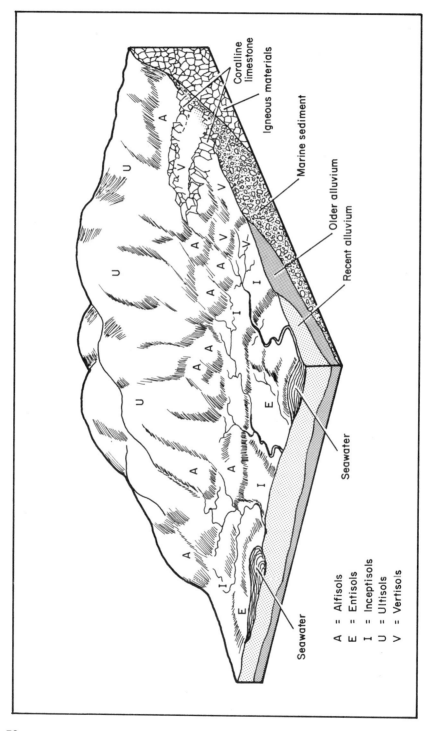

Coralline limestone

Igneous materials

Marine sediment

Older alluvium

Recent alluvium

Seawater

Seawater

A = Alfisols
E = Entisols
I = Inceptisols
U = Ultisols
V = Vertisols

Figure 3.4   Paddy soilscapes on a coastal plain physiography in the Philippines. (From Raymundo 1978)

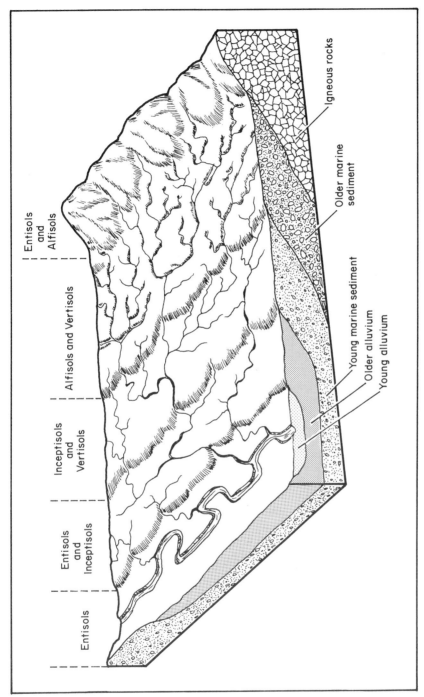

Figure 3.5 Paddy soilscapes on a broad river valley in the Philippines. (From Raymundo 1978)

Entisols and Alfisols

Alfisols and Vertisols

Inceptisols and Vertisols

Entisols and Inceptisols

Entisols

Igneous rocks

Older marine sediment

Young marine sediment

Older alluvium

Young alluvium

These characteristics develop because of long-term cultivation. The Andosols are mainly distributed on nearly flat alluvial wetlands.

Yellow soils are used for lowland rice where irrigation is available. Leaching of bases has caused the yellow soils to become acidic.

Mitsuchi (1974) reported relationships between the soil permeability and profile development in Japan's lowland rice fields without groundwater within 1.5 m of the surface. They are:

- Brown lowland soils—highly permeable.
- Gray lowland soils—moderately permeable.
- Hanging-water lowland soils—impermeable.

Recently, Matsuzaka (1978) correlated the soil groups in Japan using *Soil Taxonomy* and the FAO soil map of the world (Table 3.10). Inceptisols cover the largest rice-growing area in Japan. Other soils on which rice is grown are Ultisols, Entisols, and Histosols.

Rice soils in Japan are divided into production capability classes ranging from Class I soils with no or only a few limiting factors to Class IV with the most limiting factors. These are based on:

- Thickness of top soil.
- Gravel content of top soil.
- Permeability.
- Redox potential (Eh).
- Inherent fertility.
- Content of available nutrients.
- Presence of toxic substances (Matsuzaka 1978).

According to Kawaguchi (1973), the lowland rice soils of Japan are generally acidic with an average pH of 5.5 in the plow layer. Total phosphorus is higher than that in rice soils in South and Southeast Asia.

Because of the heavy rate of fertilizer and compost application, rice yields in Japan are one of the highest in the world. Nevertheless, for growing rice the initial soil fertility is considered highly important in Japan to the extent that there is a popular Japanese saying, "Grow paddy with soil fertility, grow barley (or wheat) with fertilizers." This is because the response of rice to fertilizer is not so great as that of barley or wheat (Takahashi 1965).

UNITED STATES   The soils in the United States on which rice is grown are either naturally wet during part of the year, or they have soil horizons that, because of their high clay content, mineralogy, and sodium content or pan characteristics, allow little downward percolation of water. The following are the main rice soils in the United States described by Flach and Slusher (1978).

Table 3.10 Area by Soil Group in Japan on which Rice is Grown and Correlation of Soil Groups of the Japanese System with the Suborders/Great Groups of *Soil Taxonomy* and the Soil Units of the *Soil Map of the World* (Adapted from Matsuzaka 1978)

Japanese System	Rice-Growing Area (1000 ha)	Soil Taxonomy[a]	Soil Map of the World[b]
Andosols	17	Dystrandepts (Eutrandepts, Durandepts, Vitrandepts)	Humic (Mollic, Ochric, Vitric) Andosols
Wet Andosols	279	Aquic Dystrandepts	Gleyic Andosols
Gleyed Andosols	43	Andaquepts	Gleyic Andosols
Brown Forest soils	5	Dystrochrepts	Dystric (Eutric) Cambisols
Gray Upland soils	79	Haplaquepts	Dystric (Eutric) Gleysols
Gley Upland soils	40	Haplaquepts	Dystric (Eutric) Gleysols
Red soils	0.4	Hapludults	Orthic Acrisols
Yellow soils	148	Hapludults (Aquic Hapludults)	Orthic (Gleyic) Acrisols
Brown Lowland soils	145	Udifluvents, Haplaquepts	Eutric (Dystric) Fluvisols, Eutric (Dystric) Gleysols
Gray Lowland soils	1061	Haplaquepts	Eutric (Dystric) Gleysols
Gley soils	882	Haplaquents, Haplaquepts	Eutric (Dystric) Gleysols
Muck soils	74	Saprists (?)	Dystric (Eutric) Histosols (?)
Peat soils	113	Fibrists (?)	Dystric (Eutric) Histosols (?)

[a]USDA 1975.
[b]Unesco 1974.

75

Alfisols.   About 50% of the rice is grown in Aqualfs, which are the wet Alfisols. Crowley soils are by far the most important single soil series for rice in the United States. Crowley soils are somewhat poorly drained. In California 23% of rice-growing areas are on Alfisols.

Inceptisols.   Inceptisols make up 21% of the soils used for rice in the southern United States.

Vertisols.   About 11% of the rice soils in the southern United States are Vertisols. They are in two soil series on the lower coastal plain in Texas. Vertisols are clayey soils with high shrink-swell potential.

Vertisols provide the second largest rice land area in California. The soils do not have a shallow water table, but their fine textures and their smectitic (montmorillonitic) clay mineralogy make them ideally suited for rice culture.

Mollisols.   About 8% of the soils in the southern United States are Mollisols. They differ from Vertisols, which may have a similar surface soil, but do not show strong evidence of shrinking and swelling.

In California about 45% of the rice grown is on Mollisols, mostly on Sacramento soils, which are very fine smectitic, thermic vertic Haplaquolls.

Problem Soils on Which Lowland Rice is Grown

Interest in the vast areas of marginal land in the tropics and subtropics that could grow rice if suitable technology is available is rapidly increasing. In South and Southeast Asia alone about 90 million hectares of land suited to rice production lie idle largely because of soil toxicities (Table 3.11). In these marginal lands, soils are problematic and modern rice varieties and their technology are not always suitable.

Saline Soils

An estimated 150 million hectares of current and potential rice lands in the tropics and subtropics are affected by salinity (Massoud 1974). Salinity is the main constraint to high rice yields in deltas, estuaries, and coastal fringes in the humid tropics. It also causes a serious problem in arid and semiarid areas (Ponnamperuma 1977).

In general, soil salinity is caused by presence or by intrusion of seawater, or by surface evaporation of soil water initially high in salt content. Some saline land is found along every coast, but the extent of affected areas varies greatly. Generally, salt accumulation increases in the drier climates and diminishes strongly in an equatorial climate without a pronounced dry season. Extremely saline deltas are found in arid areas.

Rice is generally reported as a moderately salt-tolerant crop but no rice variety

Table 3.11 Problem Soils in South and Southeast Asia (Adapted from
Ponnamperuma and Bandyopadhya 1980)

Country	Area (millions of hectares)			
	Saline Soils	Alkali Soils	Acid Sulfate Soils	Peat Soils
Bangladesh	2.5	0.5	0.7	0.8
Burma	0.6	—	0.2	—
India	23.2	2.5	0.4	—
Indonesia	13.2	—	2.0	16.0
Kampuchea	1.3	—	0.2	—
Malaysia	4.6	—	0.2	2.4
Pakistan	1.1	9.4	—	—
Philippines	0.4	—	—	—
Thailand	1.5	—	0.7	0.2
Vietnam	1.0	—	1.0	1.5
Total	49.4	12.4	5.4	20.9

can withstand high salinity throughout its growth cycle. Soil solutions high in
sodium chloride with specific conductance values of 6–10 mS/cm (6–10
mmho/cm) are harmful to rice plants and cause as much as 50% decrease in rice
yield. In tidal areas, the specific conductance values change from day to day
depending on tidal regime.

Inland saline soils generally have a pH > 7. They are low in nitrogen, low in
available phosphorus, but well supplied with potassium. Coastal saline soils may
be eutric (fertile), dystric (infertile), calcaric (calcareous), thionic (sulfidic and
sulfatic), or histic (organic). Saline-sodic soils and calcaric saline soils are
deficient in available zinc; thionic saline soils may have problems of aluminum
toxicity and iron toxicity as well as phosphorus deficiency; histic saline soils are
deficient in major nutrients as well as copper and zinc, and have other soil
problems (Table 3.12). Hydrology and relief are important in determining the
suitability of saline lands for rice production (Ponnamperuma and Bandyopadhya 1980).

Sodic (Alkali) Soils

Sodicity (alkalinity) is caused by excessive exhangeable sodium and occurs on
millions of hectares of irrigated or irrigable land in the Indo-gangetic plain in
India and Pakistan as well as in large areas in the Middle East and Africa. Sodic
soils are less common than saline-sodic soils (saline-alkali soils) (Ponnamperuma
1977). Harmful high alkalinity is related primarily to the presence of sodium
carbonate and sodium bicarbonate. Accumulation of these salts in the soil and

Table 3.12 Some Saline Soil Problems for Rice (From Ponnamperuma and Bandyopadhya 1980)

Kinds of Saline Soil	Accessory Growth-Limiting Factors
Inland saline soils	High pH; deficiencies of zinc, nitrogen, and phosphorus
Acid saline soils	Aluminum toxicity; iron toxicity; phosphorus deficiency
Coastal acid sulfate soils	Iron and aluminum toxicities; phosphorus deficiency; deep water
Neutral and alkaline coastal soils	Zinc deficiency; deep water
Coastal organic soils	Deficiencies of nitrogen, phosphorus, zinc, copper, molybdenum; toxicities of iron, hydrogen sulfide, and organic substances; deep water; *Pyricularia oryzae* (Cav.), *Helminthosporium oryzae* Breda de Haan

soil water occurs when groundwater with high content of bicarbonate ions evaporates.

In strongly alkaline soils, sodium carbonate and bicarbonate become prevalent salts, and sodium will replace calcium in the clay complex, with concurrent precipitation of calcium carbonate (sodic soils). This leads to the formation of sodium clays, which are strongly dispersed and hence, highly impermeable to water.

Rice cultivation is suggested for reclamation of highly sodic soils, but because of their poor permeability and relatively low mobility of exchangeable sodium, the alkaline soils generally do not improve under lowland rice culture (Moormann and van Breemen 1978). There have been efforts to amend sodic soils in India and Pakistan, for example, by applications of gypsum, pyrite, or acid.

Acid Sulfate Soils

Actual and potential acid sulfate soils cover 5 million hectares in South and Southeast Asia, mainly in Thailand, Vietnam, and Indonesia. There are about 3.7 million hectares of acid sulfate soils in Africa and about 2 million hectares in Venezuela (van Breemen 1980).

Rice is grown on most of about 800,000 ha of acid sulfate soils in Thailand and about 500,000 ha of acid sulfate soils in Vietnam (Moormann and van Breemen 1978). Brammer (1978) reported acid sulfate soils occurring near the coast on the Chittagong coastal plain in Bangladesh and in the southwestern part of the Ganges Tidal floodplain. For Burma, Kyuma and Takaya (1976) reported acid sulfate soils at two places on the banks of the Irrawaddy river.

Toxicities of acid sulfate soils for lowland rice are mainly due to high level of aluminum and perhaps acidity per se at the early stages of rice growth. Excess iron is harmful in young acid sulfate soils after reduction (Moormann and van Breemen 1978). Although the pH of surface soils may vary between 3.5 and 4.5 before flooding, after flooding the pH may sometimes rise to 6.0–6.5 making it possible to grow lowland rice. However, phosphorus fertilization is essential to get high rice yields (van Breemen 1980).

Acid sulfate soils developed from reduced, nonacid sediments, high in pyrite and low in calcium carbonate are the so-called potential acid sulfate soils. Less seriously affected acid sulfate soils on which rice is currently grown belong to sulfic subgroups of Tropaquepts and Haplaquepts.

Histosols

Histosols are soils characterized by a high organic matter content in the major part of the profile down to 80 cm. Histosols are highly acid, with a pH between 3.5 and 4.0. According to Moormann and van Breemen (1978), the major areas of Histosols in the world's rice-growing zone occur in the high rainfall equatorial part of Southeast Asia, East and West Malaysia, the islands of Sumatra and Kalimantan, and West Irian in Indonesia. Ponnamperuma (1977) estimated 20–30 million hectares of Histosols in Indonesia.

In Bangladesh, small areas of Histosols occur in perennially wet basins and also in some deep depressions (*haors*) in the upper Meghna floodplain (Brammer 1978). Histosols are found to a lesser extent in the Mekong Delta of Vietnam and the Ganges Delta. Rice is grown on the thinner Histosols containing relatively high proportions of mineral material.

Histosols vary in chemical properties, profile characteristics, and hydrology. The main mineral stresses are deficiencies of nitrogen, phosphorus, potassium, zinc, copper, and molybdenum. Salinity, strong acidity, and hydrogen sulfide toxicity are other possible stresses (Ponnamperuma 1977). Land preparation of Histosols is difficult.

Wet soils that are high in organic matter in the top soil layers do not provide the rice plant with proper anchorage and make the crop vulnerable to lodging. Drainage minimizes some of the problems on Histosols but reclamation of these soils is expensive. Therefore, Histosols are currently not used widely for rice production. If rice is grown, surface drainage is usually necessary to ensure any yield at all. Drainage decreases toxicity due to dissolved organic substances and increases the availability of most nutrients, presumably as a result of increased oxidation and mineralization of organic matter (van Breemen 1980). Artificial drainage may cause rapid oxidation and mineralization of the organic matter of Histosols causing subsidence of the soil surface.

UPLAND RICE SOILS

The characteristics of soils on which upland rice is grown are nonspecific in respect to soil texture, pH, organic matter content, slope, and soil fertility variations (De Datta and Feuer 1975).

According to Moormann and Dudal (1965) soil texture affects the moisture status of a soil more than any other property except topography. That makes texture particularly important in upland rice fields, which have no bunds to hold moisture. For upland soils, it is important to consider texture of surface and subsoils. Subsoil texture is perhaps the most important because it can serve as a moisture reservoir.

The textures of upland rice soils vary greatly. For example, loam and sandy soils are typical of the slightly elevated areas of Thailand's foothills and flat lands, but most upland rice in the hills is grown on clayey and clay loam soils.

Upland Rice Soils of Southeast Asia

In Southeast Asia, the major soils of the sloping, unterraced land where most upland rice is grown are Ultisols and Alfisols (Table 3.4). Upland rice soils in Southeast Asia are mostly acidic with soil pH varying from 4.5 to 5.8 (De Datta and Feuer 1975).

Ultisols

Ultisols are characterized by clay translocation to the subsoil and intensive leaching with depletion of bases. Ultisols are common, particularly in Sumatra and Kalimantan, Indonesia and in Thailand. These soils have a sandy surface texture, and a more clayey subsoil. The base saturation is low, and the clay type is 1 : 1. With Ultisols the clayey subsoil is commonly closer to the surface than the subsoil of Alfisols, which makes Ultisols more subject to erosion during heavy rainfall; the surface soil becomes saturated, causing runoff of excess water from sloping areas. Many Ultisols have nutrient problems for rice that are difficult to correct.

Alfisols

Alfisols are characterized by translocation of silicate clays to the subsoil without excessive depletion of bases and without a mollic epipedon (a dark-colored, well structured, deep surface horizon).

Alfisols have more clay in the B than in the A horizons and base saturation is high. Alfisols are most common in very dry zones. They are easy to till because of their typically granular surface soil structure. These soils have a high capacity to retain rainfall because the more clayey B horizon retards percolation losses.

Oxisols

Oxisols show extreme weathering of most aluminosilicate minerals, an absence of translocation of silicate clays, and a low activity of the kaolinite type clay fraction.

Oxisols cover minor upland rice-growing areas in Southeast Asia. They are found in lower areas in Sarawak, Malaysia (USDA 1966) and in the mountainous Ban Me Thout area of southern Vietnam (Tung 1973). Oxisols are the most important rice soils in Java; they are locally known as Latosols.

Oxisols in Southeast Asia are found on materials derived from highly weathered rock and have few textural changes throughout the profiles. Oxisols are high in clay content—mostly kaolin with oxides of iron and aluminum—and usually have high infiltration rates. Oxisols are low in all available nutrients.

Upland Rice Soils in West Africa

In West Africa, upland rice is the major system of rice culture. The West African soils have low capacity to hold available moisture and droughts of short duration significantly reduce rice yields (Moormann 1973).

Because upland rice depends entirely on rainfall for water supply in the freely drained soils, it is limited to high rainfall areas with good rainfall distribution. The freely drained soils, as classified by the FAO system (Table 3.4), include Nitosols, Acrisols, Luvisols, Ferralsols, and Cambisols. Because areas with hydromorphic soils receive supplemental water through surface runoff, rice production is not limited to only the most favorable rainfall distribution patterns. The main soils of this type are Entisols, Ultisols, and Vertisols.

Hydromorphic soils in West Africa are considered suitable for upland rice. Moormann (1973) reported marked correlation between the depth of groundwater and the growth and production of upland rice. He contended that surface soils with medium to sandy texture overlying a subsoil of finer texture are often considered best for upland rice because they are easy to till.

In West Africa, a number of level, fine-textured soils have been classified as particularly suitable for rice culture. Riquier (1971), using the FAO classification system, considered Gleyic Cambisols, Humid Gleysols, and Eutric Fluvisols as suited for rice where climatic factors are favorable. Orthic Ferralsols (Table 3.4) are considered marginally suited for rainfed rice. According to Moormann (1973) subunits such as Gleyic Cambisols, Humid Gleysols, and Eutric Fluvisols represent a large proportion of potential rice areas in humid parts of West Africa.

The inherent fertility levels and chemical composition of soils often explain yield differences and even cultural practices (De Datta and Feuer 1975). A review of fertility levels of West African rice soils by Kang (1973) indicated that the infertile Oxisols in areas under shifting cultivation cannot sustain continuous rice culture without added fertilizer.

Upland Rice Soils in Latin America

Most of Latin America's upland rice (about 75% of the total rice-growing area in that continent) is grown on well drained soils, principally Oxisols, particularly in *campo cerrado* of Brazil, the *llanos orientales* of Colombia, and parts of the Amazon jungle.

Some Alfisols and Ultisols are found in Sao Paulo, Minas Gerais, and Goias states of Brazil. All those soils are dry and well drained; most are clayey, with low levels of available moisture because of their excellent structures and high levels of iron, and aluminum oxides.

The Oxisols generally have a pH between 4 and 5, aluminum saturation of more than 60%, low levels of available phosphorus, and a cation exchange capacity of 2–8 meq/100 g. The Ultisols have kaolinitic type of clays. On Oxisols and Ultisols rice responds to nitrogen, phosphorus, potassium, calcium, and in some cases zinc and iron. Lime is sometimes necessary, depending on soil pH and the rice variety, and is usually applied as calcium and magnesium fertilizer. Farmers prefer to rotate rice and pasture on the Alfisols.

Oxisols are particularly important for upland rice around Goiania and Anapolis of Brazil. Upland rice yields in the *campo cerrado* and at the *llanos orientales* are about 1 t/ha. The principal factors that limit yield seem to be moisture stress and low fertility.

In the Amazon jungle area, where rice is grown on well drained soils, principally on highly acid Ultisols, cassava, maize, and other crops are intercropped with rice in a shifting cultivation system (De Datta and Feuer 1975). These soils generally have a higher range of available water than do the Oxisols of the *campo cerrado,* although their topsoils are coarser. The soil pH levels are low, aluminum saturation is high, and phosphorus content is low. On the Amazon Ultisols rice is grown for a few years without fertilizer but on ashes left after the vegetation is burned.

Oxisols under savanna vegetation are most prevalent in the *campo cerrado* (Feuer 1956, Cline and Buol 1973) and the *llanos orientales* of eastern Colombia. Upland rice is commonly grown on soils that have some weatherable minerals and are able to support semideciduous forests. In general, soils under savanna are not protected by vegetation at the beginning of the rainy season, and most losses of nutrients are due to truncation of the surface layers (van Wambeke 1975).

In terms of management of Oxisols, van Wambeke suggested that tillage practices should be developed to increase the water intake capacity of the soils and take advantage of rainfall, reduce runoff, and control erosion. Under intensive management systems, correction of nutritional deficiencies by fertilizers is essential.

In Peru, Ultisols and Alfisols predominate in both well drained and poorly drained areas and upland rice is grown on both acid and fertile alluvial soils, which are not subjected to permanent flooding (Sanchez and Nureña 1972).

Upland rice can also be found on poorly drained soils throughout Latin America. They range from Vertisols in Cuba and Andosols (volcanic ash soils) on the pacific coast of Guatemala to excellent alluvial soils on the Pacific coast of Costa Rica and poorly drained Ultisols, Oxisols, Inceptisols, and Alfisols in Peru and Brazil. In many of those areas, rainfall is high and drought is generally absent. Yields as high as 7 t/ha are fairly common with lowland varieties grown under upland rice culture.

RICE SOILS AND THEIR FERTILITY CONSIDERATIONS

Rice is grown on all soil Orders. Some are especially important, but on others, like the Aridisols, Histosols, and Spodosols, rice is seldom grown (Moormann and Veldkamp 1978). In general, particularly in Asia, the wet (aquic) suborders and subgroups are most extensively used.

The sources of the natural supply of plant nutrients in soil are the soil constituents, rain and irrigation water, the flooded rice culture, rice-based cropping systems, recycling of crop residues and other organic materials, and biological nitrogen fixation. Information on the natural sources of plant nutrients and the scope of their efficient use in rice farming will greatly facilitate the development of efficient use of the natural sources of plant nutrients and chemical fertilizers (Patnaik 1978).

In humid tropical Asia, lowland rice production depends on the availability of sufficient water of adequate quality to satisfy the requirements of rice. All kinds of soils can be used for rice cultivation if water conditions are favorable. Soils in most countries in Asia can sustain rice yields of 1–2 t/ha without chemical fertilizers but with good water supply and good agronomic management. In fact, most soils in the Philippines can sustain at least a 3 t/ha yield without chemical fertilizer. Some soils in the Mindanao region in the Philippines produce yields of 5–6 t/ha without fertilizer.

In Asia, nitrogen fertility of soils is the major source of the nutrient for lowland rice. For example, in a recent study by Koyama et al. (1973) in Thailand more than 60% of the nitrogen taken up by the rice plants at harvest was derived from mineralization of soil organic matter. The amount estimated was a little less than that in a Japanese rice soil with low fertility (Kawaguchi and Kyuma 1977).

In seasonally wet, acid soils in Bangladesh, Brammer and Brinkman (1977) reported anomalously low clay contents in the surface horizons. This may be explained by the cyclic process of soil reduction and eluviation of cations alternating with oxidation and acidification. The process, termed ferrolysis (Brinkman 1970), is known to occur in temperate and humid tropical climates, where there is a seasonal alternation between wet and dry seasons. In Northeast Thailand, ferrolysis can explain the low cation-exchange capacity of surface soils reported by Kawaguchi and Kyuma (1969). The natural fertility level is usually low; the soils are moderately to strongly acid, with high proportions of exchangeable aluminum particularly in the surface and shallow subsoil horizons. In the upper horizons, both organic matter content and percentage of free iron are generally low (Brinkman 1978). In these soils rice can be grown—one crop per year in the rainy season, or two with irrigation.

In tropical Africa, Alfisols and Ultisols dominate the landscape. The principal constraints of Alfisols to rice production are susceptibility to erosion, low available water retention, low reserves of nitrogen, phosphorus, copper, zinc, and sulfur, and the development of acidity with continuous cultivation. For Ultisols the constraints are low nutrient retention, low reserves of nitrogen, phosphorus, potassium, calcium, manganese, copper, and zinc, and problems associated with soil acidity (Moormann and Greenland 1980).

In most areas of West Africa, but more especially in the areas with less than 1500–1700 mm annual rainfall with an unreliable distribution, the occurence of drought stress is one of the main production-limiting factors. In the hydro-morphic (phreatic) soils—soils that receive supplementary water from surface flow and from interflow—drought stress is less severe or, in the most favorable cases, not present at all. In those soils, rice can be grown with proper management. Phreatic rice lands are mainly found in the Aquic (wet) Suborders of Entisols, Inceptisols, and the Aquic Subgroups of the already mentioned soil Orders.

The often sandy texture of African soils may be considered as a restraint to productivity. Because most of the soils in the wetter zone of West Africa and Central Africa are derived from either intermediate to acid crystalline rocks or from various sedimentary rocks, they predominantly give rise to soils with a poor potential for rice. Unlike in Asia, the development of rice-based cropping systems, with continuous land use, is difficult for pluvial rice lands (Moormann and Veldkamp 1978).

In Latin America, Oxisols are the most important rice soils followed by Ultisols and Alfisols. Favorable rainfall, topography, drainage, and physical characteristics make Oxisols and Ultisols suitable for rice and other crops, but low fertility, high soil acidity, and deficiencies of phosphorus have restricted their use.

REFERENCES

ASTM (American Society for Testing and Materials). 1958. *Procedures for testing soils.* ASTM Committee D-18 on soils for Engineering Purposes. ASTM. Philadelphia, Pennsylvania. 544 pp.

Blake, G. R. 1965a. Bulk density. Pages 374–390 *in* American Society of Agronomy. *Methods of soil analysis. Part* I. Monogr. 9. Madison, Wisconsin.

Blake, G. R. 1965b. Particle density. Pages 371–373 *in* American Society of Agronomy. *Methods of soil analysis. Part* I. Monogr. 9. Madison, Wisconsin.

Bodman, G. B., and J. Rubin. 1948. Soil puddling. *Soil Sci. Soc. Am. Proc.* 13:27–36.

Brammer, H. 1978. Rice soils of Bangladesh. Pages 35–55 *in* International Rice Research Institute. *Soils and rice.* Los Baños, Philippines.

Brammer, H., and R. Brinkman, 1977. Surface-water gley soils in Bangladesh: environment, landforms and soil morphology. *Geoderma* 17:91–109.

Brinkman, R. 1970. Ferrolysis, a hydromorphic soil forming process. *Geoderma* 3:199–206.

Brinkman, R. 1978. Ferrolysis: chemical and mineralogical aspects of soil formation in seasonally wet acid soils, and some practical implications. Pages 295–303 *in* International Rice Research Institute. *Soils and rice*. Los Baños, Philippines.

Briones, A. A. 1979. Physical properties of paddy soils. INSFFER (International Network on Soil Fertility and Fertilizer Evaluation for Rice) Lecture series, 16 March 1979, International Rice Research Institute, Los Baños, Philippines. (unpubl. mimeo.)

Cline, M. G., and S. W. Buol. 1973. Soils of the Central Plateau of Brazil and extension of results of field research conducted near Planaltina, Federal District, to them. *Agron. Mimeo* 73-13, Cornell University, Ithaca, New York. 43 pp.

De Datta, S. K., and R. Feuer. 1975. Soils on which upland rice is grown. Pages 27–39 *in* International Rice Research Institute. *Major research in upland rice*. Los Baños, Philippines.

Dixon, J. B., and S. B. Weed, eds. 1977. *Minerals in soil environments*. Soil Science Society of America, Madison, Wisconsin. 972 pp.

Dommergues, Y. R., L. W. Belser, and E. R. Schmidt. 1978. Limiting factors for microbial growth and activity in soil. *Adv. Microbiol. Ecol.* 2:49.

Dudal, R. 1958. Paddy soils. *Int. Rice Comm. Newsl.* 7(2):19–27.

Dudal, R., and F. R. Moormann. 1964. Major soils of Southeast Asia: their characteristics, distribution, use and agricultural potential. *J. Trop. Geogr.* 18:54–80.

Feuer, R. 1956. An exploratory study of the soils and agricultural potential of the soils of the future federal district in Central Plateau of Brazil. PhD dissertation, Cornell University, Ithaca, New York. 432 pp.

Flach, K. W., and D. F. Slusher. 1978. Soils used for rice culture in the United States. Pages 199–214 *in* International Rice Research Institute. *Soils and rice*. Los Baños, Philippines.

Foth, H. D., and L. M. Turk. 1972. *Fundamentals of soil science*. John Wiley and Sons, Inc., New York. 454 pp.

Hayashi, S., and C. Furusaka. 1979. *Studies on* Propionibacterium isolated from paddy soils, Antonie van Leeuwenhoek. 45:565–574.

Hoppe, G. 1977. Analysis of actively metabolizing bacterial populations with autographic method. Pages 179–197 *in* G. Rheinheimer, ed. *Microbial ecology of a brackish water environment*. Springer-Verlag.

IRRI (International Rice Research Institute). 1978. *Soils and rice*. Los Baños, Philippines. 825 pp.

Kang, B. T. 1973. Soil fertility problems in West Africa in relation to rice production. Paper presented at the seminar on rice-soil fertility and fertilizer use, 22–27 January 1973, West Africa Rice Development Association (WARDA), Monrovia, Liberia.

Kanno, I. 1956. A scheme for soil classification of paddy fields in Japan with special reference to mineral paddy soils. *Bull. Kyushu Agric. Exp. Stn.* 4:261–273.

Kanno, I. 1962. A new classification system of rice soils in Japan. *Pedologist* 6:2–10.

Kawaguchi, K. 1973. The description and classification of the soils of Japan. Pages 1–54 *in* Soils of the ASPAC region 6. *ASPAC Tech. Bull.* 15.

Kawaguchi, K., and K. Kyuma. 1969. *Lowland rice soils in Thailand*. Reports on research in Southeast Asia. Natural Science Ser. N-4. The Center for Southeast Asia Studies, Kyoto University, Japan. 270 pp.

Kawaguchi, K., and K. Kyuma. 1974. Paddy soils in tropical Asia. 2. Description of material characteristics. *Southeast Asia Stud.* 12:177–192.

Kawaguchi, K., and K. Kyuma. 1977. *Paddy soils in tropical Asia. Their material nature and fertility.* The University Press of Hawaii, Honolulu. 258 pp.

Koyama, T., C. Chammek, and N. Niamsrichand. 1973. Nitrogen application technology for tropical rice as determined by field experiments using ^{15}N tracer technique. *Trop. Agric. Res. Cent. Tech. Bull.* 3. 79 pp.

Kyuma, K. 1978. Mineral composition of rice soils. Pages 219–235 *in* International Rice Research Institute. *Soils and rice.* Los Baños, Philippines.

Kyuma, K., and Y. Takaya. 1976. Interim report of the field survey on geomorphology and soil of paddy lands in the Irrawaddy delta. Rangoon, Burma.

Massoud, F. I. 1974. Salinity and alkalinity as soil degradation hazards. FAO/UNEP Expert Consultation on Soil Degradation. 10–14 June 1974, FAO, Rome.

Matsuguchi, T., and B. Tangcham. 1974. Free-living nitrogen fixers and acetylene reduction in tropical paddy field. Pages 180–188 *in* International Society of Soil Science. *Transactions of the 10th International Congress of Soil Science.* Vol. IX.

Matsuzaka, Y. 1978. Rice soils of Japan. Pages 163–177 *in* International Rice Research Institute. *Soils and rice.* Los Baños, Philippines.

McCarthy, D. F. 1977. *Essentials of soil mechanics and foundations.* Reston Publ. Co., Inc., Virginia. 505 pp.

Mitsuchi, M. 1974. Pedogenic characteristics of paddy soils and their significance in soil classification [in Japanese, English summary]. *Bull. Natl. Inst. Agric. Sci.* Ser. B, 25:29–115.

Miura, K., and T. Yoshida. 1972. Microflora of the rice root zone in submerged Maahas clay soil. *Kalikasan (Philipp. J. Biol.)* 1:182–196.

Moormann, F. R. 1973. General assessment of land on which rice is grown in West Africa. Paper presented at the seminar on soil fertility and fertilizer use, 22–27 January 1973, West Africa Rice Development Association (WARDA), Monrovia, Liberia.

Moormann, F. R. 1978. Morphology and classification of soils on which rice is grown. Pages 255–272 *in* International Rice Research Institute. *Soils and rice.* Los Baños, Philippines.

Moormann, F. R., and R. Dudal. 1965. Characteristics of soils on which paddy is grown in relation to their capability classification. Soil Survey Rep. 32. Land Development Department, Bangkok. 22 pp. (Also FAO DO. IRC/SF-64/4, mimeo.)

Moormann, F. R., and D. J. Greenland. 1980. Major production systems related to soil properties in humid tropical Africa. Pages 55–77 *in* International Rice Research Institute. *Priorities for alleviating soil-related constraints to food production in the tropics.* Los Baños, Philippines.

Moormann, F. R., and N. van Breemen. 1978. *Rice: Soil, water, land.* International Rice Research Institute. Los Baños, Philippines. 185 pp.

Moormann, F. R., and W. J. Veldkamp. 1978. Land and rice in Africa: constraints and potentials. Pages 29–43 *in* I. W. Buddenhagen and G. J. Persley, eds. *Rice in Africa.* Academic Press, London.

Murthy, R. S. 1978. Rice soils of India. Pages 3–17 *in* International Rice Research Institute. *Soils and rice.* Los Baños, Philippines.

Panabokke, C. R. 1978. Rice soils of Sri Lanka. Pages 19–33 *in* International Rice Research Institute. *Soils and rice.* Los Baños, Philippines.

Patnaik, S. 1978. Natural sources of nutrients in rice soils. Pages 501–519 *in* International Rice Research Institute. *Soils and rice.* Los Baños, Philippines.

Ponnamperuma, F. N. 1977. *Screening rice for tolerance to mineral stresses.* IRRI Res. Pap. Ser. 6. 21 pp.

Ponnamperuma, F. N. 1978. Electrochemical changes in submerged soils and the growth of rice. Pages 421–441 *in* International Rice Research Institute. *Soil and rice.* Los Baños, Philippines.

Ponnamperuma, F. N., and A. K. Bandyopadhya. 1980. Soil salinity as a constraint on food production in the humid tropics. Pages 203–216 *in* International Rice Research Institute. *Priorities for alleviating soil-related constraints to food production in the tropics.* Los Baños, Philippines.

Raymundo, M. E. 1978. Rice soils of the Philippines. Pages 115–133 *in* International Rice Research Institute. *Soils and rice.* Los Baños, Philippines.

Riquier, J. 1971. Note on the suitability of most African soils for rice cultivation. Pages 6–10 *in* United Nations Development Programme. Notes on the ecology of rice and soil suitability for rice cultivation in West Africa. Food and Agriculture Organization, United Nations, Rome.

Rojanasoonthon, S. 1978. Rice soils of Thailand. Pages 73–85 *in* International Rice Research Institute. *Soils and rice.* Los Baños, Philippines.

Sanchez, P. A. 1976. *Properties and management of soils in the tropics.* John Wiley and Sons, Inc., New York. 618 pp.

Sanchez, P. A., and M. A. Nureña. 1972. *Upland rice improvement under shifting cultivation systems in the Amazon Basin of Peru.* North Carolina Agric. Exp. Stn. and Peruvian Ministry of Agriculture Cooperating Tech. Bull. 210. 20 pp.

Soepraptohardjo, M., and H. Suhardjo. 1978. Rice soils of Indonesia. Pages 93–113 *in* International Rice Research Institute. *Soils and rice.* Los Baños, Philippines.

Takahashi, J. 1965. Natural supply of nutrients in relation to plant requirements. Pages 271–293 *in* International Rice Research Institute. *The mineral nutrition of the rice plant.* Proceedings of a symposium at the International Rice Research Institute, February, 1964. The Johns Hopkins Press, Baltimore, Maryland.

Takai, Y. 1969. The mechanism of reduction in paddy soil. *JARQ* 4:20–23.

Tung, T. C. 1973. Soils of the ASPAC region. Part 7. South Vietnam. *ASPAC Food Fert. Technol. Cent. Tech. Bull.* 16. 29 pp.

Unesco (United Nations Educational, Scientific, and Cultural Organization). 1974. FAO-Unesco soil map of the world. 1:500,000. Vol. 1, Legend, Paris. 59 pp.

USDA (United States Department of Agriculture). 1966. *A classification of Sarawak soil. Kuching.* Department of Agriculture Soil Survey Division. Tech. Pap. 2.

USDA (United States Department of Agriculture) Soil Conservation Service, Soil Survey Staff. 1975. *Soil taxonomy: a basic system of soil classification for making and interpreting soil surveys.* USDA Agric. Handb. 436. U.S. Government Printing Office, Washington, D.C. 754 pp.

Van Breemen, N. 1980. Acidity of wetland soils, including Histosols, as a constraint to food production. Pages 189–202 *in* International Rice Research Institute. *Priorities for alleviating soil-related constraints to food production in the tropics.* Los Baños, Philippines.

Van Wambeke, A. 1975. Management properties of Oxisols in savanna ecosystems. Pages 364–371 *in* E. Bornemisza and A. Alvarado, eds. *Soil management in tropical America*. North Carolina State University, Raleigh, North Carolina.

Watanabe, I., and W. L. Barraquio. 1979. Low levels of fixed nitrogen required for isolation of free-living N_2-fixing organisms from rice roots. *Nature* 277:565–566.

Watanabe, I., and C. Furusaka. 1980. Microbial ecology of flooded rice soils. *Adv. Microb. Ecol.* 4:125–168.

Yoshida, T. 1975. Microbial metabolism of flooded soils. Pages 83–122 *in* E. A. Paul and A. D. McLaren, eds. *Soil biochemistry*. Marcel Dekker, Inc., New York.

Yoshida, T. 1978. Microbial metabolism in rice soils. Pages 445–463 *in* International Rice Research Institute. *Soils and rice*. Los Baños, Philippines.

4

Chemical Changes
In Submerged Rice Soils

Rice grows in soils with moisture regimes that range from the submerged lowland to the water-deficient upland and with nutrient transformation processes that vary with the moisture regimes. Several physical, physicochemical, and biochemical changes that accompany submergence or waterlogging are important in determining a soil's suitability for rice production.

It is important to understand the unique properties of flooded soils in order to manage soil, fertilizer, and moisture regimes and to maximize rice production in a given environment.

CHEMICAL CHANGES IN SUBMERGED SOIL

The important chemical and electrochemical changes in flooded soils are:

- Depletion of molecular oxygen.
- Chemical reduction of the soil, or a decrease in redox potential.
- Increase in pH of acid soils and decrease in pH of calcareous and sodic soils.
- Increase in specific conductance.
- Reduction of Fe(III) to Fe(II) and Mn(IV) to Mn(II).
- Reduction of NO_3^- and NO_2^- to N_2 and N_2O.
- Reduction of SO_4^{2-} to S^{2-}.
- Increase in supply and availability of nitrogen.
- Increase in availability of phosphorus, silicon, and molybdenum.
- Decrease in concentrations of water-soluble zinc and copper.
- Generation of carbon dioxide, methane, and toxic reduction products such as organic acids and hydrogen sulfide (Ponnamperuma 1972, 1976).

The extent of these changes is determined by the chemical and physical properties of the soil, and by water regime and temperature. Detailed reviews are available on various physical, physicochemical, and biochemical changes of a flooded soil: Shioiri and Tanada (1954), Ponnamperuma (1955, 1972), Chang (1971), Patrick and Mikkelsen (1971).

Depletion of Oxygen

When a soil is submerged, water replaces the air in the pore spaces. Except in a thin layer at the soil surface, and sometimes a layer below the plow sole, most soil layers are virtually oxygen-free within a few hours after submergence. Under these conditions, soil microorganisms use oxidized soil constituents and some organic metabolites in place of molecular oxygen as electron acceptors in their respiration, causing reduction of the soil. The anaerobic condition influences the availability of several plant nutrients and the production of toxic substances in the soil. Thus, the properties of a flooded soil are substantially different from those of a well drained soil.

In a well drained soil there is enough oxygen available from the atmosphere to supply the needs of microorganisms and higher plants. Flooding the soil changes this condition drastically. Air movement through the floodwater is restricted and the soil no longer has an adequate supply of oxygen. Rice, however, is able to exploit the chemical benefits of soil submergence because its roots receive oxygen through aerenchyma in the shoot system and lysigenous channels in the roots (van Raalte 1941, Arikado 1959, Armstrong 1971). Many oxidation-reduction systems in the soil of importance to the nutrition of rice plants are affected by the anaerobic conditions that result from flooding. In the absence of oxygen in the flooded soil system, facultative and true anaerobic organisms become active. Organic matter decomposition is slower and less complete in anaerobic than in aerobic soils.

Characteristics of the Oxidized and Reduced Zones

The differentiation of a submerged soil (a lake mud) into two distinct zones as a result of limited oxygen penetration was first described by Pearsall and Mortimer (1939) and first thoroughly investigated by Mortimer (1941, 1942) using a special assembly of platinum electrodes. Pearsall and Mortimer (1939) found that the surface layer of marsh soils contains oxidized forms of iron, inorganic nitrogen, and sulfur, whereas the underlying layer contains reduced forms of those elements. The thickness of the oxidized surface layer is determined by the supply of oxygen at the soil surface and the consumption rate of oxygen in the soil.

The two layers of a submerged soil can be characterized by the differences in the oxidation-reduction or redox potential. The various inorganic and organic redox systems in the soil contribute to this potential (Ponnamperuma 1965). Patrick and Delaune (1972) reported that the apparent thickness of the oxidized layer differs according to distribution of the various components of the profile, with sulfur indicating the thickest oxidized zone, manganese indicating the thinnest oxidized zone, and iron showing intermediate thickness. The thickness of the oxidized layer increases with duration of flooding.

Decrease in Redox Potential

After an aerobic mineral soil is submerged, it undergoes reduction and its redox potential (Eh) drops to a fairly stable value of +0.2 to –0.3 V depending on the soil, but the redox potential in the surface water and the first few millimeters of topsoil remains at +0.3 to +0.5 V (Ponnamperuma 1972). The bulk of the root zone of a submerged soil is reduced, but the subsoil and spots in the reduced matrix, as well as streaks corresponding to root channels, may be oxidized.

Redox potential influences:

- Oxygen concentration in the soil (which in turn influences redox potential).
- pH.
- Availability of phosphorus and silicon.
- Concentration of Fe^{2+}, Mn^{2+}, Cu^+, and SO_4^{2-} directly and those of K^+, NH_4^+, Ca^{2+}, Mg^{2+}, Cu^{2+}, Zn^{2+}, $B(OH_4)^-$, and MoO_4^{2-} indirectly.
- The generation of organic acids, ethylene, mercaptans, organic sulfides, and hydrogen sulfide (Ponnamperuma 1978).

Optimum Eh or pE for Rice

The decrease in Eh or pE (pE = Eh/0.0591) brought about by soil submergence and its secondary physicochemical effects has positive and negative effects on rice growth. The benefits are increases in the supply of nitrogen, phosphorus, potassium, iron, manganese, molybdenum, and silicon.

The disadvantages are losses of nitrogen by denitrification; decrease in availability of sulfur, copper, and zinc; and production of substances that interfere with nutrient uptake or that poison the plant directly. Ponnamperuma (1978) suggested that the optimum level of reduction may be in the range of Eh 10–120 mV, or pE 0.2–2.0 at a soil solution pH of 7.0.

Changes in pH

Within a few weeks after submergence the pH of acid soils increases and the pH of sodic and calcareous soils decreases (Ponnamperuma et al. 1966). Thus, the pH of most acid and alkaline soils converge between 6 and 7 after flooding (Fig. 4.1).

The change of soil pH after submergence has been attributed to several factors,

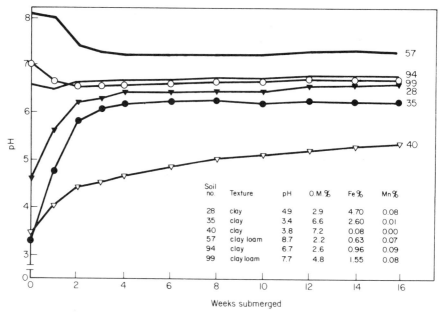

Figure 4.1   Kinetics of the solution pH of six submerged soils. (From Ponnamperuma 1976)

such as change of ferric to ferrous iron, accumulation of ammonium, change of sulfate to sulfide, and change of carbon dioxide to methane under reducing conditions. Ponnamperuma (1967) and Ponnamperuma et al. (1969), by applying the thermodynamic approach to both pure chemical systems and soil systems, concluded that the change of pH in submerged soils is regulated fundamentally by three systems:

- Na_2CO_3–CO_2–H_2O for sodic soils.
- $CaCO_3$–CO_2–H_2O for calcareous soils.
- $Fe(OH)_3$–Fe^{2+} system for ferruginous soils.

The first two systems control the decrease of pH in alkaline soils. The third, an oxidation-reduction system, regulates increase of pH and its stabilization at 6.5–7.0, particularly in acid soils.

The rate and degree of the pH changes depend on soil properties and temperature. Organic matter and active iron contents largely determine the pH changes of acid soils. If acid soils are low in organic matter or active iron, or high in acid reserves such as in acid sulfate soils, they may not attain a pH of 6.0 even after months of submergence (Fig. 4.1).

In sodic and calcareous soils, organic matter magnifies the decrease in pH. Low temperature retards pH changes in both acid and alkaline soils. Thus, the pH values of reduced soils generally remain close to 7.0 and have a narrower range than pH values of oxidized soils.

The stabilization of the soil pH at neutrality after submergence has several effects on rice growth:

- Adverse effects of low or high pH per se are minimized.
- Excess aluminum and manganese in acid soils are rendered harmless.
- Iron toxicity in acid soils is lessened.
- Availability of phosphorus, molybdenum, and silicon is increased.
- Mineralization of organic nitrogen is favored.
- Organic acids decompose.
- Lime is seldom necessary (Ponnamperuma 1976).

The concentration of ferrous iron in the soil is very sensitive to pH changes. An increase of one unit of pH from 6.25 to 7.25 will decrease the concentration of ferrous iron by one hundred times. Assuming the solid phase of Fe(II) in reduced soils is $Fe_3(OH)_8$, which has a solubility product of 6.4×10^{-18}, the dependence of Fe^{2+} concentration in the soil solution on pH is as follows (Chang 1971):

pH	Fe^{2+} (ppm)
6.25	1114
6.50	352
6.75	111
7.00	35
7.25	11
7.50	3.5

Changes in Specific Conductance

Specific conductance of the solution of most soils increases after submergence. Based on studies with 150 lowland rice soils, Ponnamperuma (1976) reported that the initial conductances are highest in the saline soils and lowest in the leached acid Ultisols and Oxisols. Redman and Patrick (1965) studied specific conductance of 26 soils using 1:1 soil-water suspensions. Submergence for 30 days increased the specific conductance of the 26 soils studied. The two soils that decreased in specific conductance after submergence had the highest initial nitrate nitrogen and the decrease of conductance was due to loss of nitrate through denitrification.

The specific conductance of the 26 soils 30 days after submergence was highly correlated to the original organic matter content of the soil (Fig. 4.2). That was because organic matter upon decomposition produces bicarbonate and organic

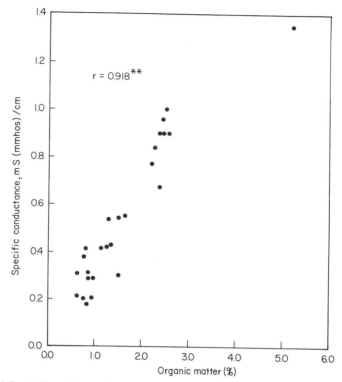

Figure 4.2   Effect of organic matter content on specific conductance of 26 submerged soils. (Adapted from Redman and Patrick 1965)

ions and serves as an energy source for the reduction of insoluble inorganic compounds to soluble ionic forms.

Reduction of Fe(III) to Fe(II)

After submergence of a soil, Fe(III) oxide hydrates are reduced to Fe(II) compounds. As a result, the soil color changes from brown to gray, and large amounts of Fe(II) enter into solution phase. The concentration of water-soluble iron may vary from 0.1 ppm shortly after submergence to as high as 600 ppm (Fig. 4.3). In acid sulfate soils the concentration may be as high as 5000 ppm within a few weeks after submergence (Ponnamperuma 1976).

Based on their studies with 26 soils, Redman and Patrick (1965) reported that the ferrous iron concentration after flooding ranged from 567 ppm to 2230 ppm. Ferrous iron concentration after submergence was positively correlated with clay and organic matter, with r values of .625 for clay and .674** for organic matter.

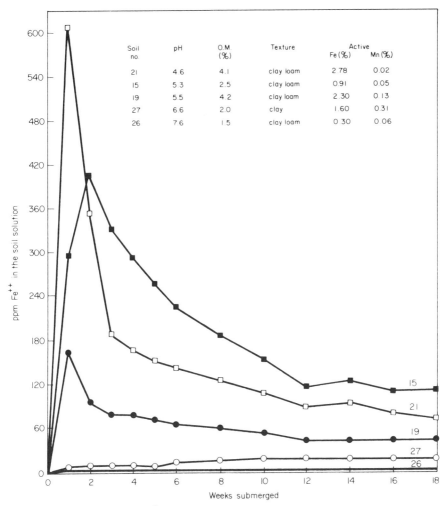

Figure 4.3   Kinetics of Fe^{2+} in the solutions of five flooded soils. (From Ponnamperuma 1976)

Other factors that affect ferrous iron concentration in submerged soils are nature and content of the Fe(III) oxide hydrates, pH of the soil, and temperature. Strongly acid soils with adequate amounts of organic matter and reactive iron oxides can build up toxic concentrations of ferrous iron and thereby cause iron toxicity problems in submerged Ultisols, Oxisols, and acid sulfate soils in the tropics (Ponnamperuma et al. 1973).

Iron toxicity may also occur in acid sandy soils and in peat soils low in active iron, as in degraded paddy soils. Low temperatures ($< 20°C$), by bringing about late but high and persistent concentrations of water-soluble iron, may cause iron toxicity in soils. At 25–35°C, toxic high concentrations are short-lived (Cho and Ponnamperuma 1971). Criteria for the iron toxicity hazard are:

· pH of the dry soil.
· Amount of reserve acidity.
· Reactivity and content of Fe(III) oxide hydrates.
· Soil temperature.
· Soil organic matter content.
· Percolation rate.
· Interflow from adjacent areas (Ponnamperuma 1976).

Reduction of Mn(IV) to Mn(II)

In flooded soils, the reduction of the higher oxides of manganese, chiefly Mn(IV), takes place almost simultaneously with denitrification (Ponnamperuma 1965). The reduction may result from these compounds functioning as:

· Electron acceptors in the respiration of microorganisms.
· Chemical oxidants of reduction products.

Manganese is more readily reduced and rendered soluble than iron. The release of manganese into the soil solution, therefore, precedes that of iron.

The kinetics of manganese reduction varies markedly from soil to soil. Soils with a high content of manganese, regardless of pH and organic matter content, show steep increases in Mn^{2+} during the first weeks of submergence and a slow decline thereafter.

Manganese mobilization in soils is markedly increased after flooding due to the reduction of manganic compounds to more soluble forms as a consequence of the anaerobic metabolism of soil bacteria. The amount of extractable (in normal sodium acetate solution of pH 2.8) manganese before and after submergence is closely correlated, with an r value of .755** (Redman and Patrick 1965).

Supply and Availability of Nitrogen

The availability of nitrogen in flooded soils is higher than in nonflooded soils. Even though organic matter is mineralized at a slower rate in anaerobic soils than in aerobic soils, the net amount mineralized is greater because less nitrogen is immobilized. The availability of nitrogen in flooded soils increases with increases in nitrogen content of the soil, soil pH, temperature, and duration of previous desiccation of the soil (Ponnamperuma 1965).

Availability of Phosphorus, Silicon, and Molybdenum

Availability of phosphorus, silicon, and molybdenum in a soil increases after flooding.

Phosphorus

The availability of soil phosphorus increases after soil submergence, mainly due to reduction of ferric phosphate to ferrous phosphate, although other changes such as hydrolysis of aluminum phosphate and the dissolution of calcium phosphate resulting from the accumulation of carbon dioxide are also involved. The ready availability of ferrous phosphate is illustrated by the experimental results from Taiwan (Table 4.1).

The increase in solubility of phosphorus, however, is low in Ultisols and Oxisols high in active iron (Fig. 4.4).

Redman and Patrick (1965) reported an average increase of 21% in extractable (in 0.1 N hydrochloric acid) phosphorus due to submergence, but there was no appreciable increase in extractable phosphorus in soils that released less than 1800 ppm iron.

Silicon

The concentration of silicon in the solutions of submerged soils increases slightly after flooding and then decreases gradually, and after several months the concentration may be lower than at submergence.

Silicon in water is present as the monomer $Si(OH)_4$, whose solubility is independent of pH in the range 2.0–8.0. The increase in silicon solubility is not due, therefore, to the increase in pH following soil reduction (Ponnamperuma 1965).

The increase in silicon concentration after flooding may be due to its release following reduction of hydrous oxides of Fe (III)-sorbing silicon and to the action of carbon dioxide on aluminosilicates. The subsequent decrease may be the result of combination with aluminosilicates, following the decrease in P_{CO_2}. Reversible changes in solubility of silicon appear to be associated with oxidation-reduction of iron (Ponnamperuma 1972).

Molybdenum

The concentration of water-soluble molybdenum in the soil increases after flooding, presumably as a result of desorption following reduction of ferric oxides (Ponnamperuma 1976).

Table 4.1 Aluminum-, Iron-, and Calcium Phosphates (ppm of P) in Soils With and Without Growing Rice (Adapted from Chang 1965)

| | Slate Alluvial Soil | | | | | | Sandstone and Shale Alluvial Soil | | | | | |
| | Al-P | | Fe-P | | Ca-P | | Al-P | | Fe-P | | Ca-P | |
Replication	Bl[a]	Pl[b]	Bl	Pl	Bl	Pl	Bl	Pl	Bl	Pl	Bl	Pl
1	29	27	80	74	300	295	251	265	357	273	50	52
2	26	25	71	71	300	315	265	252	353	243	48	46
3	26	28	71	78	295	320	254	244	341	283	52	46
4	29	26	80	74	315	315	243	260	342	307	50	48
5	29	26	83	74	340	320	266	241	361	274	50	50
Average	28	26	78	75	310	309	255	252	351	276	50	48

[a]Bl = Blank, without rice.
[b]Pl = Planted with rice.

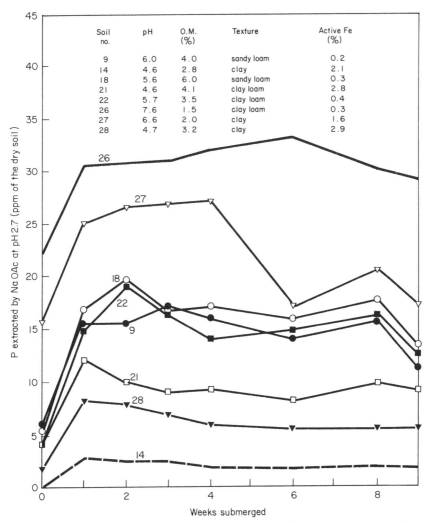

Figure 4.4   Kinetics of P extractable in pH 2.7 NaOAc in some flooded soils. (From Ponnamperuma 1976)

Water-Soluble Zinc and Copper

A decrease in concentrations of water-soluble zinc and copper is one of the few disadvantages of flooding soils for rice. Zinc deficiency is widespread in many rice-growing countries in the tropical and temperate regions (see Chapter 10).

Production of Toxins

Since the initial work on *Akiochi* soils (degraded paddy soils) was first reported from Japan (Mitsui 1955), numerous studies have been made on hydrogen sulfide in flooded soils. Hydrogen sulfide in submerged soils results from sulfate reduction and anaerobic decomposition of organic matter. In normal soils, hydrogen sulfide is rendered harmless by precipitation as ferrous sulfide, but in soils high in sulfate and organic matter and low in iron it may harm rice plants.

Organic acids (mainly acetic and butyric acids with small amounts of formic, propionic, and lactic acids) and the organic reduction products produced in anaerobic soils harm rice plants.

NUTRIENT TRANSFORMATIONS IN SUBMERGED SOILS

To increase the fertilizer efficiency in lowland rice, it is imperative to understand the transformations of nutrient elements that occur in submerged soils. Submergence causes changes in the properties of the soil because of physical reactions between the soil and water and the biological and chemical processes set in motion as a result of excess water. The most important change is the conversion of the root zone of the rice plant from an aerobic environment to an anaerobic or near-anaerobic environment.

Nitrogen Transformations

The special conditions prevailing in a flooded soil cause considerable modification of nitrogen transformation processes (Fig. 4.5). Most of the transformations involve processes performed by microorganisms.

Soil nitrogen occurs primarily in organic combination in the soil. The breakdown of organic matter leading to the release of ammonium ions to the soil solution proceeds at a slower rate in a flooded soil than in a nonflooded soil.

The nitrogen supply for lowland rice comes largely from:

· Ammonium and nitrate nitrogen present when the soil is flooded.
· Nitrogen mineralized from soil organic matter and plant residue under flooded conditions.
· Nitrogen fixed by algae and heterotrophic bacteria.
· Nitrogen added as fertilizer.

Organic nitrogen forms are a potential reserve of nitrogen for rice only after the organic form has been converted to the inorganic form.

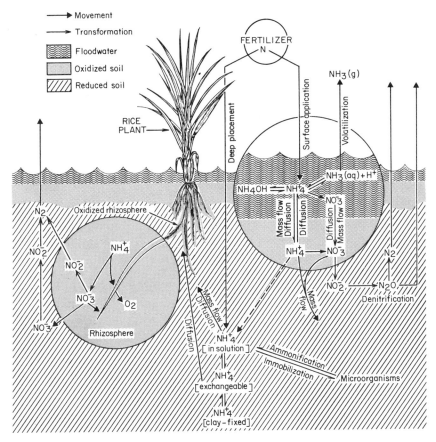

Figure 4.5 Nitrogen transformations in submerged rice soils. (From Savant and De Datta 1980, unpublished)

Inorganic Nitrogen

The inorganic nitrogen in the soil constitutes only a part of the total. However, it is from this fraction that plants derive the greater part of their nitrogen.

EXCHANGEABLE AMMONIUM Most of the inorganic nitrogen present in reduced soils is water soluble or adsorbed on the exchange complex. Exchangeable ammonium (NH_4^+) represents the form of nitrogen readily exchanged on the external surfaces of soil particles by suitable extractants, for example, $2N$ KCl.

The inorganic nitrogen regime in flooded soils is essentially characterized by the accumulation of ammonia. This occurs because the mineralization of organic nitrogen does not proceed beyond the ammonium stage in the absence of oxygen, because oxygen is necessary for microbial conversion of ammonium to nitrate.

NITRITE   The nitrite form of nitrogen may accumulate under certain circumstances in aerobic alkaline soils and its production is attributed to the oxidation of ammonium to nitrite by specific groups of microorganisms—*Nitrosomonas* and *Nitrococcus*. Nitrite occurs in flooded soils as an intermediate in nitrification and denitrification. The levels of nitrite in flooded soils, including those fertilized with nitrate, are generally in the 0–3 ppm range.

NITRATE   In submerged soils, nitrate rapidly disappears, presumably through denitrification, leaching, and plant uptake.

FIXED AMMONIUM   Fixed ammonium in soils has been described as that held within the lattice structures of silicate minerals. It is neither water soluble nor readily exchanged. From 14 to 78% of the total nitrogen in some tropical soils exists in this form.

NITROUS OXIDE AND ELEMENTAL NITROGEN   Nitrous oxide and elemental nitrogen can be formed in flooded soils. Nitrate under deficient oxygen levels replaces oxygen as the final electron acceptor in the respiration of many facultative anaerobes forming nitrous oxide and elemental nitrogen.

Organic Nitrogen

Although distribution of various nitrogenous organic compounds in flooded soils is not well documented, it may not be very much different from that in normal, well drained soils. Numerous forms of organic nitrogen are found in surface soils.

The methods used for fractionation of organic nitrogen in soils have been based largely on studies involving identification and estimation of the nitrogen compounds released by treatment of soils with mineral acids. This is done by refluxing the soil with $6N$ HCl. Following hydrolysis, analysis of the hydrolyzate and residue has revealed the presence of compounds containing organic nitrogen.

AMINO ACIDS   Hydrolysis of soils has shown that 20–40% of the total nitrogen in most surface soils is in the form of amino acid-N.

AMINO SUGARS   About 5–10% of the soil nitrogen in most surface soils is represented by amino sugar nitrogen in the form of hexosamines.

PURINE AND PYRIMIDINE BASES   The evidence on the occurrence and amount of purine and pyrimidine derivatives indicates that these bases do not account for more than 1% of the surface soil nitrogen.

Nitrogen Mineralization in Flooded Soils

Fertilized rice obtains 50–80% of its nitrogen requirement from the soil; unfertilized rice obtains an even larger proportion, mainly through mineralization of organic sources. Thus, rice depends primarily on the mineralization of organic sources for its nitrogen (Koyama 1975, Broadbent 1979).

The mineralization process includes hydrolysis of proteins to polypeptides and amino acids, with subsequent deamination resulting in the formation of ammonia. Most of the complex variety of nitrogen-containing compounds that may be present in the paddy soils are ultimately converted to ammonia. However, the breakdown of organic matter that leads to the liberation of ammonium ions into the solution phase proceeds at a much slower rate in a flooded soil than in a nonflooded soil.

Research in Japan indicates that several pretreatments have remarkable effects on the mineralization of native organic nitrogen in flooded soils. These include:

- Air-drying the soil between the successive rice crops.
- Increasing the soil temperature in flooded soil.
- Raising the soil reaction to pH 9.0, by addition of weak alkaline solution (Mitsui 1955).

Broadbent (1979) listed soil temperature, moisture level, wetting and drying, quantity of organic matter in the soil, clay content, organic amendments, and others as important factors in mineralization of organic nitrogen.

Quantities of nitrogen mineralized from organic sources during the growing period of a lowland rice crop are larger than for a similar period of an upland crop. The greater part of the nitrogen mineralized during a season appears as ammonia within 2 weeks after submergence if the temperature is favorable and the soil is not strongly acid or very deficient in available phosphorus. The amount of ammonia produced during the first 2 weeks of submergence may range from 50 to 350 ppm N on the basis of the dry soil. The soil solution may contain 2–100 ppm N depending on texture and organic matter content of the soil (Ponnamperuma 1972).

Rain

The amount of nitrogen added to a soil by rainwater is small, probably in the range of 5–14 kg/ha.

Losses of Nitrogen

Nitrogen is removed from the soil by plant uptake but it is not considered a loss mechanism. Nitrogen losses in the soil occur mainly from denitrification, ammonia volatilization, leaching, and surface runoff. In addition, immobilization and ammonium fixation make nitrogen temporarily unavailable to the rice crop but do not cause loss of nitrogen from the soil system.

Denitrification

Substantial losses of nitrogen from the soil have long prompted studies on denitrification in flooded rice soils. In 1916, Harrison and Aiyar reported that 10–95% of the total volume of gas evolved from flooded rice soils in India was nitrogen but only some of it came from denitrification. The authors analyzed the gas escaping as bubbles from the surface of flooded Indian rice soils in a study of the nature and extent of the change in gas composition caused by submergence. Besides nitrogen, methane was also found.

Although Pearsall and Mortimer (1939) were the first to report profile differentiation in the submerged soils of lake mud, it was Shioiri and Mitsui (1935) and Shioiri (1941) that gave details on the profile differentiation. They observed considerable losses of nitrogen from applied ammonium sulfate as the period of incubation proceeded. The profile differentiation of a flooded rice soil is in the surface layer, corresponding to a few millimeters or a centimeter or so. This layer is considered to be the oxidized zone where microorganisms live aerobically; below that layer lies the reduced soil layer constituting the main part of the furrow slice where microorganisms live anaerobically (Pearsall and Mortimer 1939, Mitsui 1955). In the rhizosphere of the rice plant, however, the soil zone in close proximity to the rice roots is distinctly oxidized and, therefore, constitutes a condition similar to the surface layer.

Some believe that part of the nitrogen present in a flooded soil is reduced to ammonia, but others have found that little of the nitrate in a flooded soil reaches the ammonia form (De and Sarker 1936, Broadbent and Stojanovic 1952).

In a flooded soil, denitrification of applied nitrogen fertilizer can be substantial. De and Sarker (1936) reported losses of 30–40% of the added inorganic fertilizer starting about 7–12 days after addition of NO_3^-–N fertilizer.

The magnitude of loss of applied nitrogen has been estimated as 20–40% in India, 30–50% in Japan, 37% in California and 68% in Louisiana in the United States, and 25% in the Philippines.

In a study of nitrogen loss by using ^{15}N-labeled nitrogen Patrick and Tusneem (1972) differentiated profiles of oxidation-reduction (redox) potential for the flooded soil exposed to the atmosphere. Figure 4.6 shows those profiles for samples incubated for 1 and 4 weeks. Oxygen penetrated only a short distance into the flooded soil before being reduced. Denitrification loss was very likely a result of the formation of a thick oxidized layer in which more ammonium nitrogen could be nitrified.

Patrick and his associates (1972, 1974, 1976, 1977, 1978) studied extensively the mechanisms and kinetics of the nitrification-denitrification process in controlled laboratory experiments without plants. The highlights of their work are summarized in Fig. 4.7.

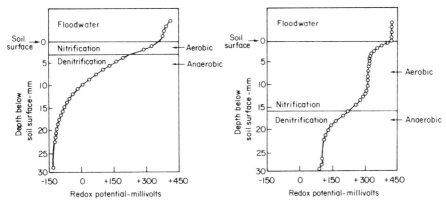

Figure 4.6 Redox potential (at pH 7) of flooded soil incubated 1 week (left) and 4 weeks (right). Potential was measured with platinum electrode advancing downward through soil at rate of 2 mm/hour. A soil zone with redox potential greater than about +220 mV is considered to be aerobic and will support nitrification, whereas a zone with redox potential below about +220 mV is considered to be anaerobic and will support denitrification. (From Patrick and Tusneem 1972)

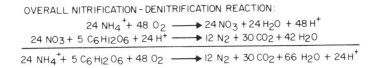

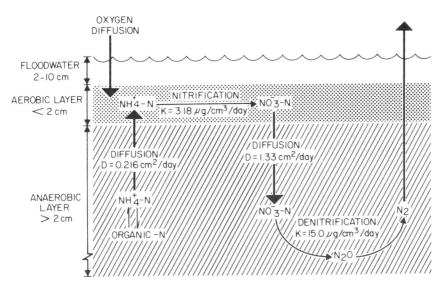

Figure 4.7 Nitrification-denitrification reaction and kinetics of the related processes controlling nitrogen loss from aerobic-anaerobic layer of a flooded soil system. (From Patrick and Reddy 1977)

105

The biological oxidation of NH_4^+ to NO_3^- ($k = 3.18$ μg N/cm^3 per day) takes place in the surface aerobic layer. This results in the development of a concentration gradient of NH_4^+–N across the aerobic layer and anaerobic layer, and causes the NH_4^+ present in the anaerobic layer to diffuse ($\bar{D} = 0.216\,cm^2/day$) into the aerobic layer where it is nitrified. The NO_3^-–N formed diffuses back into the anaerobic layer ($\bar{D} = 1.33$ cm^2/day) where it is readily denitrified ($k = 15.0$ μg N/cm^3 per day). The slow rate of upward diffusion of NH_4^+ and also the slow rate of nitrification are probably the two limiting factors in nitrogen loss through the process of sequential nitrification-denitrification in the flooded surface soil. These studies were, however, conducted in experimental systems without any water movement. Natural field systems may be much more dynamic.

FACTORS AFFECTING DENITRIFICATION   Various incubation and other studies to date envisage that many factors directly or indirectly influence the denitrification process in the anaerobic surface soil layer. Those factors that affect the denitrification process are:

1   Soil factors
    · pH and pE
    · Temperature
    · Organic matter (nature and amount)
    · Nitrate–N content and nitrification rate
    · Submergence period
    · Degree of puddling or aggregation
    · Activity of denitrifiers
    · P_{O_2}
    · Soil fertility
2   Other factors
    · Floodwater regime
    · Fertilizer nitrogen management
    · Pesticide use
    · Presence of plants

Oxygen Concentration.   Experiments using labeled ^{15}N have shown that nitrogen loss from a flooded soil occurs when the flooded soil is exposed to oxygen. Nitrogen loss was generally related to the thickness of the aerobic layer, even though appreciable loss occurred at 5 and 10% oxygen concentrations where the aerobic layer was relatively thin (Patrick and Gotoh 1974).

Alternate Drying and Wetting.   Severe nitrogen loss occurs in soils subjected to alternate draining (aerobic) and flooding (anaerobic) (Wijler and Delwiche 1954, MacRae et al. 1968). Nitrogen loss through sequential nitrification and denitrification is especially high in soils planted to lowland rice where water management practices in irrigated rice, or irregular rainfall distribution in rainfed paddies, cause frequent draining and reflooding. Reddy and Patrick (1975) reported a greater nitrogen loss in a 2-day aerobic-anaerobic cycle than in longer cycles (Table 4.2). A subsequent study (Reddy and Patrick 1976) indicated that besides nitrification during the aerobic period and denitrification during the anaerobic period, NO_2^--N formation and chemical decomposition may also be involved in nitrogen loss, especially for the more frequent changes in aerobic-anaerobic condition.

pH and pE. Stanford et al. (1975) found no correlation between the initial aerobic soil pH (within the range from 5 to 8) and potential denitrification. This is explainable on the basis of the kinetics of pH in submerged soils (Ponnamperuma 1972). The critical soil pE for ammonia reduction seems to be between 1.7 and 5.1 (Gotoh 1973). The stabilized pH (6.5 to 7.2) and pE (-1 to 3) values of the lowland flooded soils are generally highly conducive to denitrification (Ponnamperuma 1972, 1977).

The soil microorganisms mediating denitrification seem to function best near pH 7.

Komatsu et al. (1978) obtained data that indicate a possibility of participation of a Fe^{3+}–Fe^{2+} system in denitrification in a flooded soil system.

Soil Temperature. The denitrification rate in anaerobically incubated Philippine rice soils varied with soil type (Fig. 4.8) and increased markedly as the temperature rose from 5 to 45° C (Ponnamperuma 1977).

Level of Nitrate Present. A few researchers have considered that the rate of denitrification is independent of the NO_3^--N concentration, suggesting that it followed zero-order reaction kinetics (Broadbent and Clark 1965, Patrick 1960, Keeney 1973). Others have reported denitrification following the first-order reaction kinetics, indicating its dependence on the NO_3^- concentration (IRRI 1965, Ponnamperuma 1972, Bowman and Focht 1974, Stanford et al. 1975, Kohl et al. 1976). Reddy et al. (1978) attempted to elucidate this phenomenon on the basis of the NO_3^- diffusion from the overlying floodwater layer to the soil layer. They concluded that where nitrogen loss due to denitrification involves NO_3^- diffusion, the denitrification appears to be a first-order reaction rather than zero-order. In a water-sediment system, Van Kessel (1977) found that denitrification rate is dependent on NO_3^- concentration in the overlying water, approximating first-order reaction at lower concentration and gradually becoming independent of the NO_3^- concentration (zero-order reaction) at higher NO_3^- content.

Organic Matter Content of Soil. Because denitrifying bacteria need organic matter as a source of carbon (and nitrogen) for their activity, the nature and amount of organic matter in submerged lowland soils largely determine denitrification at a given temperature and nitrate concentration.

The capacity of Iowa surface soils for denitrification when submerged has been

Table 4.2 Changes in Total Nitrogen as a Result of Alternate Aerobic and Anaerobic Treatments (From Reddy and Patrick 1975)

Length of Aerobic Period (CO_2-free air) (days)	Length of Anaerobic Period (days)	Number of Complete Cycles During Incubation Period	Total Nitrogen ($\mu g/g$)[a]		
			At End of Experiment	Net Loss	Loss (%)
2	2	32	704	226	24.3
4	4	16	717	213	22.9
8	8	8	730	200	21.5
16	16	4	744	186	20.0
32	32	2	762	168	18.0
64	64	1	810	120	12.9
128		Completely aerobic	866	64	6.9
	128	Completely anaerobic	925	5	0.5

[a]At the beginning of the experiment soils in all treatments contained 930 μg N/g of soil.

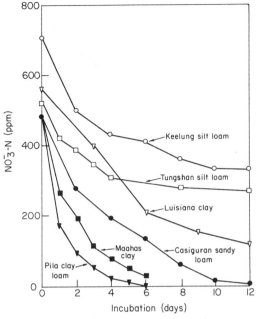

Figure 4.8   Kinetics of denitrification in six submerged rice soils of the Philippines at 35°C. (From Ponnamperuma 1977)

highly and significantly correlated ($r = .99**$) with their content of soluble or readily decomposable organic matter (Burford and Bremner 1975). Some of the recent studies of Bremner's group (Blackmer and Bremner, unpublished work cited by Bremner 1977) indicate that the ratio of nitrogen oxide to nitrogen gas in the gases produced during denitrification is influenced by molecular weight and water solubility of organic substances added to the soil. As expected, the water-soluble organic substances with low molecular weight have much greater effect on denitrification than high molecular weight substances with low water solubility.

In general, loss of nitrogen through denitrification may not be a serious problem in tropical lowland rice fields, such as those in India or Burma, primarily due to their low organic matter content (Kawaguchi and Kyuma 1977) and eventually due to weak development of reducing conditions (Yamada 1975). Earlier work of Bremner and Shaw (1958) indicated little or no gaseous loss of nitrogen from incubated waterlogged soils containing <1% organic carbon. Broadbent and Stojanovic (1952) observed that the loss of nitrogen was greatest when a low oxygen content (0.1%) was combined with a high level of organic matter (2% maize leaves).

Activity of Denitrifiers.   Takai and Uehara (1973) studied changes in bacterial flora in the surface and underlying layers of an incubated flooded soil system with progressing submergence time. In the early stage of flooding, the NO_3^- disappeared rapidly and was accompanied by a pronounced increase in denitrifiers. There was no marked increase in nitrifiers in the aerobic layer. At the middle stage, as the oxidized layer developed, a considerable increase in nitrifiers in the surface layers occurred. The denitrifiers' population was maintained at a high level and as a result NO_3^--N did not accumulate. At the later stage, the number of nitrifiers was sufficiently high but the number of denitrifiers dropped, presumably due to a decrease in substrate concentration.

Submergence Period. Using ^{15}N-tagged NH_4^+-N, Patrick and Reddy (1976) observed that $^{15}NH_4$ content of a flooded Crowley silt loam decreased and a buildup of N_2 became evident after 15 days. They noticed rapid conversion of NH_4^+ to N_2 after 30 days. It is apparent, therefore, that nitrogen loss from the flooded soil due to the sequential nitrification-denitrification would become significant only after 1–2 weeks of flooding. This period is normally required for the differentiation of the discernible oxidized layer overlying the reduced layer.

Ammonium Diffusion

Experiments with ^{15}N have shown that more NH_4^+-N is lost from a flooded soil than is actually present in the aerobic soil layer at any one time (Tusneem and Patrick 1971). Apparently, NH_4^+-N diffuses from the anaerobic soil layer to the aerobic soil layer where it undergoes nitrification and denitrification. Subsequent studies by Reddy et al. (1976) suggest that diffusion of NH_4^+-N from the anaerobic soil layer to the aerobic soil layer accounts for more than 50% of the total NH_4^+-N loss with the remainder lost from NH_4^+-N originally present in the aerobic layer.

Ammonia Volatilization

Ammonia, a readily identifiable product of nitrogen mineralization, is formed continuously in the soil and water of a flooded paddy. The ready release of ammonia from organic matter decomposing in the absence of oxygen and the high pH associated with anaerobic decomposition favor ammonia volatilization from flooded soils that have had large amounts of organic matter added. Use of nitrogen fertilizers on rice also contributes large concentrations of dissolved NH_4^+-N salts. Ammonium-form fertilizers may dissociate directly or, like urea, may decompose by catalytic hydrolysis to produce NH_4^+ ions in water. Ammonium ions, loosely bound to water molecules, predominate in water at a pH below 9.2. With increasing hydroxyl-ion concentrations in the water, ionized NH_4^+ increasingly converts to nonionized ammonia, which may escape from the water as a gas.

   Ammonia volatilization loss may occur from floodwater on a soil moderately to slightly acid, although losses are usually highest on alkaline soils. The photosynthetic and respiratory activity of submerged aquatic biota, their biomass, and factors affecting their growth play a prominent role in regulating water pH by the reaction:

$$photosynthesis$$

$$nCO_2 + nH_2O \rightleftharpoons (CH_2O)_n + nO_2$$

$$respiration$$

Aquatic organisms control water pH over a wide range in a primarily aqueous carbonate equilibrium system

$$CO_2 \text{ (atm.)} + H_2O = H_2CO_3 \rightleftharpoons H^+ + HCO_3^- \rightleftharpoons CO_3^= + H^+$$

When water pH rises above 7.4, ammonia volatilization losses may be appreciable. Water pH values increase by midday to values as high as pH 9.5–10 and decrease as much as 2–3 pH units during the night. When 60 kg N/ha was broadcast on paddy water on a neutral Maahas clay (Aquic Tropudalf),the pH of the floodwater varied from 7.5 to 9.0 between 0600 and 1600 hours daily when the level of solar radiation was high and photosynthesis by the aquatic biota was active. Free carbon dioxide declined in the system after 0600 hours, reached a low point from 1200 to 1800 hours, then increased until 2200 hours. The percentage of mole fraction of carbonic acid decreased in proportion to the free carbon dioxide content of the water. A similar effect was noted in the HCO_3^- and CO_3^{2-} ion fraction (Fig. 4.9).

Paddy water on acid Luisiana clay (Typic Tropaquept) did not develop an active algae population and the diurnal pH variations were reduced. The consequences of the high pH induced by an active biota and ramifications of dissociation of carbon dioxide in the floodwater pertaining to an ammonium-nitrogen source are apparent (Mikkelsen et al. 1978).

FACTORS AFFECTING AMMONIA LOSS BY VOLATILIZATION   The volatilization of ammonia from a soil is a function of the various properties of the soil system involved, including moisture content, soil pH, cation exchange capacity (CEC), exchangeable cations, texture, lime content, temperature, and atmospheric conditions above the soil. Agronomic variables such as rates and sources of nitrogen fertilizers and time, method, and depth of application are also important.

Moisture Content.   The importance of water and rate of water loss from the soil in the ammonia volatilization process is emphasized by several workers. However, Chao and Kroontje (1964) showed that the rate of ammonia volatilization and of water evaporation from some soils follows different functions and is not related to soil water or the relative humidity of air flowing over a soil surface.

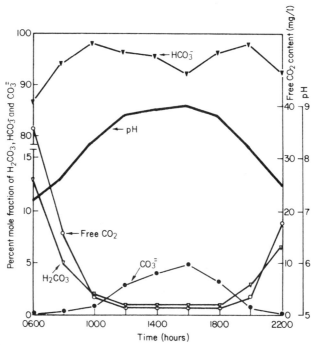

Figure 4.9 Changes in the pH and components of the carbonic acid system of rice floodwater 3 days after nitrogen fertilization on Maahas clay. IRRI, 1976 wet season. (From Mikkelsen et al. 1978)

Soil pH. It is recognized that soil pH may influence ammonia volatilization and that the higher the soil pH, the higher the potential losses. Up to about pH 9, ammonia concentration increases by a factor of 10 per unit increase of pH.

Cation Exchange Capacity. The CEC of the soil influences the degree of ammonia volatilization (Lehr and van Wesemael 1961). The higher the CEC, the lower the ammonia volatilization losses. For example, in a 1979 study at IRRI, nitrogen loss from urea fertilizer was shown to increase as the CEC of the soil decreased. At all CEC values, the percent nitrogen loss as ammonia from surface-applied urea was higher than the broadcast and incorporated urea (Fig. 4.10). The peak ammonia volatilization loss occurred between the first and second day when the nitrogen content in floodwater was also highest. Placement of urea supergranules in the San Manuel sandy loam soil with low CEC caused nitrogen loss of 6.5%. The loss increased gradually with time from the day of application. In a coarse-textured soil, slow-release nitrogen fertilizer is better than placement of supergranules for minimizing nitrogen losses as ammonia.

Lime Content. Lime content of the soil influences the ammonia volatilization losses. Some studies suggest that even in acid soil there could be a loss of

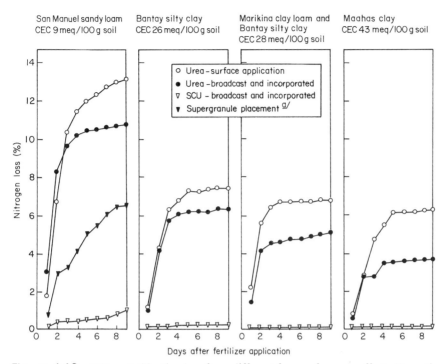

Figure 4.10 NH_3 volatilization loss from different forms of urea as affected by cation exchange capacity and texture of soil. Large drum experiment. IRRI, 1979 late dry season. [a]Supergranule placement at 10–12-cm depth gave negligible loss of nitrogen as NH_3 in silty clay, clay loam, and clay soils.

ammonia through volatilization. Larger losses have been observed with urea than with ammonium sulfate (Mikkelsen et al. 1978, Vlek and Craswell 1979). Topdressed urea is hydrolyzed to ammonium carbonate, which decomposes to yield ammonia. Mikkelsen and De Datta (1979) reported a 19.2% N loss through ammonia volatilization occurred when urea was applied either broadcast and incorporated (basal), or surface broadcast (Fig. 4.11). Although San Manuel sandy loam soil is acidic, the pH of the floodwater ranged from 8.0 to 9.0 with urea and pH 7.7–8.5 with ammonium sulfate, explaining the high ammonia losses. In addition, the CEC of San Manuel sandy loam soil was low (9 meq/ 100 g soil), which contributed to high ammonia losses.

Temperature and Atmospheric Conditions.   The high temperatures and solar radiation that prevail in the dry season of tropical Asia increase ammonia losses through volatilization. Dry season losses are also increased if the algal population in the floodwater becomes dense (Mikkelsen et al. 1978). In the wet season, however, the amount of nitrogen applied to rice is low and the losses are lower because of relatively lower solar energy and lower temperature than in the dry season.

MAGNITUDE OF AMMONIA VOLATILIZATION LOSSES IN FLOODED SOIL   Ammonia volatilization losses in flooded soils range from negligible to almost 60% of applied nitrogen (Table 4.3). The differences in these values are due to many factors, but probably mostly due to differences in the measurement techniques. Most of those studies were made in the laboratory where conditions may differ considerably from those in the field. Among the various techniques used or suggested, estimating volatilization by micrometeorological considerations is probably the most unbiased, but difficult to set up and use accurately.

In total, the following physicochemical and biological factors (or parameters), and others, in a flooded soil-water-atmosphere system simultaneously influence ammonia volatilization:

1   Soil parameters
    · Soil pH and pE
    · Salinity (EC) and alkalinity (SAR)
    · Calcium carbonate content
    · Cation exchange capacity (CEC)
    · Predominant exchangeable ions
    · Buffering capacity
    · Partial pressure of carbon dioxide (P_{CO_2})
    · Microbial activity
2   Floodwater parameters
    · pH
    · P_{CO_2}
    · Concentration of ammonia [$NH_3(aq) + NH_4 + NH_4OH$]
    · Total alkalinity
    · pH-buffering capacity
    · Temperature
    · Water movement or turbulence
    · Depth
    · Algal growth or activity
    · Concentration of phosphorus
    · Use of pesticides
3   Atmospheric parameters
    · Air temperature
    · Solar radiation
    · Wind speed
    · Partial pressure of ammonia (P_{NH_3})
4   Other parameters
    · Nitrogen management
    · Water management
    · Plant canopy

Table 4.3 Ammonia Volatilization Loss from Fertilizer-Amended Lowland Paddies

Soil	Season	Methodology	Fertilizer N Added Form	Rate (kg N/ha)	NH$_3$-N Loss (%)	Reference
Maahas clay loam (pH 7.1) Philippines	1973 wet season	Field study, closed system with opening for air pressure and acidified glass wool trap	(NH$_4$)$_2$SO$_4$	100	4	Ventura and Yoshida (1977)
			(NH$_2$)$_2$CO	100	8	
Maahas clay Philippines	1976 dry season	Field study, open and closed system	(NH$_4$)$_2$SO$_4$	40–60	30–59	Bouldin and Alimagno (1976) IRRI (1977)
Maahas clay (pH 7) and Luisiana clay (pH 5.7) Philippines	1977 wet and dry seasons	Greenhouse and field studies, closed system with air exchange and acid trap	(NH$_4$)$_2$SO$_4$	90	0.25–6.8	Mikkelsen and De Datta (1979)
			(NH$_2$)$_2$CO	30–90	1.0–20	Mikkelsen et al. (1978)
Clay (pH 7.0–7.5) Thailand	1971–1972 dry season	Field study, closed system with air exchange and acid trap	(^{15}NH$_4$)$_2$SO$_4$	50–100	0.8–14.0	Wetselaar et al. (1977)
Black clay (Vertisol) (pH 8.1) India	1976 wet season	Field study, static closed system and acid trap	(^{15}NH$_2$)$_2$CO	100	9.7	Krishnappa and Shinde (1978)
Maahas clay (pH 6.7) Philippines	1978 wet season	Field study, chemical and micrometeorological technique	(NH$_4$)$_2$SO$_4$	40–80	5.1–10.6	Freney et al. (1981)

AGRONOMIC PRACTICES TO MINIMIZE AMMONIA VOLATILIZATION LOSSES   The magnitudes of direct ammonia volatilization losses are affected by agronomic practices such as sources of nitrogen and time and method of its application.

The least ammonia volatilization losses are reported when fertilizer nitrogen (urea or ammonium sulfate) is placed in the reduced soil layer (10–12 cm deep) or slow-release nitrogen fertilizers such as sulfur-coated urea and isobutylidene diurea (IBDU) are broadcast and incorporated (Fig. 4.11). Mudball placement has allowed less than 1% loss of total nitrogen through ammonia volatilization (Mikkelsen et al. 1978).

Some of the factors affecting ammonia volatilization losses and management practices that minimize losses have been reviewed by Mikkelsen and De Datta (1979).

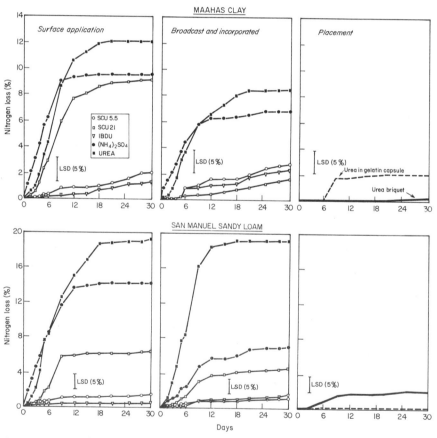

Figure 4.11   Effect of nitrogen sources and methods of application on ammonia volatilization loss in neutral Maahas clay and acid San Manuel sandy loam in an outdoor drum experiment. IRRI, 1978 dry season. (From Mikkelsen and De Datta 1979)

Immobilization of Nitrogen

The microbial processes of nitrogen mineralization and immobilization occur simultaneously in soil. Mineralization replenishes the available nitrogen supply in soil and immobilization depletes or temporarily depletes it. In any soil system, there is a continuous interchange between inorganic and organic forms of nitrogen. The immobilization (assimilation) of inorganic nitrogen by micro-organisms is rapid during periods of active cell proliferation and slow when microbial activity is low.

The nitrogen requirements of the microorganisms in a flooded soil are generally lower than those of the microbial population of a nonflooded soil (Acharya 1935a,b,c). For example, in a California field study on the decomposition of rice straw in flooded soil, Williams et al. (1968) concluded that the nitrogen requirement for the decomposition of rice straw was one-third (0.5 vs. 1.5%) the average concentration of nitrogen required for aerobic decomposition of plant residue.

Two reasons are cited for the low microbial population in flooded soils:

- The anaerobic metabolism is inherently less efficient than aerobic metabolism in providing energy for synthesis of new cells.
- Actinomycetes and fungi, which convert from 15 to 40% of substrate carbon, are inactive in submerged rice soils due to exclusion of oxygen (Broadbent 1978).

Broadbent summarized various studies reporting the critical nitrogen content of rice straw at which no net immobilization occurs when incorporated into flooded soil. These ranged from 0.47 to 1.17% N in the rice straw, depending on degree of maturity of the rice. The apparent discrepancies are due to differences in the methods used to measure immobilization.

The factors that affect immobilization are water regime, soil temperature, kind and amount of organic matter, rate of nitrogen applied, and nitrification rate. Because ammonium is the preferred form of nitrogen for soil microorganisms, more will be immobilized in a soil where ammonium persists than in one where it is quickly depleted. Information on the nitrogen immobilization process in flooded soils is meager and scattered.

Studies at IRRI suggest that immobilization of nitrogen is about 20% of the nitrogen added for a Philippine rice soil (Maahas clay). That rate will vary among different soils (Yoshida and Padre 1975).

Leaching of Nitrate and Ammonium Nitrogen

Nitrate produced in the oxidized surface layer of a flooded soil moves easily by diffusion and percolation into the underlying reduced layer, where it is rapidly denitrified.

Ammonium nitrogen is much less subject to leaching from the soil than nitrate because of its adsorption on the cation-exchange complex. Losses of nitrogen by leaching occur mainly in coarse-textured soils with low CEC. In soils where the primary source of negative charges is organic rather than inorganic, the ammonium ion is also subject to leaching by percolation water (Harada and Kutsuna 1955). Some volcanic ash soils have low ammonium adsorbing power.

Loss of ammonium by leaching is more severe in flooded soil than in nonflooded soil because:

- Ammonium is not as likely to accumulate in a well drained soil as in the flooded soil.
- Reduction reactions in a flooded soil produce ferrous and manganous ions, which displace ammonium from the exchange complex to the soil solution where it is more subject to leaching.
- The constant head of standing water in a flooded soil results in greater downward percolation of the soil solution than occurs in a nonflooded soil (Patrick and Mahapatra 1968).

Ammonium Fixation by Clay Minerals

Among the important transformation processes that involve ammoniacal nitrogen applied to the soil is that of ammonium fixation. Many researchers have observed that quantitative recovery of added ammonium could not be obtained in some soils, even when soils were extracted immediately after addition of the ammonium.

Early research suggested that ammonium retention was due to clay fixation. Subsequent studies revealed that some of the ammonium is fixed so tightly that it is resistant to extraction, even after prolonged boiling with $6N$ HCl. Recovery of tracer nitrogen was invariably less than that of total ammonium.

The well-known ability of 2:1 type clay minerals to entrap (fix) ammonium ions between the silica sheets, often accompanied by a contraction in the interlayer spacing, has received limited attention in rice soils, probably because fixation is often associated with drying to moisture contents below those usually encountered in lowland rice culture (Broadbent 1978). However, there are some soils developed from volcanic ash in areas of high rainfall that have predominantly amorphous colloidal hydrated oxides of aluminum and iron and also allophane in some cases. These soils also fix ammonium.

With regard to the magnitude of ammonium fixation in flooded soils, reports are contradictory. Some investigators believe immobilization is less in flooded than in nonflooded soils. Others report more ammonium retention in flooded than in nonflooded soil (Shimpi and Savant 1975).

When ammonium and potassium ions were added to the soil at different concentrations and order of applications, the two cations were fixed in nearly equivalent proportions but ammonium was fixed preferentially to potassium when both were added (Nommik 1965). According to Nommik (1965) smectite does not fix ammonium under moist conditions. Nevertheless, many soils maintained in the wet condition do fix some ammonium in a form that is slowly replaced by cations such as calcium, magnesium, and sodium, but not by potassium.

From the above controversies, Broadbent (1978) concludes that the line of demarcation between exchangeable and fixed ammonium is somewhat superficial, and some of these discrepancies can be resolved if one realizes that plants and microorganisms are able to extract and use this so-called fixed ammonium.

Among the factors that affect ammonium retention or fixation by minerals and soils are pH, moisture content, clay mineral content, nature of clay minerals, organic matter, and presence of cations such as potassium ions. However, in a recent study by Sahrawat (1978), the ammonium fixation capacity of 12 soils was not correlated with pH, organic matter, or clay content but significantly correlated with active iron ($r = .61*$). Iron oxides in combination with organic matter and allophanic materials may cause ammonium fixation.

The ammonium fixing capacity of one Philippine smectitic soil increased as the rate of ammonium application was increased but the percent recovery of ammonium remained more or less constant.

The availability of clay-fixed ammonium to rice has not been evaluated adequately but preliminary experiments by Manguiat (1976) suggest that defixation may occur more rapidly under flooding than has been reported in unsaturated soils (Allison et al. 1953). In some soils, ammonium fixation may be a factor in decreasing nitrogen loss by retarding the rate of nitrification.

The literature cited above suggests that ammonium can be fixed by both clay minerals and amorphous materials in soils and may be rendered unextractable by a rice crop even shortly after application. The mechanism of fixation by these materials is different and not clearly understood. Research with isotopically labeled nitrogen is needed to assess the importance of ammonium fixation in both lowland and upland rice.

Surface Runoff

Information is limited on the quantity of nutrient losses by runoff from lowland rice fields. Takamura et al. (1977) in Japan made a balance sheet of nitrogen applied to rice fields and reported that 13–16% of that applied was lost with surface drainage.

Singh et al. (1978) from studies in the Philippines reported that more than 10% of the applied nitrogen was lost in surface runoff from lowland rice fields. Their results further indicated that in the wet season the total amount of water lost through surface drainage varied from 140 to 2040 mm, with an average loss of 980 mm. The net change in the nitrogen content of the water varied from a loss of 5.6 kg N/ha to a gain of 2.6 kg N/ha depending on fertilizer and water management practices. Most of the nitrogen outflow from rice fields was in the form of ammonium nitrogen with very little loss of nitrate nitrogen.

Nitrogen losses by runoff in drainage can be substantially reduced by incorporating the nitrogen fertilizers and holding the irrigation or rainwater on the field for 5 days after fertilizer application.

Accretion of Nitrogen

Accretion of nitrogen takes place mainly by application of fertilizer and organic manure (see Chapter 10), from rain, through urea hydrolysis, and by biological nitrogen and rhizosphere nitrogen fixation.

Urea Hydrolysis

Urea is fast replacing ammonium sulfate as the most important nitrogen fertilizer for rice. The rice plant can take up urea directly (Mitsui and Kurihara 1962) but this mode of uptake is probably not important. Most of the urea nitrogen is taken up by the rice crop after urea is hydrolyzed to $(NH_4)_2CO_3$. The hydrolysis site is the soil rather than the floodwater.

The rate of urea hydrolysis has a profound effect on many transformation processes of nitrogen. Urea, relative to ammonium, is only weakly adsorbed by soil colloids. Therefore, losses by leaching and runoff would be high if hydrolysis of urea is slow. Researchers such as Delaune and Patrick (1970) have shown that urea is hydrolyzed as rapidly in flooded as in nonflooded soils. Studies at IRRI found that urea hydrolysis is rapid in flooded tropical soils.

Recently, Sahrawat (1978) reported that urease activity was highest in the alkaline Maahas clay and lowest in an acid sulfate soil.

Biological Nitrogen Fixation

For centuries rice has been grown in tropical Asia without significant inputs of chemical fertilizer, but harvests have remained at 1–2 t/ha. At IRRI, 31 successive crops of rice were harvested from plots that had no fertilizer added, and yield remained at 3–4 t/ha. Recently, Watanabe et al. (1977) summarized data from Japan, Thailand, Indonesia, and the Philippines and reported that a single rice crop can take from 37–113 kg N/ha from sources other than mineral nitrogen fertilizer.

Biological nitrogen fixation may partly explain the long-term nitrogen fertility of flooded rice soils. An indication of biological nitrogen fixation is clearly shown in the results of many long-term fertility experiments carried out for several decades at Japanese agricultural experiment stations (Yamaguchi 1979). Rice yields did not decrease in most of the experiments, in which no nitrogenous fertilizer was supplied, and no significant decrease in soil nitrogen was recognized.

The principal agents of biological nitrogen fixation for rice are:

· Blue-green algae.

- The water fern, azolla, in association with blue-green algae (*Anabaena azollae* Strassburger).
- Nonsymbiotic nitrogen-fixing bacteria around the rice plant's root zone (in the rhizosphere).
- Nonsymbiotic bacteria in the bulk of anaerobic soil (Watanabe et al. 1977).

BLUE-GREEN ALGAE   The prolific development of algae in rice fields is a common phenomenon and these algae may contribute greatly to the maintenance of soil fertility. In the tropics, many species, belonging to the genera *Tolypothrix, Nostoc, Schizothrix, Calothrix, Anoboenopsis,* and *Plectonema,* are abundant.

Numerous trials, mainly in India, have shown 15–45 kg N/ha per crop fixed by blue-green algae in unfertilized rice fields. Extensive field trials in India indicate that about 30% of the cost of commercial nitrogen fertilizers could be reduced by algal supplementation. Work done at the Indian Agricultural Research Institute, New Delhi, and elsewhere, suggests that in areas where commercial fertilizers are not used, algal application can provide as much as 30 kg N/ha per crop; in the presence of fertilizers, the benefits are complementary (Venkatraman 1977).

Application of phosphatic fertilizers, and lime, stimulate algal growth and nitrogen fixation.

AZOLLA   The water fern *Azolla* occurs in warm temperate and tropical regions throughout the world. *Azolla* is also known as water velvet and mosquito fern in English, *Lu Ping* and *Ho Ping* in Chinese, *Akaukikusa* in Japanese, and *Beo hoa dau* in Vietnamese (Lumpkin 1977).

The blue-green algae *Anabaena azollae* lives in association with *Azolla* and fixes nitrogen during all phases of the fern's life cycle. There are several species of azolla, but *Azolla pinnata* R. Br. is most common in tropical Asia. When used intensively, azolla may contribute 0–100 kg N/ha to a rice crop (3–4 crops of azolla per crop of rice).

Azolla is extensively used to supply soil nitrogen in Vietnam (300,000–400,000 ha) in the Red River Delta. The azolla is used only for the early rice crop planted in January or in April–May; it is not used for the main crop planted in June–July when temperatures are too high for profitable azolla production. The optimum range is 20–22°C. There is normal growth at 15–30°C, and azolla is killed below 7°C and above 42°C.

Farmers in China have also used azolla extensively to supply nitrogen to rice. Figure 4.12 shows azolla being grown in a rice field and unused swampy areas. Generally, a small amount of complete fertilizer (NPK) is applied to improve yields of azolla.

RHIZOSPHERE NITROGEN FIXATION   Reports by Yoshida and Ancajas (1973) suggested that up to 63 kg N/ha per crop is fixed in the rice root zone. Subsequent reports by Watanabe et al. (1977) and Watanabe (1978) suggest that the nitrogen fixing activity at the root zone is rather low and constant, at about

Figure 4.12   Azolla is grown in rice fields and swampy areas in China.

0.052 kg/ha per day, a low figure in terms of total nitrogen fixed in a soil-floodwater system (Fig. 4.13). After most of the nitrogen fixing activity of the algae was accounted for, the remaining activity was ascribed to the root zone of rice. Nitrogen fixing activity of the rice rhizosphere started to increase at 4 weeks after transplanting and attained its maximum after the heading stage.

Ratoons from rice stubble showed higher nitrogen fixing activity than intact plants before harvest or than stubble in which ratooning was prevented (Watanabe et al. 1977).

NONSYMBIOTIC BACTERIA IN ANAEROBIC SOIL   Blue-green algae appear in the paddy at early stages of rice growth and disappear at later stages (Takai and Wada 1966, Watanabe 1978). On the other hand, the nitrogen fixing activity in the rice rhizosphere increases as the plants grow older. The highest nitrogen fixing activity appears to result when root zone fixation is combined with paddy water and algae (Table 4.4).

Table 4.4   Effect of Removal of Floodwater and Algae on Nitrogen Fixation in Paddy Field (From Watanabe et al. 1978)

Treatment	C_2H_4 Formed[a] (μmol/24 hours)
Control (with paddy water and algae)	117.6 ± 7.2
Without paddy water and algae	18.6 ± 4.5

[a]$\pm$ standard error of mean.

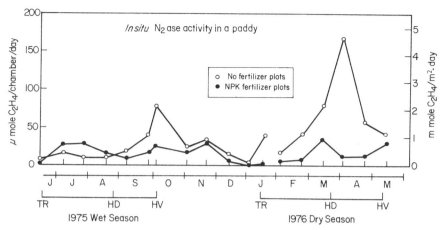

Figure 4.13   Seasonal variation of *in situ* nitrogenase activity in IRRI long-term fertility experiments plots. Growth periods of rice (IR26) from transplanting are: TR: tranplanting; HD: heading; HV: harvesting. In NPK fertilizer plots, 30 kg P_2O_5 and 30 kg K_2O were applied before transplanting; 60 kg N/ha as ammonium sulfate (all basal incorporated) in the wet season and 140 kg N/ha (100 kg N basal incorporated and 40 kg N topdressed at the panicle initiation stage) were applied. (Adapted from Watanabe et al. 1977)

The relative importance of nitrogen fixation by blue-green algae, azolla, and rhizosphere bacteria should be clearly understood before their practical significance is appreciated. The relative ease of establishing azolla, however, makes it somewhat attractive for certain tropical rice culture systems in Asia.

Transformation of Soil Phosphorus in Flooded Soils

The transformation processes of phosphorus in flooded soils are greatly different from those in nonflooded soils. The processes also differ for different water regimes. The understanding of these transformation processes is important for phosphorus fertilization of rice, which grows in water regimes ranging from continuous flooding to alternate flooding and drying.

Availability and fixation of phosphorus are two opposing chemical reactions that largely determine the phosphorus status in soils.

Availability of Phosphorus

The phosphorus supplying capacity of flooded rice soils is higher than that of nonflooded soils. Flooded rice frequently does not show response to phosphorus addition even though crops show a large response when grown on the same soil when it is dry.

The availability of phosphorus is closely related to the degree of soil reduction. In a Taiwan soil (Chiang 1963), the contents of available phosphorus and iron increased proportionately with the decrease in Eh. However, the increase in available phosphorus and iron is not so closely related to the increase in pH. The main reactions involved in this change of availability are:

- Reduction of insoluble ferric phosphate ($FePO_4 \cdot 2H_2O$) to more soluble ferrous phosphate [$Fe_3(PO_4)_2 \cdot 8H_2O$].
- Hydrolysis of aluminum and iron phosphate at the higher soil pH that occurs after submergence.
- The dissolution of the apatite because of the higher CO_2 pressure in the soil solution.
- Desorption of phosphorus from clay and oxides of aluminum and iron.

The other mechanisms of phosphorus release in a flooded soil postulated by various researchers and summarized by Patrick and Mahapatra (1968) include:

- Release of occluded phosphate by reduction of hydrated ferric oxide coating.
- Displacement of phosphate from ferric and aluminum phosphate by organic anions.

Among these reactions, the reduction of ferric phosphate to ferrous phosphate appears to be dominant. Chang (1965), by directly measuring the quantity of phosphorus absorbed from each chemical form of inorganic phosphorus, demonstrated that iron phosphate is the main source of phosphorus absorbed by rice under submergence in acidic and calcareous soils. Shapiro (1958) reported that an increase in available phosphorus by flooding as measured by A values of phosphorus (available soil phosphorus in relation to applied phosphorus fertilizer) is considerable in an acid soil rich in iron phosphate, but there is no increase in a calcareous soil and muck (organic) soil low in iron phosphate (Table 4.5). He concluded that the increase in the availability of soil phosphorus with submergence is mainly due to reduction, rather than hydrolysis.

From the above studies, it can be concluded that calcium phosphate is not an important source of available phosphorus, either in acidic soils low in calcium phosphates, or in calcareous soils. The above findings can be explained on the basis of specific surface activity. Calcium phosphate is more concentrated in the fraction of coarser soil particles (sand and silt) and thus has smaller specific surface activity. Iron and aluminum phosphates are more concentrated in the finer fraction (clay) and thus have greater surface activity.

It is believed that phosphates existing in a more dispersed state, such as iron phosphates, are more available to plants, irrespective of their chemical form (Chang 1976). With submergence, crystallized iron phosphate tends to change into colloidal iron phosphate through solution-precipitation. Juo and Ellis (1968) found that colloidal iron phosphate has a larger surface area than its crystallized counterpart and, therefore, the availability of the former is greater. It appears that the increase in availability of iron phosphate is partly due to the

Table 4.5   Effects of Flooding on the Availability of Soil P in Acid, Calcareous, and Organic Soils (Adapted from Shapiro 1958)

Soil	Dry Weight of Plant Tops (g/pot)	A Value[a] (kg P/ha)
Chester loam (pH 6.7)		
Flooded	23.5	56
Unflooded	12.2	10
Millville silty clay loam (pH 8.0)		
Flooded	2.2	405
Unflooded	1.3	432
Muck soil (pH 6.0)		
Flooded	14.5	32
Unflooded	8.5	33

[a]Available soil phosphorus in relation to applied phosphorus fertilizer.

change in the solid phase from crystalline form to the more easily soluble colloidal form.

Changes in the availability of iron phosphate in soils as influenced by submergence and drying can be postulated as follows (Chang 1971):

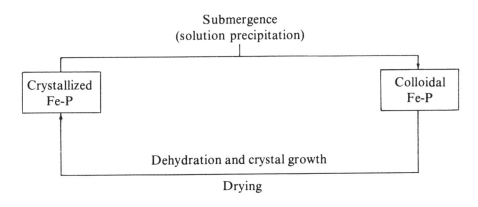

Different reactions of soil phosphorus in flooded and nonflooded soils have been attributed to the association of the phosphate ion with the ferrous iron instead of ferric iron that is usually associated with nonflooded soils (Mitsui 1955). Subsequently, Davide (1960) reported that with $CaH_4(PO_4)_2$ as a phosphorus source, response of rice was almost the same in flooded and nonflooded soils, but with iron phosphate, rice showed much better response on flooded soils than on nonflooded soils. He concluded that the beneficial effects of flooding on phosphorus availability depend on the intensity of reduction and the iron content of the soil.

In another study with five flooded soils, increased phosphorus availability after flooding was apparently due to the hydrolysis of $AlPO_4$ and the reduction of $FePO_4$. After prolonged periods of flooding, phosphorus becomes less available, probably due to higher fixation (Patrick and Mahapatra 1968).

Islam and Elahi (1954) demonstrated the reversion of ferric to ferrous iron under flooding. For laboratory incubation there was a progressive reduction of Fe^{3+} to Fe^{2+} and an increase in readily soluble phosphate in a lateritic soil. The addition of oxidizable materials, such as green manure, increased the soil reduction process and thereby increased the phosphorus availability of soil.

Among the various fractions, reductant-soluble phosphate is of special significance in flooded soils. This fraction consists of phosphate occluded in hydrated ferric oxide that is unavailable in nonflooded soils. Patrick and Mahapatra (1968) using an acid soil from Louisiana, United States, suggested Fe-P as the most dominant form under flooding (Fig. 4.14).

With alternate wetting and drying, which is the case in most rainfed rice-growing areas and in some irrigated areas in the Asian tropics, the percentage of phosphate in the aluminum form decreases and the percentage in the iron form increases.

Drying a soil subsequent to flooding generally decreases the solubility of both soil and added phosphorus and increases phosphorus fixing capacity, thus decreasing the solubility of phosphorus.

Phosphorus Fixation

The phosphorus fixing capacity of soils is considered an important soil parameter for recommendation of adequate levels of phosphorus for the desired fertilizer response.

Although the mobility of phosphorus in flooded soils may be greater than that in nonflooded soil, the soluble phosphorus applied to flooded soils is fixed on the surface of the solid phase of the soil (Chang 1976). Lin et al. (1973) analyzed the phosphorus content of soil samples obtained from plots in a long-term fertility

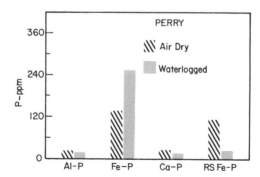

Figure 4.14 Transformation of inorganic phosphorus in a soil as a result of waterlogging. (From Patrick and Mahapatra 1968)

experiment (48 years of application of phosphorus) and found that 58% of the applied phosphorus was retained in the surface soil and subsoil.

The relative amounts of phosphorus fixed by aluminum, iron, or calcium depend on the exposed reactive surface area of the respective cations of the solid phase and the concentration of these cations in the soil solution. With time, both aluminum and calcium phosphates transform to iron phosphate. The rate of transformation increases in the presence of water. Therefore, it is reasonable to infer that the aluminum and calcium phosphates on the surface are readily transformed into iron phosphate. Thus, in a flooded soil, the available phosphorus is probably stored as surface iron phosphate (Chang 1976).

Studies at IRRI indicate that phosphorus fixation in flooded rice soils is rapid in acid and neutral soils. The amount of added phosphorus remaining in solution (^{32}P-labeled superphosphate) is lowest within 4 days of equilibration in acid Ultisols and slightly higher in Alfisols. The fixation of phosphorus is considerably slower in slightly alkaline soils (De Datta et al. 1966). Soils containing hydrated iron and aluminum oxides, halloysites, and allophanes fix phosphorus in both upland and lowland soils.

Kinetics of Water Soluble Phosphorus

The increase in concentration of water-soluble phosphorus after soil submergence is appreciable and is markedly affected by soil properties (Fig. 4.15). The increases in concentration of water-soluble phosphorus after flooding are highest in sandy calcareous soil low in iron, moderate in acid sandy soil low in iron, small in the nearly neutral clay, and least in the acid ferralitic clay (Ponnamperuma 1972). This phenomenon could possibly result from a secondary reaction between the dissolved phosphorus and the iron compound. Patrick and Khalid (1974) found that anaerobic soils release more phosphorus into soil solutions low in phosphorus and sorb more phosphorus from solutions high in phosphorus, than do aerobic soils. They attributed this phenomenon to the greater surface area of the gellike (colloidal) ferrous compounds in a flooded soil following soil reduction.

Transformation of Potassium in Flooded Soils

The effect of flooding on the chemistry and availability of soil potassium has not been studied adequately. Potassium is present in soils in four forms, which are in dynamic equilibrium as follows (Su 1976):

| soluble K (instantly available K) | $\rightleftharpoons$ | exchangeable K (easily mobilizable reserve of available K) | $\rightleftharpoons$ | nonexchangeable K (slowly mobilizable stock of available K) | $\leftarrow$ | mineral K (semi-permanent reserve) |

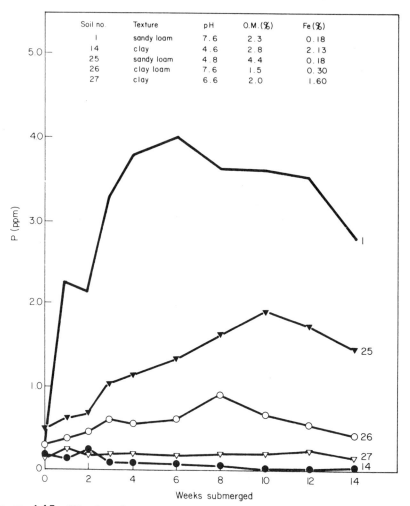

Figure 4.15 Kinetics of water-soluble phosphorus in five submerged soils. (From Ponnamperuma 1976)

Potassium is absorbed by plants after it has been transported to the root surface. In a submerged soil, *in situ,* convective transport of potassium may play a major role, whereas diffusive transport plays a minor one.

Increasing Soluble Potassium and Leaching

With flooding, soluble ferrous ions and manganous ions increase and exchangeable potassium is then displaced into the soil solution. The increase in soluble potassium after flooding is closely related to the ferrous ion content of the soil solution. In the field, with adequate drainage, leaching of soil potassium may be considerable. Studies in Taiwan summarized by Chang (1971) suggest that losses of potassium by leaching could be substantial. If the displaced potassium cations (and other basic ions) are lost by leaching, or by diffusion to the surface and loss in surface runoff, soils become acid when oxidized (Brinkman 1970).

Fixation and Release of Fixed Potassium

The mechanism governing potassium fixation and release is not clearly understood but it is generally known to be affected by the nature of soil colloids, wetting and drying, freezing and thawing, and the presence of lime.

Only about 1–2% of the total amount of potassium in an average mineral soil is readily available and exists as potassium in the soil solution or as exchangeable potassium sorbed on the soil colloids. Under certain soil conditions, added potassium is fixed by soil colloids and is not readily available to the rice plant. The nonexchangeable form is in dynamic equilibrium with the available forms and, therefore, acts as an important reservoir of slowly available potassium.

Clays of the 2:1 type (vermiculite, hydrous mica, and beidellite) can fix potassium readily and in large amounts (McLean 1978). Studies in Taiwan indicate that the fixation of applied potassium is low in Latosols, more in a slightly acidic sandstone and shale alluvial soil, and high in calcareous and shale mudstone soils (Chang and Feng 1958/1959). It was also known in Taiwan that rice plants can absorb a larger percentage of the total absorbed potassium from the nonexchangeable form under flooding than they can under nonflooded conditions. Vigorous growth of rice in a flooded soil probably increases extraction of potassium from the soil.

Chang (1971) suggested that under flooding with continuous removal of soluble and exchangeable potassium, the release of fixed potassium is favored. The correlation of the response of rice to H_2SO_4-extractable potassium from several soils was better than with exchangeable potassium, suggesting exchangeable potassium may not be entirely indicative of the available potassium status due to the rapid release of nonexchangeable potassium from the soil.

An intensity-quantity relationship has been used for studying potassium availability to plants grown in an unsaturated soil moisture regime (upland conditions). However, such soils differ considerably from submerged rice soils. The ionic composition of the reduced soil solution is dominated by Fe^{2+} and Mn^{2+} ion species. This situation poses a problem in application of the intensity-quantity relationships for potassium to flooded rice soils.

Potassium Dynamics of Lowland Rice Soils

Traditionally, potassium fertility of soils is studied using acid or NH_4OAc-extractable potassium. Recent IRRI studies on the potassium dynamics of rice soils used an electroultrafiltration (EUF) technique that has shown promise in upland soils (Nemeth 1979).

Electroultrafiltration is a combination of electrodialysis and ultrafiltration techniques and can be used to determine the amount of plant-available nutrients in soils. Soils are extracted with deionized water under an external electrical field. By adequate variation of voltage (50, 200, or 400 V) and timing (0–35 min), the total extractable nutrients can be separated into water-soluble and exchangeable forms with different bonding energies. Extraction and fractionation is done automatically by the instrument.

For the determination of soil potassium dynamics, two EUF parameters are of importance:

- The EUF-K fraction I (potassium in the extract obtained after 10 min of EUF—the first 5 min at 50 V and the next 5 min at 200 V) and

- The total EUF-K fractions (sum of all potassium fractions obtained after 35 min of EUF—the first 5 min at 50 V, the next 25 min at 200 V, and the last 5 min at 400 V).

The EUF-K fraction I represents the potassium concentration in the soil solution (intensity factor) whereas the total EUF-K fractions represent the total amount of the effectively available potassium (quantity factor).

These EUF-K parameters, particularly the total EUF-K fractions, have been closely related to the grain yields and have been better indicators of potassium availability than the exchangeable potassium technique (Fig. 4.16). However, the EUF-K fraction I clearly underestimated the potassium availability in a Maligaya silty clay loam and overestimated it in a Pili clay. This was because the

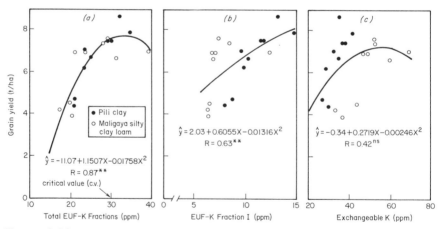

Figure 4.16 Relationship between total EUF-K fractions (a), EUF-K fraction I (b), or exchangeable K (c) and absolute grain yields of IR26 (1977 dry season) at the Bicol and Maligaya stations in the Philippines. The soil samples were taken from the NP- and NPK-plots after harvest of the 1977 dry season crop. Residual fertilizer potassium has accumulated on the potassium-treated plots as a result of the long-term potassium application since 1968, which amounted to 60 and 90 kg K_2O/ha per season (two crops per year).

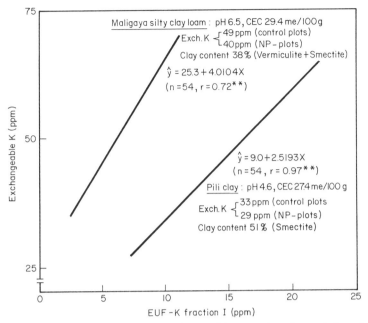

Figure 4.17 Relationship between exchangeable potassium (quantity) and EUF-K fraction I (intensity) for Pili clay and Maligaya silty clay loam in the Philippines. The potassium buffer capacity is indicated by the slope of the regression line of exchangeable potassium on the EUF-K fraction I.

potassium availability in soils is controlled not only by the soil solution potassium but also by its buffering by the soil particles, which was distinctly different for the two soils studied (Fig. 4.17).

The soil solution potassium (intensity = I) is maintained by the exchangeable potassium (quantity = Q) in a dynamic equilibrium. The higher dQ/dI or the higher potassium buffer capacity in the Maligaya silty clay loam than in the Pili clay indicates that during the active growing season, the potassium concentration in the soil solution will be depleted faster in the Pili clay than in the Maligaya silty clay loam. This explains the higher optimum potassium concentration in the soil solution in the Pili clay than in the Maligaya silty clay loam (Fig. 4.16b). The optimum potassium concentration in the soil solution can thus be lower if the potassium buffer capacity of the soil is higher.

The difference in the potassium buffer capacities of both tested soils was due mainly to the differences in the clay mineralogy. The vermiculite-bound exchangeable potassium in the Maligaya silty clay loam is more strongly bound than the entirely smectite-bound exchangeable potassium in the Pili clay. Regardless of the differences in the clay mineralogy, the total EUF-K fractions made a satisfactory assessment of the potassium availability in both soils and correlated closely with the grain yield (Fig. 4.16a). Maximum grain yields were obtained when the total EUF-K fractions after harvest prior to wetland preparation were above 30 ppm.

The EUF results suggest that potassium dynamics of lowland rice soils can be evaluated better by the EUF technique than the conventional method of exchangeable potassium (Wanasuria et al. 1980).

Antagonistic Action Between Potassium and Iron

An excessive concentration of ferrous iron in the soil solution will, under certain conditions, combine with the potassium salts in the soil to form sparingly soluble double salts consisting of K_2SO_4, $FeSO_4$, and H_2O in various proportions, and thus decrease the availability of potassium. It is also possible that excessive absorption of iron may retard the absorption of potassium. This probably explains a physiological disorder of rice in a poorly drained soil in northeastern Taiwan that can be corrected by application of potassium fertilizer (Chang 1971).

Reactions of Zinc in Flooded Rice Soils

Mikkelsen and Kuo (1977) developed a simplified diagram illustrating dynamic equilibria of zinc (Fig. 4.18). It shows that rice receives zinc from the soil solution and the exchangeable and adsorbed solid phase, including the soil organic fraction.

Unlike the redox elements iron and manganese, the concentrations of zinc in soil solution generally decrease after flooding, although zinc may undergo a temporary increase immediately after flooding (IRRI 1971, Mikkelsen and Kuo 1976, 1977); the zinc levels equilibrate around 0.3–0.5 μM (Forno et al. 1975). Although no definite explanation can be offered, Patrick and Reddy (1977) suggested that the decreased zinc concentration may be due to:

- Precipitation of Zn $(OH)_2$ as a result of increased pH after flooding.
- Precipitation of $ZnCO_3$ due to CO_2 accumulation resulting from organic matter decomposition.
- Precipitation of ZnS under very reduced soil conditions.

The mobility of zinc is affected by pH, adsorption, clay, and organic matter. Other soil parameters, such as percentage of clay and organic matter, and calcium and phosphorus status are also associated with total soil zinc.

A wide variety of chemical reactions in soils undoubtedly influence the availability of zinc to rice. For example, high manganese concentrations antagonize zinc absorption and translocation. Calcium and magnesium may also affect zinc uptake, but to a lesser extent. Giordano et al. (1974) showed the order of interference of zinc absorption and translocation to be:

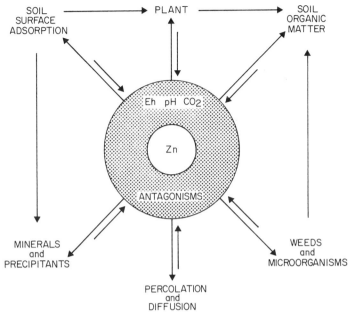

Figure 4.18   Dynamic equilibria of soil zinc and plant uptake. (From Mikkelsen and Kuo 1977)

- Depression of zinc uptake: Fe = Cu >>Mg > Mn > Ca
- Depression of zinc translocation: Fe = Cu >> Ca > Mn

Another factor that may influence the chemistry of zinc in flooded soils is phosphorus. Although considerable research has been done to determine how cations and anions affect zinc in plants, the mechanism involved in these reactions remains highly speculative.

Although the effect of flooding on the chemistry of zinc in soil and on plant absorption is not yet clear, some soil constituents have been implicated (Mikkelsen and Kuo 1977). The reversible pH change, oxidation-reduction, and carbon dioxide partial pressures may play an important role in regulating zinc uptake by rice. The reversible pH change of the flooded soil, where the pH tends to increase in acid soil and decrease in alkaline soils, undoubtedly alters the zinc equilibrium concentration in the soil solution. Because the solubility of zinc minerals and zinc sorbed by soil colloidals is pH-dependent (higher at higher pH), an increase in the pH of an acid soil when flooded will tend to decrease the zinc concentration in the solution. Therefore, the uptake of zinc by rice will decrease. In the alkaline soil, however, zinc uptake increases as the pH decreases after submergence. The reversible pH change could partially explain why zinc uptake was lower in acid or higher in alkaline soil after flooding (Mikkelsen and Kuo 1977).

Zinc deficiency is aggravated by addition of organic matter. Yoshida and Tanaka (1969) reported that normal rice plants became zinc deficient with the addition of 0.5% of cellulose to the soil. The cellulose additions enhance soil reduction and carbon dioxide accumulation as a result of anaerobic decomposition of organic matter (IRRI 1971). Experiments of Forno et al. (1975) and Mikkelsen and Brandon (1972) also showed that addition of readily decomposable organic materials may aggravate zinc deficiency and reduce zinc uptake by rice plants. The role of organic matter can, therefore, be related to the soil reduction and accumulation of carbon dioxide.

The formation of insoluble sphalerite (ZnS) has been considered a factor controlling zinc availability for rice (IRRI 1971). With the release of carbon dioxide when organic materials decompose, carbonate and bicarbonate ions in the soil and floodwater increase. The amounts of carbonate and bicarbonate ions present are a function of partial pressure of carbon dioxide, pH of the solution, and ionic strength of the soil solution. Flooded soils, especially alkaline or calcareous soils, possess all the essential characteristics for the formation of high levels of bicarbonate ions. There is a strong coincidence between high carbon dioxide partial pressures and the appearance of zinc deficiency in alkaline soils.

The rate of zinc absorption by rice roots is about twice that of wheat, which explains greater depletion of zinc in the vicinity of rice roots and creation of a concentration gradient in which rice becomes zinc deficient faster than wheat. However, some rices are relatively more tolerant of zinc deficiency than others.

Factors Affecting Zinc Availability to Rice

In addition to the physical and chemical reactions in soils, various crop management factors combine to influence the availability of zinc to rice. Among the most important are:

- The native zinc content of the soil.
- Soil pH.
- Organic matter.
- Flooding.
- Carbon dioxide partial pressure (P_{CO_2}).
- Bicarbonate ions.
- Organic acids.
- Various natural interactions.
- Environmental effects.
- Water quality.

The soil factors affecting zinc availability to rice are shown in Fig. 4.19. These factors are discussed in detail by Mikkelsen and Kuo (1977).

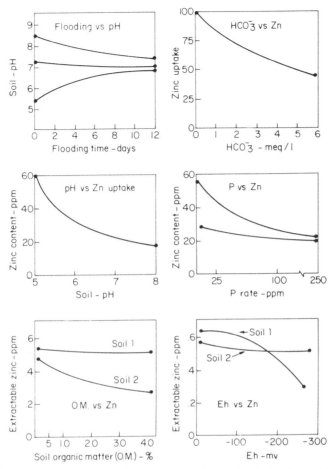

Figure 4.19   Factors affecting zinc availability to rice. (Adapted from Mikkelsen and Kuo 1977)

Transformation of Sulfur in Flooded Soils

The main transformations of sulfur in flooded soils are the reduction of sulfate to sulfide and conversion of organic sulfur to hydrogen sulfide. The hydrogen sulfide formed may react with heavy metal ions in soil (such as Fe^{2+}, Zn^{2+}, and Cu^{2+}) to give insoluble sulfides. As a result, the availability of these metal elements may be reduced. Because Fe^{3+} reduction to Fe^{2+} precedes SO_4^{2-} reduction, there usually will be Fe^{2+} present in the soil solution by the time hydrogen sulfide is produced so that hydrogen sulfide will be converted to insoluble FeS. This reaction protects microorganisms and higher plants from toxic effects of hydrogen sulfide. In muck soils and sandy soils low in iron, however, where iron is inactivated by the complex formation with organic matter, FeS formation may not take place. As a result, hydrogen sulfide toxicity to the rice plant is possible.

The kinetics of water-soluble sulfate (which is a measure of sulfate reduction) in flooded soils depends on soil properties. In neutral and alkaline soils, concentrations as high as 1500 ppm SO_4^{2-} may be reduced to zero within 6 weeks after flooding. Acid soils first show an increase in water-soluble SO_4^{2-}, then a slow decrease spread over several months. The initial increase in SO_4^{2-} concentration is due to the release (following increase in pH) of SO_4^{2-}, which is strongly sorbed at low pH by clay and hydrous oxides of iron and aluminum. Sulfate reduction proceeds slowly in flooded acid sulfate soils, despite their strong acidity. Lime accelerates the reduction considerably (Ponnamperuma 1972).

Rice, like other plants, absorbs sulfur primarily in the form of sulfate and the reduction of sulfate to sulfide in submerged soils reduces the availability of sulfur. However, rice takes up sulfur which is oxidized as SO_4^{2-} on the root surface (Engler and Patrick 1975).

The reduction of sulfate in flooded soils has three implications for rice culture: the sulfur supply may become insufficient, zinc and copper may be immobilized, and hydrogen sulfide toxicity may arise in soils low in iron (Ponnamperuma 1972).

Transformation of Silicon in Flooded Soils

Silicon occurs in soils as crystalline and amorphous silica, as silicates, as silicon adsorbed or coprecipitated with hydrous oxides of Al^{3+}, Fe^{3+}, and Mn^{4+}, and as silicon dissolved in soil solution. The dominant silicon form in soil solution of flooded soils is $Si(OH)_4$, which is in equilibrium with quartz (SiO_2). Silica concentration in a silica-water system is independent of pH over most of the pH range (2–8) (Iler 1979). Patrick and Reddy (1978) obtained similar results for flooded soils.

The concentration of silicon in the solutions of flooded soils increases slightly after flooding and then decreases gradually; after several months of submergence, the silicon concentration may be lower than at the start (Ponnamperuma 1972). The increase is probably due to the release of adsorbed and occluded silica from hydroxides of iron and aluminum (McKeague and Cline 1963). Further, the concentration of silicon in the soil solution increases with soil reduction, apparently because of the reduction of Fe(III) oxide hydrates that sorb silicon (Ponnamperuma 1965, 1978).

Ponnamperuma (1965) showed the increase in solubility of SiO_2 with time of submergence and indicated marked differences among soils in soluble SiO_2 content before and after flooding. Soils with a high content of organic matter, regardless of pH, gave the highest increases. Decomposing rice straw with its high silicon content may also contribute to the increased silicon content of the soil solution in flooded soils (Patrick and Reddy 1978).

Factors Affecting the Silicon Content in the Soil

The concentration of $Si(OH)_4$ in equilibrium with amorphous silica at $25°\,C$ is 120–140 ppm as SiO_2 and is independent of pH in the 2–9 range. The concentrations of silicon in natural waters are much less than 120 ppm, and in solutions, decrease with increase in pH up to about pH 8 (Jones and Handreck 1967). The low concentrations of silicon in natural waters have been attributed to the sorption of silicon by hydrous oxides of Fe(III) and Al and to recombination of silicon with aluminum silicates.

The solubility of silicic acid in the soil solution is sometimes directly related to the availability of the soil phosphorus to plants (Elawad and Green 1979).

Transformations of Copper, Boron, and Molybdenum in Flooded Soils

Very little is known of the transformation processes involved with copper, boron, and molybdenum in flooded soils. In a recent summary, Patrick and Reddy (1977) suggest that the chemistry of copper in flooded soils may be similar to zinc.

Boron concentration seems to remain more or less constant after submergence of rice soil.

Molybdenum concentration in rice soils was found to increase after submergence (Ponnamperuma 1975), possibly due to increased pH.

CHEMICAL KINETICS AND SOIL FERTILITY

Flooding brings about drastic changes in a soil, affecting many chemical and biochemical processes that affect the availability and loss of nutrients. The rates and magnitudes of these changes vary widely among soils. They are higher in soil low in nitrate and manganese dioxide or high in organic matter (Ponnamperuma 1965).

Availability of Nutrients

Soil properties, duration of submergence, and temperature strongly influence the concentration of:

- Water-soluble NO_3^-.
- Exchangeable and water-soluble NH_4^+.
- Total and water-soluble Fe^{2+} and Mn^{2+}.
- Water-soluble K^+, Ca^{2+}, Mg^{2+}, SO_4^{2-}.
- Water-soluble Cu, Zn, and Mo.
- Carbon dioxide and organic acids (Ponnamperuma 1977).

Small amounts of nitrate may be present in the soil surface and the supernatant water at the time of flooding and throughout submergence. Nitrate is easily absorbed by rice and nitrate disappearance through denitrification or leaching represents the loss of a valuable nutrient.

The period of intensive reduction, production of ammonia, buildup of high concentration of CO_2, Fe^{2+}, Mn^{2+}, and organic reduction products occurs in the first 2–6 weeks after flooding. Also during this period the availability of phosphorus, silicon, and potassium increases markedly. This means that the duration of submergence of a soil prior to planting is an important variable in the productivity of flooded soils.

REFERENCES

Acharya, C. N. 1935a. Studies on the anaerobic decomposition of plant materials. I. The anaerobic decomposition of rice straw (*Oryza sativa*). *Biochem. J.* 29:528–541.

Acharya, C. N. 1935b. Studies on the anaerobic decomposition of plant materials. II. Some factors influencing the anaerobic decomposition of rice straw (*Oryza sativa*). *Biochem. J.* 29:953–960.

Acharya, C. N. 1935c. Studies on the anaerobic decomposition of plant materials. III. Comparison of the course of decomposition of rice straw under anaerobic, aerobic and partially aerobic conditions. *Biochem. J.* 29:1116–1120.

Allison. F. E., M. Kefauver, and E. M. Roller. 1953. Ammonium fixation in soils. *Soil Sci. Soc. Am. Proc.* 17:107–110.

Arikado, H. 1959. Comparative studies on the development of the ventilating system between lowland and upland rice plants growing under flooded and upland soil conditions. *Fac. Agric. Mie Univ. Bull.* 19:1–10.

Armstrong, W. 1971. Radial oxygen losses from intact rice roots as affected by distance from the apex, respiration and waterlogging. *Physiol. Plant.* 25:192–197.

Bouldin, D. R., and B. V. Alimagno. 1976. NH_3 volatilization from IRRI paddies following broadcast applications of fertilizer nitrogen. Terminal report as visiting scientist at IRRI. 51 pp. (unpubl. mimeo.)

Bowman, R. A., and D. D. Focht. 1974. The influence of glucose and nitrate concentrations upon denitrification rates in sandy soils. *Soil Biol. Biochem.* 6:297–301.

Bremner, J. M. 1977. Role of organic matter in volatilization of sulfur and nitrogen from soils. *Proc. Symp. Soil Organic Matter Stud.* II:229–239.

Bremner, J. M., and K. Shaw. 1958. Denitrification in soil. I. Methods of investigation. *J. Agric. Sci.* 51:22–39.

Brinkman, R. 1970. Ferrolysis, a hydromorphic soil forming process. *Geoderma* 3:199–206.

Broadbent, F. E. 1978. Nitrogen transformation in flooded soils. Pages 543–559 *in* International Rice Research Institute. *Soils and rice.* Los Baños, Philippines.

Broadbent, F. E. 1979. Mineralization of organic nitrogen in paddy soils. Pages 105–118 *in* International Rice Research Institute. *Nitrogen and rice.* Los Baños, Philippines.

Broadbent, F. E., and F. Clark. 1965. Denitrification. Pages 344–359 in W. V. Bartholomew and F. E. Clark, eds. *Soil nitrogen.* American Society of Agronomy, Madison, Wisconsin.

Broadbent, F. E., and B. F. Stojanovic. 1952. The effect of partial pressure of oxygen on some soil nitrogen transformations. *Soil Sci. Soc. Am. Proc.* 16:359–363.

Burford, J. R., and J. M. Bremner. 1975. Relationships between the denitrification capacities of soils and total, water-soluble and readily decomposable soil organic matter. *Soil Biol. Biochem.* 7:389–394.

Chang, S. C. 1965. Application of phosphorus fractionation to the study of the chemistry of available soil phosphorus. *Soils Fert. Taiwan*: 1–15.

Chang, S. C. 1971. *Chemistry of paddy soils.* ASPAC Food Fert. Technol. Cent. Ext. Bull. 7. 26 pp.

Chang, S. C. 1976. Phosphorus in submerged soils and phosphorus nutrition and fertilization of rice. Pages 93–116 *in* ASPAC Food and Fertilizer Technology Center. *The fertility of paddy soils and fertilizer applications for rice.* Taipei, Taiwan.

Chang, S. C., and M. P. Feng. 1958/1959. Potassium in soils of Taiwan. *Soils Fert. Taiwan*: 1–10.

Chao, T. T., and W. Kroontje. 1964. Relationships between ammonia volatilization, ammonia concentration and water evaporation. *Soil Sci. Soc. Am. Proc.* 28: 393–395.

Chiang, C. T. 1963. A study of the availability and forms of phosphorus in paddy soils. 1. The interrelationship between available phosphorus and soil pH, Eh. *Soils Fert. Taiwan*: 61.

Cho, D. Y., and F. N. Ponnamperuma. 1971. Influence of soil temperature on the chemical kinetics of flooded soils and the growth of rice. *Soil Sci.* 112:184–194.

Davide, J. G. 1960. Phosphate studies in flooded soils. Unpublished PhD dissertation, North Carolina State College, Raleigh, North Carolina. 167 pp.

De, P. K., and S. N. Sarker. 1936. Transformation of nitrate in waterlogged soils. *Soil Sci.* 42:143–155.

De Datta, S. K., J. C. Moomaw, V. V. Racho, and G. V. Simsiman. 1966. Phosphorus supplying capacity of lowland rice soils. *Soil Sci. Soc. Am. Proc.* 30:613–617.

Delaune, R. D., and W. H. Patrick, Jr. 1970. Urea conversion to ammonia in waterlogged soils. *Soil Sci. Soc. Am. Proc.* 34:603–607.

Elawad, S. H., and V. E. Green, Jr. 1979. Silicon and the rice plant environment: a review of recent research. *Il Riso* 28:235–253.

Engler, R. M., and W. H. Patrick, Jr. 1975. Stability of sulfides of manganese, iron, zinc, copper, and mercury in flooded and nonflooded soil. *Soil Sci.* 119:217–221.

Forno, D. A., C. J. Asher and S. Yoshida. 1975. Zinc deficiency in rice. II. Studies on two varieties differing in susceptibility to zinc deficiency. *Plant Soil* 42:551–563.

Freney, J. R., O. T. Denmead, I. Watanabe, and E. T. Craswell. 1981. Ammonia and nitrous oxide losses following applications of ammonium sulfate to flooded rice. *Aust. J. Agric. Res.* 32:37–45.

Giordano, P. M., J. C. Noggle, and J. J. Mortvedt. 1974. Zinc uptake by rice, as affected by metabolic inhibitors and competing cations. *Plant Soil* 41:637–646.

Gotoh, S. 1973. Reduction processes in waterlogged soils with special reference to transformation of nitrate, manganese and iron. *Bull. Kyushu Agric. Exp. Stn. Jpn.* 16:669–714.

Harada, T., and K. Kutsuna. 1955. Cation exchanges in soils. *Bull. Natl. Inst. Agric. Sci., Jpn.* Ser. B, 5:1–26.

Harrison, W. H., and P. A. S. Aiyar. 1916. The gases of swamp rice soils. IV. The source of the gaseous soil nitrogen. *Mem. Minist. Dep. Agric. India, Chem. Ser.* 5:1–31.

Iler, R. K. 1979. *The chemistry of silica. Solubility, polymerization, colloid and surface properties, and biochemistry.* John Wiley & Sons, New York. 866 pp.

IRRI (International Rice Research Institute). [1965] *Annual report 1964.* Los Baños, Philippines. 335 pp.

IRRI (International Rice Research Institute). 1971. *Annual report for 1970.* Los Baños, Philippines. 265 pp.

IRRI (International Rice Research Institute). 1977. *Annual report for 1976.* Los Baños, Philippines. 418 pp.

Islam, M. A., and M. A. Elahi. 1954. Reversion of ferric iron to ferrous iron under waterlogged conditions and its relation to available phosphorus. *J. Agric. Sci.* 45:1–2.

Jones, L. H. P., and K. A. Handreck. 1967. Silica in soils, plants, and animals. *Adv. Agron.* 19:107–149.

Juo, A. S. R., and B. G. Ellis. 1968. Chemical and physical properties of iron and aluminum phosphates and their relation to phosphorus availability. *Soil Sci. Soc. Am. Proc.* 32:216–221.

Kawaguchi, K., and K. Kyuma. 1977. *Paddy soils in tropical Asia: their material nature and fertility.* University Press of Hawaii. 258 pp.

Keeney, D. R. 1973. The nitrogen cycle in sediment-water systems. *J. Environ. Qual.* 2:15–29.

Kohl, D. H., F. Vithayathil, P. Whitlow, G. Shearer, and S. H. Chien. 1976. Denitrification kinetics in soil systems: the significance of good fits of data to mathematical forms. *Soil Sci. Soc. Am. Proc.* 40:249–253.

Komatsu, Y., M. Takagi, and M. Yamaguchi. 1978. Participation of iron in denitrification in waterlogged soil. *Soil Biol. Biochem.* 10:21–26.

Koyama, T. 1975. Practice of determining potential nitrogen supplying capacities of paddy soils and rice yield [in Japanese]. *J. Sci. Soil Manure, Jpn.* 46:260–269.

Krishnappa, A. M., and J. E. Shinde. 1978. The fate of an initial ^{15}N pulse under field conditions of the tropical flooded rice culture. All-India Coordinated Rice Improvement Project. Working Pap. 41. Presented at the 4th Research coordination meeting of the joint FAO/IAEA/GSF coordinated programme on N residue held at Piracicaba, Brazil, 3–7 July 1978a. 29 pp. (unpubl. mimeo.)

Lehr, J. J., and J. C. van Wesemael. 1961. The volatilization of ammonia from lime-rich soils [in Dutch]. *Landbouwkd.* Tijdschr. 73:1156–1168.

Lin, C. F., T. S. Lee Wang, A. H. Chang, and C. Y Cheng. 1973. Effects of some long-term fertilizer treatments on the chemical properties of soil and yield of rice. *J. Taiwan Agric. Res.* 22:241–262.

Lumpkin, T. A. 1977. Azolla: morphology of the symbiosis. Paper presented at a Saturday seminar, 3 September 1978, International Rice Research Institute, Los Baños, Philippines. (unpubl. mimeo.)

Macrae, I. C., R. R. Ancajas, and S. Salandanan. 1968. The fate of nitrate nitrogen in some tropical soils following submergence. *Soil Sci.* 105:327–334.

Manguiat, I. J. 1976. Tracer studies on nitrogen transformations in a flooded soil-plant system. Unpublished PhD dissertation, University of California, Davis. 109 pp.

McKeague, J. A., and M. G. Cline. 1963. Silica in soils. *Adv. Agron.* 15:339–396.

McLean, E. O. 1978. Influence of clay content and clay composition on potassium availability. Pages 1–19 *in* Potash Research Institute of India. *Potassium in Soils and Crops.* New Delhi.

Mikkelsen, D. S., and D. M. Brandon. 1972. Zinc deficiency in the soil-water-rice plant system. Proc. 14. Rice Tech. Working Group. p. 81–82.

Mikkelsen, D. S., and S. K. De Datta. 1979. Ammonia volatilization from wetland rice soils. Pages 135–156 *in* International Rice Research Institute. *Nitrogen and rice.* Los Baños, Philippines.

Mikkelsen, D. S., and S. Kuo. 1976. Zinc fertilization and behavior in flooded soils. Pages 170–196 *in* ASPAC Food and Fertilizer Technology Center. *The fertility of paddy soils and fertilizer application for rice.* Taipei, Taiwan.

Mikkelsen, D. S., and S. Kuo. 1977. *Zinc fertilization and behavior in flooded soils.* Commonwealth Bureau of Soils. Commonwealth Agricultural Bureau Spec. Publ. 5. 59 pp.

Mikkelsen, D. S., S. K. De Datta, and W. N. Obcemea. 1978. Ammonia volatilization losses from flooded rice soils. *Soil Sci. Soc. Am. J.* 42:725–730.

Mitsui, S. 1955. *Inorganic nutrition, fertilisation, and soil amelioration for lowland rice.* 2nd ed. Yokendo Ltd., Tokyo. 107 pp.

Mitsui, S., and K. Kurihara. 1962. The intake and utilization of carbon by plant roots from ^{14}C-labelled urea. IV. Absorption of intact urea molecule and its metabolism in plant. *Soil Sci. Plant Nutr.* (Tokyo) 8:219–225.

Mortimer, C. H. 1941. The exchange of dissolved substances between mud and water in lakes. *J. Ecol.* 29:280–329.

Mortimer, C. H. 1942. The exchange of dissolved substances between mud and water in lakes. *J. Ecol.* 30:147–201.

Nemeth, K. 1979. The availability of nutrients in the soil as determined by electroultrafiltration (EUF). *Adv. Agron.* 31:155–188.

Nommik, H. 1965. Ammonium fixation and other reactions involving a nonenzymatic immobilization of mineral nitrogen in soil. Pages 198–258 *in* W. V. Bartholomew and F. E. Clark, eds. *Soil nitrogen.* American Society of Agronomy, Madison, Wisconsin.

Patrick, W. H., Jr. 1960. Nitrate reduction rates in a submerged soil as affected by redox potential. Pages 494–500 *in* International Society of Soil Science. *Transactions of 7th International Congress of Soil Science.* Vol. II.

Patrick, W. H., Jr., and R. D. Delaune. 1972. Characterization of the oxidized and reduced zones in flooded soil. *Soil Sci. Soc. Am. Proc.* 36:573–576.

Patrick, W. H., Jr., and S. Gotoh. 1974. The role of oxygen in nitrogen loss from flooded soils. *Soil Sci.* 118:78–81.

Patrick, W. H., Jr., and R. A. Khalid. 1974. Phosphate release and sorption by soils and sediments: effect of aerobic and anaerobic conditions. *Science* 186:53–55.

Patrick, W. H., Jr., and I. C. Mahapatra. 1968. Transformation and availability to rice of nitrogen and phosphorus in waterlogged soils. *Adv. Agron.* 20:323–359.

Patrick, W. H., Jr., and D. S. Mikkelsen. 1971. Plant nutrient behavior in flooded soil. Pages 187–215 *in* Soil Science Society of America. *Fertilizer technology and use.* Madison, Wisconsin.

Patrick, W. H., Jr., and C. N. Reddy. 1978. Chemical changes in rice soils. Pages 361–379 *in* International Rice Research Institute. *Soils and rice.* Los Baños, Philippines.

Patrick, W. H., Jr., and K. R. Reddy. 1976. Nitrification-denitrification reactions in flooded soils and water bottoms: dependence on oxygen supply and ammonium diffusion. *J. Environ. Qual.* 5:469–472.

Patrick, W. H., Jr., and K. R. Reddy. 1977. Fertilizer nitrogen reactions in flooded soils. Pages 275–281 *in* Society of the Science of Soil and Manure, Japan. *Proceedings of the international seminar on soil environment and fertilizer management in intensive agriculture (SEFMIA), Tokyo-Japan, 1977.* Tokyo.

Patrick, W. H., Jr., and M. E. Tusneem. 1972. Nitrogen loss from flooded soil. *Ecology* 53:735–737.

Pearsall, W. H., and C. H. Mortimer. 1939. Oxidation reduction potentials in waterlogged soils, natural waters and muds. *J. Ecol.* 27:483–501.

Ponnamperuma, F. N. 1955. The chemistry of submerged soils in relation to the growth and yield of rice. PhD dissertation, Cornell University, Ithaca, New York. 208 pp.

Ponnamperuma, F. N. 1965. Dynamic aspects of flooded soils. Pages 295–328 *in* International Rice Research Institute. *The mineral nutrition of the rice plant.* Proceedings of a symposium at the International Rice Research Institute, February, 1964. The Johns Hopkins Press, Baltimore, Maryland.

Ponnamperuma, F. N. 1967. A theoretical study of aqueous carbonate equilibria. *Soil Sci.* 103:90–100.

Ponnamperuma, F. N. 1972. The chemistry of submerged soils. *Adv. Agron.* 24:29–96.

Ponnamperuma, F. N. 1975. Micronutrient limitations in acid tropical rice soils. Pages 330–347 *in* E. Bornemisza and A. Alvarado, eds. *Soil management in tropical America.* North Carolina State University, Raleigh, North Carolina.

Ponnamperuma, F. N. 1976. *Specific soil chemical characteristics for rice production in Asia.* IRRI Res. Pap. Ser. 2. 18 pp.

Ponnamperuma, F. N. 1977. *Physicochemical properties of submerged soils in relation to fertility.* IRRI Res. Pap. Ser. 5. 32 pp.

Ponnamperuma, F. N. 1978. Electrochemical changes in submerged soils and the growth of rice. Pages 421–441 *in* International Rice Research Institute. *Soils and rice.* Los Baños, Philippines.

Ponnamperuma, F. N., T. Attanandana, and G. Beye. 1973. Amelioration of three acid sulphate soils for lowland rice. Pages 391–406 *in* International Institute for Land Reclamation and Improvement. *Acid sulfate soils.* Proceedings of the international symposium on acid sulfate soils, 13–20 August 1972, Wageningen, The Netherlands. Publ. 18, Vol. II. Wageningen.

Ponnamperuma, F. N., R. U. Castro, and C. M. Valencia. 1969. Experimental study of the influence of the partial pressure of carbon dioxide on the pH values of aqueous carbonate systems. *Soil Sci. Soc. Am. Proc.* 33:239–241.

Ponnamperuma, F. N., E. Martinez, and T. Loy. 1966. Influence of redox potential and partial pressure of carbon dioxide on the pH values and the suspension effect of flooded soils. *Soil Sci.* 101:421–431.

Reddy, K. R., and W. H. Patrick, Jr. 1975. Effect of alternate aerobic and anaerobic conditions on redox potential, organic matter decomposition and nitrogen loss in a flooded soil. *Soil Biol. Biochem.* 7:87–94.

Reddy, K. R., and W. H. Patrick, Jr. 1976. Effect of frequent changes in aerobic and anaerobic conditions on redox potential and nitrogen loss in a flooded soil. *Soil Biol. Biochem.* 8:491–495.

Reddy, K. R., W. H. Patrick, Jr., and R. E. Philipps. 1976. Ammonium diffusion as a factor in nitrogen loss from flooded soils. *Soil Sci. Soc. Am. J.* 40:528–533.

Reddy, K. R., W. H. Patrick, Jr., and R. E. Philipps. 1978. The role of nitrate diffusion in determining the order and rate of denitrification in flooded soil. I. Experimental results. *Soil Sci. Soc. Am. Proc.* 42:268–272.

Redman, F. H., and W. H. Patrick, Jr. 1965. *Effect of submergence on several biological and chemical soil properties.* La. Agric. Exp. Stn. Bull. 592. 28 pp.

Sahrawat, K. L. 1978. Nitrogen transformations in rice soils. Paper presented at a Saturday seminar, 14 January 1978, International Rice Research Institute, Los Baños, Philippines. (unpubl. mimeo.)

Savant, N. K., and S. K. De Datta. 1980. Nitrogen transformations in wetland rice soils. (unpubl. manuscript)

Shapiro, R. E. 1958. Effect of organic matter and flooding on availability of soil and synthetic phosphates. *Soil Sci.* 85:267–272.

Shimpi, S. S., and N. K. Savant. 1975. Ammonia retention in tropical soils as influenced by moisture content and continuous submergence. *Soil Sci. Soc. Am. Proc.* 39:153–154.

Shioiri, M. 1941. Denitrification in paddy soils [in Japanese]. *Kagaku* 11:1–24. (Translated in English by Masanori Saito, 1977.)

Shioiri, M., and S. Mitsui. 1935. The effect of soil stirring on the fate of ammonia under waterlogged soil condition. [in Japanese]. *J. Sci. Soil Manure, Jpn.* 9 (complement):46–48.

Shioiri, M., and T. Tanada. 1954. *The chemistry of paddy soils in Japan.* Ministry of Agriculture and Forestry, Tokyo. 45 pp.

Singh, V. P., T. H. Wickham, and I. T. Corpuz. 1978. Nitrogen movement to Laguna Lake through drainage from rice fields. Paper presented at the 9th annual scientific meeting of the Crop Science Society of the Philippines, 11–13 May 1978, Iloilo City, Philippines.

Stanford, G., R. A. Vander Pol, and S. Dzienia. 1975. Denitrification rates in relation to total and extractable soil carbon. *Soil Sci. Soc. Am. Proc.* 39:284–289.

Su, N. R. 1976. Potassium fertilization of rice. Pages 117–148 in ASPAC Food and Fertilizer Technology Center. *The fertility of paddy soils and fertilizer application for rice.* Taipei, Taiwan.

Takai, Y., and Y. Uehara. 1973. Nitrification and denitrification in the surface layer of submerged soils. I. Oxidation-reduction condition, nitrogen transformation and bacterial flora in the surface and deeper layer of submerged soils [in Japanese]. *J. Sci. Soil Manure Jpn.* 44:463–470.

Takai, Y., and H. Wada. 1966. Chemical changes and microorganisms in paddy field. Pages 45–72 *in* Furusaka, ed. *Soil microorganisms.* Iwanamishoten, Tokyo.

Takamura, Y., T. Tabuchi, and H. Kubota. 1977. Behaviour and balance of applied nitrogen and phosphorus under rice field conditions. Pages 342–349 *in* Society of the Science of Soil and Manure, Japan. *Proceedings of the international seminar on soil environment and fertility managment in intensive agriculture (SEFMIA), Tokyo-Japan, 1977.* Tokyo.

Tusneem, M. E., and W. H. Patrick, Jr. 1971. Nitrogen transformations in waterlogged soil. La. Agric. Exp. Stn. Bull. 659. 75 pp.

Van Kessel, J. F. 1977. Factors affecting the denitrification rate in two water-sediment systems. Water Res. 11:259–267.

Van Raalte, M. H. 1941. On the oxygen supply of rice roots. *Ann. Bot. Gard. Buitenzorg.* 51:43–57.

Venkatraman, G. S. 1977. *Blue-green algae (A biofertilizer for rice).* Indian Agricultural Research Institute, Division of Microbiology, Bulletin. 8 pp.

Ventura, W. B., and T. Yoshida. 1977. Ammonia volatilization from a flooded tropical soil. *Plant Soil* 46:521–531.

Vlek, P. L. G., and E. T. Craswell. 1979. Effect of nitrogen source and management on ammonia volatilization losses from flooded rice-soil systems. *Soil Sci. Soc. Am. J.* 43:352–358.

Wanasuria, S., K. Mengel, and S. K. De Datta. 1980. Use of electroultrafiltration (EUF) technique to study the potassium dynamics of wetland soils and potassium uptake by rice. International Rice Research Institute. (unpubl.)

Watanabe, I. 1978. Biological nitrogen fixation in rice soils. Pages 465–478 *in* International Rice Research Institute. *Soils and rice.* Los Baños, Philippines.

Watanabe, I., K. K. Lee, and B. V. Alimagno. 1978. Seasonal change of N_2-fixing rate in rice field assayed by in situ acetylene reduction technique. I. Experiments in long-term fertility plots. *Soil Sci. Plant Nutr. (Tokyo)* 24:1–13.

Watanabe, I., K. K. Lee, B. V. Alimagno, M. Sato, D. C. Del Rosario, and M. R. De Guzman. 1977. *Biological nitrogen fixation in paddy field studied by in situ acetylene-reduction assays.* IRRI Res. Pap. Ser. 3. 16 pp.

Wetselaar, R., T. Shaw, P. Firth, J. Oupathum, and H. Thitipoca. 1977. Ammonia volatilization from variously placed ammonium sulphate under lowland rice field conditions in central Thailand. Pages 282–288 *in* Society of the Science of Soil and Manure, Japan. *Proceedings of the International seminar on soil environment and fertilizer management in intensive agriculture (SEFMIA), Tokyo-Japan, 1977.* Tokyo.

Wijler, J., and C. C. Delwiche. 1954. Investigations on the denitrifying process in soil. *Plant Soil* 5:155–169.

Williams, W. A., D. S. Mikkelsen, K. E. Mueller, and J. E. Ruckman. 1968. Nitrogen immobilization by rice straw incorporated in lowland rice production. *Plant Soil* 28:49–60.

Yamada, Y. 1975. Behavior of nitrogen in soil and its effects on plant growth. Pages 116–124 *in* East-West Food Institute. *Proc. Fert. INPUTS (Increasing Productivity Under Tight Supplies) Projects.* Honolulu, Hawaii.

Yamaguchi, M. 1979. Biological nitrogen fixation in flooded rice field. Pages 193–204 *in* International Rice Research Institute. *Nitrogen and rice.* Los Baños, Philippines.

Yoshida, S., and A. Tanaka. 1969. Zinc deficiency of the rice plant in calcareous soil. *Soil Sci. Plant Nutr. (Tokyo)* 15:75–80.

Yoshida, T., and R. R. Ancajas. 1973. The fixation of atmospheric nitrogen in the rice rhizosphere. *Soil Biol. Biochem.* 5:153–155.

Yoshida, T., and B. C. Padre , Jr. 1975. Effect of organic matter application and water regimes on the transformation of fertilizer nitrogen in a Philippine soil. *Soil Sci. Plant Nutr. (Tokyo)* 21:281–292.

5

Morphology, Growth, and Development of the Rice Plant

The most comprehensive among early studies of the morphology of the rice plant was that of Juliano and Aldama (1937). Other reports of interest are those of Kuwada (1910), Santos (1933), and Morinaga and Fukushima (1934). A widely adopted description of the plant morphology and varietal characteristics is by Chang and Bardenas (1965).

STRUCTURE OF THE RICE GRAIN

The rice fruit is a caryopsis in which the single seed is fused with the wall (pericarp) of the ripened ovary, forming a seedlike grain. The caryopsis is enveloped by the lemma and palea (Fig. 5.1). Among cultivars, the rice caryopsis varies widely in shape and in size. The mature rice seed is harvested as a covered grain (rough rice) in which the caryopsis is enclosed.

Anatomy of the Rice Grain

The main components of the rice grain are the hull, caryopsis, endosperm, and embryo. Figure 5.1 details the composition of each component.

Hull

The rice caryopsis is surrounded by a hull (husk) composed of two modified leaves, the palea and a larger lemma. The palea and lemma are held together by hooklike structures (Bechtel and Pomeranz 1978). The cells of the mature hull are highly lignified and brittle, with high concentrations of silica in the hull cells, presumably in the outer epidermal cells.

146

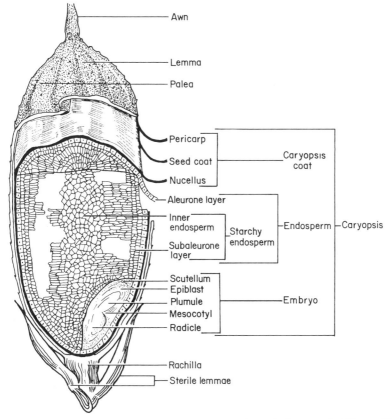

Figure 5.1   Structure of the rice grain. (Adapted from Juliano and Aldama 1937, Juliano 1980)

Caryopsis Coat

Surrounding the endosperm of mature rice caryopsis, but inside the hull, are three distinct layers that make up the caryopsis coat—the pericarp, seed coat, and nucellus.

The seed coat is beneath the pericarp. Abutting the seedcoat cuticle is another thick cuticle of the crushed nucellar cells.

Endosperm

The endosperm consists of:

- An aleurone layer which forms the outermost layer of endosperm tissue. The number of aleurone layers present is dependent on location in the grain.

variety, and environmental factors. The aleurone layer is rich in phosphorus, magnesium, and potassium (Tanaka et al. 1974).

- A starchy endosperm, which consists of thin-walled parenchyma cells usually radially elongated and heavily loaded with compound starch granules and some protein bodies (Juliano 1972).

Juliano and Bechtel (1980) divided starchy endosperm into a subaleurone, located just beneath the aleurone, and a central region consisting of the rest of the starch endosperm.

Embryo

The embryo or germ is extremely small and is located on the ventral side of the caryopsis. It contains the embryonic leaves (plumule) and embryonic primary root (radicle), which are joined by a very short stem (mesocotyl). The plumule is enclosed by a cylinderlike protective covering, the coleoptile, and the radicle is ensheathed by a mass of soft tissue, the coleorhiza. The outer side of the embryo is enclosed by the aleurone layer. The coleoptile is surrounded by the scutellum and the epiblast. The vascular trace of coleoptile is fused with lateral parts of the scutellum (Juliano 1972).

MORPHOLOGY OF THE RICE PLANT

Cultivated rice is an annual grass with round, jointed culms, rather flat leaves, and terminal panicles.

Vegetative Organs

The vegetative organs consist of roots, culms, and leaves. A branch of the plant bearing the root, culm, leaves, and often a panicle, is called a tiller.

Roots

The rice plant has a fibrous root system. There are two kinds of roots:

- *Seminal roots* that grow out of the radicle and are temporary in nature.
- *Secondary adventitious roots* that are freely branched and produced from the lower nodes of the young culm. These roots replace the seminal roots.

Culms

The culm, or stem, is made up of nodes and internodes in alternate order. The node bears a leaf and a bud, which may grow into a tiller, or shoot. The mature internode is hollow and finely grooved. At an early growth stage, what is often referred to as the culm is composed mostly of leaf sheaths and is not the true culm, which is very short at that stage.

Tillers grow out of the main culm in an alternate order. The primary tillers grow from the lowermost nodes and give rise to secondary tillers. These in turn give rise to a third group, called tertiary tillers (Fig. 5.2).

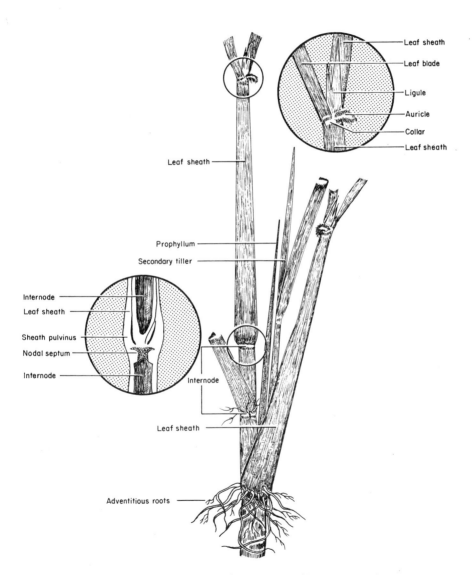

Figure 5.2   Parts of a primary tiller and its secondary tiller. (Adapted from Chang and Bardenas 1965)

Leaves

The leaves are borne at an angle on the culm in two ranks, one at each node.

- The *blade*, or the expanded part of the leaf, is attached to the node by the leaf sheath.
- The *leaf sheath* is the lower part of the leaf, originating from the node and enclosing the internode above it and sometimes the leaf sheaths and blades of the succeeding internodes.
- The *flag leaf* is the uppermost leaf below the panicle.
- The *auricles* are on either side of the base of the blade in pairs of small and earlike appendages (Fig. 5.2).
- The *ligule* is a papery triangular structure just above the auricles.

Floral Organs

The floral organs of the rice plant are modified shoots consisting of a panicle and spikelets.

Panicle

The panicle is a determinate inflorescence on the terminal shoot (Fig. 5.3). The extent to which the panicle and a portion of the uppermost internode extend beyond the flag leaf sheath determines the exsertion of the panicle. Varieties differ in degree of exsertion and environment can modify the extent of exsertion.

Spikelets

A spikelet is the unit of the panicle, and consists of two sterile lemmas, the rachilla, and the floret (Fig. 5.4). The rachilla is the small axis between the rudimentary glumes (the sterile lemmas) and the fertile floret. The floret includes the lemma, palea, and the enclosed flower.

- The *lemma* is the hardened, five-nerved bract of the floret partly enclosing the palea. It bears an *awn*, a filiform extension at different lengths from the keel (middle nerve) of the lemma.
- The *palea* is a hardened, three-nerved bract of the floret and fits closely to the lemma. It is similar to the lemma but narrower.
- The *flower* consists of six stamens and a pistil, with the perianth represented by the lodicules. The six stamens are composed of two-celled anthers borne on slender filaments. The pistil contains one ovule.

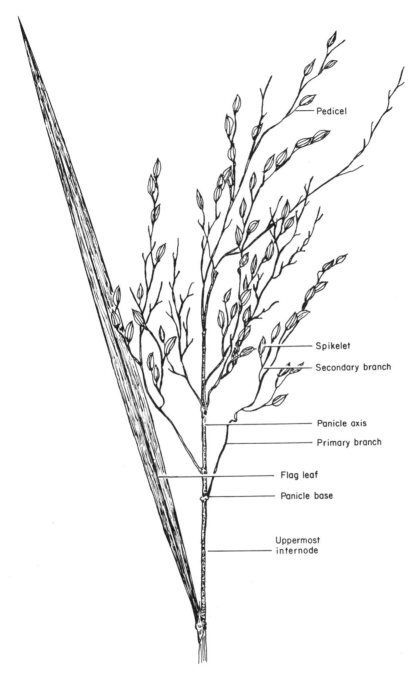

Figure 5.3   Component parts of a panicle (partly shown in this illustration). (Adapted from Chang and Bardenas 1965)

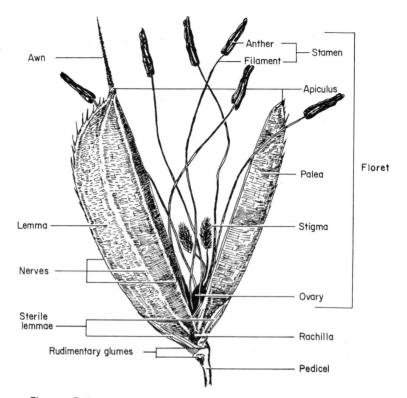

Figure 5.4   Parts of a spikelet. (From Chang and Bardenas 1965)

Grain

The grain is the ripened ovary, with the lemma, palea, rachilla, sterile lemmas, and the awn, if present, firmly adhered to it. The dehulled rice grain (caryopsis) with a brownish pericarp is called brown rice; dehulled rice grain with a red pericarp is red rice.

GERMINATION, GROWTH, AND DEVELOPMENT OF THE RICE PLANT

The development of the rice plant may be divided into three phases:

- The vegetative phase, which runs from germination to panicle initiation.
- The reproductive stage, which runs from panicle initiation to flowering.
- A ripening phase, which runs from flowering to full maturity.

These main phases, however, may be further divided into physiologically distinct stages or periods. Figure 5.5 shows the detailed divisions of the main phases of the rice crop with different durations.

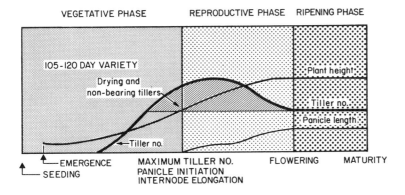

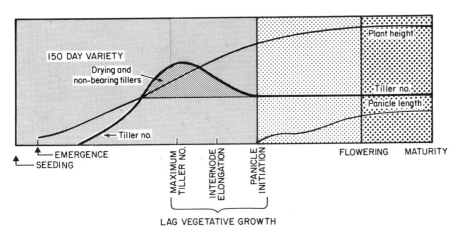

Figure 5.5 In short-duration, photoperiod-insensitive varieties (105–120 days), maximum tiller number, panicle initiation, and internode elongation occur almost simultaneously. In long-duration varieties (150 days), there is a so-called lag vegetative period during which maximum tillering, stem elongation, and panicle initiation occur in succession.

Vegetative Stage of Crop Development

Germination of Seed

Rice seeds germinate by pushing the radicle through the coleorhiza. The coleoptile that encloses the young leaves emerges as a tapered cylinder. The coleoptile later ruptures at the apex and the primary leaf emerges. Figure 5.6 shows parts of young seedlings germinated under light and darkness.

In warm, moist conditions, the grains of nondormant varieties can germinate immediately after ripening. In dormant varieties, a period of time (depending upon the variety) must elapse before germination can occur. A heat treatment ($50°C$ for 4–5 days), mechanical dehulling, or chemicals (such as HNO_3) can be

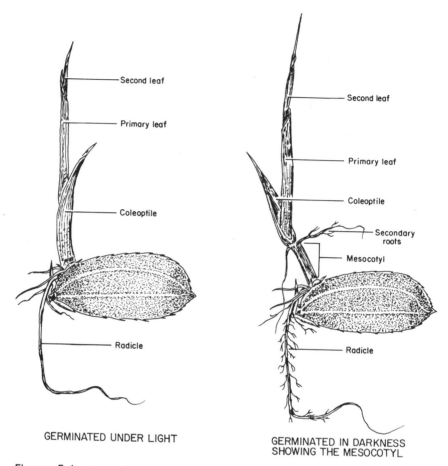

Figure 5.6   Parts of young seedlings. (Adapted from Chang and Bardenas 1965)

used to break the dormancy of freshly harvested seeds. Many tropical rice
varieties have a period of dormancy, which prevents the panicles from
germinating when in contact with water particularly when the crop is lodged
during the ripening stage.

When a rice grain germinates in an aerated environment, such as a well drained
soil, the sheath (coleorhiza) enveloping the radicle in the embryo protrudes (Fig.
5.6) before the radicle. If it germinates in water, the cylinderlike structure
(coleoptile) emerges ahead of the coleorhiza.

The radicle breaks through the coleorhiza shortly after the latter appears. This
is followed by the formation of two or more seminal roots, all of which develop
lateral branches.

Emergence and Growth of Seedlings

In the tropics, the first leaf usually emerges 3 days after sowing pregerminated seeds (pregermination is usually accomplished by soaking for 24 hours and incubating for 48 hours).

The seedling stage includes the period from emergence until just before the appearance of the first tiller. During this stage, the seedling develops seminal roots and absorbs most of the endosperm. By the 10th day, two more leaves should be fully developed. Leaves continue to develop at the rate of one every 3–4

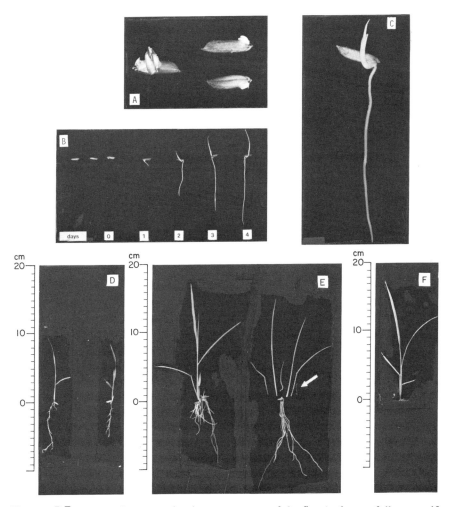

Figure 5.7   Stages from germination, emergence of the first leaf up to full-grown 18-day-old seedlings ready for transplanting.

days during the early stages. Adventitious roots that form the fibrous root system rapidly replace the temporary seminal roots. Dissected seedlings show six leaves: three fully developed, two developing, and one dead. Leaves are numbered according to the order in which they appear. Figure 5.7 shows germination, emergence of seedlings, and the full grown 18-day-old seedlings.

The tillering stage follows the seedling stage, and starts with the appearance of the first tiller from the axillary bud in one of the lowermost nodes. Tillers displace a leaf as they grow and develop.

After emergence of primary tillers (large arrow in Fig. 5.8A) they begin to form secondary tillers (small arrow). This happens at about 30 days of age with IR36, an early-maturing (105 days from seed to maturity), photoperiod-insensitive semidwarf variety. At that stage, the plant rapidly increases in length and tillers actively (Fig. 5.8B). Besides the primary and secondary tillers, new tertiary tillers start to appear as the plant becomes taller and larger.

The growth of tertiary tillers is in two stages:

- MAXIMUM TILLERING STAGE   The increase of tertiary tillers continues up to a certain point designated as the maximum tiller number stage. At this stage, tillers have increased in number to the point that it is difficult to pick out the main culm. After the maximum-tiller-number stage, some tillers die and the number of tillers declines and levels off (Fig. 5.5). The plants stop producing secondary tillers after the tertiary tillers (third group of tillers) are produced.

- STEM ELONGATION STAGE   This stage begins before panicle initiation in long-growth-duration varieties and it usually occurs during the later part of the

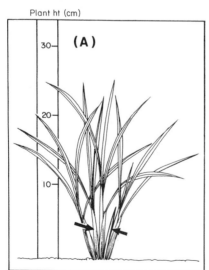
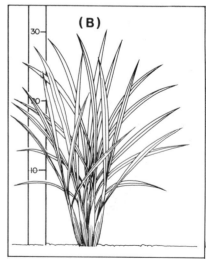

Figure 5.8 Stages of crop growth showing: (A) Emergence of primary tillers (large arrow) through beginning-to-form secondary tillers (small arrow); (B) increase in plant height and tiller number.

tillering stage. In short-growth-duration varieties, stem elongation and panicle development occur simultaneously.

Reproductive Stage of Crop Development

The reproductive stage begins just before or just after the maximum tillering stage, depending on variety and environment. The reproductive stage is marked by the initiation of a panicle primordium of microscopic dimensions in the growing shoot.

Panicle Initiation

The panicle initiation stage begins when the primordium of the panicle has differentiated and becomes visible. In a short-duration variety (105 days from seed to maturity), the panicle primordium starts to differentiate at about 40 days after seeding and becomes visible 11 days later (visual panicle initiation) as a white feathery cone 1.0–1.5 mm in length (Fig. 5.9).

Panicle initiation occurs first in the main culm and follows in the tillers in an uneven pattern. In long-duration varieties (135–160 days), the basal internodes have elongated considerably (as much as 1 m) before panicle initiation. If water is

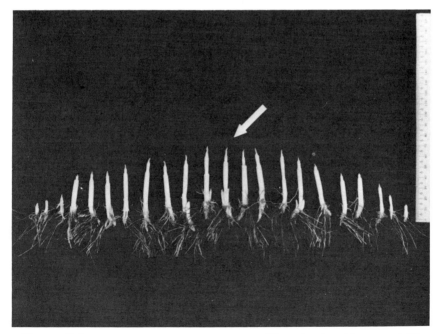

Figure 5.9   The primordia of panicles have differentiated and become visible.

limiting, panicle initiation may be delayed. This often occurs in rice direct-seeded in a nonpuddled soil.

Panicle Development

During the panicle development stage, the spikelets become distinguishable and the panicle extends upward inside the flag leaf sheath. The panicle continues to develop slowly. When it has grown to a length of 5 cm (about 7 days after panicle becomes visible in a dissected specimen), the spikelet primordia differentiate and the number of spikelets is determined. During this first part of the reproductive stage, yield is adversely affected by any stress exerted on the plant.

BOOTING   Booting is the latter part of the panicle development stage. About 16 days after visual panicle initiation, the sheath of the flag leaf swells. This swelling of the flag leaf sheath is called booting. Senescence of leaves and unproductive (nonpanicle-bearing) tillers become noticeable at the base of the plant.

When the collar of the flag leaf and that of the preceding leaf are at the same level, as shown in the center of Fig. 5.10, meiosis is occurring in the spikelets located at the midregion of the growing panicle.

HEADING   The booting stage is followed by the emergence of the panicle (heading) out of the flag leaf sheath.

FLOWERING   Anthesis (blooming or flowering) begins with protrusions of the first dehiscing anthers in the terminal spikelets. At the time anthesis is occurring, the panicle is erect in shape. The panicles flower beginning at the top, middle, and lower thirds, occurring in the 1st, 2nd, and 3rd day after panicle exsertion (heading) in a tropical environment (Fernandez et al. 1979).

Flowering occurs about 25 days after visual panicle initiation regardless of variety. Flowering continues successively until most spikelets in the panicle have bloomed.

Pollination and Fertilization

Rice is highly self-pollinated. The florets open from 0900 to 1500 dependent on variety and weather. They open early on bright days and late on humid and cloudy days. The stamens elongate and the anthers move out of the flowering glumes as pollen is shed. The pollen grains fall on the pistil, a feathery structure, through which the pollen tube of the germinating grains will extend into the ovary (Fig. 5.11). The lemma and palea then close.

Ripening Stages of Crop Development

The rice grain develops after pollination and fertilization. Grain development is a continuous process and the grain undergoes distinct changes before it fully

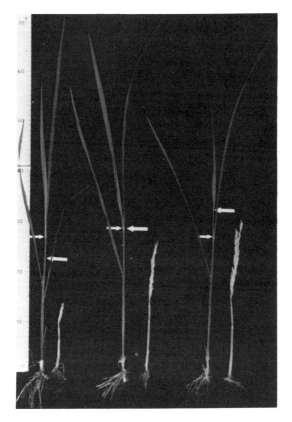

Figure 5.10   Meiosis in the spikelets takes place when the collar of the flag leaf and that of the preceding leaf are at the same level as shown in the center.

Figure 5.11   At the time of anthesis, the panicle is erect and the following developments take place: (A) The florets of a rachis open in the morning in most cultivars; (B) The stamens elongate and the anthers move out of the flowering glumes as pollen is shed. The florets then close; (C) The pollen on the pistil, seen here as a feathery structure through which the pollen tube will extend into the ovary.

Figure 5.12   Development of the rice grain after flowering through the various stages to the mature grain.

matures. In the tropics, the ripening stage (from flowering to maturity) takes 25–35 days regardless of variety. In temperate countries, such as Japan, southern Australia, and the United States, ripening takes 45–60 days.

Ripening involves three stages:

- In the *milk grain stage,* the contents of the caryopsis (the starch portion of the grain) are first watery but later turn milky in consistency. When held upright, the top of the panicle during the milk stage will bend gently in an arc. The content of the grain is a white liquid that can be squeezed out.
- In the *dough grain stage* the milky portion of the grain turns first into a soft, and later a hard dough.
- In the *mature grain stage* grain color in the panicles begins to change from green to yellow. The individual grain is mature, fully developed, and is hard and free from green tint. The mature grain stage is complete when 90–100% of the filled spikelets have turned yellow. The panicle arches further with the exception of a few still green spikelets and all grains are yellow and hard. At this time, senescence of the upper leaves including the flag leaves is noticeable. In some varieties, the culm and upper leaves may remain green even when the grains have ripened. Figure 5.12 shows the development of the rice grain after flowering.

Development Stages of the Rice Plant

The growth and development stages of the rice plant differ under different climatic and cultural conditions. Based on experience in Texas, Stansel (1975) developed simplified time ranges for each development stage of the rice plant (Fig. 5.13). The time ranges presented are for the very early- and early-maturing

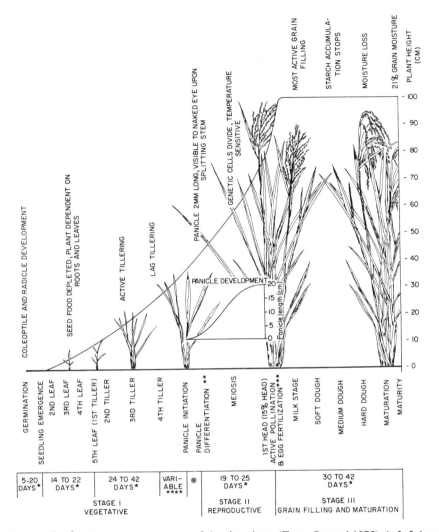

Figure 5.13   Development stages of the rice plant. (From Stansel 1975) * 3–5 days.
*Under warm conditions use the lower number of days and for cool conditions use the
larger number of days. **The reproductive stage begins with panicle differentiation,
which can be seen only by splitting the stem lengthwise. At this stage 30% of the main
culms sampled have a panicle 2 mm or longer. ***Stage III begins when 50% of the florets
are pollinated. ****Variable time—0–25 days (dependent upon variety).

photoperiod-insensitive varieties. Although the time ranges represent a warm or
cool weather combination for the Texas rice belt, the basic time ranges should be
applicable in areas with a similar environmental regime.

Recently, Zadoks et al. (1974) proposed a decimal code for the growth stages
of cereals that may be applicable to rice with some modifications.

PLANT AGE AND LEAF DEVELOPMENT

The development stage of the rice plant can be determined by the number of leaves it bears. The leaves develop consecutively and live for a short period. From germination to heading, the numbered leaves developing from the main culm are generally less in number for a short-duration variety than a long-duration variety.

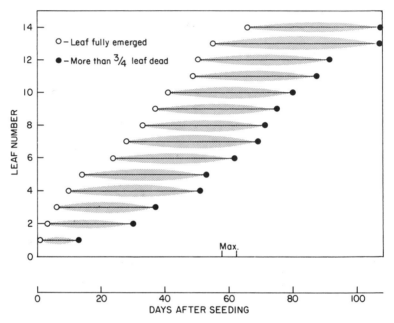

Figure 5.14   Leaves in the rice plant develop consecutively and live for a short period. In early-maturing IR36, the first leaf is bladeless and is considered leaf zero. This leaf dies in 10–12 days. At flowering, five to six leaves are still active and during grain ripening, only two or three remain active.

The leaves are named as the first, second, third leaf, and so on, in the order they emerge, and the number of fully developed leaves on the main culm is an indication of the physiological growth stage of a given variety. Using early-maturing IR36, Fernandez et al. (1979) determined the relative time of full emergence and life duration of 14 leaf blades throughout the stages of the rice plant (Fig. 5.14). In IR36, the first leaf, which is bladeless and is considered leaf zero, dies in 10–12 days. The other leaves last 25–35 days except the last two leaves, which may last longer. At flowering, five or six leaves remain active. During grain ripening, only two or three leaves remain active.

Synchronization in Leaf Development

During vegetative plant growth, there is a close relationship between the appearance of each tiller and the emergence of leaves on the main culm. For example, simultaneously with the appearance of the fourth leaf (4/0) from the main culm, the first leaf of primary tiller No. 1 (1/1) emerges. Simultaneously with the appearance of the fifth leaf, sixth leaf, ... (5/0, 6/0, ...) on the main culm, the second leaf, third leaf, ... (2/1, 3/1 ...) emerges from primary tiller No. 1. Leaves represented by (2/1, 3/1, ...) and (5/0, 6/0, ...) are called the synchronously-emerged leaves. Matsubayashi et al. (1963) reported that each of the respective leaves, both on the main culm and tillers, develops at a definite interval, and that all the leaves from tillers develop in parallel with the development of leaves on the main culm.

TILLERING CAPACITY AND LEAF AREA INDEX EFFECT ON YIELD

In monsoonal Asia, high-tillering rices are very desirable for transplanted or direct-seeded rice. Cultivars with improved plant type and high tillering capacity can be planted at a wide range of spacings and still produce an adequate number of tillers per unit area. The tiller number is positively or negatively correlated with grain yield depending on the rice cultivar and crop environment (Kawano and Tanaka 1968). It is well established that nitrogen application markedly increases the tillering of rices (Fig. 5.15).

The total leaf area of a rice population is a factor closely related to grain production because the total leaf area at flowering greatly affects the amount of photosynthates available to the panicle. It is known that 75–80% of the carbohydrates in the grain are photosynthesized after flowering (Ishizuka and Tanaka 1953, Welbank et al. 1968, and Yoshida and Ahn 1968).

In the tropics, Yoshida and Parao (1976) showed a close correlation between grain yield and leaf area index (LAI) at heading (Fig. 5.16). Similar results were obtained with IR8 in northern Australia (Basinski and Airey 1970). Later studies suggest that except for Peta, a lodging-susceptible variety, grain yields of all other rices in the dry season increase as the plant density and LAI values increase (Fig. 5.17). Peta lodged heavily and the yields were low at most plant densities and at most LAI values. High plant densities and nitrogen levels increased the spikelet number and LAI value of IR8 (Fig. 5.18). But with increased spikelet number and a high LAI, filled-spikelet percentage decreased particularly during the wet season. Murata (1969) reported high LAI values as desirable for increasing spikelet number, an important component of physical capacity for grain yield.

An important barrier to raising grain yields at high LAI values is the resulting decrease in filled-spikelet percentage. To increase yields beyond currently attainable levels, cultivars must be found that have a higher percentage of filled-

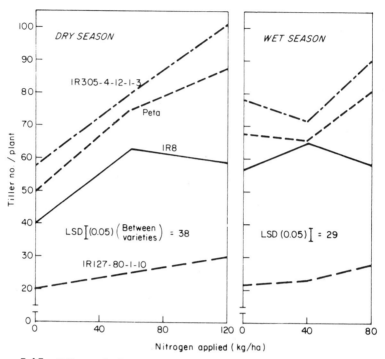

Figure 5.15 Effects of nitrogen level on the tillering capacity of rice cultivars at flowering (100 × 100 cm spacing). IRRI, 1969 dry and wet seasons. (From Fagade and De Datta 1971)

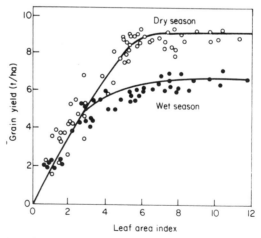

Figure 5.16 Relationship between leaf area index at heading and grain yield of IR8 in wet and dry seasons, 1966–1971. (From Yoshida and Parao 1976)

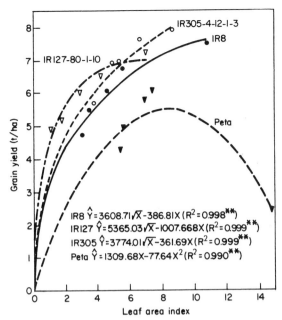

Figure 5.17   Relationship between leaf area index at flowering and yield of rice at 120 kg N/ha (average of five plant densities). IRRI, 1969 dry season. (From Fagade and De Datta 1971)

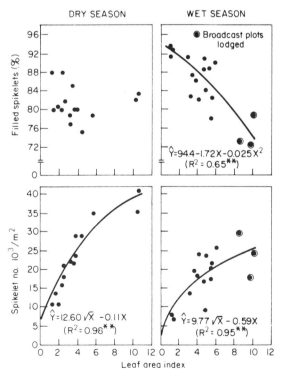

Figure 5.18   Effects of leaf area index on the number of spikelets and the percentage of filled spikelets of IR8. IRRI, 1969. (Adapted from Fagade and De Datta 1971)

Table 5.1   Morphological Characters Associated with High Yielding Potential of Rice Varieties (Adapted from Yoshida et al. 1972)

Plant Part	Desirable Characters	Effect on Photosynthesis and Grain Production
Leaf	Thick	Associated with more erect habit. Higher photosynthetic rate per unit leaf area.
	Short and small	Associated with more erect habit. Even distribution of leaves in a canopy.
	Erect	Increases sunlit leaf surface area, thereby permitting more even distribution of incident light.
Culm	Short and stiff	Prevents lodging.
Tiller	Upright (compact)	Permits greater penetration of incident light into canopy.
	High tillering	Adapted to a wide range of spacings; capable of compensating for missing hills; permits faster leaf area development (transplanted rice).
Panicle	Low sterility or high ripening percentage at high nitrogen rates	Permits use of larger amounts of nitrogen.
	High grain-to-straw ratio (high harvest index)	Associated with high yields.

spikelets at higher spikelet number and LAI values. If such cultivars are found, broadcast seeding will raise grain yields still further because broadcast-seeded plants produce high spikelet numbers more readily than the transplanted ones. For broadcast seeding, the varieties must have better lodging resistance than IR8, however (Fagade and De Datta 1971).

Table 5.1 summarizes certain varietal characters probably related to the high-yield potential of rice varieties. Three major characters are considered important for obtaining high yields:

· Short and stiff culms.
· Erect leaves.
· High tillering capacity.

Short and stiff culms make the rice plant more resistant to lodging. A major advance was made in the mid-1960s when a semidwarf gene was effectively introduced into tropical rice varieties.

A close association between erect leaves and high yield potential has been shown in the past. There is a need for real understanding of the physical meaning of erect leaves in terms of light use by a plant community. Yoshida et al. (1972) have discussed it in relation to plant response. The usefulness of erect leaves is more pronounced at high light intensities than at low light intensities. In rice, the upper three leaves export their assimilation products to the grains during the ripening period. Thus, erect leaves, which increase exposure of leaf surface to sunlight, must be important in increasing yield.

In transplanted rice, limited leaf-area development due to wide spacing may reduce grain yield. For such conditions, high- and early-tillering varieties have a definite advantage. Further, high tillering capacity gives the plant greater ability to compensate for missing hills that may be caused by poor stand establishment, insects, and diseases. In fact, Hayashi (1969) found that many high-yielding varieties in Japan are short, erect-leaved, and high tillering. Similar results have been reported from many studies in the tropics (Yoshida et al. 1972).

In temperate countries such as Australia and the United States, where direct seeding is the only method used to plant rice, high tillering capacity of rice varieties is not essential. In fact, many farmers in Australia get 8–9 t/ha yield with one or two tillers per plant. However, panicle density often exceeds $700/m^2$.

In China also, where transplanting is a major planting method, high tillering capacity of rice is not considered essential. Low tillering capacity is compensated by higher number of seedlings per hill. These results suggest that the most important factor is to get high panicle or spikelet number, or both, per unit area. This should be attainable partly through breeding and partly through improved management practices.

GROWTH PATTERNS IN RICE

In the diverse climatic environment in which rice is grown, the patterns of growth also vary.

Growth has two aspects:

- Dry matter production (quantitative change).
- Phasal development (qualitative change during the developmental stage).

Figure 5.19 shows an example of growth process described by Tanaka et al. (1964).

Phasal development is the sequential development of the three growth phases; the condition of the plant during the vegetative phases determines the tiller number, which is also the potential number of panicles. It also determines the

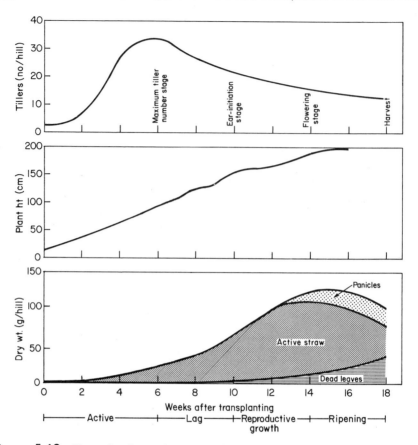

Figure 5.19 Example of growth process of tropical rice (Peta in the rainy season at IRRI). (From Tanaka 1976)

condition of leaves that function during the reproductive phase. The plant's condition during the reproductive phase determines the number and size of spikelets, and also the status of the leaves that contribute to ripening (Tanaka 1976).

Grain production, which is the final product of growth and development, is controlled by dry matter production during the ripening phase. The dry matter production in turn is controlled by two factors:

- The potential ability of the population to photosynthesize (the source).
- The capacity of spikelets to accept the photosynthates (the sink) (Tanaka 1972).

The sink is composed of:

- The panicle number per unit field area (determined during vegetative stage).

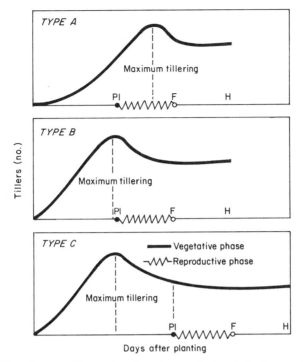

Figure 5.20   Diagram illustrating three types of phasal development: PI = panicle primordia initiation stage; F = flowering stage; H = harvest. (Adapted from Tanaka 1976)

- The spikelet number per panicle (determined during reproductive stage).
- The filled-spikelet percentage (determined during reproductive and ripening stages).
- The size of the individual spikelets (determined during reproductive stage).

Tanaka et al. (1964) classified three types of growth patterns from the standpoint of phasal development (Fig. 5.20). The types vary according to the combinations of varieties, environmental conditions, and cultural practices. Details on these growth patterns are described by Tanaka (1976).

RICE GROWTH IN DIFFERENT ENVIRONMENTS

Although selection of varieties is made by considering the environmental condition in which the crop is going to be grown, the growth patterns vary considerably.

Long-duration varieties of rice may be considered a necessity under certain rainfall or photoperiod regimes in the low-latitude areas. In the high-latitude areas, because of a limited length of season with suitable temperature regime, long-duration varieties cannot be grown.

In the high latitude areas, vegetative growth is slow and long, whereas in low-latitude areas it is vigorous and short. Because of this, the possibility is greater for a medium-duration variety to assume Type B patterns in the high-latitude areas and Type C patterns in low-latitude areas (Fig. 5.20).

For a long time, the Type C growth pattern was common in the tropics. Therefore, despite high total dry matter production, the grain yields were low due to low harvest index. However, with the introduction of modern varieties in the tropics, rice follows the Type A (Fig. 5.20) growth pattern that maintains a high growth rate throughout growth due to a favorable plant type, and produces high LAI and high grain yield.

Growth Duration in Relation to Yield

In temperate Asia, the growth duration of rice can be manipulated in a limited way by manipulating cultural practices. For example, in Japan where rice is mostly machine transplanted, seedbed technology has been considerably improved. Farmers in Japan extend the rice season by the use of temperature-protected seedbeds. This led to use of longer-duration varieties and to significant yield increase (Ishizuka et al. 1973). Rice yields in Japan have increased steadily due to improvements in variety and cultural practices. The rice growing season has been shifted to earlier in the spring by using nonseasonal varieties. Transplanting is done when the temperature is lower.

In the tropics, where temperature is favorable for year-round rice culture, there appears to be an optimum growth duration for high grain yields. For example, the growth period of short-duration plants (of less than 100-day

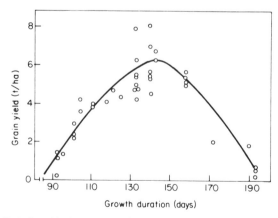

Figure 5.21   Relationship between growth duration and grain yield. The spacing used was 25 × 30 cm and the nitrogen level was 40 kg/ha. IRRI, 1963. (From Vergara 1970)

maturity) grown under normal field conditions usually does not permit the production of sufficient leaf area to result in production of larger number of panicles with well filled spikelets.

Results of tests by Vergara (1970) indicate that under certain environmental conditions (14° N), the best yields are from varieties that mature in 130–140 days (Fig. 5.21). Optimum growth duration could be attained earlier by closer spacing or higher nitrogen levels.

With increased emphasis on increased cropping intensity in both irrigated and rainfed areas, there is a considerable demand for rices with growth duration around 100 days. With the use of those shorter-duration varieties, it is possible to raise the productivity per hectare per day even if the individual crop yield is somewhat lower than for varieties with 130 days or longer. Thus, growth manipulation should be designed to efficiently use given resources such as water, fertilizer, solar radiation, and temperature.

REFERENCES

Basinski, J. J., and D. R. Airey. 1970. Nitrogen response of IR8 rice at Coastal Plains Research Station in northern Australia. *Aust. J. Exp. Agric. Anim. Husb.* 10:176–182.

Bechtel, D. B., and Y. Pomeranz. 1978. Implications of the rice kernel structure in storage, marketing and processing: a review. *J. Food Sci.* 43:1538–1542.

Chang, T. T., and E. A. Bardenas. 1965. *The morphology and varietal characteristics of the rice plant.* IRRI Tech. Bull. 4. 40 pp.

Fagade, S. O., and S. K. De Datta. 1971. Leaf area index, tillering capacity, and grain yield of tropical rice as affected by plant density and nitrogen level. *Agron. J.* 63:503–506.

Fernandez, F., B. S. Vergara, S. Yoshida, L. D. Haws, N. Yapit, and O. Garcia. 1979. Growth and development stages of the rice plant: *An early maturing dwarf in the tropics.* International Rice Research Institute. Los Baños, Philippines. (unpublished manuscript)

Hayashi, K. 1969. Efficiencies of solar energy conversion and relating characteristics in rice varieties. *Proc. Crop Sci. Soc. Jpn.* 38:495–500.

Ishizuka, Y., and A. Tanaka. 1953. Biochemical studies on the life history of rice plants. II. Synthesis and translocation of organic constituents. *J. Sci. Soil Manure, Jpn.* 23:113–116.

Ishizuka, Y., Y. Shimazaki, A. Tanaka, T. Satake, and T. Nakayama. 1973. *Rice growing in a cool environment.* Food Fert. Technol. Cent., ASPAC, Taipei, Taiwan. 98 pp.

Juliano, B. O. 1972. The rice caryopsis and its composition. Pages 16–74 *in* D. F. Houston, ed. *Rice chemistry and technology.* American Association of Cereal Chemists, Inc., St. Paul, Minnesota.

Juliano, B. O. 1980. Preparation and properties of rice starch. *In* R. L. Whistler, E. F. Paschall, and J. N. Bemuller, eds. *Starch chemistry and industry.* Academic Press, New York. (in press)

Juliano, B. O., and D. B. Bechtel. 1980. Composition and structure. *In Chemical Rubber Corporation (CRC) handbook series in agriculture.* Section G. Processing and utilization. Plant proteins—cereal grains. 4. Rice. CRC Press. (in preparation)

Juliano, J. B., and M. J. Aldama. 1937. Morphology of *Oryza sativa* Linnaeus. *Philipp. Agric.* 26:1–134.

Kawano, K., and A. Tanaka. 1968. Studies on the interrelationships among plant characters in rice. I. Effect of varietal difference and environmental condition on the correlation between characters. *Jpn. J. Breed.* 18:75–79.

Kuwada, Y. 1910. A cytological study of *Oryza sativa* L. *Bot. Mag. Tokyo* 24:267–281.

Matsubayashi, M., R. Ito, T. Tsunemichi, T. Nomoto, and N. Yamada. 1963. *Theory and practice of growing rice.* Fuji Publishing Co., Ltd., Tokyo. 502 pp.

Morinaga, T., and E. Fukushima. 1934. Cyto-genetical studies on *Oryza sativa* L. I. Studies on the haploid plant of *Oryza sativa. Jpn. J. Bot.* 7:73–106.

Murata, Y. 1969. Physiological responses of nitrogen in plants. Pages 235–259 *in* American Society of Agronomy and Crop Science Society of America. *Physiological aspects of crop yield.* Madison, Wisconsin.

Santos, J. K. 1933. Morphology of the flower and mature grain of Philippine rice.; *Philipp. J. Sci.* 52:475–503.

Stansel, J. W. 1975. The rice plant—its development and yield. Pages 9–21 *in* Texas Agricultural Experiment Station in cooperation with U.S. Department of Agriculture. *Six decades of rice research in Texas.* Res. Monogr. 4.

Tanaka, A. 1972. *The relative importance of the source and the sink as the yield-limiting factors of rice.* ASPAC Food Fert. Technol. Cent. Tech. Bull. 6. 18 pp.

Tanaka, A. 1976. Comparisons of rice growth in different environments. Pages 429–448 *in* International Rice Research Institute. *Climate and rice.* Los Baños, Philippines.

Tanaka, K., T. Yoshida, and Z. Kasai. 1974. Distribution of mineral elements in the outer layer of rice and wheat grains, using electron mircroprobe x-ray analysis. *Soil Sci. Plant Nutr. (Tokyo)* 20:87–91.

Tanaka, A., S. A. Navasero, C. V. Garcia, F. T. Parao, and E. Ramirez. 1964. *Growth habit of the rice plant in the tropics and its effect on nitrogen response.* IRRI Tech. Bull. 3. 80 pp.

Vergara, B. S. 1970. Plant growth and development. Pages 17–37 *in* University of the Philippines College of Agriculture in cooperation with the International Rice Research Institute. *Rice production manual.* Los Baños, Philippines.

Welbank, P. J., K. J. Witts, and G. N. Thorne. 1968. Effect of radiation and temperature on efficiency of cereal leaves during grain growth. *Ann. Bot. N. S.* 32:79–95.

Yoshida, S., and S. B. Ahn. 1968. The accumulation process of carbohydrate in rice varieties in relation to their response to nitrogen in the tropics. *Soil Sci. Plant Nutr. (Tokyo)* 14:153–161.

Yoshida, S., and F. T. Parao. 1976. Climatic influence on yield and yield components of lowland rice in the tropics. Pages 471–494 *in* International Rice Research Institute. *Climate and rice.* Los Baños, Philippines.

Yoshida, S., J. H. Cock, and F. T. Parao. 1972. Physiological aspects of high yields. Pages 455–469 *in* International Rice Research Institute. *Rice breeding.* Los Baños, Philippines.

Zadoks, J. C., T. T. Chang, and C. F. Konzak. 1974. A decimal code for the growth stages of cereals. *Weed Res.* 14:415–421.

6

Varietal Development of Rice

Information gathered from many sources allows the reconstruction of a series of events that led to the cultivation of the Asian cultivated rice (*Oryza sativa* L.) and the planting of the African cultigen (*O. glaberrima* Steud.)

TAXONOMY, ORIGIN, AND EARLY CULTIVATION

The genus *Oryza* belongs to the tribe *Oryzeae* in the family Gramineae. About 20 valid species are distributed chiefly in the humid tropics of Africa, South and Southeast Asia, southern China, South and Central America, and Australia (Chang 1976). Cultivated rice belongs to the genus *Oryza* and its most important species is *O. sativa*. *Oryza glaberrima*, grown sporadically in some West African countries, is gradually being replaced by *O. sativa*.

Among many botanists who studied *Oryza,* Roschevicz (1931) was the first one to postulate that the center of origin of section Sativa, Rosch., to which *O. sativa* L. and *O. glaberrima* Steud. belong, was in Africa. Sampath (1962) and Oka (1964) considered that *O. perennis* Moench was the common progenitor of both the Asian and African cultivated rice. Porteres (1956) suggested that the common progenitor was a rhizomatous and floating form, but no mention of name was made. *Oryza nivara* Sharma et Shastry, an annual wild form from central India (Sharma and Shastry 1965), appears to be the immediate progenitor of the Asian cultivated rice *O. sativa*. Three ecogeographic races of *O. sativa* are recognized: indica, japonica, and javanica (Fig. 6.1). Indica rices are indigenous to the humid regions of the Asian tropics and subtropics. The japonicas are limited to temperate zones and subtropics. The temperate race was differentiated in China and therefore japonicas are also known as sinicas or keng (Chang 1964, 1976). Javanicas are mainly grown in parts of Indonesia.

Cultivation of rice in many humid parts of tropical and subtropical Asia probably began about 10,000 years ago. India may have had the earliest date of cultivation because wild rices were abundant there. The domestication process first took place in China. Cultural practices for domestication, such as puddling and transplanting, were first developed in North and Central China and later transmitted to Southeast Asia. Lowland culture preceded upland culture in

173

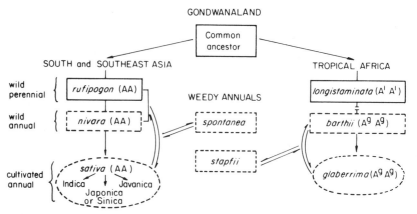

Figure 6.1 Evolutionary pathway of the two cultivated species of rice. Taxa boxed by solid lines are wild perennials. Taxa boxed by broken lines are annuals. Arrow with solid line indicates direct descent. Arrow with broken line indicates indirect descent. Double arrows indicate introgressive hybridization. (Adapted from Chang 1976)

China, but in many hilly areas of Southeast Asia, upland rice cultivation preceded lowland. Planting methods progressed from shifting cultivation to direct sowing in permanent fields, then to transplanting in bunded fields (Chang 1976).

Domestication of rice in Asia may have occurred independently and at about the same time in many places along a broad belt extending from the Ganges Plains below the eastern foothills of the Himalayas, across northern Burma, northern Thailand, Laos and Vietnam, to Southwest and South China (Chang 1964, 1976). There were likely several routes by which the Asian rices were introduced into other parts of the world. South Asia was undoubtedly the main source of indica varieties later found in ancient Persia and various parts of Africa. The japonica type spread from China to Korea and Japan.

Europeans could have obtained rices through ancient Persia, Central Asia, or directly from China. Latin American countries obtained rices largely from Spain and Portugal. The early rices in the United States came from the Malagasy Republic, Europe, and the Far East. Details on the dispersal of the rice cultivars have been summarized by Lu and Chang (1980).

The introduction of *O. sativa* in Africa was primarily through visitors from Malayo-Polynesia a few centuries B. C. Another possible route of introduction was from Sri Lanka and Indonesia via Oman and then on to Somalia, Zanzibar, and Kilua (Carpenter 1978). *Oryza glaberrima* is indigenous to tropical West Africa and probably originated there about 1500 B. C. (Porteres 1956). *Oryza glaberrima* evolved from the wild annual rice, usually named either *O. breviligulata* Chev. et Roehr. or *O. barthii* A. Chev. (Porteres 1956, Carpenter 1978).

RICE VARIETAL IMPROVEMENT

Since the dawn of agriculture, rice in tropical Asia has been primarily a monsoonal crop grown from June to December in the tropics north of the equator and from November to April south of the equator. In parts of South India and Sri Lanka the main crop is grown during the northeast monsoon during September to February. Earlier, varieties grown during the monsoon had a long maturity period (160–200 days) and a limited number of varieties grown in the off season in Sri Lanka, eastern India, Bangladesh, Philippines, West Malaysia, and Indonesia had a shorter maturity period (90–130 days). Most of the latter varieties, for example the *aus* and *boro* varieties of northeastern India and Bangladesh, and the *kuruvai* and *kars* of South India, had little sensitivity to photoperiod changes. Thus, length of growing season was an important varietal specification. Most varietal collections were limited to varieties grown in the lowlands of the monsoonal tropics (Parthasarathy 1972).

Early Breeding Work in Tropical Asia

Rice breeding at the beginning of the century was based on selection among farmers' varieties, which had been selected for local adaptation and preferred grain quality for many centuries. The selection was limited to purification by removal of off-types in the varieties popular with farmers. The next step was mass selection in such varieties. Following that, cross fertilization for combining specific traits in different varieties was attempted. Introductions from neighboring countries and regions played an important role through direct use by farmers and effective utilization by breeders in hybridization programs.

Early rice breeding in tropical Asia was done at a single experiment station without testing of varieties in widely different environments. Workers in Indonesia, notably H. Siregar, were probably the first group who made selections from common hybrid materials at the central experiment station at Bogor and each of the six regional stations in Java. That led to the evolution of varieties adapted to the whole of Java and its different soil types and climates.

In India, rice breeding work was initiated in 1911 in East Bengal (now Bangladesh). In 1929, the Imperial (now Indian) Council of Agricultural Research (ICAR) aided breeding work in various states. A detailed account of rice breeding work in India is given by Ghosh et al. (1960). The breeding objectives of India primarily included development of varieties with different growth duration, yielding ability, grain quality, tolerance for flooding, salinity, drought, and resistance to lodging and blast (Parthasarathy 1972).

Intensive selection from local populations in different parts of India led to the identification of several varieties that served as the foundation of a modern breeding program for developing varieties with high and stable yields and resistance to major diseases and insects. Examples of those varieties, selected through continuous exposure to plant stress caused by climatic, edaphic, and biologic conditions are:

- Yield potential: Latisail, T141, T-90, GEB 24
- Disease and insect resistance: TKM6, Ptb 18, Ptb 21, Ptb 33, Eswarakora
- Tolerance for drought: N22, Dular
- Tolerance for salinity and alkalinity: Pokkali, SR 26B
- Tolerance for flood: FR 13A, FR 43B

In Burma, a traditionally rice-exporting country, 19 improved varieties were distributed in the early 20th century in lower Burma and eight in upper Burma (Grant 1932). The breeding objectives included yield, particular grain types, and milling quality.

Sri Lanka is divided into dry and humid regions according to the amount of rainfall received. Only indica rices were used in breeding varieties highly responsive to fertilizer. The cross between Murungakayan 302 and Mas (from Indonesia) gave rise to H-4 (red rice) and H-5 (white rice). H-5 was considered fertilizer-responsive and resistant to blast.

The early varieties grown in Indonesia belonged to the two groups, the *tjereh* (indica) and the *bulu* (or javanica). On the islands of Bali and Lambok, where rice culture was highly specialized, *bulus* are still grown. *Bulus* are also grown on the mountain terraces in Banaue, Philippines and in the mountain areas of Taiwan (Chang 1976). The *bulus* as a group are distinguished from japonicas and indicas by low tillering, tall plants, stiff straw, long panicles, long awns, large and bold nonshattering grains, and low sensitivity to photoperiod.

Thailand specialized early in growing varieties with high-quality, long, slender grains. These varieties still form the bulk of Thailand's exported rice. The strict demands of the export trade for high-quality grain naturally imposed some restrictions on achieving other objectives in breeding. In the Central Plain and Northeast Thailand, nonglutinous varieties of the grain type mentioned earlier were grown. During the late 1950s, an intensive breeding program for blast resistance was initiated. Also, five floating varieties needed for the North, the Northeast, and the Central Plain were released.

Early Breeding Work in Temperate Regions

In Japan, rough rice yield has increased from 3.7 t/ha in 1916 to 5.0 t/ha in 1966 (Institute of Developing Economies 1969). In 1927, a rice breeding effort was organized to develop rice varieties suitable to 12 ecological conditions. During the last 60 years, rice breeders in Japan have placed major emphasis on breeding for high yield with heavy fertilizer application; blast resistance; resistance to the Akiochi soil problem, bacterial leaf blight, and stripe virus; short culms for lodging resistance; high tillering; cold tolerance in low-temperature regions; and improved grain quality (Okabe 1972). From 1927 to about 1949, the Konosu Experiment Station was responsible for making crosses followed by F_2 selections for distribution to regional breeding stations. Today, the breeding network consists of many small breeding stations in different environments. In 1962, 12 ecological regions were consolidated into four macro-ecological districts: a cold-weather district (Hokkaido), a cool-weather district (northern parts of Honshu), a warm-weather district (southwestern Honshu), and a hot-weather district (Kyushu, Shikoku, and Sato Inland sea coastal areas). Since 1962, new breeding objectives in Japan included suitable plant characters for direct seeding, early maturity, and wide adaptability.

In the United States, cooperative rice breeding studies were initiated in 1909 with the establishment of Louisiana's Crowley Station. Rice breeding research in the United States from 1909 to 1961 are covered by Jones (1936) and Adair et al. (1973).

In 1931, the United States Department of Agriculture initiated coordinated rice breeding programs with state agricultural experiment stations in Louisiana, Texas, California, and Arkansas. There was continuous attention to development of varieties with a reasonably wide maturity range within each grain type (short, medium, and long grains) (Johnston et al. 1972). Some examples of varieties developed during the early years include Colusa (short grain) in 1917, Zenith (medium grain) in 1936, and Rexoro (long grain) in 1928. Century Patna 231 developed in 1951 was an early-maturing, long-grain variety with moderate height that was adapted to high fertilizer application and combine harvesting.

In 1970, only about 8.8% of the rice-growing area in the United States grew short-grain varieties such as Caloro and Colusa. About 41.5% was sown to medium grain varieties such as Arkrose, Calrose, CS-M3, Nato, Nova 66, Saturn, and Zenith. The remaining area was sown to long-grain varities such as Belle Patna, Bluebelle, Bluebonnet 50, Dawn, Della, Rexoro, Starbonnet, Toro, and TP 49. Descriptions of these varieties are given by Adair et al. (1973).

Evolution of the Ponlai Rices

During the early period of the Japanese occupation of Taiwan, which began in 1894, several Japanese introductions were extensively tested but none was outstanding in performance (Huang et al. 1972). At a May 1926 Japan Rice Production Conference in Taipei, the Japanese varieties planted in Taiwan were named *ponlai* rice, meaning heavenly rice and synonymous with *horai* in Japanese. After 1926, varietal work emphasized blast resistance. From 1931 to 1943, crosses in Taiwan led to new *ponlai* varieties with grain qualities equal to, and yielding abilities superior to, the varieties introduced from Japan. Taichung 65 was the most prominent pre-World War II variety that could be profitably grown in both the first and second crop seasons. It yielded well on a wide range of soil types. These improved *ponlai* varieties were also early maturing, which provided opportunity to practice multiple cropping in Taiwan. The photoperiod-

and temperature-insensitive *ponlai* rices also produced record yields in West Africa and India (Chang 1967, Shastry 1966).

Indica × Japonica Hybridization Program

The concept of indica × japonica hybridization, with its more clearly defined objectives and systematic planning, represents a transition from early to modern rice breeding. In the United States, indica × japonica crosses led to a number of commercial varieties during the late 1930s and early 1940s. As a sequel to the recommendations of the International Rice Commission (IRC) Working Party, an indica × japonica hybridization program was initiated in 1951 under the auspices of Food and Agriculture Organization (FAO) of the United Nations. This project represented the first international effort toward multicountry cooperative breeding in rice. The FAO project also focused, for the first time, on use of the genetic potential of the rice plant to capitalize on added fertilizers under tropical conditions. All the countries of tropical Asia participated in the project by sending the seeds of their best indica varieties for crossing with japonicas at the Central Rice Research Institute (CRRI), Cuttack, India, which was selected as the center for making the crosses and the growing of the F_1 plants. F_2 seeds from the crosses were dispatched to participating countries for further selection work. Simultaneously, ICAR funded a similar program for the different states of India.

Only India and Malaysia distributed early-maturing, nonseasonal commercial varieties derived from the project. In India, ADT 27, which was suitable for the early monsoonal season, replaced the earlier varieties ADT 3 and ADT 4, in the Tanjore delta. In Malaysia, Malinja and Mashuri had the preferred grain quality and were found adapted to the second-crop season in the irrigated areas in Wellesley province (Parthasarathy 1972). Even today, Mashuri is widely grown in many parts of tropical Asia where there is poor soil and poor water control in the monsoonal season. Lack of dormancy and susceptibility to blast, however, limit Mashuri's yield.

Improving the Plant Type

Rice breeders have continually worked to develop high-yielding varieties of rice that respond to high rates of nitrogen fertilizer (Baba 1954). Nagai (1958) reported, "in countries like Japan where high yield is aimed at by intensive culture with heavy application of fertilizers, the leading varieties are required to possess heavy tillering ability with short culm and resistance to lodging." The relationship between morphological characteristics and grain yield has received considerable attention since the pioneering work of Tsunoda (1962).

The introduction of semidwarf genes into rice and wheat varieties in the early 1960s increased spectacularly the yielding ability of these crops largely because of increased resistance to lodging. The relationship of improved plant type to yielding ability is described by Beachell and Jennings (1965) and McDonald (1978).

IRRI began its operations with the important breeding objectives of development of high-yielding , short, sturdy-strawed rice varieties that would resist lodging even with high rates of fertilizer application (Beachell and Jennings 1965, Beachell et al. 1972). In 1969, Beachell and Khush summarized the traits that were objectives of IRRI's breeding work as nitrogen responsiveness, resistance to diseases, resistance to insects, photoperiod insensitivity, grain dormancy, high milling yield, superior cooking and eating quality, slow leaf senescence, and tolerance for cold weather.

Some of the traits are essential for all improved rice. Others are essential in a certain country, or in a given region of a country. For example, high nitrogen responsiveness should be desirable everywhere. Other traits, such as early vegetative vigor and relatively high tillering capacity to be competitive with weeds are essential in the rainfed transplanted rice culture but not essential under direct seeding with irrigation. Also, grain dormancy, resistance to leaf damage during typhoons, slow leaf senescence, and ability to withstand drought and flood are essential in some regions in the tropics but not essential in temperate regions such as in Australia, the United States, or Japan.

Development of High-Yielding Semidwarf Rice Varieties

The following are examples of modern semidwarf varieties that revolutionized rice culture in tropical, subtropical, and temperate Asia. Semidwarfism as used here refers to a reduction in plant height controlled primarily by a single major (recessive) gene.

Taichung Native 1

One of the significant events in the history of varietal improvement of rice was the development of a high-yielding semidwarf indica variety, Taichung Native 1. It resulted from the cross of Dee-geo-woo-gen and Tsai-yuan-chung made in 1949 by breeders at the Taichung District Agricultural Improvement Station, Taiwan. The precise origin of the semidwarf Dee-geo-woo-gen is not known, although it was grown by Taiwanese farmers prior to 1951 (Huang 1956). Taichung Native 1 was selected and named in 1956 but its official seed multiplication and distribution began in 1960 (Huang et al. 1972). Outside Taiwan, the value of Taichung Native 1 and some ponlai varieties was not fully recognized until after the mid 1960s (Chang 1967).

Guang-chang-ai

Breeders at China's Academy of Agricultural Sciences in Guang-dong province crossed Ai-zai-zan with Guang-chang 13 in 1956 to develop the semidwarf Guang-chang-ai. In 1961, it was released for general use (IRRI 1978a). Guang-chang-ai was the first short-statured high-yielding variety successfully developed by crossbreeding in China (Shen 1980). Guang-chang-ai is about 90-cm tall—more than 10 cm shorter than Guang-chang 13—and has a vigorous root system and tillering capacity and erect leaves. Its excellent lodging resistance makes it withstand typhoon conditions well.

Ai-jio-nan-t'e

Also in Guang-dong province, a semidwarf plant was selected from a field of Nan-t'e 16 during 1956. The selection was named Ai-jio-nan-t'e and introduced into neighboring provinces. It attained a planted area of 733,000 ha during 1965 in areas south of the Yangtze River (Shen 1980).

Guang-hwai-ai

The semidwarf varieties were widely accepted and by 1965 most of the first crop in Guang-dong province was planted to them. At about the same time (1963) a dwarf variety, Kuang-er-ai, was released for the second crop of the two-rice-crop system. It was further crossed with a locally adapted second-crop variety, and an improved variety, Guang-hwai-ai, was released for the second crop. It has narrow leaves, and is photoperiod-sensitive (IRRI 1978a).

IR8

In 1962, among 38 crosses made at IRRI, the most successful was the 8th cross, which involved Peta, a tall and heavy-tillering, disease-resistant indica variety of Indonesia, and Dee-geo-woo-gen, the semidwarf Chinese variety. IR8-288-3 was one of the promising lines identified in the F_4 generation of the cross. In March 1965, IR8-288-3 was first planted in the yield nursery and produced a computed yield of 6.6 t/ha (Chandler 1969). In the 1966 dry season at IRRI, it showed outstanding fertilizer nitrogen response with a high yield of 9.4 t/ha and outyielded Taichung Native 1 at all fertilizer nitrogen levels (Fig. 6.2). Taichung Native 1 had, until then, been considered the highest yielding indica variety. In the same season, IR8-288-3 produced 10.3 t/ha in a replicated experiment at IRRI, the highest grain yield then reported for any variety in replicated experiment in the tropics (De Datta et al. 1966). Seeds of IR8-288-3 were sent to 60 sites throughout the world and from mid-1965 through 1966 it was tested extensively in India, Malaysia, Pakistan, Philippines, and Thailand.

Based on its yield ability and wide adaptability as an irrigated lowland crop IRRI named IR8-288-3 as IR8 in November 1966, and released it for commercial cultivation. Figure 6.3 shows IR8 with its two parents Peta and Dee-geo-woo-gen.

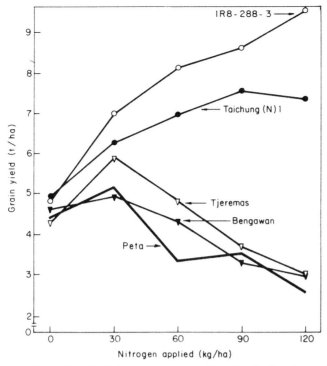

Figure 6.2　Effect of levels of nitrogen on the grain yield of indica rice varieties. IRRI, 1966 dry season. (Adapted from De Datta et al. 1968)

Figure 6.3　IR8-288-3 was developed from a cross with tall indica variety Peta and semidwarf Dee-geo-woo-gen. In 1966 it was named as IR8.

CURRENT BREEDING PROGRAMS

In the tropics, despite the introduction of modern varieties, the average rice yield is still between 1.5 and 2.0 t/ha. Yields are 5.0 t/ha or more in temperate rice-growing countries such as Japan, Republic of Korea, United States, Australia, Spain, and Italy. Much of the yield difference is because rice in the tropics is grown under widely different agroclimates, and under adverse growing conditions. Furthermore, the more numerous and more diverse insects and diseases cause more severe damage to the rice crop in the tropics than in temperate regions.

Breeding Programs in the Tropics

Current breeding programs for the tropics and temperate regions must consider issues such as grain quality and resistance to, or tolerance for, the diverse environmental regimes rice is exposed to.

Genetic Evaluation and Utilization (GEU) Program

In 1973, IRRI initiated a Genetic Evaluation and Utilization (GEU) program based on interdisciplinary and problem-oriented rice breeding (Brady 1975). Plant breeders were teamed with problem-area scientists such as plant pathologists, entomologists, cereal chemists, soil scientists, plant physiologists, and agronomists. Each team member contributes specialized knowledge to the joint effort to identify, screen, and cross diverse rices. The GEU goal is incorporation of the best characteristics into nutritious rice varieties that resisted or tolerated environmental and biological enemies of the rice plant (Fig. 6.4).

   Major GEU problem areas include:

- Agronomic characteristics.
- Grain quality.
- Resistance to diseases.
- Resistance to insects.
- High nutrient levels (including protein).
- Tolerance for drought.
- Tolerance for adverse soils.
- Tolerance for deep water.
- Tolerance for temperature extremes.

An integral part of the GEU program is linkage with national rice improvement programs in rice-producing countries, particularly in Asia. The four interrelated components of the GEU program are:

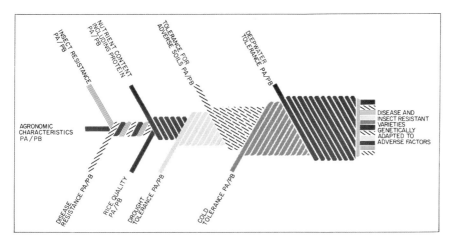

Figure 6.4 The building of modern rice varieties. Disease and insect resistance and superior agronomic characteristics are the core for all varieties developed through the Genetic Evaluation and Utilization (GEU) program. Plant breeders (PB) and problem area scientists (PA) work together to incorporate the genetic ability to withstand other production constraints. (Adapted from Brady 1975)

- Germ plasm collection, conservation, and evaluation.
- Research by interdisciplinary teams.
- Development of improved germ plasm.
- Distribution, evaluation, and exchange of germ plasm.

Germ Plasm Collection, Conservation, and Evaluation

It is difficult to determine when and where systematic germ plasm collection was initiated. However, the earliest record of the rice varietal diversity in the cultivated rice in India was recorded by Roxburgh (1832). In India, the earliest and most extensive collections of germ plasm were maintained at Coimbatore, Dacca (now in Bangladesh), Raipur, Karjat, and Kanpur (Seetharaman et al. 1972). With the establishment of CRRI in India, national rice collection began. Between 1967 and 1971, the Indian Agricultural Research Institute (IARI) and All-India Coordinated Rice Improvement Project (AICRIP) collected 6730 accessions from northeastern India and provided valuable germ plasm for varietal improvement. This collection provided excellent sources of genes for various traits. Part of this collection was from the State of Assam and was later called the Assam Rice Collection.

In the People's Republic of China, 30,000 local varieties were assembled in the late 1950s and many of the varieties were used for varietal improvement (Shen 1980).

An early phase of the rice improvement program in the United States benefited from a sizeable collection of foreign introductions (Adair et al. 1973).

International Cooperation in Rice Genetic Resources Conservation and Identification

In 1961, IRRI began to assemble rice accessions with the assistance of rice researchers in Asia, the U.S. Department of Agriculture, and FAO. By the end of 1962, IRRI had acquired 6867 accessions from 73 countries (Chang 1972). IRRI's collection of *O. sativa* accession is now 60,000; the collection of *O. glaberrima* has also increased steadily. IRRI staff directly participated in the field collection activities of Bangladesh, Burma, Indonesia, Kampuchea, Philippines, Sri Lanka, and Vietnam. Workers in seven other countries added 29,700 samples to the collection (Chang 1980).

Recognizing the serious implications of the rapid disappearance of genetic resources in the diverse rice germ plasm, the International Board for Genetic Resources and IRRI cosponsored the 1977 Rice Genetic Conservation Workshop. Participants from 18 countries agreed on monitoring of field collection efforts and the need to promote international and institutional collaboration on the collection, characterization and conservation of rice genetic resources (IRRI 1978b).

Figure 6.5 shows different operations at IRRI relating to collection and utilization of germ plasm. IRRI's Rice Genetic Resources Laboratory opened in 1977. In its long-term storage facilities, seed samples are stored in vacuum sealed cans at $-10°$ C and should remain viable for at least 50 years, perhaps 100 years. In the medium-term storage a $4°$ C temperature is maintained and seed samples should remain viable for about 25 years. Seeds stored in the short-term storage area ($20°$ C) should remain viable from 3 to 5 years. The viability of the seed samples of different control varieties is checked every six months. As an added precaution, a 15-g duplicate of each completely recorded accession is sent to the U. S. National Seed Laboratory at Fort Collins, Colorado.

Contacts with other international institutes and organizations, such as the International Institute of Tropical Agriculture (IITA), the West Africa Rice Development Association (WARDA), and the Institut de Recherches Agronomiques Tropicales et des Cultures Vivrierès (IRAT) provide for collecting indigenous rice germ plasm in Africa. IITA's germ plasm collection and conservation program are summarized by Sharma and Steele (1978).

Progress in GEU Research

Research progress to serve rice growers in problem areas has been steady and significant since the GEU concept was formalized in 1973. The GEU concept is gradually being adopted in national programs in Asia.

AGRONOMIC CHARACTERISTICS Increasing the yield potential of rice by developing rices with desirable plant type is the major objective of all breeding programs. For irrigated areas, the semidwarf plant type exemplified by IR8 with short stature (about 100 cm), sturdy stem, high tillering ability, lodging

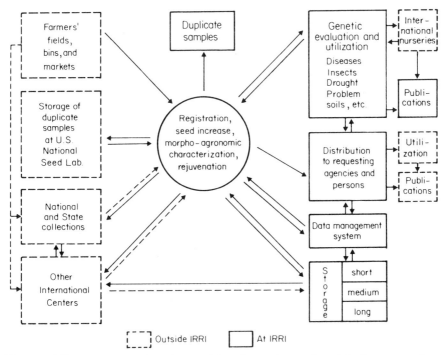

Figure 6.5 IRRI operations in acquiring, increasing seed, cataloging, preserving, evaluating, and utilizing rice germ plasm with the collaboration of other agricultural research centers. (From Chang 1980)

resistance, and dark green erect leaves still represents the most efficient combination of plant characters essential to high fertilizer responsiveness and high grain yields.

In many rainfed lowland areas, varieties with intermediate plant stature (about 120 cm) are more popular than the shorter-statured varieties. Rices intermediate or moderately tall also appear more desirable for upland conditions (Chang et al. 1972). Consequently, intermediate stature and moderate tillering ability are being introduced into lines with erect leaves and sturdy stems that are known to be resistant to the major insects and diseases (Khush and Coffman 1977).

With increased emphasis on crop intensification came a growing demand for early-maturing rices (90–100 days from seed to maturity) for irrigated and rainfed areas. Several early-maturing varieties have been released: IR28 and IR30 in 1975; IR36 in 1976; and IR50 in 1979 by the Government of the Philippines; BR7 in 1977 by Bangladesh; and Ratna in 1970, Palman 579 in 1972, and Pusa 2-21 in 1975 by India all mature in 105–110 days.

GRAIN QUALITY Next to yield, grain quality is the most important factor considered by plant breeders. If consumers do not accept the taste, texture, aroma, or appearance of a newly developed variety, its usefulness is greatly impaired. In the developed countries and in the rice-exporting countries, physical appearance of the grain is often more important than grain yield. In the developing countries, grain quality takes greater importance as the countries become more prosperous and self-sufficient in rice. Grain size and shape, milling recovery of head rice, absence of white belly, appropriate amylose content, gelatinization temperature, gel consistency, and aroma are important factors in the development of a successful rice variety.

Grain appearance is largely determined by endosperm opacity, the amount of chalkiness, either on the dorsal side of the grain (white belly) or in the center (white center), and the condition of the "eye" or pit left in the embryo when the grain is milled. Rice samples with damaged eyes have poor appearance and low market acceptability. Similarly, the greater the chalkiness the lower the market value.

Cooking and eating qualities are largely determined by the properties of the starch, which makes up 90% of milled rice. Several tests are conducted in standard laboratories to determine the cooking characteristics of rice varieties. These characteristics are gelatinization temperature, amylose content, gel consistency, grain elongation, and aroma. Details on these characteristics are given by Juliano (1979).

High-grain-quality rice has different meaning in different countries, depending on consumer preference or preference in the international market. In Thailand, good quality rice is defined as that with a long, slender, translucent grain that produces a fluffy, tender, cooked product (Kongseree 1979). A cursory examination of the old recommended varieties reveals that most have slender, clear grain with a length of more than 7 mm. Many varieties have moderately high amylose content. The popularity of Khao Dawk Mali 105, a low-amylose rice in Thailand, may be more the result of its aroma than its amylose content.

Although association between long grain and slender shape, chalkiness and bold shape, grain dormancy and late maturity, and high protein and light grains has been detected in several instances, no serious barrier in obtaining various combinations of desirable traits in a breeding program is envisaged (Chang and Somrith 1979). The ultimate test of a rice variety is in the market place where consumers generally decide on the basis of physical appearance. It is important to translate the physicochemical properties and eating quality characteristics used in the laboratory into visual terms that can be easily used by consumers. Details on these characteristics are given by Juliano (1973, 1979).

DISEASE RESISTANCE   Rice is often cultivated year-round in the hot, humid tropics. These conditions encourage rapid buildup of pathogenic organisms (Ou 1972). The breeding of resistant varieties is the most practical way to control diseases.

Sources of resistance to important diseases of rice have been identified at IRRI and elsewhere. Numerous rices with resistance to blast, tungro virus, and bacterial blight have been identified. However, one strain of *O. nivara* is the only source of resistance to grassy stunt virus that has been identified (Ling et al. 1970).

The sources of resistance, often tall donor parents, were crossed with an improved-plant-type parent, and the improved plant types segregating with resistance to the diseases in question were selected. The improved-plant-type lines with resistance to different diseases were intercrossed to obtain lines with multiple resistance. At IRRI, resistance to as many as five diseases has been combined (Khush and Beachell 1972, Khush and Coffman 1977). The disease-resistance component of the IRRI GEU program shares these materials with the national programs in Asia and other rice-producing areas of the world. Genetic improvement of rices with resistance to diseases is reviewed by Khush (1977).

INSECT RESISTANCE   Rice farmers in tropical Asia generally use small amounts of insecticides, and then only when crop damage is visible. But when a farmer observes damage, it is often too late to control the insect. Such insecticide treatment is also ineffective if the insect transmits a virus disease to the rice crop.

The initial success of IR20, an early insect-resistant variety, showed that farmers quickly adopt rice varieties that are resistant to insect pests. Not only does insect resistance stabilize yields in farmers' fields, it also lowers production costs.

Devastation of the rice crop by insects, however, remains a constant threat in Asia. Many of the popular rice varieties lack resistance to the brown plant-hopper. Other insects, still of minor economic importance, have the potential to become a serious menace. And the development of the new brown planthopper biotype—a variation within the species making it capable of overcoming a rice plant's resistance to attack—remains a somber threat to rice growers throughout South and Southeast Asia. Nevertheless, clearcut cases of host resistance to several insect species harmful to rice have been speedily incorporated into varieties with improved plant type. However, improved varieties with resistance to several important insect species are still not available. Details on the resistance of rice to insects are given by Pathak (1972) and Khush (1977).

Except in resistance to whorl maggot and sheath blight, the modern genetic materials are definitely superior to the traditional varieties. These modern pest-resistant lines are made available to rice scientists worldwide through the International Rice Testing Program (IRTP).

PROTEIN CONTENT   Rice protein is one of the most nutritious of all cereal proteins. It is relatively rich in lysine (about 4% of the protein fraction), a commonly limiting essential amino acid of cereals. However, the protein content of milled rice is relatively low (about 7% at 14% moisture). Increasing the protein content of rice would mean an increased supply of protein in rice-based diets.

Efforts at IRRI to improve rice protein content go back to 1966, when screening of the world germ plasm collection to identify high-protein varieties started. IR480-5-9 was identified as a high-protein improved line through screening of advanced breeding lines (De Datta et al. 1972). All improved IRRI breeding lines are now screened for protein content.

Breeding for high-protein rices includes conventional methods as well as reinforcing genetic factors for high protein content in a single genotype, a breeding scheme called "long-cycle recurrent selection" (Nanda and Coffman 1979).

Several factors have impeded success in developing high-protein breeding lines with high yield potential.

- Protein content in rice has a complex genetic system (Chang and Somrith 1979).
- Protein content is greatly influenced by environment.
- Heritability values for protein content are low (Suprihatno 1976).
- There appears to be quadratic relationship between grain yield and protein content.

Grain yield and protein content may be improved simultaneously up to a point, beyond which an increase in protein results in a decrease in grain yield (Fig. 6.6).

Weather and cultural practices such as rate of fertilizer application, water management, and weed control also affect levels of protein. These factors complicate selective breeding for higher protein. Nevertheless, the high-protein line IR480-5-9 yields well under disease-free conditions and produces high-protein grains on farmers' fields in the Philippines (Table 6.1).

DROUGHT TOLERANCE Rice is normally grown in the tropics during periods of high rainfall. But the time and intensity of rainfall varies markedly, and the crop is often subjected to periods of severe moisture stress. A major objective of GEU research is to develop rice varieties that will tolerate drought and then recuperate quickly when the rains come.

Drought-tolerant varieties are most needed by upland rice farmers. Drought tolerance is also important for rainfed lowland rice, the yields of which are often limited by unseasonal drought. Even deepwater rices must have some drought tolerance because in some countries the crop is direct seeded into dry soil long before flooding occurs.

Mass screening techniques for drought tolerance have been developed (Chang

Table 6.1 Grain Yield and Brown Rice Protein Content of IR8 and IR480-5-9 Grown as an Upland Crop in Farmers' Fields in Two Villages in Batangas, Philippines, 1973 Wet Season (IRRI 1974)

Village	Yield (t/ha)		Protein Content (%)	
	IR8	IR480-5-9	IR8	IR480-5-9
Santo Tomas	3.8	4.5	7.9	10.7
Cuenca	4.1	5.2	8.2	11.7

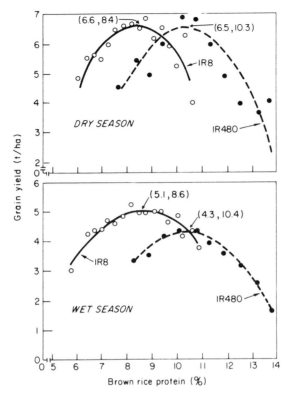

Figure 6.6 Estimated protein-yield thresholds of IR8 (from 964 observations) and IR480-5-9 (from 538 observations) for the dry and wet seasons at IRRI, 1968–1972. Figures in parentheses are yield and protein content, respectively. (From Gomez and De Datta 1975)

et al. 1974, De Datta and O'Toole 1977) to evaluate field tolerance for drought at different growth stages. Major differences for drought tolerance have been noted among rice varieties. For example, varieties Salumpikit and Pinursigi from the Philippines, Nam Sagui 19 and KU 86 from Thailand, ARC 10372 from India, and DJ 29 and DV 110 from Bangladesh are highly drought tolerant.

Many varieties tolerant of drought in the field have a high proportion of long, deep, and thick roots, which reach into the subsoil for water during moisture stress (Chang et al. 1972). Others are semidwarf modern rices such as IR36 and IR442-2-58, which do not have deep root systems but have good levels of drought tolerance.

TOLERANCE FOR ADVERSE SOIL FACTORS   Millions of hectares of land potentially suitable for growing rice remain uncultivated because of soil toxicities caused by salt, alkali, acid, or organic matter. Additionally, vast areas have deficiencies of zinc, phosphorus, and iron, or an excess of iron, aluminum, and manganese that limit rice yields. At IRRI, and in some national programs, rice germ plasm is screened to identify varieties with natural tolerance to these soil problems.

DEEPWATER AND FLOOD TOLERANCE In vast areas of South and Southeast Asia, the water level during the growing season is too deep for the high-yielding, semidwarf varieties. The traditional varieties grown in deepwater and floating rice areas are sensitive to photoperiod, which allows them to mature after the rainy season is over and the floodwater had receded. Early generation lines from IRRI's GEU and other national programs such as in Thailand, Bangladesh, India, and Indonesia are extensively evaluated where the problems exist (IRRI 1977).

TEMPERATURE TOLERANCE Temperature extremes are important limiting factors in many rice growing countries. Germ plasm tolerant of low and high temperatures have been identified and are being evaluated in various areas affected by low or high temperatures.

Specific breeding lines from the GEU programs are screened for tolerance for high or low temperatures.

Distribution, Exchange, and Evaluation of Germ Plasm

All crosses recommended by different GEU teams are made at one place and grown in one nursery. All F_2 materials are grown together. Materials in the F_2 populations, pedigree nurseries, and yield trials are screened by problem-area scientists. Figure 6.7 shows the flow of materials within the IRRI GEU program

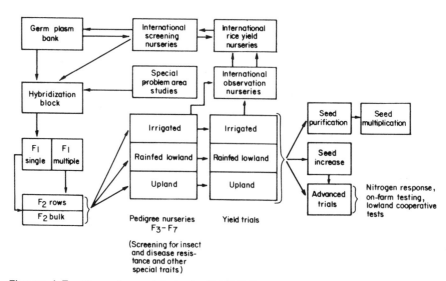

Figure 6.7 Flow of materials in the IRRI GEU program and to national programs through international nurseries and back to IRRI Germ Plasm Bank. (Adapted from Khush and Coffman 1977)

and to national programs. The IRTP, which IRRI coordinates, channels genetic materials into a network of many national programs.

Integration and Exchange of Data

All data on GEU materials are first recorded in common computer-printed fieldbooks. The data are later stored as computerized records and are available to rice scientists worldwide. A computerized data management system, initiated by IRRI in 1976, provides an information retrieval system that gives instant access to, and retrieval of, desired information from the data bank.

National Rice Improvement Programs in the Asian Tropics

During the last decade, impressive gains have been made by the national rice improvement programs in Asia. The following are examples of national breeding programs where major accomplishments have been made:

India

Introduction of the semidwarf variety Taichung Native 1 to India in the 1960s signaled a new era in India's rice breeding programs. At about the same time, national rice improvement efforts were reinforced with the initiation of a countrywide testing network under AICRIP. The introduction of the semidwarf rice enriched the variety trials with several promising breeding lines of a productive plant type, which had been generated at the national institutes and at several breeding stations in different states. Some of the native traditional tall varieties proved to be excellent sources of genetic material for developing semidwarfs suited to diverse situations and stresses.

Jaya, one of the earliest semidwarf varieties developed in India (released 1968) created a significant impact in the rice production in the country. It is a photoperiod-insensitive variety developed from the cross Taichung Native 1/T141. It produced high and stable yields in multilocation tests as well as in on-farm demonstrations. Jaya matures 1 week earlier than IR8 and has similar grain type. It covered a significant area in several states in India and was also released by a few other countries (Ivory Coast and Vietnam).

During the 10-year period after release of Jaya, more than 100 varieties were named in India, including those released centrally and by different states. Some of those released were:

- For early maturity: Cauvery (TN1/TKM6), Pusa 2-21 (IR8/TKM6), Ratna (TKM6/IR8), Rasi or IET 1444 (TN1/CO29).
- For medium maturity: Vijaya (TN1/T141).
- For long-duration photoperiod sensitivity: Rajarajan (CO 25/IR8).
- For gall midge resistance: Phalguna (IR8/Siam 29), Kakatia (IR8/W1263).

To meet farmers' needs for diverse environments and stresses, the following have been considered important breeding objectives in India:

- A wide range in maturity, including photoperiod sensitivity.
- Grain quality to include slender rice, aromatic rice, rice suitable for parboiling, and so on.
- Tolerance for moisture stress, including drought and deep water.
- Tolerance for soil problems, including mainly salinity, alkalinity, and iron toxicity.
- Tolerance for low temperature.
- Resistance to major diseases (blast, bacterial leaf blight, tungro virus, brown spot, sheath blight, and stem rot).
- Resistance to major insect pests (stem borer, gall midge, brown planthopper, thrips).

Bangladesh

The Bangladesh national rice improvement program was considerably strengthened by establishment in 1970 of the East Pakistan Rice Research Institute now called the Bangladesh Rice Research Institute (BRRI). An intensive rice breeding program was initiated in 1966 with the introduction of 303 germ plasm accessions to develop modern rices (Zaman 1980). The program aimed to develop a well-adapted variety with good response to fertilizer, resistance to diseases and insects, and good cooking and market quality. Some examples of BRRI semidwarfs that showed outstanding performance are:

- Mala or BR2, a selection from IRRI advanced generation (IR272-4-1-2) materials released in 1971 as a *boro* (dry season) and *aus* (early summer broadcast) variety.
- BR3 (IR506-133-1/Latisail) released in 1973 for *aus,* transplanted *aman* (monsoon crop) and *boro.*
- BR4 (IR20/IR5-114-3-1), developed specifically for the transplanted *aman* crop of Bangladesh. It can be planted in 15–18 cm of standing water.

Later releases were BR6 (IR28), BR7 (IR2053-87-3-1), and BR9 (IR272-4-1-2-J/IR8). Bangladesh has about 2 million hectares of deepwater and floating rice-growing areas. Some improvements in varieties for those areas have been made but success is limited. Salinity tolerance is an important trait needed for rices grown in Bangladesh's coastal areas.

Sri Lanka

The climatic and edaphic diversity of Sri Lanka's rice-growing regions imposes the need for varieties with photoperiod insensitivity and different maturity classes ranging from 3 to 5.5 months as well as those that are photoperiod-sensitive and mature in the 5.5–6 months range.

Varieties with improved plant type and semidwarf growth habit have been bred and released for cultivation since 1970. Those occupy nearly 60% of the cultivated area. Among them, those bred in the early 1970s were BG34-8 (3-month duration), BG 94-1 (3.5-month duration), BG 11-11, Ld 66 and BG 90-2 (4.5-month duration), and BG 3-5 (5.5–6-month duration, photoperiod-sensitive). High yielding ability, fertilizer responsiveness, and disease resistance were their desirable attributes but they grossly lacked resistance to insects, particularly the brown planthopper and gall midge, which have assumed major importance within the last few years. Resistance to those insects has been incorporated in the latest varieties released: BG 400-1 (4.5-month duration) and BG 276-5 (3-month duration).

The adoption of the new semidwarf varieties was rapid, particularly during 1971–1974. These varieties now occupy the highly productive areas characterized by adequate irrigation water, fertile soils, and high solar radiation; purelines and other varieties (H-4, H-7, H-8, and 62-355) cover the rest of the rice lands in the less productive areas, which are beset by excesses or deficiencies of water, poor soils, unfavorable temperatures, low solar radiation, and so on.

Breeding of varieties suitable for the less productive areas has produced varieties with 4.5-month duration such as BW 78 and BW 100. These were bred for the bog and half-bog soils of the low-country wet zone where iron toxicity is a major constraint to rice production.

Indonesia

The climatic and edaphic factors that limit rice production in different areas of Indonesia are varied and complex. Introduced varieties from other countries, which provide specific dwarfing genes, resistances and tolerances, help to make up essential characteristics needed for breeding area-specific modern varieties. Varieties such as Peta, Bengawan, Sigadis, and Syntha were used in the development of the improved-plant-type varieties. Pelita 1-1 and Pelita 1-2 (IR5/Syntha), bred from varieties of wide genetic base, have been developed for various growing conditions in lowland, upland, high elevation, and tidal swamp areas.

Thailand

Since 1966, new semidwarf types have been used extensively in crosses with the tall, photoperiod-sensitive, recommended Thai varieties in an effort to combine virus resistance and stiff straw with greater responsiveness to fertilizer. From those efforts, RD1 and RD3 (Leuang Tawng × IR8) were released in 1969. RD1 and RD3 are photoperiod-insensitive, nonglutinous, resistant to tungro, have long translucent grains, and mature in 120–130 days (Awakul 1972).

In 1979, the Department of Agriculture approved the release of several new rice varieties. Two of those, RD17 and RD19, represent Thailand's first deepwater rice varieties of hybrid origin. Both varieties are selections from BKN6986, the cross between the semidwarf IR262 and the floating Pin Gaew 56. RD17 and RD19 possess dwarfing genes, which confer shallowwater plant heights ranging from 110 to 150 cm, depending on the environment. By forming longer-than-usual internodes to a maximum plant height of about 180 cm, they can stand not-too-abrupt flooding to a one-meter water depth during the vegetative phase. Both varieties are moderately susceptible to bacterial leaf blight, have long grains, and acceptable cooking quality:

- RD 17 (BKN6986-66-2) matures in 140–150 days irrespective of the photoperiod prevailing in Thailand. It can therefore be grown in both the dry and the wet seasons.
- RD 19 (BKN6986-147-2) is photoperiod sensitive and will flower mid-November in Thailand, if planted early. Harvest is mid-December. This variety is for the vast areas immediately east and north of Bangkok, where traditionally tall, photoperiod-sensitive varieties are grown in the wet season. RD 19 has the distinction of being Thailand's first photoperiod-sensitive semidwarf rice variety.

Philippines

Before 1960, practically all commercial lowland rice varieties grown in the Philippines were tall, weak-strawed, lodging susceptible, and late maturing.

Before IR8 was introduced in the Philippines in 1966, BPI-76 (Fortuna/ Seraup Besar 15) was one of the major varieties with intermediate stature and yielding ability. The original photoperiod-sensitive BPI-76 was released in 1960. There were several selections made from it that were essentially insensitive to photoperiod.

The most significant variety developed in the Philippines was C4-63 (Peta × BPI-76). In 1970, a green-base C4-63 was released to replace the original seed stocks of C4-63. C4-63 has about 125-day growth duration, is photoperiod-insensitive, intermediate-statured, and moderately high-tillering. It has long grains with a 1-month grain dormancy period, high head rice recovery, and intermediate amylose content (Cada and Escuro 1972). Because of its good eating quality (intermediate amylose content) it has spread in Indonesia, Malaysia, and Burma.

Rice Improvement in Africa

The breeding objectives in Africa are high yield potential, yield stability, desirable agronomic characteristics, high grain quality, blast resistance, and drought tolerance.

Among the desirable agronomic characteristics, a plant height of 100–120 cm is considered optimal. In certain regions with a bimodal rainfall pattern, such as in western Cameroon, two rice crops can be grown per year. For that, an early- or medium-growth-duration variety is required. In other areas, the growth duration of rice varieties must be suitable for different cropping systems (Jacquot 1978). Upland rice improvement programs in West Africa are discussed by Abifarin et al. (1972). For lowland rice, introduced rice varieties such as IR5 and IR20 are most widely planted. To develop modern varieties for Africa, Buddenhagen (1978) and Virmani et al. (1978) proposed that they should be matched to important ecosystems in a country.

About 91% of tropical Africa's rice is cultivated in Malagasy Republic, Sierra Leone, Nigeria, Guinea, Ivory Coast, Tanzania, Liberia, Mali, and others. Most of the hybridization and selection of segregating materials in anglophone Africa has been in Sierra Leone, Nigeria, and Liberia, with a little in Tanzania (Virmani et al. 1978). The success of upland rice in terms of the number of varieties has come primarily from selection among African varieties and hybridization between African and exotic varieties. The latter has been practiced in Nigeria and Sierra Leone. The most popular rice varieties grown as upland are LAC 23, ROK 3, and Faya, which are in fact selections from farmers' varieties.

For the francophone countries in Africa, most of the varietal improvement programs were carried out by IRAT, particularly the improvement of upland rice. Historically, the National Institute for the Development of the Congo (now INEAL, Zaire) began collecting rice ecotypes in 1933 and produced such varieties as R66 and OS6 (Jacquot 1978). OS6 is extensively grown in West Africa. Moroberekan, a local variety from the northern Ivory Coast, is grown extensively in West Africa under upland rice culture.

Rice Improvement in South America

Three examples can be given for varietal improvement of rice in South America. They are lowland rice in Colombia and Surinam and upland rice in Brazil.

Colombia

In 1969, about 6.6 million hectares were planted to rice in Latin America. The most successful breeding program was developed in Colombia with cooperation between Centro Internacional de Agricultura Tropical (CIAT) and Instituto Colombiano Agropecuario (ICA). That program developed superior varieties with high yield potential, improved plant type, strong seedling vigor, moderately heavy tillering, semidwarf stature and erect leaves, early maturity (90–120 days), resistance to blast and sheath blight, resistance to *hoja blanca* virus and its insect vector *Sogatodes orizicola* (Muir), and good milling and cooking quality.

From the CIAT-ICA collaboration, the first variety, CICA-4, was developed from the segregating population of IR930-31-1-1B. CICA combines the initials of CIAT and ICA. CICA-4 is resistant to *hoja blanca* and highly resistant to

Sogatodes. It is susceptible to blast, however (Rosero 1972). IR22 also has become an important variety in Colombia.

CICA-9 was released in 1976 and is now grown not only as lowland rice in Colombia but as upland rice in Costa Rica and Panama. It has higher yield potential and greater disease resistance than previously used varieties. It is also somewhat taller than the other Colombian varieties. CICA-9 and other varieties being developed in Colombia promise to make a sizable contribution to higher rice yields in Latin America.

Surinam

A number of varieties developed in Surinam have unique morphological and grain characteristics. Many of these rices are about 130 cm tall, have dark green, long, narrow leaves and are generally low tillering with relatively good straw strength. The grains are very long. Some examples of Surinam varieties are Alupi, Awini, Cauponi, Apuri, and Ciwini. During 1976, Surinam's 21st rice variety, Diwani (IR454-1-17-1-1/SML Washabo), was released. Diwani is 80–90 cm tall, medium tillering, and resistant to lodging, blast, and brown spot (Lieuw et al. 1977).

Brazil

For upland rice-growing areas in Brazil, the Instituto Agronomico at Campinas (IAC) has developed several rices with tall stature and good drought tolerance. Examples are IAC 1246, IAC 47, and IAC 25. The latter matures about 10 days earlier than the other two and escapes the drought period locally known as *veranico* (see Chapter 2).

Breeding Programs in the Temperate Regions

In temperate Asia, impressive gains in rice yields have been made through varietal improvement.

Japan

Development of the variety Hoyoku was the most significant event in breeding for high rice yields in Japan. Hoyoku was developed at the Kyushu Agricultural Experiment Station in 1961 by crossing Jukkoku (female parent) with Zensho 26 (male parent) (Kariya 1966). Jukkoku has a short stem and is highly resistant to lodging but susceptible to blast and bacterial leaf blight. Zensho 26 is resistant to bacterial leaf blight. Hoyoku has a short stem, is highly responsive to fertilizer, and its development may be considered of similar importance to the development of IR8 in the tropics.

Table 6.2 Plant Characters, Grain Yield, and Yield Components of a New (Hoyoku) and Old (Norin 18) Rice Variety (Adapted from Ito and Hayashi 1969)

Attributes	Hoyoku	Norin 18
Growth duration after transplanting (days)	130	134
Total number of leaves on main stem	17.6	16.2
Length of culm (cm)	72	87
Length of panicle (cm)	19	22
Panicles per m^2 (no.)	378	259
Grains per panicle (no.)	86	95
Grains per m^2 (no.)	32,500	24,600
Weight of 1000 grains (g)	28.0	30.4
Yield of rough rice per m^2 (g)	730	645

In 1962, another high-yielding variety, Kokumasari, was released (Shigemura 1966). Hoyoku and Kokumasari spread rapidly and by 1964 they completely replaced Norin 18, the leading variety, and others that had occupied almost all rice land of northern Kyushu. Table 6.2 compares plant characters and yield components of Hoyoku and Norin 18.

Since the development of Hoyoku, only minor improvements have been made in increasing yield potential of Japanese varieties through breeding; most of the improvements in yields have been through improvements in management practices. Shigemura (1966) reviewed yearly average rice yields for each of Japan's prefectures for the past 100 years and found that the high-yielding rice-growing area had shifted from southwestern areas to the northwestern districts of Tohoku and Hokuriku. Toriyama (1979) summarizes details on rice breeding programs and the current status of varietal improvement in rice in Japan.

China

Breeders in China work closely with researchers in the production teams, brigades, and communes. Methods used are introduction, pureline selection, hybridization, radiation treatment, and haploid breeding using anther culture. Hybridization is done by county agricultural institutes and provincial agricultural academies.

Most of the rice improvement work in China has been to improve agronomic characteristics. High yields and short growth duration have been emphasized. Most of the japonica varieties in the northern regions and the second-crop japonica varieties in the central regions mature in 150 days. Any reduction in growth duration of those varieties without sacrificing yield would be extremely beneficial. Short growth duration has also been a major breeding objective because early-maturing varieties were essential for the multiple cropping systems developed in China. Chinese breeders bred many early-maturing indica varieties such as Guang-lu-ai 4 and Ai-nan-tsao 1, which are grown widely in central and southern China. At present, first-crop rice varieties Zhen-zhu-ai, Guan-lu-ai 4, Xian-fang, Er-jiu-ging, Lu-shuang 1011 and second-crop rice varieties Bao-xuan 2, Nan-jing 11, and Dong-ting-wuan-xian are widely grown in the southern region (Shen 1980).

To increase the yield potential of rice, the Chinese have put major emphasis on reduction of plant height as a way to decrease lodging. The japonica varieties of northern and central China have been selected progressively for shorter height during several decades and most are now between 80 and 110 cm in height. They are highly responsive to nitrogen fertilizer and have good lodging resistance.

High tillering capacity is not considered an important rice plant trait in China. Apparently, the large number of seedlings per hill (more than 10) used in the Chinese production system offsets any disadvantage of low-tillering varieties.

The Chinese consider photoperiod sensitivity an important trait for the second crop of rice in the two-rice-crop system or for the third crop in the three-rice-crop systems in the south. With increased intensity of cropping, tolerance for extreme temperatures in rice has become an essential feature in breeding work (IRRI 1978a). There were also efforts to develop rice varieties resistant to blast, planthoppers, leafhoppers and stem borers.

Other methods of varietal development include hybrid breeding and wide crosses (crossing of rice with sorghum and other grasses). Efforts are made to develop high yielding varieties with high photosynthetic efficiency.

Republic of Korea

One of the remarkable success stories of rice production is that of Korea for 1965–1975. Although Korea has had a long history of breeding and testing programs, in 1965 a cooperative rice research program was initiated with IRRI. One of the first crosses made at IRRI and tested in Korea was between a Japanese variety, Yukura, and the short-statured Taichung Native 1. The F_1 progeny was then crossed with IR8. The most promising selection from that triple cross was IR667-98, which was named Tong-il in 1972. In subsequent years, several newer Tong-il type varieties have been developed from the indica-japonica crossing program.

Tong-il and the related semidwarf varieties were resistant to blast and stripe virus but susceptible to insects. Of the semidwarf varieties, Milyang 30 is resistant to many destructive insects (Choi 1978). By 1978, about 76% of Korea's rice-growing areas were planted to the new varieties. Figure 6.8 shows remarkable gain in rice yields over the years, making Korea's the highest national average in the world. In 1979, the area planted to the Tong-il semidwarfs have probably dropped somewhat due to a blast disease problem that affected large areas in 1978. The blast outbreak was attributed to heavy fertilizer applications and a new race of blast that attacked only the newer rices.

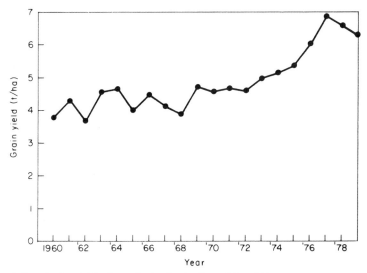

Figure 6.8   Grain yield of rough rice in the Republic of Korea, 1960–1979. (Adapted from Office of Rural Development, Korea, 1979, and Ministry of Agriculture and Forestry, 1979, unpublished data)

United States

Varietal development in the United States is the joint responsibility of each rice-growing state and the U.S. Department of Agriculture. Incorporation of the semidwarf gene in the U.S. rice varieties has been slow, however.

LOUISIANA   The primary objective of the Louisiana varietal improvement program is to develop high-yielding semidwarfs superior to the current commercial varieties. Hybridization of appropriate stocks followed by selection of resulting progenies has, in fact, resulted in development of a considerable number of high-tillering, nitrogen-responsive lines with desirable plant type and resistance to some diseases. Progress, however, has been slow because crosses with IR8, Taichung Native 1, and the other semidwarf accessions used for their yield and plant type resulted in undesirable traits such as susceptibility to stem rot, a major cause of lodging in Louisiana (McIlrath et al. 1975). In a 1977 regional uniform nursery, the yield of the semidwarf variety LA 110 (8.3 t/ha) exceeded all other Louisiana's entries in the short-stature group (McIlrath et al. 1978).

TEXAS   The variety Labelle, released in 1972, covered about 65% of the Texas rice area by 1974. The current breeding objectives include development of semidwarf lines with high-yielding ability, initiation of crosses with potential for development of long-grain rice with resistance to blast, and development of varieties of all grain types resistant to sheath blight, brown spot, stem rot, and kernel smut (Bollich and Scott 1975).

ARKANSAS   There are about 14 varieties commercially available that produce satisfactory grain yields in Arkansas. Examples are Starbonnet, Bonnet 73, Lebonnet, and Labelle. Rice varieties grown in Arkansas comprise three maturity groups: very short season (100–115 days); short season (116–130 days); and midseason (131–155 days). Those varieties further divide into short-, medium-, and long-grain rices.

Acceptable varietal performance is measured by high grain and milling yields, suitable maturity, resistance to lodging and diseases, desirable grain type, cooking and processing characteristics, and consumer preference (Huey 1977). Performance of rice varieties in Arkansas during 1972–1975 were summarized by Johnston et al. (1976). In 1978, the leading varieties in Arkansas were Starbonnet (38% of area) and Lebonnet (30% of area) (Huey 1979).

CALIFORNIA   California annually produces about 25% of the rice in the United States. The yield per hectare is 20% greater than the other rice-producing states. Varietal improvement has provided significant contribution to that yield advantage. The breeding objectives in California include short stature for greater lodging resistance, early maturity to facilitate harvest before fall rains, and smooth hulls to reduce dust during harvesting and drying.

A series of steps were taken to incorporate these features into Calrose, a high-yielding variety with good cold tolerance and milling and cooking qualities (Rutger and Peterson 1976). In 1969, Calrose seeds were exposed to radioactive cobalt-60. In 1971, selections were made for short stature and early maturity mutants in the first segregating generation (M_2). By 1972, genotypes were identified that were true breeding for short stature (Fig. 6.9). The most promising short stature selection, designated D7, was similar to Calrose but 25 cm shorter.

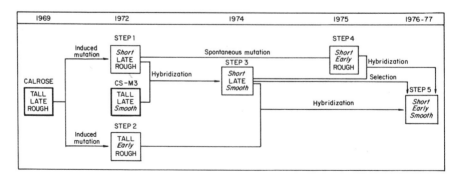

Figure 6.9   Step-by-step additions of short stature, smooth hulls, and early maturity into California rice germ plasm are delineated. Desirable attributes are italicized. (From Rutger and Peterson 1976)

Table 6.3 Agronomic and Quality Characteristics of the Short-Stature Genotype D7 Compared with the Tall Check Variety CS-M3; Yield Data Are the Means of 19 Trials in 1973–1975 (Adapted from Rutger and Peterson 1976)

Character	Variety D7	Variety CS-M3
Yield, at 14% moisture (t/ha)	8.3	7.9
Plant height (cm)	85	120
Lodging (%)	18	45
Seedling vigor score[a]	3.96	4.00
Days to heading	112	111
Brown rice kernel length (mm)	5.80	5.96
Head rice yield (%)	69.1	68.8[b]
Amylose (%)	18.4	18.9[b]

[a] 1 = poor, 5 = excellent vigor.
[b] Average of CS-M3 and Calrose from the same trials.

Table 6.3 shows a significant advantage of D7 when compared with a tall rice CS-M3. In 1976, D7 was released as Calrose 76 (Rutger and Peterson 1976).

In 1978, twenty of the California breeding lines and the semidwarf variety M9 yielded significantly greater than the commercial tall variety Earlirose. ESD-7, later released as M-101, (Rutger et al. 1979) yielded about 10% higher than Earlirose in 2 years of statewide testing. Figure 6.10 shows the nitrogen responsiveness of ESD7-1, which is mainly because of higher lodging resistance. It is a widely adapted, highly promising, short-statured rice, with medium grain that matures about 5 days earlier than Earlirose. Head rice yields of the line may be less than desirable, particularly if it is harvested at low moisture contents (University of California, Davis and USDA 1978).

Genetic studies by Foster and Rutger (1978a) showed that the most useful sources of short stature have been the D7 mutants selected from the tall, adapted variety Calrose, and derivatives of the semidwarf Taichung Native 1 or similar materials. Each source is conditioned by a single recessive gene, with additional genes of small effects introduced from indica sources (IRRI 1965, Foster and Rutger 1978b).

Australia

The need to develop high-quality, long-grain varieties that are better adapted to high fertility and the cool temperatures of southern New South Wales, and which will yield as well as the currently used japonicas, has been discussed by McDonald (1978). The japonica varieties, such as Calrose and Kulu, cultivated in New South Wales are among the world's most tolerant of low temperature during

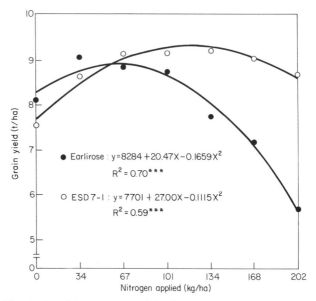

Figure 6.10 Grain yield response of Earlirose and ESD 7-1 to preplant nitrogen; Sorrenti Ranch, San Joaquin County, 1978. (Adapted from University of California, Davis and U.S. Department of Agriculture 1978)

reproduction. The ultimate solution may be to develop early-maturing varieties that escape the low temperature regime. Improvement in lodging resistance by shortening the culm length is also a major objective in breeding rices for New South Wales. Semidwarfs developed in California are being used for incorporating lodging resistance into japonica varieties grown in New South Wales.

BREEDING METHODS AND PROCEDURES

The breeding objectives and the facilities available to the breeders have an important bearing on the method used.

Breeding Methods

The selection method of plant breeding has been used from time immemorial to select new varieties from old varieties that may have passed down from generation to generation of farmers. Other methods have been developed as science has advanced and new technology has become available.

Pure Line Selection

For a pure line breeding program to be successful, the base population or the original variety must have genetic variability because selection can act on heritable differences only. For pure line breeding, a large number of plant selections are made from the genetically variable original population. As a second step, progeny rows from the individual plant selections are grown for initial evaluation. As a final step, promising selections are compared with each other and the parent variety in the replicated yield trials and the highest yield lines are released as a pure line variety.

Mass Selection

The main difference between mass selection and the pure line selection method is that a number of plants, rather than just one, are selected to make up the new variety. The varieties developed by this method include fewer genotypes than the parent population but more than the "single genotype" of varieties developed by the pure line selection. The number and variability of types included depend upon the variability of the original population and the intensity of selection practiced.

In some countries, mass selection finds its greatest use in purification of existing varieties in connection with pure seed programs. In South and Southeast Asia, mass selection can be profitably employed in purifying old varieties of rice that are under cultivation and in maintaining the genetic purity of new, high-yielding, improved plant type varieties being developed.

Hybridization

As plant breeding has progressed and more of the natural variability existing within the populations of rice as a self-pollinated species has been exploited, it has become increasingly important for the plant breeder to generate new variability and to produce desired recombination by making artificial hybrids.

The hybrids are allowed to self-pollinate and the resulting populations are handled either by the bulk method, the backcross method, or the pedigree method.

Bulk Breeding

In Japan, the bulk method of breeding has been widely used in experiment stations because, in general, the effectiveness of selection in early generations of rice hybrids is low for quantitatively inherited traits, including yield, particularly in individual plant selections. Selection in advanced generations that have been purified somewhat is effective for such traits. For these reasons, the bulk method without plant selections in early generations seems to be more suitable to breeding for yield increase than the pedigree method, particularly on a small scale. Labor saving is also a great concern of rice breeders in Japan. Finally, procedures for shortening breeding cycles in a greenhouse are practicable only by means of the bulk method (Okabe 1972).

Despite its inherent advantages of simplicity and convenience, years of bulk breeding in tropical Asia have not produced significant gains in developing rices with high yields. At present, the bulk method as practiced earlier is not used because it does not permit concurrent screening for resistance to a number of diseases and insects (Khush 1978).

Backcross Breeding

The backcross method has not been used extensively because of a lack of suitable recurrent parents. The major disadvantage of backcrossing is that no single variety is so nearly ideal that it needs to be improved in only one character. However, at IRRI, a few backcrosses have been made in the crosses with *Oryza nivara* for grassy stunt resistance; IR8 and IR24 were used as recurrent parents. At present, several named varieties, such as IR28, IR32, IR34, IR36, IR40, and IR42, are being used as recurrent parents in a backcrossing program (Khush 1978). Backcrossing is also used to develop isogenic lines with different genes for resistance to the brown planthopper. Simple inheritance and high heritability of the character to be improved through backcrossing are the prerequisites for successful use of the method.

Pedigree Breeding

The pedigree method has been the most widely used in rice improvement. Early generations of field-selected plants can be evaluated in special tests for characters such as insect and disease resistance and grain quality. The method provides a sound basis for the discarding of undesirable lines and results in the concentration of useful material (Jennings et al. 1979).

The pedigree method is highly suitable to develop rices with resistance to insects and diseases if the resistance is governed by major genes. It is possible, therefore, to combine genes for resistance to six or seven major diseases and insects in a short period (Khush 1978).

The major disadvantage of pedigree breeding is that it requires much time to evaluate lines periodically throughout the growing season and to keep records on which selection is based at maturity. Of all breeding methods, the pedigree method requires the greatest familiarity with the material and with the relative effects of genotype and environment on character expression. For traits governed by polygenes, the pedigree method of breeding is not the most effective approach. For example, resistance to stem borers and sheath blight appears to be under polygenic control. For these two traits, Khush (1978) suggested a diallel selective mating system proposed by Jensen (1970).

Rapid Generation Advance

With rapid generation advance, which is a modified form of the bulk method, it is possible to grow three or four generations in a year (Goulden 1939, Okochi et al. 1958). No selection is practiced during that period. To develop insect and disease resistant lines, the bulk population at F_5 or F_6 is exposed to the disease or insect pressure and individuals with better level of resistance are identified and grown in progeny rows for further evaluation.

Mutation Breeding

Variability caused by induced mutations is not essentially different from variability caused by spontaneous mutations during evolution. X-rays, gamma rays, and neutrons are effective ionizing radiations for mutation induction. Ethyl-methane sulfonate (EMS) is one of several chemical mutagens that has been used.

The direct use of mutations is a valuable supplementary approach to plant breeding, particularly when it is desired to improve one or two easily identifiable characters in an otherwise well adapted variety.

Induced mutations offer three advantages (IAEA 1971):

1 They are capable of conferring specific improvements to varieties without otherwise significantly affecting their performance, and the time required for such specific improvement is shorter than if only hybridization is used.
2 They represent the only possible method of creating a character that is not found in the natural population and their use is often the easiest and quickest method if the desired character is part of an undesirable genotype.
3 They offer a method of breaking tight linkages, producing translocations for gene transfer.

Mutational changes of agronomic importance that have been reported are plant height, flowering and maturity period, resistance to diseases and insects, and increased protein content in rice. One of the earlier reports on the use of mutagenic agents to induce artificial variability in rice in relation to X-ray induced mutants is GEB 24 in India (Ramiah and Parthasarathy 1938).

Some of the successful varieties developed by mutation breeding are as follows (Mikaelsen 1980):

Variety	Country and Date
Reimei	Japan, 1966
Jagannath	India, 1969
IRAT 13	Ivory Coast, 1974
Calrose 76	USA, 1976

Variety	Country and Date
Hokuriku-100	Japan, 1976
M-7	USA, 1977
RD-6	Thailand, 1977

Hybrid Breeding

The phenomenon of heterosis or hybrid vigor has been exploited commercially in maize, pearl millet, and sorghum, and its possible use in wheat and rice is being explored. Essential prerequisites to a successful hybrid breeding program are:

- Presence of true F_1 superiority or heterobeltiosis; that is, the F_1 hybrid is superior to the better parent, not only superior to the mean of the two parents, as in heterosis.
- Availability of efficient cytoplasmic-genetic male-sterile and fertility restorer lines.
- Ability of the male-sterile lines to show satisfactory seed set through cross-pollination.

HETEROSIS AND HETEROBELTIOSIS Many rice crosses have been reported to show heterosis but some specific cross combinations have also shown significant true F_1 superiority or heterobeltiosis (Davis and Rutger 1976, Lin and Yuan 1980). The extent of the F_1 yield advantage over the better parent has been found to range from 10 to 210% in experiments using limited population size and noncommercial planting densities. However, results from China, which are based on large production plots and commercial planting densities, have indicated clearly that hybrid varieties had a yield advantage of 20–30% over the best conventionally bred varieties. The highest yield obtained from an F_1 hybrid was 12 t/ha (Chang 1979a).

CYTOPLASMIC-GENETIC MALE STERILITY AND FERTILITY RESTORATION A cytoplasmic-genetic male-sterility system depends on cytoplasmic and genetic factors. Plants carrying a particular type of cytoplasm in combination with recessive allele(s) of pollen fertility gene(s) in the nucleus would produce aborted pollen but would produce seeds if cross-pollinated, indicating that ovule fertility is normal. The F_1 offspring from a cytoplasmic-genetic male-sterile plant would be pollen-sterile if the pollen parent contributed the same recessive allele(s) of gene(s) in the nucleus as in the male-sterile plant. Such a pollen parent is called a maintainer line because it enables maintenance of the sterility of a male-sterile plant. On the other hand, if a pollen parent contributed dominant allele(s) of gene(s) responsible for pollen fertility the F_1 offspring from a cytoplasmic-genetic male-sterile plant would be pollen-fertile. Such a pollen parent is called restorer line because it enables restoration of the pollen fertility in the offspring of a male-sterile plant.

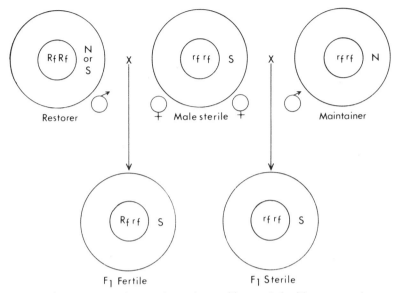

Figure 6.11 Cytoplasmic-genetic male sterility and fertility restoration system. S = sterility-inducing cytoplasm; N = normal cytoplasm; Rf = dominant allele for pollen fertility in the nucleus; rf = recessive allele for pollen fertility in the nucleus.

The mechanism of cytoplasmic-genetic male-sterility in a plant is explained in Fig. 6.11. The cytoplasmic-genetic male sterility is also often designated as cytoplasmic male sterility because the male-sterile and its corresponding male-fertile maintainer line differ only with regard to cytoplasm; the genetic constitution of the two is identical.

Several cytoplasmic male-sterile lines have been identified or developed, or both, in indica as well as japonica rices (Katsuo and Mizushima 1958, Shinjyo 1969, Athwal and Virmani 1972). In China, a number of agronomically superior cytoplasmic male-sterile lines, such as Er-Chiu-Nan 1A, Zhen-Shan 97A, V20A, V41A, Yar-Ai-Zhao A, Chen-Bao 1A, Lien-Tong-Tsao A, and others, have been developed for use in the hybrid breeding program (Lin and Yuan 1980). Two IRRI varieties, IR24 and IR26, have been identified as effective restorers for these cytoplasmic male-sterile lines and used extensively in the development of commercial hybrids.

NATURAL OUTCROSSING AND SEED SET FOR HYBRID SEED PRODUCTION The ease and cost of production of hybrid seed for rice will determine to what extent the hybrid breeding approach can be used in other rice-growing countries in the world. The floral structure of rice is not well adapted to cross-pollination and less than 1% natural outcrossing is observed normally. However, on male-sterile plants 33–45% seed set has been obtained in large-scale hybrid seed production plots in China. The highest frequency of cross-pollination and seed set (74%) was observed in the Hunan province during early September when the daily mean temperature was 22°C or above (Chang 1979a).

Research on the "three lines" (cytosterile line, maintainer, and restorer) of hybrid rice in China was started in 1964 but the major breakthrough in hybrid seed production was not realized until 1974. In 1978, the area growing hybrid rice in China covered 4.2 million hectares (Shen 1980). Other reports (Lin and Yuan 1980, Shen 1980, Chang 1979a) suggest that China has effectively used hybrid rice breeding as an important tool for varietal improvement and that the method is feasible for developing new varieties.

Varietal Improvement Through Tissue Culture

Techniques have been developed to regenerate rice plants from cultured cells (Bajaj 1980). Diploid callus can be induced by plating surface-sterilized seeds on chemically defined nutrient media. Haploid callus can be obtained from uninucleate pollen by plating immature rice anthers on media containing the basic nutrients for growth of plant cells and supplemented with growth regulators, including 2,4-D and naphthalene acetic acid (NAA).

Diploid cells can be treated with mutagenic agents and screened *in vitro* for mutants tolerant of herbicides, salinity, mineral deficiencies and toxicities, and other yield constraints.

Haploid cells generated through anther culture give rise to homozygous plants. When the anthers are obtained from F_1 hybrids, the regenerated plants have the same inherent genetic variability found in segregating F_2 populations from conventional crosses. About 40% of the plants derived from haploid callus are diploid through spontaneous doubling of the chromosomes when the plant is differentiated. Additional diploid, homozygous plants can be obtained by treating the growing point of haploid plantlets with mutagens. Because the plants obtained through anther culture are homozygous, they give rise to homozygous breeding lines that can immediately be screened for disease and insect resistance, agronomic characteristics, and yield potential. Outstanding lines can be rapidly advanced through conventional seed multiplication programs because there will be no segregation in succeeding generations.

Three difficulties have been encountered with rice tissue culture (Chaleff 1980):

1 The morphogenetic capacity of the callus cultures may be lost rapidly during continued maintenance in culture.

2 Only a small portion of cultured anthers give rise to callus

3 Some of the regenerated plants may be albino; green plants from callus range from 5 to 90%.

In China, some new varieties such as Mu-hua 1 (Heilongjiang/Sin-siu Shanghia), and Wan-dan 7 (Guangxi) have been developed from pollen culture of hybrids of varietal and intersubspecific crosses (Shen 1980).

Breeding Procedures

The most important factor in developing a successful breeding program is to be systematic. Naturally, a rice breeding program must aim to achieve its objectives but systematic program organization should permit maximum efficiency in screening and generation advances.

Growing of Donor Parents and Crossing

Each growing season, donor parents should be grown for hybridization. A broad-based, high-volume crossing program is essential in any rice improvement effort. Types of crosses are:

- SINGLE CROSSES   A single cross is the hybridization of one variety or line with another variety or line.
- BACKCROSSES   A backcross is the cross of an F_1 to one of its parents.
- TOPCROSSES   A topcross (3-way cross) is the cross of an F_1 with a third variety or line.
- DOUBLE CROSS   A double cross is a cross of two F_1 hybrids.

Details on operation and procedure in a hybridization program are given by Jennings et al. (1979).

F₂ Populations

The intense interplant competition caused by differential tillering ability and sizes of neighboring plants in early segregating generations is a critical factor that affects the choice of breeding methods. To minimize this problem, Jennings et al. (1979) suggest multiple crosses involving both tall and dwarf parents and production of a large number of F_1 plants from these multiple-cross combinations. The resulting F_1 will have both tall and short phenotypes; only the latter are advanced to the F_2 in the field.

F₂ populations from the individual F_1 plants of topcrosses or double crosses should be grown separately and selections should be made only from agronomically suitable crosses. To develop insect and disease resistant varieties, wherever feasible most of the F_2 populations should be grown without insecticides and fungicides to ensure insect and disease pressures (Khush 1978).

Although there is no ideal F_2 size for a single-cross population, a useful rule is to make them as large as possible, especially for the wider crosses. Larger size of an F_2 population becomes imperative with emphasis on important recessive traits, such as dwarf stature, and also to permit a wide spectrum of recombinations.

Pedigree Nurseries

The pedigree nursery consists of progenies of single plant selections made in F_2–F_5 generations. The pedigree nurseries are the heart of a rice improvement program. Good management and good judgment are more essential in handling pedigree nurseries than in any other phase of operation.

From a 25- to 30-g seed sample, 5 g of seed may be planted in a single 5-m row in early generations and 3 rows per line in later generations. The remaining seeds are then sent to individual specialists for screening for resistance to various insects and diseases and tolerance for drought or other stress problems.

Yield Trials

Selected F_5, F_6, or late generation rows, should be evaluated for yield in replicated trials. Promising materials are then tested for yield potential and nitrogen responsiveness by agronomists collaborating in the program.

Multilocation Testing

An important and integral part of plant breeding is testing of the breeding products, as well as breeding sources, in diverse agroclimates. This provides an opportunity for the material to be exposed to various biological, climatic, and edaphic stresses and will not only help identify potential varieties and potential genes, but will also reduce the time needed for development of varieties. Multilocation testing can be achieved through multicountry, multidisciplinary, collaborative research.

Seed Programs

In any varietal improvement program it is essential to have an effective program on seed multiplication, testing, certification, and distribution. Considerable attention must be given to a seed program if the farmer and the consumer are to benefit from improved varieties.

Seed Certification

Four classes are usually included in the seed certification program:

1 *Breeder seed* is the seed directly controlled by the originating or sponsoring plant institution, or individual, and the source for the production of seed for the certified classes.
2 *Foundation seed* is generated from the progeny of breeder seed. Foundation seed is handled to maintain genetic purity and identity. Production must be acceptable to the certifying agency.

3 *Registered seed* is the progeny of breeder seed or foundation seed handled by a procedure acceptable to the certifying agency to maintain satisfactory genetic purity and identity.

4 *Certified seed* is the progeny of breeder, foundation, or registered seed, which is handled to maintain satisfactory genetic purity and identity in a manner acceptable to the certifying agency.

Breeder Seed Production

Breeders are often faced with the problem of having identified several excellent lines without being certain which should be named as a variety. The breeder seed procedure allows the breeders to simultaneously multiply seed of the leading candidates and narrow them to the best one or two. The following steps are suggested for multiplying breeder seed in tropical Asia (Jennings et al. 1979):

- Six hundred panicles are selected from each line individually from either yield or regional trials.
- Each panicle is then inspected in the laboratory for uniformity of grain types and then threshed separately.
- To speed multiplication, the material is transplanted (wherever possible) in the single-plant hills to facilitate roguing. Occasional off-type plants are destroyed during frequent inspection of the material.
- To avoid mixture, extreme care is exercised when the field seed is bulk harvested, dried, cleaned, and bagged. About one ton of seed is harvested from the 600 panicles.
- About 300 kg seed of the line is stored in an air-conditioned room as breeder seed, to be used for subsequent production of foundation seed. Seed viability remains excellent for 3–5 years if stored under low temperature and relative humidity. The breeder seed is carefully separated and labeled to prevent mixture with other seed lot. If questions are raised about varietal characteristics, the breeder seed is the ultimate source for rechecking.

In the United States, the production of breeder and foundation seed is an integral part of the cooperative rice-breeding projects of U.S. Department of Agriculture and the State Agricultural Experiment Stations (Adair et al. 1973). The steps followed for initial testing are similar to those just described for tropical Asia. The minor differences are as follows:

- Once seeds are checked for uniformity, they are stored in a small envelope.
- The next year, the seed from each panicle is sown in single rows 1.2–6 m long and 30–60 cm apart. In some cases, there are 90-cm alleys between ranges of rows to facilitate careful inspection and roguing.
- The block of panicle rows of each variety is isolated from those varieties similar in maturity so as to eliminate natural crossing and subsequent segregation for off-types.

- In the early seedling stage, off-type plants or rows are identified and removed immediately or tagged for removal during harvesting. Albino seedlings are identified and removed at an early stage.

- During the ripening period, careful observations are made to detect variations in plant type, plant height, panicle type, pubescence, color of apex or apiculus, and grain type.

- If the entire group of panicle rows appears uniform, the rows are harvested in bulk. However, if considerable variations exist due to genetic segregation, then it may be necessary to make further selection of rows for purification. This may be done by selection from within a block or rows or individual rows that appear to be uniform in appearance and that typify the variety being increased.

Foundation Seed Production

Once dormancy is broken, all or part of the breeder seed can be planted to produce foundation seed. The area planted depends on the amount of foundation seed desired. With 700 kg of seed, 10–20 ha of foundation seed can be planted, depending on the planting method used. Twenty hectares should produce 100 t of foundation seed, enough to transplant 2500 ha or to directly grow 800 ha for production of registered seed (Jennings et al. 1979).

Foundation seed fields should be rogued several times during the last part of the growing season. In the United States, roguing of foundation seed is facilitated by a space left every few meters by stopping up one or more of the holes in the farm-type grain drill used for seeding. Management practices should be followed that minimize lodging.

The release of foundation seed to growers is usually handled through a committee, seed council, or similar organization that allots the seed to carefully selected growers. Sometimes, foundation seed is distributed directly to the growers from the State Agricultural Experiment Station.

Registered and Certified Seed Production

The production and certification of registered and certified seed are not part of the breeding program. The details on cleaning, grading, and processing seed rice are described by Adair et al. (1973).

According to Jennings et al. (1979), it requires at least 5 years from the time crosses are made until seed of new varieties are available to the farmers, assuming that two generations can be advanced in a year. This is possible only if growing conditions permit year-round rice culture. In temperate regions, only one crop of rice can be grown a year but an alternate generation can be grown in a favorable climatic environment.

Types of Seed Production Programs

Probably all seed programs have started with government participation. It is, however, desirable to strengthen private-sector production program for initiating seed multiplication. In the United States and Europe, private enterprises are preferred to handle production of seeds.

Semi-Government Seed Production

In some countries, the government establishes a national agency to produce, process, and distribute seed. In general, such agencies contract seed production to selected, qualified farmers, who operate as autonomous units and are usually financed by the government bank credits.

Seed Growers Associations

In some countries, seed production is handled by a seed growers association, which is a rapid and efficient method of handling seed multiplication and marketing of quality seeds.

Table 6.4 Standards[a] for the Certification of Rice Seed (Adapted from Huey 1977)

	Standard for Each Class		
Factor	Foundation	Registered	Certified Blue Tag
Pure seed (minimum)[b]	98%	98%	98%
Other varieties	none	none	2 seeds/lb (4.4 seeds/kg)
Other crop seeds (maximum)	none	none	2 seeds/lb (4.4 seeds/kg)
Field bindweed, hemp sesbania (coffeebean), northern jointvetch (curly indigo), nutsedge, crotalaria	none	none	none
Red rice (maximum)	none	none	1 seed/2 lb (1.1 seeds/kg)
Other noxious weeds (maximum)	none	none	1 seed/2 lb (1.1 seeds/kg)
Total weed seed (maximum)	0.03%	0.03%	0.08%
Inert material (maximum)	2%	2%	2%
Germination (minimum)	80%	80%	80%
Moisture (maximum)	14%	14%	14%

[a]Official Standards for Seed Certification in Arkansas, Arkansas State Plant Board.
[b]Percent of sample by weight.

Seed Quality and Treatment

Good quality rice seed is the seed that has fully matured, has been properly dried and stored, is free of weed seeds, and has a high percentage of viable seed. Rice seed should be treated with a fungicide for control of seedling diseases. Table 6.4 shows an example of standards for certified seed that is regulated by the Arkansas State Plant Board in the United States.

A main factor is that seed growers must receive adequate remuneration for their efforts, while the interests of farmers are also protected. Seed growers, extension workers, and farmers must understand that pure seeds rarely deteriorate per se, although new pest problems can reduce or shorten the usefulness of an improved variety (Chang 1979b).

Details on the seed production technology and seed science are summarized by Feistritzer (1975) and Copeland (1976).

REFERENCES

Abifarin, A. O., R. Chabrolin, M. Jacquot, R. Marie, and J. C. Moomaw. 1972. Upland rice improvement in West Africa. Pages 625–635 in International Rice Research Institute. *Rice breeding.* Los Baños, Philippines.

Adair, C. R., C. N. Bollich, D. H. Bowman, N. E. Jodon, T. H. Johnston, B. D. Webb, and J. G. Atkins. 1973. Rice breeding and testing methods in the United States. Pages 22–75 in USDA Handb. 289. *Rice in the United States: varieties and production.* Washington, D.C.

Athwal, D. S., and S. S. Virmani. 1972. Cytoplasmic male sterility and hybrid breeding in rice. Pages 615–620 in International Rice Research Institute. *Rice breeding.* Los Baños, Philippines.

Awakul, S. 1972. Progress in rice breeding in Thailand. Pages 167–170 in International Rice Research Institute. *Rice breeding.* Los Baños, Philippines.

Baba, I. 1954. Breeding of rice variety suitable for heavy manuring. Absorption and assimilation of nutrients and its relation to adaptability for heavy manuring and yield in rice variety. Pages 167–184 in Reports for the fifth meeting of the IRC working party on rice breeding, Japan.

Bajaj, Y. P. S. 1980 Implication and prospects of protoplast, cell, and tissue culture in rice improvement program. Pages 103–134 in International Rice Research Institute. *Innovative approaches to rice breeding:* Selected Papers from the 1979 International Rice Research Conference. Los Baños, Philippines.

Beachell, H. M., and P. R. Jennings. 1965. Need for modification of plant type. Pages 29–35 in International Rice Research Institute. *The mineral nutrition of the rice plant.* Proceedings of a symposium at the International Rice Research Institute, February, 1964. The Johns Hopkins Press, Baltimore, Maryland.

Beachell, H. M., and G. S. Khush. 1969. Objectives of the IRRI rice breeding program. *SABRAO Newsl.* 1:69–80.

Beachell, H. M., G. S. Khush, and R. C. Aquino. 1972. IRRI's international breeding program. Pages 89–106 in International Rice Research Institute. *Rice breeding.* Los Baños, Philippines.

Bollich, C. N., and J. E. Scott. 1975. Rice breeding developments in Texas. *Rice J.* 78(7):68–69.

Brady, N. C. 1975. Rice responds to science. Pages 62–96 *in* A. W. A. Brown, T. C. Byerly, M. Gibbs, A. San Pietro, eds. *Crop productivity—research imperatives.* Michigan Agricultural Experiment Station and Charles F. Kettering Foundation, USA.

Buddenhagen, I. 1978. Rice ecosystems in Africa. Pages 11–27 *in* Academic Press. *Rice in Africa.* London.

Cada, E. C., and P. B. Escuro. 1972. Rice varietal improvement in the Philippines. Pages 161–166 *in* International Rice Research Institute. *Rice breeding.* Los Baños, Philippines.

Carpenter, A. J. 1978. The history of rice in Africa. Pages 3–10 *in* Academic Press. *Rice in Africa.* London.

Chaleff, R. S. 1980. Tissue culture in rice improvement: An overview. Pages 81-91 *in* International Rice Research Institute. *Innovative approaches to rice breeding*: Selected Papers from the 1979 International Rice Research Conference. Los Baños, Philippines.

Chandler, R. F., Jr. 1969. Improving the rice plant and its culture. *Nature* 221:1007–1010.

Chang, T. T. 1964. *Present knowledge of rice genetics and cytogenetics.* IRRI Tech. Bull. 1. 96 pp.

Chang, T. T. 1967. The genetic basis of wide adaptability and yielding ability of rice varieties in the tropics. *Int. Rice Comm. Newsl.* 16(4):4–12.

Chang, T. T. 1972. International cooperation in conserving and evaluating rice germ plasm resources. Pages 177–185 *in* International Rice Research Institute. *Rice breeding.* Los Baños, Philippines.

Chang, T. T. 1976. The origin, evolution, cultivation, dissemination and diversification of Asian and African rices. *Euphytica* 25:425–441.

Chang, T. T. 1979a. Hybrid rice. Pages 173–174 *in* J. Sneep and A. J. Th. Hendriksen, eds. *Plant breeding perspectives.* PUDOC, Wageningen, The Netherlands.

Chang, T. T. 1979b. The Green Revolution's second decade. *SPAN* 22 (1):2–3.

Chang, T. T. 1980. The rice genetic resources program of IRRI and its impact on rice improvement. Pages 85–105 *in* International Rice Research Institute and Chinese Academy of Agricultural Sciences. *Rice improvement in China and other Asian countries.* Los Baños, Philippines.

Chang, T. T., and B. Somrith, 1979. Genetic studies on the grain quality of rice. Pages 49–58 *in* International Rice Research Institute. *Proceedings of the workshop on chemical aspects of rice grain quality.* Los Baños, Philippines.

Chang, T. T., G. C. Loresto, and O. Tagumpay. 1972. Agronomic and growth characteristics of upland and lowland rice varieties. Pages 645–661 *in* International Rice Research Institute. *Rice breeding.* Los Baños, Philippines.

Chang, T. T., G. C. Loresto, and O. Tagumpay. 1974. Screening rice germ plasm for drought resistance. *SABRAO J.* 6:9–16.

Choi, H. O. 1978. Recent advancement in rice breeding in Korea. *Korean J. Breed.* 10:201–238.

Copeland, L. O. 1976. *Principles of seed science and technology.* Burgess Publishing Co., Minneapolis, Minnesota. 369 pp.

Davis, M. D., and J. N. Rutger. 1976. Yield of F_1, F_2, and F_3 hybrids of rice (*Oryza sativa* L.). *Euphytica* 25:587–595.

De Datta, S. K., and J. C. O'Toole. 1977. Screening deep-water rices for drought tolerance. Pages 83–92 *in* International Rice Research Institute. *Proceedings, 1976 deep-water workshop, 8–10 November, Bangkok, Thailand*. Los Baños, Philippines.

De Datta, S. K., J. C. Moomaw, and R. S. Dayrit. 1966. Nitrogen response and yield potential of some varietal types in the tropics. *Int. Rice Comm. Newsl*. 15(3):16–27.

De Datta, S. K., W. N. Obcemea, and R. K. Jana. 1972. Protein content of rice grain as affected by nitrogen fertilizer and some triazines and substituted ureas. *Agron. J*. 64:785–788.

De Datta, S. K., A. C. Tauro, and S. N. Balaoing. 1968. Effect of plant type and nitrogen level on the growth characteristics and grain yield of indica rice in the tropics. *Agron. J*. 60:643–647.

Feistritzer, W. P. 1975. *Cereal seed technology. A manual of cereal seed production, quality control, and distribution*. Food and Agriculture Organization of the United Nations. Rome. 238 pp.

Foster, K. W., and J. N. Rutger. 1978a. Independent segregation of semidwarfing genes and a gene for pubescence in rice. *J. Hered*. 69:137–138.

Foster, K. W., and J. N. Rutger. 1978b. Inheritance of semidwarfism in rice. *Oryza sativa* L. *Genetics* 88:559–574.

Ghosh, R. L. M., M. B. Ghatge, and V. Subramanyan. 1960. *Rice in India*. Indian Council of Agricultural Research, New Delhi. 474 pp.

Gomez, K. A., and S. K. De Datta. 1975. Influence of environment on protein content of rice. *Agron. J*. 67:565–568.

Goulden, C. H. 1939. Problems in plant selection. Pages 132–133 *in* Cambridge Press. *Proceedings of the seventh international genetic congress*. London.

Grant, J. W. 1932. The rice crop in Burma. Its history, cultivation, marketing and improvement. *Agric. Dep. Burma Agric. Surv*. 17. 48 pp.

Huang, C. H. 1956. The present status and trend of rice varietal improvement in Taiwan [in Chinese]. *Sci. Agric. (Taiwan)* 4:216–227.

Huang, C. H., W. L. Chang, and T. T. Chang. 1972. Ponlai varieties and Taichung Native 1. Pages 31–46 *in* International Rice Research Institute. *Rice breeding*. Los Baños, Philippines.

Huey, B. A. 1977. *Rice production in Arkansas*. University of Arkansas, Division of Agriculture, and USDA Cooperating Circ. 476 (Rev.). 51 pp.

Huey, B. A. 1979. *Rice varieties in Arkansas*. University of Arkansas, Division of Agriculture, and USDA Leaflet 518.

IAEA (International Atomic Energy Agency). 1971. *Rice breeding with induced mutations*. III. Vienna. 198 pp.

Institute of Developing Economies. 1969. *One hundred years of agricultural statistics in Japan*. Tokyo, 1969. 270 pp.

IRRI (International Rice Research Institute). [1965]. *Annual report 1964*. Los Baños, Philippines. 335 pp.

IRRI (International Rice Research Institute). 1974. *Research highlights for 1973*. Los Baños, Philippines. 62 pp.

IRRI (International Rice Research Institute). 1977. *Proceedings, 1976 deep-water rice workshop, 8-10 November, Bangkok, Thailand.* Los Baños, Philippines. 239 pp.

IRRI (International Rice Research Institute). 1978a. *Rice research and production in China: an IRRI team's view.* Los Baños, Philippines. 119 pp.

IRRI (International Rice Research Institute). 1978b. *Proceedings of the workshop on the genetic conservation of rice.* Los Baños, Philippines. 54 pp.

Ito, H., and K. Hayashi. 1969. The changes in paddy field rice varieties in Japan. Pages 13-23 *in* Symposium on optimization of fertilizer effect in rice cultivation. *Trop. Agric. Res. Ser. 3.*

Jacquot, M. 1978. Varietal improvement programme for pluvial rice in francophone Africa. Pages 117-129 *in* Academic Press. *Rice in Africa.* London.

Jennings, P. R., W. R. Coffman, and H. E. Kauffman. 1979. *Rice improvement.* International Rice Research Institute. Los Baños, Philippines. 186 pp.

Jensen, N. F. 1970. A diallel selective mating system for cereal breeding. *Crop Sci.* 10:629-635.

Johnston, T. H., N. E. Jodon, C. N. Bollich, and J. N. Rutger. 1972. The development of early maturing and nitrogen-responsive rice varieties in the United States. Pages 61-76 *in International Rice Research Institute. Rice breeding.* Los Baños, Philippines.

Johnston, T. H., B. R. Wells, W. E. Hunter, and S. E. Hebry. 1976. *Performance of rice varieties in Arkansas, 1971 to 1975.* Agricultural Experiment Station, University of Arkansas, and United States Department of Agriculture Mimeo. Ser. 241. 12 pp.

Jones, J. W. 1936. Improvement in rice. Pages 415-454 *in* U.S. Government Printing Office. *United States Department of Agriculture Yearbook for 1936.* Washington, D.C.

Juliano, B. O. 1973. Recent developments in rice grain research. Pages 57-63 *in* Reports 7th working and discussion meetings, International Association of Cereal Chemists, Vienna, 1972.

Juliano, B. O. 1979. The chemical basis of rice grain quality. Pages 69-90 *in* International Rice Research Institute. *Proceedings of the workshop on chemical aspects of rice grain quality.* Los Baños, Philippines.

Kariya, K. 1966. Rice varieties, present and future. Pages 84-93 *in* Association of Agricultural Relations in Asia. *Agriculture, Asia* (English Ed.). *Development of paddy rice culture techniques in Japan.* Tokyo.

Katsuo, K., and U. Mizushima. 1958. Studies on the cytoplasmic differences among rice varieties, *Oryza sativa* L. I. On the fertility of hybrids obtained reciprocally between cultivated and wild varieties [in Japanese]. *Jpn. J. Breed.* 8:1-5.

Khush, G. S. 1977. Disease and insect resistance in rice. *Adv. Agron.* 29:265-341.

Khush, G. S. 1978. Breeding methods and procedures employed at IRRI for developing rice germ plasm with multiple resistance to diseases and insects. Pages 69-76 *in* Symposium on methods of crop breeding. *Trop. Agric. Res. Ser. 11.*

Khush, G. S., and H. M. Beachell. 1972. Breeding for disease and insect resistance at IRRI. Pages 309-322 *in* International Rice Research Institute. *Rice breeding.* Los Baños, Philippines.

Khush, G. S., and W. R. Coffman. 1977. Genetic evaluation and utilization (GEU) program. The rice improvement program of the International Rice Research Institute. *Theor. Appl. Genet.* 51:97-110.

Kongseree, N. 1979. Physicochemical properties of Thai rice varieties and methodology used in quality improvement. Pages 183–190 *in* International Rice Research Institute. *Proceedings of the workshop on chemical aspects of rice grain quality.* Los Baños, Philippines.

Lieuw Kie Song, P. A., M. J. Ido, and C. W. van Den Bogaert. 1977. Surinam releases 'Diwani'—21st rice variety. *Rice J.* 80(7):24–25.

Lin, S. C., and L. P. Yuan. 1980. Hybrid rice breeding in China. Pages 35–51 *in* International Rice Research Institute. *Innovative approaches to rice breeding:* Selected Papers from the 1979 International Rice Research Conference. Los Baños, Philippines.

Ling, K. C., V. M. Aguiero, and S. H. Lee. 1970. A mass screening method for testing resistance to grassy stunt disease of rice. *Plant Dis. Rep.* 54:565–569.

Lu, J. J., and T. T. Chang. 1980. Rice in its temporal and spatial perspectives. Pages 1–74 *in* AVI Publishing Co. *Rice: production and utilization.* Westport, Connecticut.

McDonald, D. J. 1978. Rice and its adaptation to world environments. Farret Memorial Oration, 1977. *J. Aust. Inst. Agric. Sci.* 44:3–20.

McIlrath, W. O., N. E. Jodon, and G. J. Trahan. 1975. Rice variety improvement program. *Rice J.* 78(7):50.

McIlrath, W. O., G. J. Trahan, N. E. Jodon, and C. J. Leger. 1978. Rice variety improvement research program, variety tests (a preliminary report). Pages 6–21 *in* Rice Experiment Station, Louisiana State University, and United States Department of Agriculture 70th annual progress report.

Mikaelsen, K. 1980. Mutation breeding in rice. Pages 67–79 *in* International Rice Research Institute. *Innovative approaches to rice breeding:* Selected papers from the 1979 International Rice Research Conference. Los Baños, Philippines.

Nagai, I. 1958. *Japonica rice*—its breeding and culture. Yokendo Ltd., Tokyo. 843 pp.

Nanda, J. S., and W. R. Coffman. 1979. IRRI's efforts to improve the protein content of rice. Pages 33–47 *in* International Rice Research Institute. *Proceedings of the workshop on chemical aspects of rice grain quality.* Los Baños, Philippines.

Oka, H. I. 1964. Pattern of interspecific relationships and evolutionary dynamics in *Oryza*. Pages 71–90 *in* American-Elsevier. *Rice genetics and cytogenetics.* New York.

Okabe, S. 1972. Breeding for high-yielding varieties in Japan. Pages 47–59 *in* International Rice Research Institute. *Rice breeding.* Los Baños, Philippines.

Okochi, H., T. Ota, S. Iizuka, and K. Sugiyama. 1958. A method to decrease the field area for growing bulk-populations of rice [in Japanese]. Pages 239–246 *in* K. Sakai, R. Takahashi and H. Akemine, eds. *Studies on the bulk method of plant breeding.*

ORD (Office of Rural Development). 1979. *An introduction to the rural development program.* Suweon, Korea. 66 pp.

Ou, S. H. 1972. *Rice diseases.* Commonwealth Mycological Institute, Kew, Surrey, England. 368 pp.

Parthasarathy, N. 1972. Rice breeding in tropical Asia up to 1960. Pages 5–29 *in* International Rice Research Institute. *Rice breeding.* Los Baños, Philippines.

Pathak, M. D. 1972. Resistance to insect pests in rice varieties. Pages 325–341 *in* International Rice Research Institute. *Rice breeding.* Los Baños, Philippines.

Porteres, R. 1956. Taxonomie agrobotanique des riz cultives *O. sativa* Linné. et. *O. glaberrima* Steudel. I–V. *J. Agric. Trop. Bot. Appl.* 3:341–384, 541–580, 627–700, 821–856.

Ramiah, K., and N. Parthasarathy. 1938. X-ray mutations in rice. Pages 212–213 *in* *Proceedings of the 25th Indian Science Congress, Calcutta.* III. Sect. Agric., Abstr. 12.

Roschevicz, R. J. 1931. A contribution to the knowledge of rice [in Russian, English summary]. *Bull. Appl. Bot. Genet. Plant Breed. (Leningrad)* 27(4):3–133.

Rosero M., M. J. 1972. Rice breeding in Colombia. Pages 107–114 *in* International Rice Research Institute. *Rice breeding.* Los Baños, Philippines.

Roxburgh, W. 1832. *Flora indica; or descriptions of Indian plants.* Vol. 3. Serampore, Printed for W. Thacker.

Rutger, J. N., and M. L. Peterson. 1976. Improved short stature rice. *Calif. Agric.* 30(6):4–6.

Rutger, J. N., M. L. Peterson, H. L. Carnahan, and D. M. Brandon. 1979. Registration of 'M-101' rice. *Crop Sci.* 19:929.

Sampath, S. 1962. The genus *Oryza:* its taxonomy and species interrelationships. *Oryza* 1:1–29.

Seetharaman, R., S. D. Sharma, and S. V. S. Shastry. 1972. Germ plasm conservation and use in India. Pages 187–200 *in* International Rice Research Institute. *Rice breeding.* Los Baños, Philippines.

Sharma, S. D., and S. V. S. Shastry. 1965. Taxonomic studies in genus *Oryza* L. III. *O. rufipogon* Griff, *Sensu stricto* and *O. nivara* Sharma et Shastry nom. nov. *Indian J. Genet. Plant Breed.* 25:157–167.

Sharma, S. D., and W. M. Steele. 1978. Collection and conservation of existing rice species and varieties of Africa. Pages 61–67 *in* Academic Press. *Rice in Africa.* London.

Shastry, S. V. S. 1966. Rice needs of India and ways of meeting it. *Indian Farming* 16(6):19–21.

Shen, J. H. 1980. Rice breeding in China. Pages 9–30 *in* International Rice Research Institute and Chinese Academy of Agricultural Sciences. *Rice improvement in China and other Asian countries.* Los Baños, Philippines.

Shigemura, C. 1966. Contribution of newly developed varieties to the increased production of rice in the warm districts of Japan … a case of Hoyoku, etc. *JARQ* 1:1–7.

Shinjyo, C. 1969. Cytoplasmic-genetic male sterility in cultivated rice *Oryza sativa* L. II. The inheritance of male sterility. *Jpn. J. Genet.* 44:149–156.

Suprihatno, B. 1976. Effectiveness of protein per seed as a selection criterion for high protein rice. MS thesis, University of the Philippines at Los Baños, Philippines. 77 pp.

Toriyama, K. 1979. National program of rice breeding in Japan. *JARQ* 13:1–8.

Tsunoda, S. 1962. The growth and production system for the maximum yield. *Recent Adv. Breed.* 3:89–93.

University of California and U.S. Department of Agriculture. 1978. Annual report comprehensive rice research. University of California, Davis. 91 pp. (unpubl. mimeo.)

Virmani, S. S., J. O. Olufowote, and A. O. Abifarin. 1978. Rice improvement in tropical anglophone Africa. Pages 101–116 *in* Academic Press. *Rice in Africa*. London.

Zaman, S. M. H. 1980. The importance of the IRTP in BRRI's varietal improvement work. Pages 251–259 *in* International Rice Research Institute and Chinese Academy of Agricultural Sciences. *Rice improvement in China and other Asian countries*. Los Baños, Philippines.

7

Systems of Rice Culture

The systems of growing rice and the soil and crop management practices that have evolved for each system are complex and somewhat unique. The systems were developed to suit specific environments and socioeconomic conditions of the farmers, which makes a single classification system for rice lands—one using all the production factors—extremely difficult. Instead, several classification systems based on each area's or country's rice production systems are followed.

CLASSIFICATION OF RICE CULTURE

Rice culture is classified according to source of water supply as rainfed or irrigated. Based on land and water management practices, rice lands are classified as:

- Lowland (wetland preparation of fields).
- Upland (dryland preparation of fields).

Then, according to water regime, rice lands have been classified as:

- Upland, with no standing water.
- Lowland, with 5–50 cm of standing water.
- Deepwater, with >51 cm to 5–6 meters of standing water.

Figure 7.1 diagrams the world's rice land classification by water regime and predominant rice types.
  Based on varietal type used, rice culture can be further classified into:

- Lowland rice, with plants of semidwarf to medium to tall (100 cm to 2 m) height.
- Upland rice, with plants of medium to tall (130–150 cm) height.
- Deepwater rice, with plants of medium to tall (120–150 cm tall without standing water; 2–3 m with rising water level) height.

221

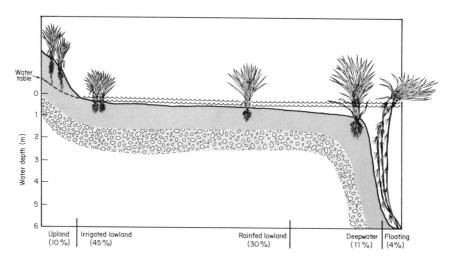

Figure 7.1 The world's rice lands classified by water regime and predominant rice types. Figures in parentheses are the percent of the world's rice-growing area that grows that type of rice culture.

- Floating rice, with tall (>150 cm tall without standing water; 5–6 meters tall with rising floodwater) plants.

Precise data are lacking on the extent of different rice cultures. In many countries, the rainfed and irrigated areas cannot be precisely differentiated because statistics on irrigated areas in tropical Asia include areas where only supplemental water is supplied during the monsoonal season. Within the cultures there exists a continuum of water conditions from very good water control to very poor water control. For South and Southeast Asia, Barker and Herdt (1979) classified rice lands into irrigated, shallow rainfed, deepwater, and upland rice-growing areas and estimated the production for each (Fig. 7.2).

Lowland and deepwater rice-growing areas can be more precisely classified to suit water regime, varietal requirement, and land preparation and stand establishment techniques.

In many rice-growing areas of the tropics, particularly in South and Southeast Asia, the year is divided into fairly distinct wet and dry seasons. In most areas, the bulk of the rice is produced in the wet season and dependence on rainfall is the most limiting production constraint for rainfed rice culture. Because rainfall received during the dry season is not enough to grow a crop of rice, rice grown in that season is entirely an irrigated crop and, because development of irrigation facilities is costly and moves slowly, the area planted to rice in the dry season is limited.

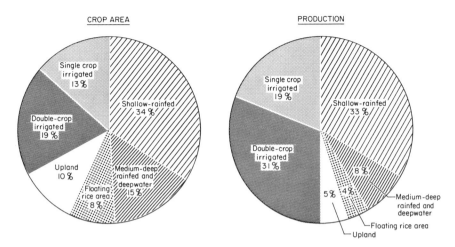

Figure 7.2   Estimate of the percentage of rice crop area and production by specified environmental complex in South and Southeast Asia, mid-1970s. (Adapted from Barker and Herdt 1979)

The principal rice-growing seasons of various rice-growing countries are shown in Fig. 7.3. Examples of growing seasons, and rice-based cropping systems in China are shown in Fig. 7.4.

In all of the rice-growing countries in Asia, lowland rice culture is the most predominant system followed. On the other hand, in Africa and Latin America, upland rice culture is the major system. For West Africa, Buddenhagen (1978) classified rice culture into four main types and eight subtypes based on climate, soil conditions, water regime, and technological level of agriculture. They are:

1 Upland
   · Dryland
   · Hydromorphic
2 Irrigated
3 Inland swamp
   · Nontoxic soil
   · Toxic soil
4 Flooded
   · Riverine shallow
   · Riverine deep
   · Boliland
   · Mangrove

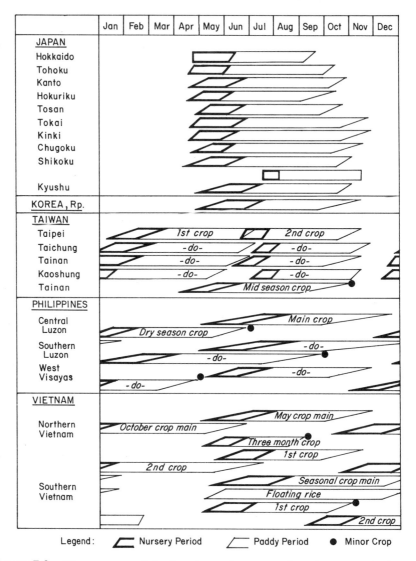

Figure 7.3   Rice seasons and rice-based cropping systems in various countries in Asia. (Adapted from Kung n.d. and Tanaka 1976)

West Africa's 1.87 million hectares of rice lands, however, equal only a little more than 50% of the rice-growing area in the Philippines and only 1.4% of the world's rice-growing area. In all of Latin America, rice is grown on 6.5 million hectares, which is less than the rice-growing area of Bangladesh or Indonesia.

The statistics on rice-growing areas make it clear that Asia is the most important rice-growing region in the world. Rainfed rice covers most rice-growing areas in the world although yield per crop is high in irrigated areas.

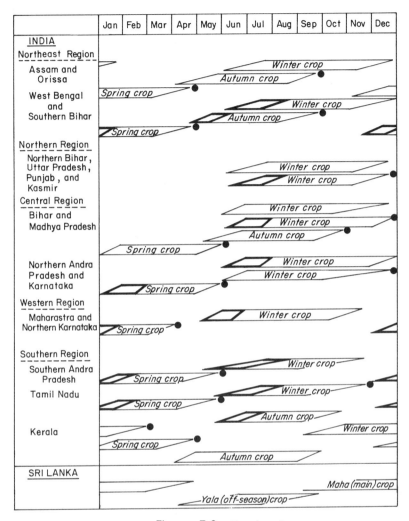

Figure 7.3   Continued.

Classification of Rainfed Rice Culture

It is difficult to classify rainfed rice lands accurately according to water depths. It is equally difficult to estimate the area under each water depth and system of rice culture.

Recently, Barker and Herdt (1979) classified rice lands of South and Southeast Asia according to water depths. For lowland rainfed rice, they used shallow rainfed (5–15 cm) and medium-deep rainfed (16–100 cm), with medium-deep rainfed further classified into intermediate-deep (15–50 cm) and semideep

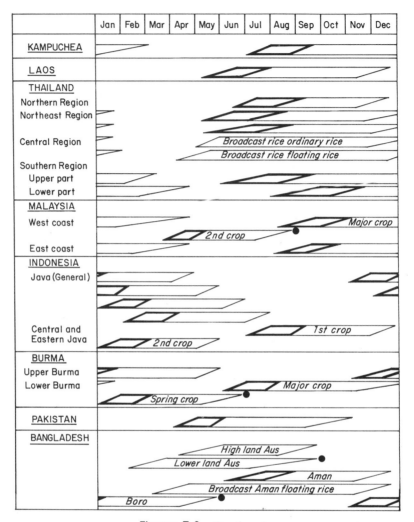

	Jan	Feb	Mar	Apr	May	Jun	Jul	Aug	Sep	Oct	Nov	Dec
KAMPUCHEA												
LAOS												
THAILAND Northern Region Northeast Region												
Central Region												
Southern Region Upper part Lower part												
MALAYSIA West coast												
East coast												
INDONESIA Java (General)												
Central and Eastern Java												
BURMA Upper Burma Lower Burma												
PAKISTAN												
BANGLADESH												

Figure 7.3   Continued

rainfed (51–100 cm). Where water depth exceeds 100 cm, they classified rice lands as deepwater.

There are distinct disadvantages to grouping intermediate-deep and semideep rice areas into medium-deep lowland areas. The rices and technology required for intermediate-deep and semideep lowland areas are distinctly different. In addition, in contrast to the sharp distinction between upland and lowland rice, the transition from lowland to deepwater is less abrupt (Krishnamoorty 1979) as is the transition between deepwater and floating rice areas. From an agronomic management viewpoint it is desirable to separate rainfed lowland rice culture,

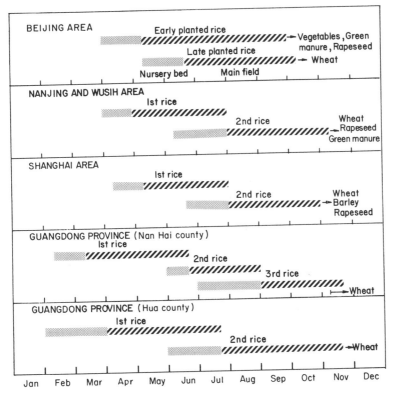

Figure 7.4   Rice-based cropping systems in rice-growing areas of the People's Republic of China. (Adapted from report on IRRI Team's visit to China 1976, and IRRI 1978)

which is primarily transplanted in a puddled soil, from deepwater and floating rice cultures, which are direct-seeded onto dry soil.

Further refinements can be made of the classification system suggested by Barker and Herdt (1979), with both deepwater and floating rice areas totally excluded from the rainfed lowland rice areas. The refinements of classification are:

Types of Rice Culture	Main Method of Planting	Maximum Water Depth (cm)
Rainfed lowland rice	Transplant	0–50
Shallow rainfed lowland rice		5–15
Medium-deep rainfed lowland rice		16–50

Types of Rice Culture	Main Method of Planting	Maximum Water Depth (cm)
Deepwater rice	Broadcast onto dry soil	51–100
Floating rice	Broadcast onto dry soil	101–600
Upland rice	Broadcast or drilled into dry soil	No standing water

With this classification, the rainfed lowland rice areas include those where maximum water depth is 50 cm. In such areas, fields are generally prepared wet and rice is transplanted in puddled soil, although direct seeding onto puddled or in dry soil is practiced in some countries.

By separating deepwater and floating rice cultures it is possible to concentrate on research specific to deepwater rice culture.

Production methodology and cultural practices followed in each system of rice culture vary greatly. In this chapter, descriptions are given for various systems of rice culture such as rainfed lowland, irrigated lowland, deepwater, floating rice, and upland rice. Under each system of rice culture, varietal requirement and cultural practices not discussed elsewhere in this book are briefly mentioned. Details on other production factors such as land preparation, water management, fertilizer management, weed control, insects and diseases and their control measures, and harvesting and postharvest operations are discussed in other chapters.

LOWLAND RICE CULTURE

In most rice-growing countries, rice is usually grown as a lowland (wetland) crop. Under this system, land is either prepared wet or dry but water is always held on the field by bunds. About 30% of the world's rice is grown as rainfed lowland and about 45% as irrigated lowland (Fig. 7.1).

Crop Establishment

For both rainfed lowland and irrigated lowland rice, the raising and handling of seedlings and plant density are generally similar.

Raising Seedlings

For lowland rice in the tropics, seedlings are grown at the time there is an adequate moisture supply. In the case of irrigated rice, seedlings can be grown any time of the year as needed. For rainfed rice, the starting of seedbeds usually conforms with the onset of monsoonal rains.

Seeds of many tropical rice varieties have a dormancy period. A dormancy period of 2–3 weeks in rice seeds is an important advantage in the tropics, especially for the wet-season crop, when high temperature and humidity at harvest would result in germination on the panicle if there is no dormancy. For seedbed seeding, farmers have to wait until the dormancy period is over or break it by heat (50° C for 5 days). Waiting until the end of the domancy period is the usual practice.

Rice seeds are soaked in water for 24 hours and then incubated for 48 hours before they are placed on the seedbed. This pregermination process, assures a quick and even start of the seedbed.

The three major methods of raising seedlings are the wet-bed, the *dapog,* and the dry-bed. The wet-bed is most widely used.

- WET-BED In the wet-bed method, pregerminated seeds are broadcast uniformly on a raised bed of puddled soil and the seedlings are ready for transplanting 20–25 days after the seed is sown. Rate of seeding is about 100 g/m^2 of seedbed, which amounts to a rate of about 50 kg/ha.

- DAPOG The *dapog* method of raising seedlings was developed in Laguna province, Philippines, and is commonly used in that region. Its use spread to other parts of the Philippines and to other Asian countries. For a *dapog* seedbed, the land is prepared as in the wet-bed method, but banana leaves or plastic sheets are used to cover the soil (or a concreted area is used as a bed). The *dapog* bed should be about 1.5 m wide, and its length will depend on the area to be planted. Banana bracts are placed on the seedbed with bamboo pegs to keep the leaves in position.

  The *dapog* method seeding rate is 3000 g/m^2 of bed, which gives a rate of 100 kg/ha, about twice that of the wet-bed method. *Dapog* seedlings are ready for transplanting 9–14 days after the seed is sown. The advantages of this method are labor savings because the bed is easily made, the short period for raising seedlings, and relative ease of transport of seedlings because a mat of seedlings can be rolled; seedlings raised in wet or dry beds have to be pulled, bundled, and tied for transport.

- DRY-BED The dry-bed method is not used extensively in the Asian tropics, although it is popular in the Philippines on the clayey soils in the rainfed lowland areas where there is insufficient water to irrigate seedbeds. Dry-bed seedlings are grown in a manner similar to that used for wet-bed seedlings, except that the soil is not puddled and drainage is provided to keep the soil moist but not inundated. In the Philippines, farmers construct a small paddy near a source of water and water the seedlings two or three times daily. The seeding rate is the same as for the wet-bed method (100 g/m^2).

In any of these three methods of raising seedlings, the seedbed should be protected from insects and diseases. Because the area to protect is small, the cost of protecting seedlings is not prohibitive.

The choice of seedbed method depends on the availability of water. Studies in the Philippines suggest similar growth characteristics and grain yield with seedlings raised by the three different methods. At maturity, the plants from all seedbeds are taller in the wet than in the dry season. Taller seedlings are desirable in the wet season because of sudden floods, which may cause damage to the rice crop during the early stage.

Age and Handling of Seedlings

An even stand containing the optimum number of optimum-age seedlings is essential for proper crop development and high grain yields.

When photoperiod-sensitive varieties are grown, it is fairly common to transplant seedlings that are 40–50 days old. However, the best age for transplanting wet-bed seedlings is 20–30 days. In general, the earlier the variety matures, the sooner the seedlings should be transplanted. Thirty-day or older seedlings recover more slowly than younger seedlings, especially if they suffer from too much stem or root injury during pulling. Such injuries reduce tillering, prolong maturity, and may reduce grain yield.

It is desirable to flood the wet and dry seedbeds one day before pulling the seedlings. Flooding softens the soil and facilitates seedling pulling. For *dapog* seedlings, the seedbed is cut into convenient sizes and the seedling mats are rolled with the roots outward. *Dapog* seedlings are ready for transplanting 9–14 days after sowing regardless of maturity days of the variety used.

Handling seedlings with care is more important for modern varieties than for traditional varieties. Irrespective of the method of raising seedlings it is important to handle the seedlings carefully to enable the transplanted seedlings to revive and resume growth in the main field as quickly as possible. This is particularly important in the rainfed rice, where early revival and rapid growth immediately after transplanting may mean avoiding damage from submergence by floods or heavy rain soon after transplanting, or from later dry weather.

Number of Seedlings Per Hill

The number of seedlings planted per hectare depends on the method of raising seedlings and plant spacing. Ordinarily, 3–4 seedlings per hill are planted from wet-bed and 6–8 per hill from *dapog* seedbeds.

It is desirable to select rice varieties with high tillering capacity to ensure adequate number of tillers per hill. In China, where rice is almost entirely irrigated, 8–10 seedlings per hill are commonly used. This is because Chinese varieties do not have high tillering capacity. Extra seedlings are, however, an expensive method of ensuring high tiller number in a given area. The cost of extra seed may discourage most tropical Asian farmers to plant a high number of seedlings per hill because of added seed cost.

Plant Spacing

Plant spacing is an important production factor in transplanted rice. Planting rice closer than necessary increases the cost of transplanting and the chances of lodging. On the other hand, spacing rice wider than necessary may result in lower yield because the number of plants in the area may be less than the optimum number needed for high yield.

In many countries in South and Southeast Asia, rice is transplanted at a random spacing. Random planting is most common in rainfed rice, particularly if traditional varieties are grown. The advantages of straight-row planting are:

- A rotary weeder can be used for weeding.
- An optimum plant population is possible, and it is easier to apply insecticides, herbicides, and topdressed fertilizers.

There is no single spacing practice best for all varieties. Optimum spacing of any variety depends on soil fertility and season of planting. A uniform stand containing an optimum plant population is essential for proper crop development and high grain yields.

In tropical Asia, rices of improved plant type and high tillering capacity can be planted at a wide range of spacings. The tiller number per unit area in a rice population is largely a function of plant density. The tiller number is positively or negatively correlated with grain yield depending on the rice variety and crop environment (Kawano and Tanaka 1968).

The effect of plant density on the yield of rice for traditional varieties (Vacchani and Rao 1959, Yin et al. 1960, Matsuo 1964) and modern varieties (Fagade and De Datta 1971) has been studied. For example, the short, lodging-resistant, photoperiod-insensitive varieties such as IR8, IR36, IR42, and IR50 should be spaced 20 × 25 cm in the wet season regardless of soil fertility. In the dry season, the tall and leafy, heavy-tillering varieties such as Peta are spaced 25 × 25 cm in relatively poor soil and 30 × 30 cm in fertile soil. In the wet season, the tall varieties are spaced 30 × 30 cm in the poor soil and 35 × 35 cm in fertile soil. Studies at IRRI suggest that the optimum spacing for the lodging-susceptible tall variety Peta is 50 × 50 cm, whereas the optimum spacing for nonlodging short variety Tainan-3 is 30 × 30 cm in the wet season.

Data on varietal response to plant density demonstrate that if there is no lodging, the yields of most varieties do not change much as the planting distance is reduced below 25–35 cm (Table 7.1). At closer spacings, the yield per plant is small but this is compensated for by greater number of plants per unit area. At distances greater than 35 cm between plants, yields of most varieties are reduced because plant population per unit area is reduced.

Most modern early-maturing rices have a short vegetative period that limits the number of panicles formed. Thus, dense planting of these varieties can overcome the limitation of a short vegetative period. With poor weed control, closely spaced rice competes better with weeds. Because of sunlight intercepted by the rice crop, plant spacing and row orientation also affect the grain yield of rice (Matsuo 1964).

Nguu and De Datta (1979) reported the effect of plant density on grain yield of rice grown with various levels of soil nitrogen. The yield of two early- maturing

Table 7.1   Grain Yield of Modern and Traditional Varieties in Relation to Plant Spacing; IRRI, 1967 (Adapted from Mabbayad and Obordo 1970)

	Grain Yield (t/ha)		
	Modern Varieties		Traditional Tall Varieties, High Tillering
Spacing (cm)	Low Tillering	High Tillering	
15 × 15	6.2	7.6	4.1
25 × 25	5.7	8.2	4.6
35 × 35	5.2	7.4	4.4
45 × 45	4.2	6.9	4.3
55 × 55	3.5	6.5	3.6

rices increased by 1.3 t/ha when the plant density increased from 0.67×10^5 to 10^6 hills per hectare (Table 7.2).

Rate of fertilizer application affected the yield response to plant density. Without fertilizer, or with low levels (60 kg N/ha), the grain yield increased either linearly or curvilinearly with increased plant density. At a high level of applied nitrogen (120 kg N/ha), grain yield increased as the plant density was increased to a certain level, beyond which the yield decreased with increased plant density (Nguu and De Datta 1979). Their results also suggested that row orientation significantly affected the rice yield. Although these experiments were with a continuously irrigated crop, results should be applicable to rainfed rice culture.

Table 7.2   Grain Yield of IR36 and IR40 Rices Planted at Different Plant Densities; IRRI, 1977 Wet Season (Adapted from Nguu and De Datta 1979)

Plant Density ($\times 10^5$ hills/ha)	Plant Spacing (cm)	Grain Yield[a] (t/ha)		
		IR36	IR40	Mean
0.67	40 × 40	2.9	3.2	3.0d
1.1	30 × 30	2.9	3.1	3.0d
2.5	20 × 20	3.5	3.6	3.6c
2.5	40 × 10	3.6	3.7	3.6c
3.3	30 × 10	3.7	3.7	3.7c
5.0	20 × 10	4.1	4.1	4.1b
5.0	40 × 5	4.1	4.3	4.2ab
6.6	30 × 5	4.1	4.1	4.1b
10.0	20 × 5	4.5	4.2	4.3a

[a]Average of seven fertilizer nitrogen treatments. Any two means in one column followed by the same letter are not significantly different at the 5% level.

Table 7.3 Standard Practice of Raising and Characteristics of Young and Medium-sized Seedlings in Japan (Adapted from Hoshino 1978)

	Young Seedlings	Medium-sized Seedlings
Sowing density (g/box)	about 200	about 100
Boxes per hectare	150–200	250–450
Amount of soil required (liters/ha)		
Bedding soil	600–800	1000–1800
Covering soil	150–200	250–450
Raising period (days)	15–20	30–45
Standard shape		
Leaves (no.)	2.0–2.5 (2.0–2.5)[a]	3.5–4.5 (3.1–3.5)
Plant height (cm)	10–15 (8–15)	15–20 (13–20)
Dry weight without roots (mg/plant)	about 10	20–30

[a]Figures in parentheses show values for Hokkaido.

Mechanical Transplanting

Almost all mechanical transplanters used in Japan transplant seedlings raised in soil. Seedlings raised by this method are young with 2–2.5 leaves and are 8–15 cm in height. These young seedlings must be transplanted 7–10 days earlier than hand-transplanted seedlings. This causes seedling submergence if irrigation water control is not good. Recently, medium-sized seedlings have become popular. Table 7.3 compares seedling characteristics for young and medium-sized seedlings.

Mat-type seedlings are now most widely used for mechanical transplanting. These are raised in a nursery box that has no partition. The roots (or soil and roots) of the seedlings take the form of a mat. Seedlings raised by this method require less labor and are easy to transplant by machine. China uses mechanical transplanters and machines that uproot seedlings before they are transplanted to speed transplanting operations (Fig. 7.5).

RAINFED LOWLAND RICE CULTURE

In tropical Asia where rainfed lowland rice is almost synonymous to rice cultivation, enough water has to accumulate in the field to soften the soil before plowing can be done. Dikes are essential for rainfed culture because the undependable water supply must be captured and controlled. The usual method of stand establishment is transplanting, although direct seeding onto puddled or dry fields is also practiced.

Figure 7.5 Seedling uprooting machine being developed in a commune factory in People's Republic of China.

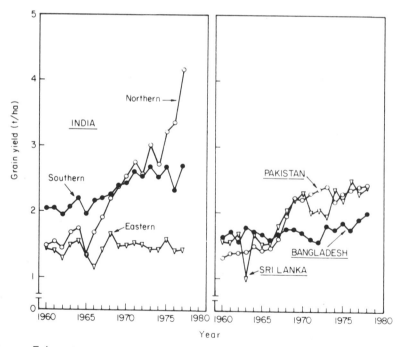

Figure 7.6 Rice yield trends in Bangladesh, Pakistan, Sri Lanka, and three regions of India, 1960–1977/78. (Adapted from Barker and Herdt 1979)

Modern rice technology has by and large bypassed many major rainfed lowland rice areas resulting generally in no yield increases on farms in those areas. For example, rice yields in eastern India (West Bengal, Bihar, Orissa, Assam, and eastern Uttar Pradesh) have not increased much since the mid-1960s when the modern rice varieties were introduced (Fig. 7.6). On the other hand where water control was good, such as in northern (Punjab) and southern India (Andhra Pradesh and Tamil Nadu), Pakistan, and Sri Lanka, rice yields increased steadily over those years, particularly after the introduction of modern varieties.

Stand Establishment Techniques

Rainfed lowland rice can be classified according to stand establishment techniques:

- Transplanted in puddled soil.
- Direct-seeded (with pregerminated seeds) on puddled soil.
- Direct-seeded on dry soil.

Transplanted Rice

Transplanted rice is the major system of rainfed rice culture in most of tropical Asia, primarily as a monsoonal crop. It has been the most important crop in the Indian subcontinent, and its success or failure means abundance or deficit (Islam 1980).

The transplanted rainfed crop is known as *kharif* (monsoon crop) in India and *aman* (an Arabic word for safety or stability) in northeastern India (West Bengal, Assam and Orissa States) and Bangladesh. In northeastern India, of about 15 million hectares of rice land, about 12.5 million hectares grow rainfed rice during the *aman* season (Mukherji 1980). In Bangladesh, of 10.4 million hectares of rice-growing area, about 5.8 million hectares grow *aman* rice (Ahmad 1980).

In Indonesia, of 8.8 million hectares of rice land, about 2.5 million hectares grow rainfed rice. The main rainfed areas are in Java, Sumatra, Kalimantan, and Sulawesi.

In Thailand, of 8.4 million hectares of rice land, 6.5 million hectares are considered rainfed, most of which is rainfed lowland. Most of the northeastern and the southern rice-growing areas grow transplanted rice during the monsoonal season.

VARIETAL REQUIREMENT   In rainfed lowland rice, water depths largely determine plant types of rice grown. In general, modern semidwarf varieties are grown in shallow rainfed lowland rice-growing areas. Early-maturing semidwarfs, such as IR36, are grown in many shallow rainfed areas in the Philippines. In most rainfed areas in South and Southeast Asia, however, taller varieties with or without photoperiod sensitivity are widely grown.

In medium-deep rainfed lowland rice-growing areas, tall varieties that are mostly photoperiod-sensitive are grown, although some tall varieties essentially photoperiod-insensitive are grown in some areas. For example, Mashuri (an indica × japonica cross), which was developed in Malaysia, is extremely popular in medium-deep rainfed areas in India and Burma (see Chapter 6).

Varieties with moderate elongating capacity, such as the IR442 type, are well suited in medium-deep rainfed areas. But these rices must have some photoperiod sensitivity to delay harvesting until the floodwater has receded. Submergence tolerance, particularly at the seedling and early tillering stages, is desirable in rices for medium-deep rainfed lowland areas.

In the region covering 15° -30° N latitudes and 70° -120° E longitudes of South and Southeast Asia, covering eastern India (Assam, Bihar, West Bengal States), Bangladesh, Nepal, parts of Burma, Laos, and Vietnam (Islam 1980), the photoperiod-sensitive varieties are grown. Their maturity is controlled by short days (less than 12 hours) and cool climate (less than 27° C) particularly during the reproductive and ripening stages of crop growth. However, photoperiod-sensitive varieties are also grown in lower latitudes of southern Asia including Malaysia and Indonesia. In northeastern India, about 87% of rice-growing area is grown to tall photoperiod-sensitive varieties or varieties with intermediate plant height (Mukherji 1980) such as Pankaj, Mashuri, and IR442 lines, which are essentially insensitive to photoperiod.

In Bangladesh, the photoperiod-sensitive variety Nazirsail is most widely grown. Photoperiod-sensitive varieties remain in the active and lag vegetative stages during the main part of the monsoon season. However, the transplanted *aman* crop is often subjected to flash floods and reduction in stand due to seedling submergence. Therefore, submergence tolerance is a desirable character (Ahmad 1980).

In Thailand, the entire Central Plains and Northeast and the southern rice-growing regions grow photoperiod-sensitive tall varieties. Figure 7.7 shows the maturity range of photoperiod-sensitive varieties in four rice-growing regions in Thailand.

Wet-Seeded Lowland Rice

In some rainfed areas in the Asian tropics, wet-seeded lowland rice culture is an important system. Wet seeding is usually by broadcasting in Sri Lanka, parts of India, Bangladesh, and the Philippines. Pregerminated seeds are broadcast onto puddled fields without much standing water. Fields are prepared wet with various degrees of puddling. Stand establishment is often poor because of poor land preparation, weed competition, and poor water control.

Farmers in Indonesia prepare the land with a hand hoe or spade. In Java, farmers may prepare the land wet or dry depending on the moisture accumulation in the field. When the amount of rain water is more than 200 mm for a

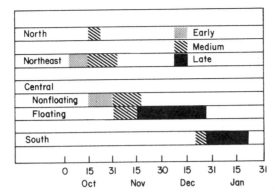

Figure 7.7   Flowering dates of photoperiod-sensitive varieties recommended by the Thai Government, according to region. (Adapted from Awakul 1980)

month, farmers generally opt for wetland preparation. Less than 200 mm of rainfall for the planting month leads to dryland preparation and seeding onto a dry or moist soil.

VARIETAL REQUIREMENT   Early-maturing rices (about 100-day duration) are preferred for wet-seeded rice. Varieties should have excellent seedling vigor and good tillering capacity. Early-maturing rices such as IR28, IR30, IR36, and Ratna (from India) have been used successfully as broadcast wet-seeded rice.

Experimental results suggest that the yield potential for direct-seeded (drilled or broadcast) rice is similar to that for transplanted lowland rice (Table 7.4). Where rainfall distribution is good and farmers can puddle the field quickly and thoroughly, it is often desirable to wet-seed rice on a puddled soil.

In some parts of the Philippines, farmers claim that they stand a good chance of double cropping with rice if the first rice crop is wet-seeded. This is due primarily to the earliness in planting for wet-seeded rice (May–June) compared to the delay for transplanted rice (July) (Roxas et al. 1978). During 1975–1979,

Table 7.4   Grain Yields of Rice as Affected by Different Planting Methods; IRRI, 1964–1966 (Adapted from Mabbayad and Obordo 1970)

Planting Method	Grain Yield (t/ha)				
	1964 Wet Season	1965 Dry Season	1965 Wet Season	1966 Dry Season	1966 Wet Season
Transplanted	2.6	5.8	5.5	5.3	4.5
Broadcast	3.1	5.6	5.1	5.3	4.8
Drilled	3.1	5.9	5.1	5.3	4.5

there was a dramatic shift to direct seeding from traditional transplanting in Iloilo province, Philippines.

Dry-Seeded Lowland Rice

Establishment of dry-seeded rice in lowland fields must be in accord with the local rainfall pattern. Field observations suggest that dry seeding after rains commence results in poor seedling emergence and low yields. Dry seeding ahead of the rainy season requires consideration of rainfall pattern and the effects of early rainfall on crop performance. For high yields, once rainfall begins it must be adequate to promote fast seed emergence and rapid early vegetative growth.

Despite the above difficulties in stand establishment, dry seeding into dry soil provides a unique opportunity to increase cropping intensity by double cropping. For many years farmers in northeastern India, Bangladesh, and Indonesia have grown rainfed lowland rice by dry seeding directly onto nonpuddled soil at the beginning of the rainy season.

This method of rice culture is known as *aus* cropping in northeastern India and Bangladesh. The *aus* (meaning early) rice is direct-seeded in March–April and harvested in July–August.

Gogorantja (also known as *gogorancah*) is a system used in certain areas in Indonesia that practice rainfed rice culture. As early as 1928, the *gogorantja* method had been used successfully in Indonesia in a multiple cropping system (Dalrymple 1971). Under this system, land is prepared during the dry season and planted at the beginning of the rainy season in October–November. In this system, rice grows as a dryland crop for the first few weeks and becomes a wetland crop as soon as rainfall is adequate to flood the field. In several regions, farmers grow a transplanted rice crop after the *gogorantja* rice.

In flood-prone areas on the north coast of Java, prolonged heavy rains cause frequent floods. By using the *gogorantja* method, flood damage is minimal because with direct dry seeding rice plants are 80–90 days old and 80–90 cm tall when floods occur.

There are disadvantages with the *gogorantja* system. Land preparation is difficult, rainfall is uncertain during the stages of land preparation and stand establishment, and weed competition is severe.

In most rice-growing areas in the Philippines, rains start at the beginning of May but not enough water accumulates for puddling the field until July or August. IRRI scientists, in 1974–1976 field tests, demonstrated that a first crop of an early-maturing rice can be started in May by seeding into dry soil at the onset of the monsoon but before the heavy rains start.

Those early planted rices are harvested in mid-August, the soils puddled and a second crop is transplanted. Seedbeds are seeded 3 weeks before the harvest of the first crop. With this system, transplanting of the second crop may come at the time of transplanting of the first crop under the traditional system (Fig. 7.8). In an IRRI study at nine sites in Central Luzon, Philippines, farmers were harvesting their second crop (transplanted) at the same time that local farmers

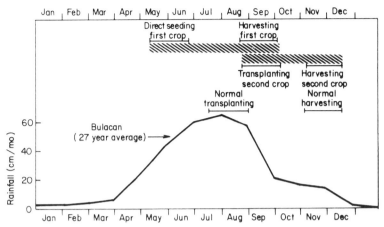

Figure 7.8 Early-maturing rice lines direct-seeded in farmers' fields in early May before heavy rains began. (Adapted from IRRI 1974)

following traditional crop cultures were harvesting their first (and only) crop (IRRI 1974). This dry seeding into bunded rainfed fields, called the *sabog tanim* system, is not a new practice in the Philippines. Farmers in the Pangasinan province who used it in a two-crop system abandoned it because of weed competition, uncertain onset of the monsoon, poor stand establishment, and drought. But with new varieties and modern weed control techniques the possibility of growing two crops of rice plus an upland crop later increased, as shown by cropping systems for the Manaoag area, Pangasinan province, Philippines (Fig. 7.9).

VARIETAL REQUIREMENT Early-maturing rices are more essential for dry-seeded rice than for transplanted rice. Varieties used should have good drought tolerance. IR36, with about 105-day duration and good drought tolerance, is an example of a good rice for direct dry seeding. Gines et al. (1978) reported a maximum yield of 7.0 t/ha with dry-seeded IR36 and 3.8 t/ha with a local variety.

TURNAROUND TIME Turnaround time, the interval between harvesting of one crop and planting of the next in a cropping sequence, is important in all two-crop rainfed cultures. In fact, delay in turnaround time is a serious impediment to intensification in a rainfed rice-based cropping system. Long turnaround time results in crops being planted too late in the wet season for dependable production. Presently, the rainfed rice–rice turnaround time averages 21 days but it varies from 5 to 37 days (Roxas et al. 1978). The method of second rice crop establishment influences turnaround time, averaging 16 days for wet-seeded fields and 26 days for transplanted fields (Magbanua et al. 1977). Any increase in

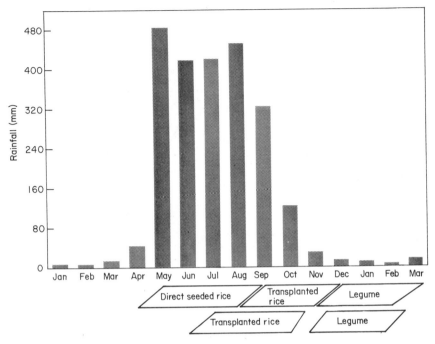

Figure 7.9 Comparison between two cropping patterns by first rice crop establishment plotted against the 24-year rainfall average, Manaoag, Pangasinan, Philippines. (Adapted from Gines et al. 1978)

the intensity of tillage causes more total tillage time for establishing a second rainfed rice crop without significantly contributing to grain yield.

IRRIGATED LOWLAND RICE CULTURE

Irrigated rice covers only 32% of the rice-growing area but produces about 50% of the rice in South and Southeast Asia. Nineteen percent of the irrigated area is double-cropped and the rest is single-cropped (Fig. 7.2). In temperate Asia— China, Japan, and Korea—most rice land is irrigated. In the United States, Europe, and Australia, the rice-growing area is entirely irrigated. In those temperate countries, rice is not grown if irrigation water is not available.

Stand Establishment Technique

In the irrigated rice-growing areas of Asia, transplanted rice is by far the most important method of crop establishment. Other systems of planting include

Table 7.5   Labor Input for Rice Production in Japan (Adapted from Hoshino 1978)

	Labor (hours/ha)					
Year	Seed Preparation	Seedling Raising	Planting	Harvesting	Other Activities	Total
1966	5	74	240	473	601	1393
1968	6	71	244	436	570	1327
1969	7	74	247	396	557	1281
1970	7	74	232	355	510	1178
1971	6	73	213	330	481	1103
1972	6	69	189	284	442	990
1973	5	66	162	260	434	927
1974	5	66	141	243	416	871
1975	5	66	122	218	404	815

direct seeding into a puddled soil, drill seeding into a dryland soil, and broadcast seeding into water.

For irrigated rice, timely preparation of the land, raising of seedlings, and transplanting are essential. In Japan, because of a steady increase in the cost of labor, the transplanting machine has been introduced. In 1968 only 14,000 transplanting machines were used in Japan, but by 1976 the number in use was more than a million, planting 2.7 million hectares (Hoshino 1978). As a result of mechanization, mainly for planting and harvesting, labor use went down from 1393 hours/ha in 1966 to 815 hours/ha in 1975 (Table 7.5).

Cultural practices vary a great deal under different systems of rice culture. Among all, the cultural practices for irrigated systems, such as fertilizer management, weed control, and insect control with insecticides, are far more efficient than those for rainfed systems. Modern varieties with semidwarf growth characteristics have done best in irrigated rice culture. The basic IR8 plant type, with a maturity period of 100–125 days, is most appropriate. In order to increase cropping intensity and productivity per hectare per day, even shorter maturity varieties are desirable.

Direct Seeding onto Puddled Soils

Direct seeding onto puddled soil is practiced in parts of India, Bangladesh, Sri Lanka, and the Philippines. For direct seeding on puddled soils the land is leveled after the soil is puddled and pregerminated seeds are either broadcast or machine drilled. With ideal conditions, it is possible to obtain similar high grain yields with rice transplanted or direct-seeded in puddled soil. To obtain high yields with direct-seeded rices, precise water management (water depths controlled with irrigation and drainage), good weed control, and optimum fertilizer management are necessary. Root anchorage is poor and lodging is more serious with direct-seeded than with transplanted rice (Castillo 1962).

VARIETAL REQUIREMENT   Varieties suitable for direct seeding rainfed rice onto puddled soil are also suitable for irrigated rice culture.

Drill Seeding into Dry Soil

Drill seeding rice into dry soil is most common in the United States and Australia where rice production is fully mechanized. Under this system the final land preparation is done with a spring-tooth or disc harrow followed by a spike-tooth harrow. This generates a mellow, firm seedbed and moisture is held near the soil surface for rapid germination (Johnston and Miller 1973). Rice is sown 3–5 cm deep with a grain drill.

In Australia, rice is grown once in 3–4 years in rotation with other crops. The rice crop is often seeded into heavily grazed pasture land without preparing the field in a method called *sod seeding*. Because of extremely high solar radiation and long days during the reproductive and ripening stages, and very few insect and disease problems, extremely high yields have been reported from sod-seeded rice.

VARIETAL REQUIREMENT   For rice drilled into dry soil, both short- (100 cm tall) and intermediate- (130 cm tall) statured rices are grown. In Australia, intermediate-statured japonicas (some from California) are widely grown. In the southern United States, where drill seeding is common, the indica × japonica crosses with intermediate grain type are grown.

Broadcast Seeding onto Dry or Moist Soils

Broadcast seeding onto dry or moist soil is usually by aircraft and the seeds are covered by harrowing. Simmons (1940) reported the method popular in Arkansas. Because it did not require much machinery, it was used extensively in new rice-growing areas. However, more seeds are required and stand establishment is poorer than with drill seeding. In Texas, rice frequently is broadcast on rough, dry seedbeds with an end-gate seeder.

VARIETAL REQUIREMENT   There is no difference between varieties for broadcast and drill seeding. However, because of a possible severe lodging problem, a broadcast-seeded rice should have a stiff straw.

Water-Seeded Rice

The practice of water seeding rice originated, and is still followed, in parts of Asia, including India, Sri Lanka, Malaysia, and Thailand. It is widely practiced in the Americas, southern Europe, Russia, and Australia.

For water seeding, precise water control is a must, and more seeds are required than for the transplant method. Good seed viability is essential. Oxygen deficiency does not appear to be a limiting factor in stand establishment of rice in water-seeded rice (Chapman and Mikkelsen 1963).

In California, water-seeded rice is the major system of rice culture. The system was started in California primarily to control barnyard grass (*Echinochloa* spp.). The usual practice in California is to soak the seed for 18–24 hours, drain for 24–48 hours, and seed by airplane into fields flooded to a depth of 7.5–15 cm. Aircraft seeding into water in California was initiated in 1929 (Johnston and Miller 1973).

On clay soils in Arkansas, a disc harrow is used to prepare the seedbed. A spring-tooth harrow follows the disc harrow to leave furrows in the seedbed to catch the seed and reduce the drift. The field is then flooded as rapidly as possible to a minimum 10–15 cm depth. The field is seeded immediately after flooding at the rate of 150 kg/ha. The 10–15-cm water level is maintained for 4–6 weeks (Hall 1960).

Similar water seeding techniques have been developed for Texas and Louisiana (Johnston and Miller 1973).

VARIETAL REQUIREMENT   A high level of lodging resistance is important for rice varieties water-seeded. Otherwise, lodging causes considerable reduction in rice yields and quality. It is also difficult to use a combine harvester in lodged fields.

Recently, several rices were developed in California with reduced height (from 130 cm to about 100 cm). Examples are the intermediate-statured Calrose (medium grain) and S-6 (short grain) and the short-statured M-7 and M-9 (medium grain). They are most popular in water-seeded rice in California (see Chapter 6). The shorter rices have better lodging resistance than the taller ones grown earlier. The shorter rices now cover at least 50% of the rice-growing areas in California.

DEEPWATER RICE CULTURE

Deepwater rice is a general term used for rice culture, or the variety that is planted, when the standing water for a certain period of time is more than 50 cm. Water depths, problems, varietal requirement and cultural practices are somewhat unique for deepwater rice-growing areas, which are in a continuum from the medium-deep rainfed lowland rice culture. For deepwater rice maximum water depths vary between 51 and 100 cm for more than half of the growth duration and sometimes there is complete submergence of the plant.

The depth of water, duration of flooding, the rate of increase in water level, temperature, turbidity, and time of occurrence, vary for different areas, so that the term deepwater may have different meanings in different countries. Figure 7.10 shows the duration, depth, and daily increase or decrease in water level in the Huntra area of Thailand where deepwater and some floating rices are grown. As in the Huntra area, water usually remains on the field in all deepwater rice areas for four months or longer. The low-lying areas, so-called "stagnant water" areas, and the tidal swamp areas are of this nature. Uncontrolled water regimes and deepwater rice (and floating rice) are found in deltas, estuaries, and river valleys

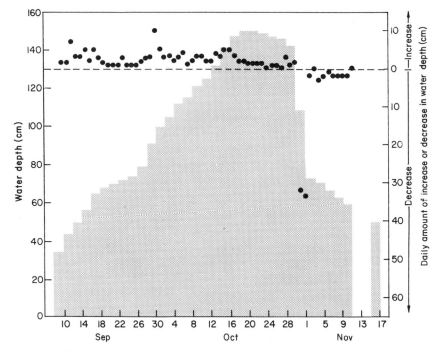

Figure 7.10 Changes in water depth during the 1969 rice-growing season at Huntra, Thailand (data supplied by K. Kupkanchanakul). The dots indicate daily amounts of increase or decrease in water level. (Adapted from Vergara et al. 1976)

of Asia in India, Bangladesh, Burma, Thailand, Vietnam, Kampuchea, and Indonesia.

India has large deepwater areas in West Bengal, Assam, Orissa, Bihar, Andhra Pradesh, and Tamil Nadu. In Africa, deepwater and floating rice-growing areas are found in Mali, Niger, Nigeria, Benin, Gambia, and Sierra Leone. Ecuador in Latin America has some areas that grow deepwater rice. It is estimated that 10 million hectares of rice land is subjected to annual floods (Table 7.6).

Problems of Deepwater Rice Culture

The main problems of deepwater rice growers are:

· Drought damage at germination and the seedling stage because the crop is direct-seeded into dry soil.
· Weed competition at the seedling stage.
· Poor stand establishment.
· High seedling mortality due to sudden flood.

Table 7.6 Summary of Deepwater and Floating Rice-growing Areas[a] (Adapted from Vergara 1977a)

Country	Area (ha)
Bangladesh	
Floating rice	2,100,000
Deepwater rice	3,300,000
Burma	
Deepwater and floating rice	486,000
India	5,500,000
Andhra Pradesh	
Deepwater and floating rice	100,000
Assam	
Floating rice	100,000
Bihar	
Floating rice	500,000
Deepwater rice	1,500,000
West Bengal	
Floating rice	20,000
Uttar Pradesh	
Deepwater and floating rice	600,000
Indonesia, Kalimantan and Sumatra	
Tidal swamp	183,000
Thailand	
Deepwater and floating rice	800,000
Vietnam	
Floating rice	500,000
Deepwater rice	
Single transplanted	1,200,000
Double transplanted	220,000
West Africa	
Gambia	
Floating rice	8,000
Mali	
Floating rice	132,000
Niger	
Floating rice	5,000
Nigeria	
Deepwater and floating rice	77,000
Sierra Leone	
Floating rice	8,000

[a]The "submerged areas" or areas where deepwater remains for only a short period of time are not included in this table.

· Submergence of rices at various growth stages.
· Lodging susceptibility.

These problems are discussed in details by Vergara (1977b).

Varietal Requirements

In the deepwater rice-growing areas, only intermediate-statured (about 130 cm), mostly traditional, varieties are grown. A few modern varieties are grown but for the deepwater area those varieties must have excellent submergence tolerance and, in some cases, moderate elongation ability.

There are many common plant characteristics needed in the different deepwater areas as well as many differences in plant type when considering a particular area. Vergara (1977c) listed six plant characters desirable for deepwater rice.

1 Good seedling vigor is essential.
2 Essentially, a plant type of the high-yielding varieties is needed with the ability to elongate if the water level increases (rapid elongation is not necessary).
3 Intermediate plant height (about 130 cm tall at 5 cm water depth) is needed. In fact, if water level is low or drought sets in, the plant height will be about 100 cm.
4 Submergence tolerance is necessary.
5 Kneeing ability is required in case elongation takes place and the plants lodge when the water recedes.
6 Photoperiod sensitivity should assure flowering at a time when plants are least vulnerable to submergence. In some areas, 150-day varieties, which are photoperiod insensitive, are adequate, but in some areas, maturity may take up to 240 days. Figure 7.11 shows an example of growth stages of deepwater rice varieties in West Bengal, India.

Cultural Practices

Dry seeding of deepwater rice is fairly common, although transplanting or double transplanting is occasionally practiced. In Bangladesh, sowing usually starts in mid-March but the advent of rains and amount of rainfall determine the actual date of sowing.

In West Bengal, India, seed is broadcast in April–May on dry soil.

In Thailand, deepwater rice is grown as a dryland crop until the full force of the monsoon sets in, which is any time from June to August. Before the onset of the monsoon, rice frequently suffers from severe drought.

In Vietnam, reseeding of deepwater rice is often done once or twice if it does not rain for several days after seeding (De Datta and O'Toole 1977). However,

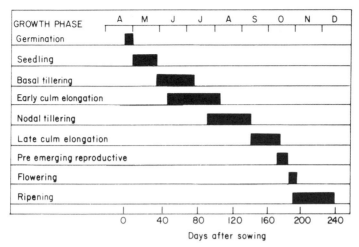

Figure 7.11 Diagrammatic representation of the relative lengths of various growth phases and subphases of rice plants in deep water. (From Datta and Banerji 1979)

farmers double-transplant the rice crop in parts of the Mekong Delta of Vietnam, where the water level ranges from 40 to 70 cm during the wet season. Farmers sow the seedbeds in June at the onset of the rainy season and pull the seedlings 30 days later to transplant them in bigger seedbeds where they grow for another 60 days. They transplant the seedlings a second time in September when the water level is as deep as 70 cm.

Little fertilizer or pesticide is applied to deepwater rice. Weeds are controlled primarily by harrowing following germination of rice and weeds. Flooding at 50–60 days after seeding provides reinfestation of some weeds, mostly of the aquatic type.

Cropping patterns of deepwater rice-growing areas vary from country to country. For example, in Bangladesh deepwater rice (and floating rice) is generally grown following an *aus* (early summer rainfed rice) or *boro* (irrigated winter rice) crop (Fig. 7.12). Other crops, particularly *rabi,* are grown in rotation with deepwater rice.

FLOATING RICE CULTURE

Water depth problems and varietal requirements are generally similar for floating and deepwater rice culture, except that the development of modern varieties and their technology is considerably more difficult for floating rice. Floating rice is grown where maximum water depth ranges between 1 and 6 m for more than half of the growth duration. In densely populated areas, floating rice is grown as a subsistence crop because no other crop will grow. These areas cover the floodplains of the Ganges and Brahmaputra in India and Bangladesh, the

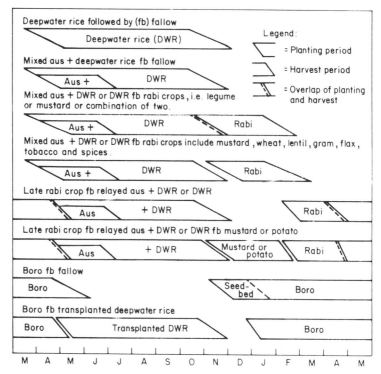

Figure 7.12 Cropping patterns in deepwater rice-growing areas in Bangladesh. *Aus* is an early summer rainfed rice; *rabi* is an irrigated winter crop; and *boro* is an irrigated winter rice. (Adapted from Hobbs et al. 1979)

Meghna in Bangladesh, the Chao Phryya in Thailand, the Irrawaddy in Burma, and the Mekong in Vietnam.

In northern Nigeria, deepwater and floating rices are grown in the floodplains of the Sokoto-Rima Valley. In Mali, floating rice is grown on the plains where the Niger River and its tributaries cause annual flooding (Martin 1974). Of about 223,000 ha of rice land in Mali, about 70% grows floating rice; this is one of the largest floating rice-growing areas in West Africa (Toure 1979).

In West Bengal, India, there are about 20,000 ha of low-lying areas where water accumulates slowly but steadily from adjoining flood areas during the monsoonal season (June–September) and increases water depths to 3 m (Mukherji 1975). In Bangladesh, it is estimated that 2 million hectares, where maximum water depths vary from 1 to 6 m, grow rice. The part of those areas that grow floating rice would be difficult to estimate.

The length of time that floodwater remains on the field varies from place to place and from year to year in the same area. In floating rice-growing areas, the duration is 3–6 months in Bangladesh, 4 months in Thailand, 3–4 months in India, 8 months in Mali, and 10 months in Niger (Vergara et al. 1976).

Problems of Floating Rice Culture

The problems of growing floating rice are similar to those described for deepwater rice, except they become more severe in floating rice-growing areas. Duration of submergence, turbidity of water, and temperature of water all affect the floating rice crop.

Varietal Requirements

Floating rice varieties have been reported to increase in plant height by as much as 20–25 cm a day. In Bangladesh, the daily increase in plant height may be as much as 60–90 cm (Chowdhury and Zaman 1970). The average increase in water rise is 3–8 cm/day in Thailand; the average increase in West Africa is 3–10 cm/day, although it may reach 50 cm/day (Bidaux 1971).

Floating rice plants have been reported to elongate as much as 6 m. The critical plant characters associated with floating rice culture are listed by Vergara et al. (1976). They are:

- They must withstand an abrupt rise in water level or total submergence for at least 1–3 weeks without losing physiological stability.
- They should have the ability to grow quickly above the water level.
- The varieties must have tillering ability before and possibly during the flood.
- They must withstand uprooting (otherwise severe winds can result in enormous waves that uproot and carry the plants great distances).
- Varieties must be photoperiod sensitive and mature between 150 and 270 days.

From these criteria, it is obvious that no modern varieties (semidwarf or intermediate-statured) can be grown as floating rice. For example, the entire floating rice-growing area in India grows photoperiod-sensitive local varieties (Table 7.7). For a long time to come, farmers in the floating rice-growing areas may need to use local varieties that are suited to the specific water and flood regime.

Cultural Practices

The ungerminated rice seeds for floating rice are normally broadcast (60–130 kg/ha) on dry soil before the advent of the rainy season. The seeds germinate as soon as rains thoroughly soak the soil. In certain areas, pregerminated seeds are broadcast on puddled soil.

The seedlings grow as a dryland crop (or at most in moist soil conditions) for 4–20 weeks before flooding occurs. Drought damage and weed competition are often limiting factors in getting an optimum stand of rice.

As the floodwater rises rice plants may elongate as much as 6 m, forming a

Table 7.7 Lowland Rice, Deepwater and Floating Rice-growing Areas in India (Adapted from Balakrishna Rao and Biswas 1979)

Type of Rice Culture	Water Depth (cm)	Lowland Area (million ha)				Kharif (Wet Season) Crop Area (%)
		Transplanted		Broadcast		
		Photoperiod Sensitive	Photoperiod Insensitive	Photoperiod Sensitive	Photoperiod Insensitive	
Shallow rainfed	5–15	3.0	2.0	nil	nil	13.27
Medium-deep rainfed	16–50	6.0	nil	4.5	nil	27.88
Deepwater	51–100	nil	nil	1.5	nil	3.98
Floating rice	>100	nil	nil	2.5	nil	6.63
Total		9.0	2.0	8.5	—	51.76

dense mat on the water surface. Branches and roots are formed on the upper nodes. Harvesting is done after the floodwater recedes but occasionally harvesting is done with boats if the crop matures before the floodwater recedes.

Floating rice yields of 0.4–1.0 t/ha are fairly common. In Thailand, floating rice yields vary between 1.5 and 2.0 t/ha. Similar yields of floating rice have been reported from many South and Southeast Asian countries.

UPLAND RICE CULTURE

The term upland rice has different meanings in different countries. Upland rice (dryland rice) refers to rice grown on both level and sloping fields that are not bunded, that are prepared and seeded under dry conditions, and that depend on rainfall for moisture (De Datta 1975). Recently, Moormann and van Breemen (1978) referred to upland rice land as pluvial rice land (see Chapter 3). It is also known as *arroz secano* in most of Latin America and *arroz sequeiro* in Brazil.

Upland Rice-Growing Areas

Upland rice is grown in Asia, Africa, and Latin America, mostly by small subsistence farmers in the poorest regions of the world. But the area planted to upland rice is nearly 10% of the world's total rice land of nearly 143.5 million hectares—about 14 million hectares.

Much of the future expansion of the world's rice land will probably be in upland rice-growing areas because most of the land suited to lowland (rainfed or irrigated) rice culture is already planted to lowland rice. Such expansion has tremendous possibility in Africa and in the *Cerrado savanna* area in central western Brazil and the Amazon basin area of South America. But in Asia, the entire tillable land suitable for upland rice culture has already been brought into production. And some steep mountainous areas that should not be tilled are being used for upland rice and other crops. An exception in Asia is the vast area in southern parts of Sumatra and Kalimantan, Indonesia, that could be brought into upland rice culture if difficult weeds such as nutsedge (*Cyperus rotundus* L.) and cogon grass or alang-alang (*Imperata cylindrica* L. Beauv.) were controlled (see Chapter 12).

Upland rice is grown under a wide range of management intensities, which vary from shifting cultivation as practiced in Malaysia, Philippines, West Africa, and Peru, to a highly mechanized crop as practiced in Brazil.

Upland rice is an important system of rice culture in South and Southeast Asia, covering 10% of the total rice area (Fig. 7.2). The largest areas of upland rice in Asia are in India, Bangladesh, Indonesia, Philippines, and Thailand. China grows very little upland rice.

Upland rice is grown in eastern India, including Assam, West Bengal, Orissa, and eastern Uttar Pradesh. However, large areas of what is classified as upland

Figure 7.13   In Indonesia, upland rice is often grown in combination with other crops such as cassava.

rice in the states of West Bengal, India and Bangladesh is in fact rainfed lowland bunded rice that is locally known as an *aus* crop. Bangladesh's upland rice-growing area is 24–35% of the total rice land (10 million hectares).

In Indonesia, of 8.8 million hectares of total area grown to rice, 1.4 million hectares are grown to upland rice. The crop is distributed in Sumatra, Java, Kalimantan, Moluccas, and others in decreasing order. Often, upland rice is grown in combination with other crops such as cassava (Fig. 7.13).

Most of Burma's 400,000 ha of upland rice (of 5.1 million hectares total rice land) is in the northern states (Shan and Kuching). In Nepal, upland rice is grown on 9% of total rice area (1.27 million hectares).

In Thailand some upland rice is grown under shifting cultivation but production is insignificant when compared with the production of 8.2 million hectares that grow mostly rainfed lowland, deepwater, and floating rices.

In the Philippines, upland rice covers about 10% of total rice-growing area (3.6 million hectares). It is grown primarily in Mindanao and the southern Tagalog and Bicol regions.

In Sri Lanka, upland rice is grown in Jaffna district, a northern state. In southern parts of Vietnam, the greatest concentration of upland rice is in the Bin Dinh and Darlac provinces (De Datta 1975).

In West Africa, upland rice is the most important system of rice culture, covering 75% of the total rice-growing area (1.87 million hectares). The important upland rice-growing African countries are Sierra Leone, Guinea, Nigeria, Ivory Coast, and Liberia.

In Latin America, of a total 6.5 million hectares of rice-growing areas, 65–75% grows upland rice. Brazil has 5.2 million hectares of rice land of which 3.5 million hectares grow upland rice. Most of Brazil's upland rice grows on small to medium-sized farms that have somewhat rolling topography. Colombia, Guyana, Panama, Ecuador, Peru, Venezuela, and several Central American countries constitute the other important upland rice-growing areas. In many Central American countries, upland rice performs and yields as well as lowland rice anywhere in the world.

The Problems of Upland Rice

All factors that limit the grain yield of lowland rice also limit the yield of upland rice, but some are more critical in the production of upland rice (De Datta and Beachell 1972). They are:

- RAINFALL DISTRIBUTION   The amount and variability of rainfall are two important constraints on production of upland rice (see Chapter 2).
- CHANGE IN NUTRIENTS IN SOIL   The forms and availability of nutrients are directly related to moisture supply in soil. These changes in nutrients under low moisture supply have profound effects on nutrition and growth of upland rice. Among the growth-limiting factors in oxidized (aerobic) soils are phosphorus deficiency, iron deficiency on neutral and alkaline soils, and manganese and aluminum toxicity on acid soils fertilized with ammonium sulfate (Ponnamperuma and Castro 1972).
- WEED COMPETITION   Weed competition in upland rice is so serious that total failure of a crop is common if weeds are not controlled (see Chapter 12).
- INCIDENCE OF BLAST   The incidence of rice blast disease is more severe for upland than for lowland rice (see Chapter 11).

Varietal Requirements

Many varieties of *Oryza sativa* or *O. glaberrima* grown as upland rice are loosely called upland rice varieties (Chang and Bardenas 1965). Most of the traditional varieties grown as upland rice are intermediate-statured or tall, with relatively long, pale green leaves, low tillering, and long, well exserted panicles (Chang and Vergara 1975). Because they are tall, most upland varieties have weak straw and are susceptible to lodging. They are also early-maturing (100–125 days) and have deeper roots than most lowland varieties (Krupp et al. 1972).

Most upland rice varieties have a fair-to-good level of drought tolerance and produce stable but low yields of about 1 t/ha (De Datta et al. 1975). Varieties that produce 3–4 t/ha as an upland rice crop have the following plant characters:

- Semidwarf to intermediate in height (100–125 cm tall).
- Medium-to-heavy tillering.

- Tolerance for and recovery from moderate drought stress.
- Resistance to lodging.
- Resistance to blast and bacterial leaf blight.

Cultural Practices

Cultural practices for upland rice have been studied far less than those for lowland rice. In most of Asia, little mechanization is used to prepare land for an upland rice crop. As soon as enough rain has fallen to permit initial land preparation, the field is plowed with animal-drawn implements (Fig. 7.14), and harrowed with a comb harrow to prepare a seedbed and firm the soil.

In Thailand, slightly elevated areas are plowed with water buffalo or cattle and then hoed. On hills the soil is hardly cultivated.

Indian and Bangladeshi farmers simply turn the soil over with country plows and pulverize it no more than 10 cm deep. A similar method is used in Sri Lanka and Nepal.

In West Africa, 98% of the rice land is prepared by hand but methods vary greatly from country to country. Under shifting cultivation, the soil is hardly tilled at all.

In most upland rice-growing areas in Brazil, large tractors are used to prepare the land and intercultivate between widely spaced rows (Fig. 7.15).

In most countries, upland rice is primarily broadcast. However, drilling is practiced in some countries (De Datta and Ross 1975). Fertilization practices are

Figure 7.14   A Filipino farmer prepares his land for upland rice with an animal-drawn wooden plow.

Figure 7.15   In Brazil, large tractors are generally used to prepare and intercultivate upland rice lands.

the same as for any other dryland crop. However, most upland rice farmers do not use fertilizers if traditional varieties are grown. In Brazil and Central American countries, because improved varieties are used, upland rice is adequately fertilized. Fertilizer response is poor in Brazil because of low moisture supply from rain and poor water holding capacity of soil.

In Central American countries, high-yielding modern varieties such as CICA-9 are grown as upland rice and have excellent fertilizer response. With adequate fertilizer, yields of 6–7 t/ha are fairly common with CICA-9 in Costa Rica.

REFERENCES

Ahmad, M. S. 1980. Plant characteristics needed for transplant *aman* rice. Pages 27–36 *in* Bangladesh Rice Research Institute. *Proceedings of the international seminar on photoperiod-sensitive transplant rice.* October 1977. Dacca.

Awakul, S. 1980. Photoperiod sensitive transplant rice in Thailand. Pages 269–275 *in* Bangladesh Rice Research Institute. *Proceedings of the international seminar on photoperiod-sensitive transplant rice.* October 1977. Dacca.

Balakrishna Rao, M. J., and S. Biswas. 1979. Rainfed lowland rice in India. Pages 87–94 *in* International Rice Research Institute. *Rainfed lowland rice: selected papers from the 1978 international rice research conference.* Los Baños, Philippines.

Barker, R., and R. W. Herdt. 1979. Rainfed lowland rice as a research priority—an economist's view. Pages 3–50 *in* International Rice Research Institute. *Rainfed lowland rice: selected papers from the 1978 international rice research conference.* Los Baños, Philippines.

Bidaux, J. M. 1971. La riziculture en eau profonde au Mali [in English, Spanish summary]. *Agron. Trop. (France)* 26(10):1100–1114.

Buddenhagen, I. W. 1978. Rice ecosystem in Africa. Pages 11–27 *in* Academic Press. *Rice in Africa.* London.

Castillo, P. S. 1962. A comparative study of directly seeded and transplanted crops of rice. *Philipp. Agric.* 45(10):560–566.

Chang, T. T., and E. A. Bardenas. 1965. *The morphology and varietal characteristics of the rice plant.* Int. Rice Res. Inst. Tech. Bull. 4. 40 pp.

Chang, T. T., and B. S. Vergara. 1975. Varietal diversity and morpho-agronomic characteristics of upland rice. Pages 72–90 *in* International Rice Research Institute. *Major research in upland rice.* Los Baños, Philippines.

Chapman, A. L., and D. S. Mikkelsen. 1963. Effect of dissolved oxygen supply on seedling establishment of water-sown rice. *Crop Sci.* 3:392–397.

Chowdhury, M. A., and S. M. H. Zaman. 1970. *Deep water rice of East Pakistan.* 13th Int. Rice Commission Working Party Rice Production and Protection. Iran, 1970. Paper IRC/pp. 70/VII/6. 20 pp.

Dalrymple, D. G. 1971. *Survey of multiple cropping in less developed nations.* USDA For. Econ. Dev. Serv. Rep. 12. 108 pp.

Datta, S. K., and B. Banerji. 1979. The influence of varying water regimes on tillering of deepwater rice and its relation to yield. Pages 233–246 *in* International Rice Research Institute. *Proceedings of the 1978 international deepwater rice workshop.* Los Baños, Philippines.

De Datta, S. K. 1975. Upland rice around the world. Pages 2–11 *in* International Rice Research Institute. *Major research in upland rice.* Los Baños, Philippines.

De Datta, S. K., and H. M. Beachell. 1972. Varietal response to some factors affecting production of upland rice. Pages 685–700 *in* International Rice Research Institute. *Rice breeding.* Los Baños, Philippines.

De Datta, S. K., and J. C. O'Toole. 1977. Screening deep-water rices for drought tolerance. Pages 83–92 *in* International Rice Research Institute. *Proceedings, 1976 deep-water rice workshop, 8–10 November, Bangkok, Thailand.* Los Baños, Philippines.

De Datta, S. K., and V. E. Ross. 1975. Cultural practices for upland rice. Pages 160–183 *in* International Rice Research Institute. *Major research in upland rice.* Los Baños, Philippines.

De Datta, S. K., T. T. Chang, and S. Yoshida. 1975. Drought tolerance in upland rice. Pages 101–116 *in* International Rice Research Institute. *Major research in upland rice.* Los Baños, Philippines.

Fagade, S. O., and S. K. De Datta. 1971. Leaf area index, tillering capacity and grain yield of tropical rice as affected by plant density and nitrogen level. *Agron. J.* 63:503–506.

Gines, H., L. Lavapiez, J. Nicholas, R. Torralba, and R. A. Morris. 1978. Dry-seeded rice: agronomic experiences in a rainfed and partially irrigated area. Paper presented at the ninth annual scientific meeting of the crop science society of the Philippines, 11–13 May, Iloilo City.

Hall, V. L. 1960. Water seeding of rice in Arkansas. *Rice J.* 63(13):13.

Hobbs, P. R., E. J. Clay, and M. Z. Hoque. 1979. Cropping patterns in deepwater areas of Bangladesh. Pages 197–213 *in* International Rice Research Institute. *Proceedings of the 1978 international deep water rice workshop.* Los Baños, Philippines.

Hoshino, S. 1978. *New developments in transplanting rice.* ASPAC Food Fert. Technol. Cent. Ext. Bull. 106. 12 pp.

IRRI (International Rice Research Institute). 1974. *Annual report for 1973.* Los Baños, Philippines. 266 pp.

IRRI (International Rice Research Institute). 1978. *Rice research and production in China: an IRRI team's view.* Los Baños, Philippines. 119 pp.

Islam, M. A. 1980. Photoperiod sensitivity: a neglected issue. pages x–xii *in* Bangladesh Rice Research Institute. *Proceedings of the international seminar on photoperiod-sensitive transplant rice.* October 1977. Dacca.

Johnston, T. H., and M. D. Miller, 1973. Culture. Pages 88–134 *in Rice in the United States: varieties and production.* USDA Agric. Handb. 289. Washington, D.C.

Kawano, K., and A. Tanaka. 1968. Growth duration in relation to yield and nitrogen response in rice plant. *Jpn. J. Breed.* 18:46–52.

Krishnamoorty, C. 1979. Rainfed lowland rice—problems and opportunities. Pages 61–71 *in* International Rice Research Institute. *Rainfed lowland rice: selected papers from the 1978 international rice research conference.* Los Baños, Philippines.

Krupp, H. K., W. P. Abilay, and E. I. Alvarez. 1972. Some water stress effects on rice. Pages 663–675 *in* International Rice Research Institute. *Rice breeding.* Los Baños, Philippines.

Kung, P. [n.d.]. Rice calendar in the Far East and South-East Asian countries (a compilation). Food and Agriculture Organization, Bangkok, Thailand. (unpubl. mimeo.)

Mabbayad, B. B., and R. A. Obordo. 1970. Methods of planting rice. Pages 84–88 *in* University of the Philippines College of Agriculture in cooperation with the International Rice Research Institute. *Rice production manual.* Los Baños, Philippines.

Magbanua, R. D., N. M. Roxas, M. E. Raymundo, and H. G. Zandstra. 1977. Testing of rainfed lowland rice cropping patterns in Iloilo, crop year 1976–1977. Paper presented at a Saturday seminar, 18 June, International Rice Research Institute, Los Baños, Philippines. (unpubl. mimeo.)

Martin, P. 1974. Rice variety improvement and selection in Mali. Pages 116–125 *in* West Africa Rice Development Association. *Rice breeding and varietal improvement seminar.* Proc. I. Monrovia, Liberia, May.

Matsuo, T. 1964. Varietal responses to nitrogen and spacing. Pages 437–448 *in* International Rice Research Institute. *The mineral nutrition of the rice plant.* Proceedings of a symposium at the International Rice Research Institute, February, 1964. The Johns Hopkins Press, Baltimore, Maryland.

Moormann, F. R., and N. van Breemen. 1978. *Rice: soil, water, land.* International Rice Research Institute, Los Baños, Philippines. 185 pp.

Mukherji, D. K. 1975. Problems of deep-water rice cultivation in West Bengal and possibilities for evolving better varieties. Pages 25–30 *in* Bangladesh Rice Research Institute. *Proceedings of the international seminar on deep-water rice, August 21–26, 1974.* Dacca.

Mukherji, D. K. 1980. Problems of photoperiod-sensitive transplant *aman* rice. Pages 150–164 *in* Bangladesh Rice Research Institute. *Proceedings of the international seminar on photoperiod-sensitive transplant rice.* October 1977. Dacca.

Nguu, N. V., and S. K. De Datta. 1979. Increasing efficiency of fertilizer nitrogen in wetland rice by manipulation of plant density and plant geometry. *Field Crops Res.* 2:19–34.

Ponnamperuma, F. N., and R. U. Castro. 1972. Varietal differences in resistance to adverse soil conditions. Pages 677–684 *in* International Rice Research Institute. *Rice breeding.* Los Baños, Philippines.

Roxas, N. M., F. R. Bolton, R. D. Magbanua, and E. C. Price. 1978. Cropping strategies in rainfed lowland area in Iloilo. Paper presented at the ninth annual meeting of the crop science society of the Philippines, 11–13 May, Iloilo City.

Simmons, C. F. 1940. *Rice production and riceland uses in Arkansas.* Arkansas Agric. Ext. Serv. Circ. 424. 16 pp.

Tanaka, A. 1976. Comparison of rice growth in different environments. Pages 429–448 *in* International Rice Research Institute. *Climate and rice.* Los Baños, Philippines.

Toure, A. I. 1979. The growing of deepwater rice and its improvement in Mali. Pages 27–33 *in* International Rice Research Institute. *Proceedings of the 1978 international deepwater rice workshop.* Los Baños, Philippines.

Vacchani, M. V., and M. V. Rao. 1959. Influence of spacing on plant characters and yield of transplanted rice. Paper presented at the seminar on "Recent Advances in Agronomy and Soil Science and their Application to Increase Crop Production," June 1959, Simla, India. Indian Council of Agricultural Research, New Delhi. 4 pp.

Vergara, B. S. 1977a. Deep-water rice. 1. Areas affected and characteristics of flood waters. Paper presented at the GEU training program, March, International Rice Research Institute, Los Baños, Philippines. (unpubl. mimeo.)

Vergara, B. S. 1977b. Deep-water rice. 2. Problems in deep-water rice culture. Paper presented at the GEU training program, March, International Rice Research Institute, Los Banos, Philippines. (unpubl. mimeo.)

Vergara, B. S. 1977c. Deep-water rice. 3. Plant type and crosses made for deep-water rice areas in 1976. Paper presented at the GEU training program, March, International Rice Research Institute, Los Baños, Philippines. (unpubl. mimeo.)

Vergara, B. S., B. Jackson, and S. K. De Datta. 1976. Deep-water rice and its response to deep water stress. Pages 301–319 *in* International Rice Research Institute. *Climate and rice.* Los Baños, Philippines.

Yin, H. Z., T. D. Wang, Y. Z. Li, G. X. Qiu, S. Y. Yang, and G. M. Shen. 1960. Community structure and light utilization of rice fields. *Sci-Sinica* 9:790–811.

8

Land Preparation
For Rice Soils

Tillage practices generally have their greatest effects on plant growth during the germination, seedling emergence, and stand establishment stages. Tillage practices that provide conditions for rapid water intake and for temporary storage of water on the soil surface or in the tilled layer help prevent water runoff. A soil surface with an uneven microrelief can store considerable water in the microdepressions on the soil surface for later intake into the soil. This mechanism has been termed depression storage (Larson 1963).

For rice, the kind and degree of land preparation are closely related to the method of planting and moisture availability, either from rain or from irrigation. The most common benefits of conventional land preparation for most crops are:

- Weed control.
- Incorporation of fertilizers.
- Increase in soil porosity and aeration.
- Mixing the soil to bring up leached deposits.
- Giving the soil a fine tilth to increase adsorption of nutrients.

The benefits cited for tillage for rice are not greatly different. However, because rice is grown in diverse land and water management systems, tillage practices for land preparation vary with the systems. Tillage operations vary according to water availability, soil texture, topography, level of resources available to the farmer, and farmer's preference for a particular type of rice culture.

This chapter focuses on land preparation and crop establishment methods for lowland (wetland preparation) and upland, deepwater and rainfed, and irrigated direct-seeded rice (dryland preparation).

LAND PREPARATION UNDER DIFFERENT SYSTEMS OF RICE CULTURE

Tillage of rice soils as wetland or dryland depends on the stand establishment technique to be followed, moisture supply, and power resources available to the farmer. The most critical factors for the growth and yield of rice are timeliness and quality of land preparation.

Wetland Tillage

In most tropical Asian countries wetland tillage is common. The traditional method of tillage for lowland rice is plowing and puddling, generally done by animal-drawn implements that frequently give poor results. Puddling has been widely adopted because it provides ease of transplanting.

Wetland Tillage in Relation to Moisture Supply

Water savings during wetland tillage provide expansion of the planted area, and minimize drought damage by timely planting of seedlings. Any advancement of planting dates increases the probability of escape from late-season dry spells.
   Wetland tillage consists of three phases:

1   Land soaking in which water is absorbed until the soil is saturated.
2   Plowing, which is the initial breaking and turning over of the soil.
3   Harrowing, during which big clods of soil are broken and puddled with water.

Together, the three phases use one-third of the total water supply in growing a rice crop. Transplanting starts at the end of land preparation.

The Puddling Process

The term puddling has several meanings. For farmers it is a simple soil operation that eases transplanting and reduces water losses through percolation. In rheological terms, the word *puddle* means a clay that can be worked to a water-impermeable stage (Rice 1943). Bodman and Rubin (1948) defined puddling as the mechanical reduction in the apparent specific volume of soil.
   Only soils with more than 20% of clay particles are prone to puddling. Puddling a soil results in the destruction of 91–100% of macropore volume and restricts porosity to only the space occupied by the water film around the clay particles. Investigations of Bodman and Rubin (1948) led to the introduction of the term *puddlability,* a factor strongly influenced by the soil moisture and energy status of the solid phase.

When clay soils are plowed and harrowed at about soil saturation, several changes take place in their structure. Among the properties of puddled soils cited by Koenigs (1963) are the marked reduction in air-filled pore volume, lowering of permeability, increase in moisture suction, lowering of resistance to raindrops, and increase in deformability. According to Jamison (1953), the increase in microporosity by puddling causes an increase in the so-called water-holding capacity of the puddled soil.

Puddling and subsequent flooding differentiate lowland rice soils chemically and pedologically from other arable soils. An important difference between a dryland and a puddled lowland soil is the presence of the reduced soil layer in the puddled soil system. The puddled layer is divided into several subhorizons. The formation of relatively impermeable layers, or plow pans, is attributed to physical compaction (at the same depth) during puddling, and to eluviation of clays and reduced iron and manganese. The plow pan is found in loamy soils that have grown rice for many years and in well-drained Latosols, but is absent in clayey soils, Vertisols, young alluvial, and calcareous soils (Moormann and Dudal 1964).

The rate of moisture loss, tillage practices, wetting and drying cycles, and temperature all influence the degree and rate of soil compaction and the formation of plow pans. The degree of compaction, however, is influenced by groundwater table and soil type. High bulk density of the soil below the plow layer after wet tillage results in less downward flow of water and less leaching of nutrients.

Advantages of Puddling

Although puddling, as practiced in much of tropical Asia, involves a great amount of labor, the method has been widely adopted primarily because of its compatibility with other components of production technology and economic conditions, which include:

- Improved weed control by primary and secondary tillage through puddling action.
- Ease of transplanting.
- Establishment of a reduced soil condition, which improves soil fertility and fertilizer management.
- Reduced draft requirements for primary and secondary tillage.
- Reduced percolation losses resulting in conservation of water from rainfall and irrigation.
- Reliability of monsoon rains by the time puddling operations have been completed (De Datta et al. 1978).

MOISTURE CONSERVATION   One important benefit of puddling is the apparent reduction of moisture losses by percolation. It is difficult, however, to accurately

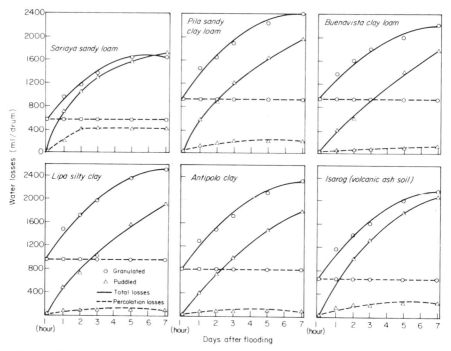

Figure 8.1 Cumulative water losses through percolation and evaporation measured in puddled and granulated soils in steel drums. (Adapted from Sanchez 1973b)

determine the difference in moisture losses from puddled and undisturbed soils. Also, puddling reduces hydraulic conductivity of soil.

Sanchez (1973b) pointed out a dramatic difference in percolation losses between puddled and granulated soils (Fig. 8.1). Granulated soils lost 30% of the water by percolation and 70% by evaporation in 7 days. The puddled soils lost 10% by percolation and 90% by evaporation during the trial. Soil moisture depletion patterns indicate that after the initial heavier loss of water in granulated soils, the trend of subsequent loss was essentially the same for all soils, with granular soils averaging 20% lower moisture contents than puddled soils.

De Datta and Kerim (1974) reported that percolation was considerably higher in a nonpuddled soil than in a puddled soil (Fig. 8.2). The nonpuddled soil received twice as much water and had three times as much seepage and percolation loss as the puddled soil. Similar results were reported by Huynh et al. (1974), who reported water requirements, including evapotranspiration, of 4.5 mm/day for puddled and 6.0 mm/day for nonpuddled soils at saturation (no standing water). For the same soils flooded to 5 cm the requirements were 7.9 and 8.1 mm/day. Puddling reduces percolation losses of moisture by decreasing aggregate cohesion thereby essentially eliminating macroporosity and increasing bulk density (Wickham and Singh 1978).

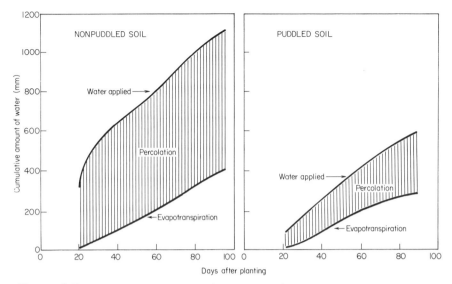

Figure 8.2   Comparison of cumulative water applied, evapotranspiration, and percolation losses in field studies of puddled and nonpuddled soils continually flooded at 5 cm. IRRI, 1972 wet season. (Adapted from De Datta and Kerim 1974)

In a field lysimeter, rice on an Alfisol underwent a 2-week drought during panicle development. Soil moisture tension rose to 55 centibars (cb) at 20 cm in the nonpuddled soil, and to 25 cb in the puddled soil (Fig. 8.3). Furthermore, soil moisture content remains higher after wetland tillage than after dryland tillage (Fig. 8.4). This indicates that rice grown on puddled soil may be less affected by drought because moisture retention is higher than in nonpuddled soil.

WEED CONTROL   Land preparation for puddling of a lowland rice soil consists of plowing and harrowing, which minimize weed growth. Some farmers in South and Southeast Asia repeat plowing and harrowing several times to reduce weed problems. Puddling aids the quick establishment and tillering of transplanted rice and results in greater competition and suppression of weed growth. In addition, transplanters trample weeds to incorporate them into the soil during the transplanting operation. This explains the generally lower weed problem in transplanted rice than in broadcast-seeded flooded rice.

Land preparation includes two main weed-control operations:

- Plowing, a deep tillage, between harvest of one crop and planting of the next crop, which turns under weed seed and rice stubble.
- Harrowing, a shallow tillage (and puddling), before rice seedlings are transplanted further incorporates weed seed and stubble into the reduced soil layer, where lack of oxygen inhibits weed seed germination. Functions of primary and secondary tillage are schematically represented in Fig. 8.5.

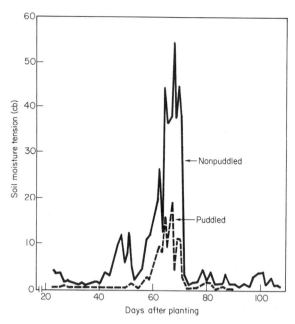

Figure 8.3 Soil moisture tension at 20-cm depth in puddled and nonpuddled rainfed soils. IRRI, 1972 wet season. (Adapted from De Datta and Kerim 1974)

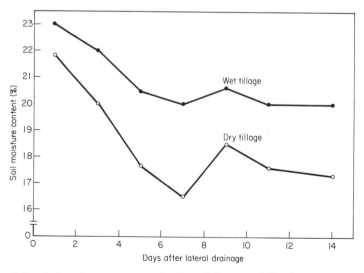

Figure 8.4 Soil moisture content after lateral drainage following wet and dry tillage. (Adapted from Curfs 1974)

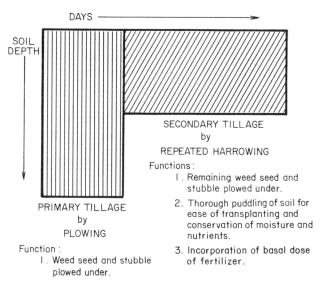

PRIMARY TILLAGE
by
PLOWING

Function :
1. Weed seed and stubble plowed under.

SECONDARY TILLAGE
by
REPEATED HARROWING
Functions:
1. Remaining weed seed and stubble plowed under.

2. Thorough puddling of soil for ease of transplanting and conservation of moisture and nutrients.

3. Incorporation of basal dose of fertilizer.

Figure 8.5 Functions of primary and secondary tillage in wetland rice field preparation time.

In discussing groups of tillage operations in relation to various functions, Kuipers (1974) suggested weed control as a main objective of each tillage operation. The number of weeds growing in association with transplanted rice declines as the number of preplanting harrowings increases (Fig. 8.6). The time between harrowings, and the number of harrowings, can be reduced substantially if postplanting weed control is practiced. Where chemical control is not practiced, tillage is the most important factor in weed control in tranplanted rice.

Minimizing weed growth is more critical for direct-seeded rice than for transplanted rice because direct-seeded rice does not get as much of a head start on weeds. The effect of level of land preparation on the number of weeds in transplanted and direct-seeded rice is shown in Fig. 8.7. Nonselective herbicides applied with or without follow-up tillage suppressed all weeds present at the time of land preparation but weeds regrew. The rate and intensity of weed regrowth were higher in herbicide treated plots than the plots that were plowed and harrowed. Direct-seeded rice had a higher weed count per unit area than transplanted rice with either zero or minimum tillage.

Tillage alone is probably the most practical method for controlling some perennial grass weeds such as *Paspalum distichum* L. Depth of plowing also affects growth of annual and perennial weeds. Land preparation alone, however, will not control all the annual and perennial weeds.

NUTRIENT AVAILABILITY The low redox potential of submerged puddled soils helps conserve water-soluble nutrients in rice soils. Accumulation of ammonium (NH_4^+) is favored, biological nitrogen fixation increases, and the availability of

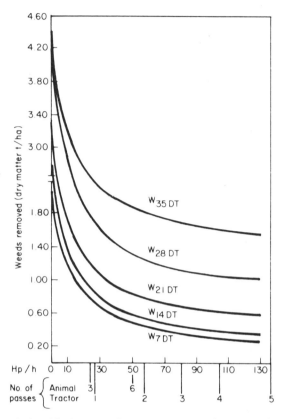

Figure 8.6 Relationship between horsepower-hours for harrowing and weeds re-moved at weeding time (av. of two varieties, IR8 and H-4). W = weeds removed at weeding time; DT = days after transplanting. IRRI, 1967 dry season. (From De Datta and Barker 1978)

phosphorus and silicon, and the solubility of iron and manganese increase. Puddling also hastens the mineralization of soil organic matter. Significant increases in NH_4^+-N production after 4 weeks of anaerobic incubation were observed in puddled soil samples by Harada et al. (1964).

Data from Sanchez (1973a) indicate that puddling decreased losses of applied nitrogen in the field and resulted in higher rice yields. When nutrient availability is evaluated in puddled soils, mud forms are best (Table 8.1).

Disadvantages of Puddling

Strong arguments have been presented for replacement of soil puddling for rice cultivation with furrow irrigation on a nonpuddled soil. A nonpuddled soil allows easier inclusion of other crops into a rice-based cropping system. The main disadvantages of puddling are high water requirement, hindrance to regeneration of soil structure, and impeded root development.

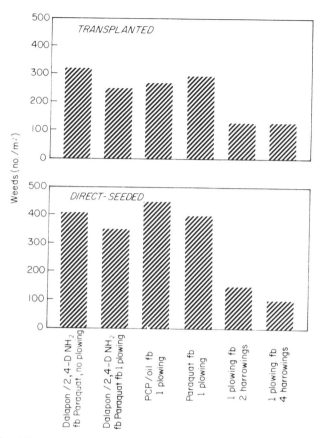

Figure 8.7 Effects of levels of land preparation on number of surviving weeds in transplanted and direct-seeded flooded rice at 40 days after planting. The data are averages for IR8 and IR22 rices. / = proprietary mixture; fb = followed by. IRRI, 1970 dry season. (Adapted from De Datta and Barker 1978)

Table 8.1 Ratings of Various Qualities of the Wet-Seedbed Soil Structure (Adapted from Koenigs 1961)

Quality	Soil Structure Rating		
	Mud	Paste	Granulated
Water percolation losses	Low	Low	High
Availability of nutrients	High	Low	Low
Availability of added nutrients	Low	Low	High
Availability of soil moisture	High	Low	High

HIGH WATER REQUIREMENT A major disadvantage of puddling is the high water requirement for land preparation. Water required for puddling a wetland field varies from 150 to 200 mm (see Chapter 9). Total water use in a nonpuddled field is about 50% of that in a wetland field despite a longer field duration of a nonpuddled planting. In a Philippine study, daily water use in a wetland field was 9.28 mm/day (including water required for land preparation), compared with 3.37 mm/day in a nonpuddled field (De Datta et al. 1973).

The high water requirement for land preparation in a puddled system is a more critical disadvantage for rainfed rice than for irrigated rice. Farmers often have to wait 1–3 months for sufficient water to allow puddling. Because of insufficient water land preparation by rainfed rice farmers is delayed at a period when all other environmental factors are favorable for planting and subsequent crop growth. This disadvantage has led some farmers in Bangladesh, northeastern India, Indonesia, and the Philippines to shift to dry seeding of rice in a nonpuddled soil.

HINDRANCE TO REGENERATION OF SOIL STRUCTURE In some parts of tropical Asia, cropping can be intensified by extending the total crop-growing season. This is made possible by regeneration of soil structure, which permits growing an upland crop following lowland rice. This is not a new concept, but the shift from a lowland puddled soil to a dryland soil requires major alteration in the physics of the soils. The conversion of a soil from the wet to dry condition is characterized by a process of improving the soil structure destroyed by puddling. The wet to dry conversion becomes a great concern in a rice-based cropping system on a clay soil. When a puddled soil dries, it may shrink and crack resulting in clumps that are fi ̣quently unworkable, and create difficulties in preparing the soil for upland crops.

Regeneration of soil structure is a major consideration in the conversion of soils from lowland paddy to a dryland field. Briones (1977) concluded that montmorillonitic clay soils with low organic matter and iron oxide contents are more difficult to convert from lowland to dryland use than kaolinitic clay with higher organic matter and iron oxide contents. This indicates that incorporation of crop residues aids regeneration of soil structure.

A lowland soil is best converted to a dryland soil by tillage at an optimum moisture content. The simplest process is to let the lowland soil dry before tillage to a moisture content below the lower limit of plasticity. For easier tillage in the wet to dry soil conversion, the soil should be worked before the maximum soil strength has developed.

IMPEDIMENT TO ROOT DEVELOPMENT Puddling creates an unfavorable rooting medium for upland crops, and puddling and subsequent continuous submergence cause several nutritional disorders of rice in tropical Asia. Furthermore, in the puddled anaerobic soil, decomposition of organic matter slows, which leads to the accumulation of toxic organic compounds. Low grain yields from puddled soils are often due to accumulation of Fe^{2+} and H_2S, especially on acid sulfate soils.

The formation of a hardpan by puddling some heavy clay lateritic soils leaves a plow layer too thin to permit adequate root development. Hardpan formation is also undesirable in saline paddy soils.

Speed of Land Preparation

For simplicity, the speed of land preparation is measured by the progress of soaking, plowing, harrowing, and transplanting of a total area. In a Philippine study, Valera et al. (1975) related the speed or duration of land preparation with the supply of water. They did not include light rains 2 or more weeks prior to the release of irrigation water in the computation of cumulative water use for land preparation. Only hard and cracked soils, which were not cropped for at least a month prior to the study, were considered.

Water supply was found to be a limiting factor in the duration of land preparation for some sites. Fifty percent of the whole Peñaranda River Irrigation System's Gapan site was transplanted in 7.5 weeks, but an additional 5.5 weeks were required to finish it (Fig. 8.8). There was a significant time gap between the first half and the second half of the system. The first half was irrigated by the upper reaches of the canal and was completely transplanted in 10 weeks. The second half, irrigated by the lower reaches of the same canal, was not transplanted any earlier than the adjacent rainfed area. For sites where water was more generously supplied, land preparation time was about 5.5 weeks (Fig. 8.8).

In the dry season Valera et al. (1975) collected data from the Upper Pampanga River Project area. Land preparation time ranged from 7.5 weeks for 50% of the area to 10 weeks for 100% of the area transplanted (Table 8.2).

WATER SUPPLIED FOR LAND PREPARATION   Savings of about 100 mm of water lost by evaporation could be achieved by reducing land preparation time to 5 weeks in many areas but shortening of land preparation time will require additional power resources. Use of tractors, for example, may speed land preparation operations. Initial moisture content of the field is another factor that may be relevant in reducing the period of land preparation.

Estimates of the water requirement for saturation and land preparation for a lowland puddled field in Malaysia are 505 mm for the wet season and 570 mm for the dry season. Data collected from a 25-ha site in the Philippines showed an average of 723 mm of water required for seepage and percolation, land soaking, and evaporation during land preparation in the wet season. Average duration of land preparation was 7.5 weeks from the time water was first turned in until 50% of the area was transplanted.

Tillage Properties of Lowland Soils

Tillage should be done within an appropriate soil moisture range because soil consistency changes widely with moisture content. There are fairly clear

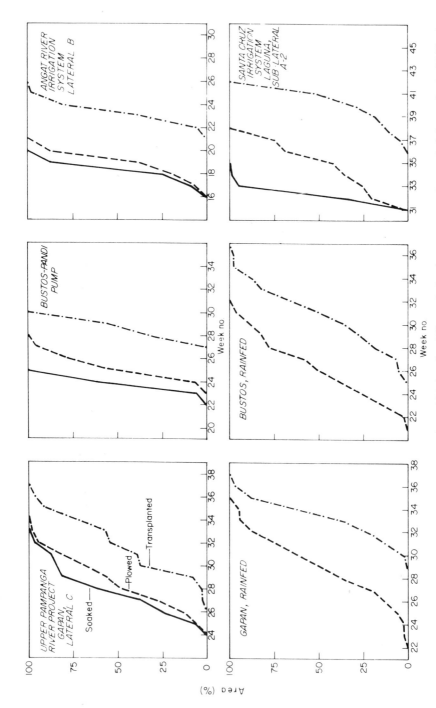

Figure 8.8 Percent of area soaked, plowed, and transplanted by date for different irrigation systems and two rainfed areas in the Philippines, 1973 wet season. (Adapted from Valera et al. 1975)

Table 8.2 Duration of Land Preparation and Amount of Water Applied to Three Upper Pampanga River Project Pilot Sites, Philippines, 1974 Dry Season (Adapted from Valera et al. 1975)

Site	Time for Land Preparation (weeks)	Total Water for Land Preparation Only[a] (mm)
Santa Arcadia		
Continuous		
50% transplanted	7	643
100% transplanted	10	690
Rotational		
50% transplanted	6.5	575
100% transplanted	9	594
Gomez		
Continuous		
50% transplanted	8	575
100% transplanted	10	596
Rotational		
50% transplanted	7.5	676
100% transplanted	9	698
Kaliwanagan		
Continuous		
50% transplanted	7.5	967
100% transplanted	10	1039
Rotational		
50% transplanted	8	1031
100% transplanted	10	1070

[a]The sum of weekly rainfall and irrigation multiplied by the percentage of area not yet transplanted.

relationships between machine performance and soil constants, and the prediction of machine performance is possible for any soil type and moisture contents. With increasing cropping intensity of rice, the tillage operation becomes difficult as the trafficability problem of machines arises.

The relationship between machine performance and soil constants (cone index and shear resistance) is important. Knowing the cone index, the trafficability of a tractor can be predicted (Table 8.3). During rotary tilling, the tractor is pushed by the action of the rotary tines, and axle torque decreases. This pushing action of the rotary tiller contributes to the trafficability of the tractor on soft soils (Kisu 1978). Among the factors cited for low frequency of use of tillage machinery are low soil-bearing capacity, lack of farm roads, cost of machines, and lack of mechanical knowledge and adequate machinery service facilities.

Table 8.3   Prediction Criteria of Trafficability of Tractor
from Cone Index[a] (From Kisu 1978)

	Trafficability		
	Easy	Possible	Impossible
Operation	Cone Index (kg/cm^2)		
Rotary tilling	> 5	3–5	< 3
Plowing	> 7	4–7	< 4
Plowing (with cage wheel)	> 4	2–4	< 2

[a]Base area of cone = 2 cm^2 (small cone). Mean value of small
cone index in the range of 0–15-cm depth is used.

Grain Yield Response to Wetland Tillage Practices

In rice-growing areas of the temperate regions where soils are not puddled, grain yields are consistently high. Yields of 5–6 t/ha are common, and farmers often harvest 8–9 t/ha. In temperate East Asia, the soil is puddled for rice and the grain yield ranges from 3.5 to 7.0 t/ha.

In tropical Asia where most rice is grown as rainfed, it is believed that a puddled field minimizes risk of yield reduction due to drought. In an experiment on a montmorillonitic clay, irrigated IR20 rice yielded 7.9 t/ha in a puddled soil, but yielded only 3.6 t/ha in a nonpuddled soil. The crop on puddled soil produced 9.0 kg grain/mm water, and the crop on nonpuddled soil produced 8.3 kg grain/mm water (De Datta et al. 1973). In a subsequent study, rice (variety Tongil) in puddled soil had about 2.5 times higher efficiency of water use (7.9 kg grain/mm water) than rice in nonpuddled soil (2.9 kg grain/mm of water) (De Datta and Kerim 1974). Similar results are reported for sandy soils in India and Nigeria.

In Bangladesh, increasing the number of plowings from two to three increased grain yield by almost 1 t/ha. Further increase in plowing did not increase grain yield (Sattar 1978).

Little has been reported on difference in grain yield between animal-drawn and power-drawn implements. From a study at the Central Rice Research Institute in India, Jacobi (1974) reported that yields after tillage with the two- and four-wheel tractor-mounted implements were higher than those after tillage by animal-drawn equipment. A yield advantage was attributed to primary tillage only in the case of animal-drawn equipment.

A survey in Central Luzon, Philippines, indicated that use of tractors in the tillage process gave an advantage in weed control but not in yield (De Datta and Barker 1978).

Tillage Problems in Double Cropping of Rice

To shorten the turnaround time between crops in double cropping of rice, mechanization in land preparation is essential. Its use can shorten the period between harvest of one crop and planting of the next. However, the between-crop tillage operation is difficult because of the trafficability problem. The clayey soils of South and Southeast Asia become hard and compact when dried but when submerged they lose bearing capacity immediately. Another problem is softening caused by the breakage of the plow sole layer by large machines. Trafficability problems can be minimized by increasing the period of soil drying between crops but that increases turnaround time. Softening of the soil can be avoided by not using large machines unless the soil is sufficiently dry to hold heavy equipment.

Dryland Tillage

The reasons cited for using dryland tillage for rice are:

- Initial crop growth is obtained from early monsoonal rainfall, which would be used for land soaking and puddling for wetland tillage.
- Labor constraints associated with seedbed preparation, land preparation, and transplanting are reduced.
- Large power units can be employed for primary and secondary tillage where capital is available and labor is lacking.
- Where a nonrice crop follows a nonpuddled rice crop, soil structure is in a more favorable state for stand establishment and root development of the following crop.
- Insect and disease buildup on alternate hosts during the period required for wetland soaking and puddling are reduced.

The major disadvantages of dryland tillage are:

- Draft power requirements are high, often beyond the reach of rice farmers in the Asian tropics.
- Early weed control requirements are comparatively exacting and critical.
- Percolation losses are comparatively high, leaving the crop more susceptible to periodic drought stress.
- The crop may be exposed to several soil insects and to blast prior to accumulation of standing water.
- Fertilizer requirements will often be higher (De Datta et al. 1978).

Many systems of rice culture are based on dryland preparation. For example, all rice-growing areas in the United States, southern Australia, most of Latin

America and West Africa, parts of tropical Asia, and many countries in Europe use dryland tillage. In some of those areas, rice is sown in dry soil and later irrigated (United States and Australia) or sown in dry soil and grown as rainfed upland rice (most of Latin America, West Africa, and parts of Asia). In the United States, the tillage is much the same as that for other small grains. In rainfed rice culture, the type of tillage done depends on whether the crop will be grown as dryland or wetland.

Land Leveling

For good water control, and also to improve quality of seedbed preparation, it is essential to have land leveled by grading. In recent years, a laser beam has been used to guide grading equipment as it cuts and fills the soil. After the cut-and-fill work, a land plane or leveler is used for the finishing touches on a level, smooth surface (Huey 1977).

Water-Seeded Rice

If the rice is to be sown in water, such as it is in California, the soil surface before flooding should be left rough, which prevents drift of seeds (Fig. 8.9). In California, rice farmers sometimes are able to prepare a good seedbed in the late fall or winter and erect levees. When that is possible, little land preparation other than harrowing is necessary before spring.

Figure 8.9   For water-seeded rice, such as in California, the land should be prepared rough before water is introduced.

The final seedbed preparation for water seeding depends on soil type. On sandy or silt loam soils, a mellow, firm seedbed similar to that for drilling should be prepared. On clay or very fine silt loam soils, the seedbed should be fairly rough with a clod size as large as 10 cm in diameter (Johnston and Miller 1973).

Drill-Seeded Rice

In the southern United States, where dryland drill seeding of rice is practiced, a mellow, firm, weed-free seedbed is desirable. The first step is plowing or disking 10 or 20 cm deep to bury the previous crop residue. The land is then leveled and a field cultivator may be used for the final seedbed preparation just prior to seeding (Huey 1977).

Reynolds (1954) reported that summer plowing frequently is practiced in Texas for fields infested with red rice (*Oryza sativa* L.) or other weeds. After summer plowing, the land is leveled and disk harrowed as needed for weed control. Land plowed in the fall is usually left rough until spring and then disk harrowed preparatory to seeding. Heavy soils plowed in the spring generally require more subsequent tillage to obtain a desirable seedbed than when plowed in the fall or early winter.

Final land preparation just ahead of drilling is usually done with a spring-tooth harrow or a disc harrow with a spike-tooth harrow behind it. A roller-packer may be used to break up clods and to firm the soil after drilling. This gives a mellow, smooth, and firm seedbed and the moisture is held near the surface so that the seed germinates soon after seeding.

If rice follows a soybean or a similar row crop, only a limited amount of land preparation may be necessary if the soil is in fairly good physical condition after the row crop is harvested.

Broadcast-Seeded Rice

If rice is to be broadcast-seeded on dry ground, the final land preparation should leave the surface rough and somewhat cloddy. The seed may be covered with a field cultivator, spring-tooth harrow, or other appropriate type of equipment. A corrugated roller, or cultipacker, is used to firm the seedbed after broadcast seeding.

Dry-Seeded Bunded Rice in Asia

Dryland tillage has been adopted over a limited area for lowland rice production in tropical Asia. However, recent technological advances may increase the area where dryland cultivation is suitable.

The most difficult problem in dryland tillage systems (see Chapter 7) is quick tillage of a dry field. In Bulacan, Philippines, it was observed that where land was prepared at the end of the previous wet season the following wet season's newly emerged rice crop was able to survive a period of drought. In contrast, a crop seeded at the same time but following land preparation at the beginning of a current wet season suffered from considerable moisture stress. Crop survival was apparently due to availability of soil moisture conserved by a soil mulch in the dry season.

Bolton and De Datta (1979) tested the conservation of soil moisture by soil mulching. They also compared the time saved by the early dry seeding of a rice crop after the dry soil mulch versus transplanting a crop after traditional wetland tillage. In plots where land preparation was completed at the end of the previous wet season, a dry soil mulch was maintained from February to the first week of May. The soil moisture tension under the mulch did not exceed 33 cb at 15-cm depth.

In contrast the soil moisture tension in weedy fallowed plots was 5 b at a depth of 15 cm at the end of the dry season despite a water table within 1 m of the soil surface during the whole period. Weed-free plots conserved some moisture, but not as much as the dry soil mulch (Fig. 8.10). It is obvious that the dry soil mulch has potential as a practice to conserve soil moisture for early direct-seeded rice.

Upland Rice

In most of Asia, little mechanization is used to prepare land for upland rice. As soon as enough rain has fallen to permit initial tillage, the field is plowed with an animal-drawn implement, then harrowed with an indigenous comb harrow. Sometimes the weed seeds are allowed to germinate for a week, and the field reharrowed.

For upland rice in Thailand, slightly elevated areas are plowed by water buffalo or cattle, and then hand hoed. On hill areas, the soil is hardly cultivated. Indonesian farmers generally prepare land for upland rice with animal-drawn plows during June–August. Indian farmers use simple country plows and pulverize the soil no more than 10 cm deep.

About 98% of upland rice land in West Africa is tilled manually because draft animals are scarce. Land preparation methods vary greatly from country to country in Latin America. In the shifting cultivation areas of Peru, for example, mature secondary forests are cut and burned during the dry months. Upland rice is then seeded by dibbling. Shifting cultivation in Peru is quite similar to the slash-and-burn methods that precede rice planting in Burma, Malaysia, and Thailand. Most upland rice farmers in Brazil use large tractors to prepare the land.

Deepwater and Floating Rice

About 15% of the world's rice lands are subject to annual floods and require deepwater and floating rice cultures (see Fig. 7.1 in Chapter 7). For those cultures, tillage and seeding are in dry soils. Land preparation is poor and drought and weed infestation are major factors affecting seedling establishment.

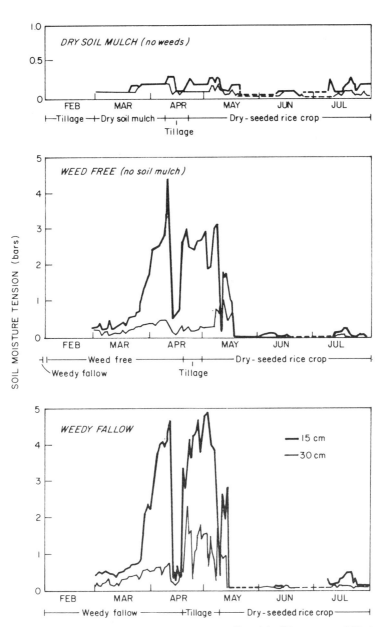

Figure 8.10   Soil moisture tension under a dry soil mulch did not exceed 33 cb at any time during the 1976 dry season. Some moisture was conserved by keeping the plots weed free, but moisture tension went higher than in the soil-mulched plots. Plots that were fallow and on which weeds were allowed to grow had 5 b of soil moisture tension at a 15-cm depth by the end of the dry season. IRRI, 1976. (Adapted from Bolton and De Datta 1979)

In Bangladesh, land preparation for deepwater rice begins in March or April. For normal seeding, as many as four plowings are common. After the plowings, the land is leveled by breaking the clods. In areas where the soil is a heavy clay, big clods remaining after leveling are broken with a long-handled wooden mallet. In some areas, rice is broadcast after the four plowings and the land is leveled, primarily to cover the seeds. If a farmer feels uncertain about the onset of the monsoon, he repeatedly plows the field primarily to dry the soil. With such tillage, seeds germinate only after the onset of the monsoon.

In the deepwater rice-growing areas in West Bengal, India, rice is broadcast in April or May after the land is plowed dry. Land preparation and seedling establishment are often poor.

In Thailand, the cropping season starts in April when farmers burn straw from the previous crop. After that, large tractors are hired to disc plow the land when it is slightly moist; most farmers with 4–7 ha hire tractors with a five-disc plow. Smaller farmers generally wait until sufficient rain has fallen and use animal-drawn implements. Rice is sown on the plowed land several weeks after the first plowing, and the land is plowed again. The last plowing destroys new growth of weeds and covers the rice seeds. Some farmers sow rice in poorly prepared fields and the crop has to compete with growing weeds (De Datta and O'Toole 1977).

Effect of Tillage On Dryland Soil

The use of machinery for tillage will become more common for upland than for lowland rice fields in tropical Asia. This is because much of the machinery used for upland crop tillage can be successfully used for upland rice.

For the dryland soil, aggregate size, moisture content, porosity and aeration, compaction, and soil temperature are important tillage-related parameters that affect crop yields.

- SOIL AGGREGATES   Tillage leaves the soil in an aggregated and loose condition, thereby improving soil structure. Incorporation of organic matter further improves soil aggregation.
- SOIL MOISTURE   Tillage breaks the capillary connection with the soil surface, which reduces evaporation of soil moisture. Excessive tillage reduces infiltrability of soil moisture.
- SOIL POROSITY AND AERATION   Tillage alters soil structure by increasing its porosity and aeration. When a bulk of the soil has large aggregate size more water is retained because of greater internal porosity.
- SOIL COMPACTION   Despite the favorable effects of tillage, it has detrimental effects over longer periods because of compaction. Repeated running over the field with heavy equipment may result in a compacted, high-bulk-density traffic pan.
- SOIL TEMPERATURE   Tillage creates a soil temperature favorable for seedling establishment. Tillage loosens the soil surface, resulting in the decrease in

thermal conductivity and heat capacity. This increases thermal insulation and resistance to water flow from moist soil layers to the atmosphere.

Effect of Soil Physical Properties on Pulverization

In the case of direct seeding of rice in dry soil, seed germination and seedling growth are greatly affected by the fineness of the soil clods. The degree of soil pulverization by tillage differs according to soil conditions such as moisture content or particle distribution (Kisu 1978). Plowing dry soils with a hard consistency will result in the formation of numerous clods. However, plowing in the correct moisture range not only achieves maximum granulation but also requires less power.

Grain Yield Response to Dryland Tillage Practices

With the advent of modern short-duration rices, the possibility of increasing rice production by growing two rice crops in rainfed fields becomes viable in regions receiving rainfall for 5–7 months or more of at least 200 mm rain per month.
  Early first crop plantings can be achieved through dry seeding. Figure 8.11

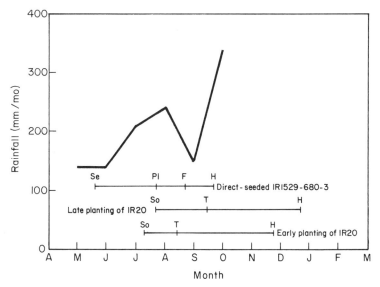

Figure 8.11 Time of seeding and harvesting of direct-seeded IRI529-680-3 and transplanting of IR20 in relation to the amount of rainfall. Se = seeding; PI = panicle initiation; F = flowering; H = harvest; So = sowing in the seedbed; T = transplanting. San Rafael, Bulacan, Philippines, 1973 wet season. (From IRRI 1974)

shows the differences in time of harvest in relation to dry seeding of IR1529-680-3 and transplanting of IR20 in a Philippine study. Under prevailing conditions, transplanting could not be done until sufficient water had accumulated in late July to allow puddling.

For most farmers of tropical Asia, mechanical drilling is not a foreseeable alternative. For the first crop of dry-seeded rice, common establishment alternatives are:

- Broadcasting on a level field.
- Broadcasting over shallow furrows and passing a spike-toothed harrow at 45° to concentrate seed in rows.
- Hand "drilling" (placing seeds in shallow furrows opened by animal-drawn implement).
- Dibbling seeds so that seedlings emerge in hills at uniform spacings.

Analysis of rainfall data indicates that a dry-seeded rice crop in two field experiments in the Philippines had completed 30–50 days of growth before 400 mm of rainfall (enough to puddle soil and transplant) had accumulated (De Datta et al. 1978). Dry seeding results in net time savings from the first crop that were passed on for growing the second crop. Similar advantages attributable to dry seeding establishment of the first crop were obtained in a study at IRRI. By delaying to transplant the first crop on a puddled soil, insufficient time remained to plant a second crop within the wet season.

Transplanting of a second crop instead of direct seeding may reduce its field duration by 2–3 weeks, and lower the risk of late season drought stress during the critical spikelet-filling period. Evidence of this effect appears in Table 8.4, in the form of higher yields and lower variability associated with transplanted second crop in comparison with direct-seeded second crops. In those studies, the number of crop failures following direct seeding was about twice the frequency of failure after transplanting (De Datta et al. 1978).

In a study of dry-season shallow and deep tillage (with straw incorporation), rice could be seeded 24 days earlier than the weedy fallow treatment—a practice generally followed by most farmers in tropical Asia (De Datta et al. 1978). The land preparation for a transplanted rice crop on puddled soil could start in early July after sufficient rain accumulated for thorough puddling. That gave harvest dates of the transplanted cultivars that were 5–8 weeks later than harvest dates of the same cultivar planted by a dry-seeding tillage system. Thus, it was possible to grow only a single crop of transplanted rice on puddled soil but in dry-seeded plots a second crop of transplanted rice was harvested (Table 8.5). Again, for reasons cited earlier, grain yields of the second crop were considerably lower than the first crop.

Table 8.4 Farmer-Cooperator Yields of First Rainfed Direct-Seeded (Broadcast) and Second Rainfed Direct-Seeded and Transplanted Rice Crops, 1976–1977 and 1977–1978 Wet Seasons, Iloilo Province, Philippines (From De Datta et al. 1978)

| | First Crop | | | Second Crop | | | | | |
| | | | | Direct-Seeded | | | Transplanted | | |
Wet Season	Observations (no.)	Yield (t/ha)	SD[a] (t/ha)	Observations (no.)	Yield (t/ha)	SD (t/ha)	Observations (no.)	Yield (t/ha)	SD (t/ha)
1976–77	41	5.3	1.5	37	2.2	1.6	4	3.5	1.4
1977–78	22	5.9	1.1	13	1.4	1.8	9	1.7	1.2

[a]Standard deviation.

Table 8.5 Grain Yields of Rainfed Bunded Rice as Influenced by Different Dry Soil Mulch and Straw Mulch Treatments; IRRI, 1977 Wet Season (From De Datta et al. 1978)

| | Grain Yield[a] (t/ha) | | | | | | |
Crop	Shallow Tillage (10 cm deep)	Deep Tillage (20 cm deep)	Straw Incorporation (5 t/ha)	Straw Mulch (5 t/ha)	Weed-Free Fallow	Weedy Fallow Dry-Seeded	Weedy Fallow Transplanted[b]
First	3.6	3.6	3.7	3.4	3.2	2.5	3.1
Second	1.6	1.6	1.8	1.4	1.7	1.1	—
Total	5.2	5.2	5.5	4.8	4.9	3.6	3.1

[a]Average over four replications and four varieties.
[b]Not included in the experiment analysis but tested in adjacent areas.

NEW TILLAGE CONCEPTS AND PRACTICES

To reduce land preparation cost without sacrificing grain yields, farmers should limit tillage to a minimum. They should select the lowest cost technique of soil tillage and replace that with lower cost operations as soon as they are available. One should, therefore, evaluate total operating costs of the farm and see how cost may be reduced by adopting a special tillage operation.

Moens (1963) divided land preparation practices for rice into three broad groups:

1 Preparative functions—tillage to save inputs like water, seeds, and plant nutrients.
2 Substitutive functions—substitution of one operation for another, for example, a chemical for a tillage operation.
3 Curative functions—difficulties encountered by one tillage operation that are corrected by a more extensive tillage.

To integrate all tillage operations, it is essential to examine each in relation to total effort. If tillage aims to avoid damage to soil structure, then one should avoid the least essential tillage. But if tillage aims to prevent weed seeds from germination, it should be limited to that objective and integrated with herbicides (Blake 1963).

Minimum and Zero Tillage

The search for substitutive functions of tillage operations has led to minimum and zero tillage practices.

The modern concept of minimum tillage dates back to the early 1950s with the introduction of the use of dalapon (2,2-dichloropropionic acid) to kill an existing grass cover prior to surface reseeding. Unfortunately, the effectiveness of the early herbicides varied with the composition of the weed flora. Those chemicals also left considerable residue in the soil. The discovery of paraquat (1,1'-dimethyl-4,4'-bipyridylium ion) and recently, glyphosate [N-(phosphono-methyl)glycine] stimulated considerable interest in reduced tillage techniques due to their broad spectrum activity and lack of toxic residues.

In the tropics, heavy rains destroy the structure of soil exposed by tillage thus reducing infiltration and leaving the soil susceptible to erosion. Minimum disturbance of soil under such conditions is desirable. As information from research on minimum and zero tillage in temperate regions has accumulated, the possibilities for research on minimum and zero tillage techniques in tropical crops have grown.

Seth et al. (1971) demonstrated in several regions of Asia that extensive tillage in rice may not be necessary if weeds are effectively controlled with preplant herbicides, and that yields could equal those following the use of traditional tillage. Tests in Malaysia have demonstrated the feasibility of using minimum cultivation techniques in several different rice crop cultures. Trials in Sri Lanka have also shown that both direct-seeded and transplanted rice can be established with minimum tillage (Mittra and Pieris 1968). Similar benefits were also suggested by Mabbayad and Buencosa (1967) from the Philippines.

Minimum and zero tillage techniques in rice involve the application of nonresidual, broad-spectrum, preplant herbicides to control weeds, and ratoon and volunteer rice. Fields are then flooded and a shallow cultivation and leveling are done (in the case of minimum tillage) followed by seeding or transplanting. The choice of herbicide and the amount used is determined by the degree of infestation and kinds of weeds. Often, a contact herbicide such as paraquat is adequate if most weeds are annual and weed density is light. When the density of broadleaf weeds and sedges is heavy, a follow-up application with MCPA (4-chloro-2-methylphenoxy acetic acid) or 2,4-D (2,4-dichlorophenoxy acetic acid) may be desirable. Where annual grasses or perennial weed species predominate, a translocated herbicide such as dalapon, or dalapon followed by a contact herbicide such as paraquat, gives better result.

Aspects of the minimum and zero tillage techniques that have only been lightly touched on are the relative roles of the preplant herbicides, tillage, and flooding. These three processes are complementary in obtaining good preplant weed control and, to a limited extent, compensatory. For example, if tillage is poor or hurried, this can be compensated for by extending the period of flooding from 3 to 7 days. This poses the question as to how the limited resources available to the rice farmer can be most effectively allocated, which will depend on the relative costs and availability of preplant herbicides, power for cultivation, and water for irrigation (De Datta and Barker 1978).

Experimental Results on Minimum and Zero Tillage Techniques in Rice

A study by Mittra and Pieris (1968) in Sri Lanka showed that paraquat application followed by flooding and cultivation provided good weed control (Table 8.6) with no adverse effect on yield. Seth et al. (1971) also found no difference in yield in comparing conventional tillage to minimum and zero tillage techniques in lowland rice where paraquat was used for preplant weed control. control.

Mabbayad et al. (1968) demonstrated that yield, panicle number, and percentage of sterile spikelets were not significantly different from soil prepared by zero, minimum, and conventional tillage. On a fertile and deep soil in the Philippines, tillage was shown to be unnecessary when rice was transplanted in land that had been sprayed with paraquat and pentachlorophenol. In those experiments, a problem sometimes encountered in obtaining good stand establishment in dense weed growth was the failure to get firm contact between

Table 8.6 Effect of Herbicide Application and Method of Land Preparation and Flooding on Weeds Growing in Association with Transplanted Rice (Adapted from Mittra and Pieris 1968)

Treatment[a]	Weeds (no./m²)		Wet Weed Weight[c] (t/ha)
	16 DS[b]	28 DS	42 DS
Conventional tillage	172	323	2.8b
Good cultivation fb flooding 3 days later	172	355	4.4b
Good cultivation fb flooding 7 days later	108	237	2.6b
Poor cultivation fb flooding 3 days later	172	312	12.4a
Poor cultivation fb flooding 7 days later	161	312	7.2ab
Paraquat (0.56 kg/ha) fb good cultivation fb flooding 3 days later	86	194	1.4b
Paraquat (0.56 kg/ha) fb good cultivation fb flooding 7 days later	75	172	2.0b
Paraquat (0.56 kg/ha) fb poor cultivation fb flooding 3 days later	161	226	4.2b
Paraquat (0.56 kg/ha) fb poor cultivation fb flooding 7 days later	129	291	6.5ab

[a]fb = followed by.
[b]DS = days after seeding.
[c]Means followed by the same letter are not significantly different at the 5% level.

seed and soil, which led to the retention of a single tillage to incorporate residual organic matter (Moomaw et al. 1968).

In trials from Malaysia, Seth et al. (1971) observed that weeds in the growing crop were generally fewer following the use of minimum tillage. However, where perennial weeds were present, continued practice of zero tillage resulted in rapid regeneration and an increased incidence of these weeds. In the same trials, weed incidence after planting was markedly reduced when a mixture of MCPA and paraquat was applied for preplant weed control.

TIME SAVING In terms of time savings with minimum tillage, Moomaw et al. (1968) reported reduction of weeding labor by 200 hours/ha compared with conventional tillage. In Indonesia, Varley (1970) noted that minimum tillage could reduce the time taken to prepare the land from 15–30 days to 5–10 days. When the preplant herbicide application was made before harvest of the previous crop, the time was reduced to as little as 2–4 days. Even with tillage time reduced from 25 days to 8 days, there was about 15% increase in return per unit area per day with continuous cropped rice.

In trials in Sri Lanka by Mittra and Pieris (1968), three rice crops were grown in the shortest possible time using conventional and minimum tillage. With minimum tillage, the interval between harvesting and transplanting the next crop was reduced to only 10 days, and production per unit area per day was increased by 22–30%.

WATER SAVING Minimum tillage, by reducing the time needed for land preparation, also reduces the amount of water used. With the usual wetland preparation time of 20–30 days, 30% of the total water needed for a crop of transplanted rice is used during land preparation. A saving of at least 20% of the total is possible by using minimum or zero tillage.

FERTILIZER MANAGEMENT At IRRI, fertilizer management practices were studied for conventional, minimum, and zero tillage. Nitrogen fertilizer efficiency, as measured by increase in grain yield, was considerably lower with zero tillage than with minimum and conventional tillage. However, the grain yield difference between minimum and zero tillage was not significant (Table 8.7). These results indicated that methods have to be developed to increase fertilizer nitrogen efficiency where minimum tillage techniques are used.

Research results with minimum and zero tillage have not, however, been

Table 8.7 Efficiency of Nitrogen Fertilizer as Influenced by Method of Land Preparation and Fertilizer Application; IRRI, 1976 Wet Season (De Datta et al. 1979)

Method of Application (56 kg N/ha)	N Fertilizer Efficiency[a] (kg rice/kg N)			
	Conventional Tillage	Minimum Tillage	Zero Tillage	Mean[b]
Split application	17	9	–6	7a
Supergranule placement	58	55	31	47b
Liquid band placement	44	47	18	34b
Mean	40b	37ab	13a	

[a]Based on comparison between mean yield of each treatment with 56 kg N/ha and mean yield of conventional tillage × split application treatment with 28 kg N/ha.
[b]Two means followed by the same letter are not significantly different at the 5% level.

consistent and it has been suggested that they may be limited to areas with no perennial weeds (De Datta et al. 1979). Recent IRRI studies indicate that minimum tillage is a dependable alternative to conventional tillage for transplanted rice, at least where difficult-to-control weeds are absent. Zero tillage should, however, be considered as a special technique for special conditions (De Datta et al. 1979).

Sod-Seeding of Rice

The extreme example of zero tillage is sod-seeding, a technique widely used by rice farmers in southern Australia. Rice is seeded directly in a well-grazed pasture sward with a sod-seeder that gives minimum disturbance of the soil. Before sowing, the pasture is heavily grazed by sheep. If the soil is too hard for sod-seeding, the field is irrigated before sowing. After sowing, the field is flush irrigated and grazed. The sheep may eat some rice tops but the plants soon recover.

Sod-seeding offers advantages of reduced cost of production, improved utilization of nitrogen fixed by leguminous pasture crops, greater utilization of pasture, less soil disturbance, less weed problem, firm ground ensuring easy harvest, and high grain yield. In conventional sod-seeding, pasture competition after sowing is eliminated by grazing or scalding, or both. In a newer technique called chemical sod-seeding, the pasture sward is eliminated with a presowing herbicide such as paraquat (Clough 1974).

Soil Compaction to Substitute for Puddling

Compaction of the soil with a roller has been suggested as a way to get the benefits of puddling without puddling (Varade and Ghildyal 1967).

Changes in Soil Physical Properties

Compaction is the increase in soil density caused by dynamic loading. On dynamic loading, the soil solids are rearranged and the compression of liquid and gas within the soil pores takes place. During compaction soil particles move to a closer state of contact, as a result of which both bulk density and porosity change. The degree of compaction depends on soil moisture, the compactive energy, the nature of the soil, and the amount of manipulation. The state of compaction affects the soil-air-water-temperature relationship profoundly and affects the physical, chemical, and biological properties of soil (Ghildyal 1978).

EFFECT ON GRAIN YIELD   In the field study, soil compaction to 1.88 g/cm^3 helped to conserve nitrogen, increased the efficiency of nitrogen fertilizer, and reduced hydraulic conductivity. Furthermore, compaction helped increase grain

Table 8.8   Effect of Puddling and Compaction on
Bulk Density of the Soil and Yield of Rice (Adapted from
Varade and Patil 1971)

Treatment	Bulk Density (g/cm^3)	Grain Yield $(g/plot)$
Control	1.54	1155
Puddling	1.71	1535
Compaction	1.88	2270

yield over that with puddling despite restricted root growth in compacted soil
layers (Table 8.8).

EFFECT ON WATER USE   Water use efficiency increases considerably after
compaction. Ghildyal (1978) reported that soil compaction to densities of 1.75
g/cm^3 reduced the water requirement of the crop considerably as compared with
puddling (Table 8.9). The primary reason for increased water-use efficiency after
compaction was due to reduced percolation losses of moisture in light-textured
soils. Compaction destroys the soil macrovoids and results in decreased
hydraulic conductivity and percolation losses.

Puddling on the other hand leaves some macrovoids intact causing lesser
reduction in hydraulic conductivity.

The factor most discouraging to adopting compaction instead of puddling a
soil, however, is that transplanting is impossible in a compacted soil. Because
transplanting into puddled soil has many distinct advantages over dibbling into
dry or moist compacted soil, transplanting will be difficult to replace in the
foreseeable future. If transplanting should be replaced it would be more
appropriate to replace it with direct seeding into dry soil without compacting it.

Table 8.9   Effect of Puddling and Compaction on Grain Yield, Water
Use, and Water Use Efficiency (Adapted from Ghildyal 1978)

Treatment	Grain Yield (t/ha)	Water Use (cm)	Water Use Efficiency (kg rice/cm of water used per ha)
Compaction 1.75 g/cm^3	8.43	215	39.2
Puddling	7.97	252	31.8
LSD (5%)[a]	0.39	70	2.31

[a]LSD = least significant difference at level given.

POWER AND ECONOMIC RESOURCES OF FARMERS AFFECTING CHOICE OF LAND PREPARATION

The power sources for tillage of rice soils are human, animal, and mechanical. The particular source or combination of sources varies from place to place.

The animal has been the principal tillage power source in Asia for thousands of years. In lowland rice culture, the water buffalo is commonly used. In dryland fields (upland, rainfed lowland, and deepwater) cattle are normally used in pairs. Water buffalo are seldom worked in pairs because they are difficult to train in pairs and because the draft in the wetland soil is light. The water buffalo works in the lowland field at a pace compatible to man and can be productively employed for 4–6 hours/day (Fig. 8.12).

Animal-power tillage equipment includes a wooden or steel plow, a wooden or steel harrow, a sled for leveling, and, in the case of upland rice, a wooden harrow for furrowing.

Human power is used in some parts of Asia, where farm size is small (less than 0.5 ha), labor is plentiful, and wage rates are extremely low (less than $0.30/day). The typical equipment is the hoe or mattock. At the other extreme, land preparation is almost completely mechanized in East Asia, where rural daily wages are high (currently about $8.00 in the Republic of Korea, $5.00 in Taiwan, and $15.00 in Japan). The principal form of mechanization in East Asia is the 10- to 15-hp tiller (De Datta and Barker 1978).

Figure 8.12   The water buffalo, such as in the Philippines, is a unique animal that can be used for land preparation of lowland fields at a work capacity of 4–6 hours/day.

Elsewhere in Asia there is considerable mechanization of land preparation but the motives are different from those in East Asia. Unlike in East Asia, tractors are not normally owned by the farmers but are generally contracted to do only the primary tillage, one of the more arduous tasks in rice production. The large tractor, 55 hp or larger, is used extensively for contract tillage in Thailand, Malaysia, and some areas in the Philippines (De Datta and Barker 1978). Annual hours of operation per tractor were reported as about 1360 for Thailand, 1040 for Malaysia, and 1400 for the Philippines. The speed of land preparation is greater with the large tractor, and it appears that the large tractor is efficient for initial land preparation if field size is not too small.

A bogging problem with large-tractor use is of great concern in irrigated areas continually planted to rice. Reports indicate that after 2–3 years of operation with the large tractor, the depth to the plow sole in paddies increased markedly, prohibiting further use of the large tractors (Nichols 1974).

The labor-, animal-, and machine horsepower-hours required for alternative power sources for primary and secondary tillage are shown in Table 8.10. Although labor requirements are reduced dramatically as more horsepower is added, the actual horsepower requirements for land preparation do not differ greatly for different power sources if a water buffalo is assumed as about equivalent to one horsepower. Thus, when labor wages are low, the cost of alternative sources of power can be approximately compared using the cost per rated horsepower. The rising cost and growing scarcity of animals (aggravated by

Table 8.10   Average Range of Power Requirements for Wetland Tillage for Rice Production, by Labor-, Animal-, and Rated Horsepower-Hours per Hectare (Adapted from Johnson 1963)

	Labor (h/ha)	Animal (h/ha)	Rated hp (h/ha)
Manpower only			
Hand hoe and spade			
Primary tillage	120–250		
Secondary tillage	120–250		
Animal power			
Animal plow—primary	50–100	50–100	
Comb harrow—secondary	60–100	60–100	
Power tiller, 6 hp			
Tractor plow—primary	20–40		100–200
Comb harrow—secondary	16–24		80–150
Power tiller, 10 hp			
Rotary tiller—primary	10–20		100–160
Rotary tiller—secondary	10–20		100–160
Tractor, 40–60 hp			
Moldboard plow—primary	2–3		75–125
Disc harrow—secondary	1–2		70–90

outbreaks of hoof-and-mouth disease) have been an important factor encouraging tractor use in some regions.

The potential benefits of mechanized land preparation include labor saving, reduction of heavy work load, increase in yield, timeliness of operation, and crop intensification. Although alternatives for employment of saved labor are limited, farmers often place a higher value on the saving of labor and the reduction of the work load than is normally realized. For example, in recent surveys in the Philippines, adoption of the tractor for tillage resulted principally in a reduction of family labor (De Datta and Barker 1978). Over a period of about 10 years, labor requirements for land preparation were reduced from 15–20 days/ha to 10 or less by mechanization.

Recent experiments in the Philippines demonstrated no apparent yield gain benefit from mechanized tillage (De Datta et al. 1978). The experiments used five combinations of land preparation equipment in different soils, with results in contrast with those from a number of farm surveys that reported higher yields from mechanized farms than from nonmechanized farms. But, in general, the studies failed to separate the effect of mechanization from the effect of fertilizer and other management and input factors (Binswanger 1978). Although more evidence is needed, the belief is that no significant yield gains are achieved by substituting mechanical for animal power.

A major benefit of mechanization that is frequently stressed is time saving, or shorter turnaround time. Reducing the time required for land preparation could have much the same effect as introducing a short-duration variety. The benefit may be in the reduced risk of damage to the second crop by drought or cold weather or in the possibility of adding one more crop per growing season.

Observations of production teams at work in the advanced communes in China suggest that they are taking advantage of timeliness in land preparation by using heavy machinery (IRRI 1978). Land preparation is mechanized in areas of China with a high population density and large agricultural labor force. The irrigation systems are highly developed making it possible to control water effectively at all times, which means operations are not slowed by delays in water delivery, or by flood or drought.

In South and Southeast Asia, neither the development of water resources nor the organization of rural labor makes it possible to take full advantage of the potential benefits from more timely tillage. In a recent survey in the Philippines, the only farmers who reported transplanting earlier as a result of mechanization were those who owned their tractors (De Datta et al. 1978). In most cases, the introduction of mechanization had reduced the time required for land preparation, but not the turnaround time. In recent survey of the literature on mechanization in South Asia (including nonrice crops), Binswanger (1978) concluded that the speed of cultivation made possible by tractors did not have an overriding advantage in the choice of cropping pattern or in crop intensification.

There are several important factors that influence the benefits and costs of, and hence the decision to use, a particular land preparation or crop establishment procedure in a given area.

In much of the rainfed rice-growing area, delayed planting results in a loss in yields due to:

- Greater probability of drought in the ripening stage, and
- A difference in environmental factors such as solar energy and temperature.

It takes about a month after the onset of the rains before there is sufficient moisture for puddling. Thus, farmers with rainfed fields typically transplant and harvest a month later than many of the farmers with irrigation. Depending on the rate of onset of the monsoons and the degree of flooding, further delays may also be encountered if conventional puddling and transplanting methods are used. It is practical in many rainfed areas to consider alternative methods of land preparation and planting.

The pattern of yield reduction will vary from year to year, place to place, and even by varietal type. An example of the types of yield reduction encountered, however, is illustrated in Fig. 8.13. The data were based on three surveys in a

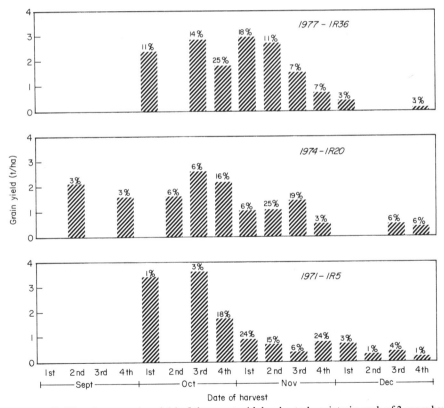

Figure 8.13   Average rice yield of the most widely planted variety in each of 3 years by date of harvest in rainfed areas of Bulacan and Nueva Ecija, Philippines, 1971–1977. (Figure above each bar refers to percent of area harvested.) (From De Datta et al. 1978)

rainfed area of Central Luzon during a 8-year period. Modern varieties were grown on nearly 70% of the farms as early as 1971. The average yield reduction was about 30 kg/day as the harvest extended into November and December. However, the introduction of IR36, an early maturing variety (100–110 days vs. 125 days for IR20 and 140 days for IR5) has had a dramatic impact on production and yield by allowing farmers to advance their harvesting date.

For land preparation in rainfed areas, the power requirement should be considered. A Philippine study determined the power requirements and rice yield differences when fields were prepared using different combinations of animal and tractor power on three irrigated and one rainfed site (Table 8.11). It cannot be assumed that rainfed areas typically require more power input per hectare. However, cone penetrometer readings of 2.46 kg/cm^2 (35 psi) were obtained at about 6-cm depth in the rainfed soil and at almost twice that depth in the irrigated soil indicating that the rainfed fields were not as well soaked. If a water buffalo works 6 hours/day, a hectare of rainfed soils can be prepared for planting in about 3 weeks. But the land prepared by water buffalo on rainfed farms in this area ranged from 0.5 to 1 ha. Thus, with available animal power, the time required to prepare a hectare may extend to a month or more (De Datta et al. 1978).

An economic survey conducted in the same rainfed village in the Philippines indicated that by employing a four-wheel tractor for primary tillage, land preparation time was reduced from 37 to 27 days. Assuming that this reduction in time could be translated into higher yields (Fig. 8.13) the value of the 300 kg additional yield ($40) is slightly higher than the custom rate for tillage of one hectare ($30). A few rainfed rice farmers in this area hired tractors when planting

Table 8.11 Alternative Land Preparation Treatments in Three Villages, Nueva Ecija Province, Philippines, 1973 Wet Season (Adapted from Orcino and Duff 1974)

| | Land Preparation Method | | | |
| | Primary | | Secondary[a] | |
Treatment	Power Source	Implement	Power Source	Implement
1	65-hp tractor	Rotary tiller	Animal	Comb harrow
2	14-hp tiller	Rotary tiller	Animal	Comb harrow
3	7-hp tiller	Moldboard plow	7-hp tiller	Comb harrow
4	Water buffalo	Moldboard plow	7-hp tiller	Comb harrow
5	Water buffalo	Moldboard plow	Animal	Comb harrow

[a]Secondary tillage consists of two passes over the field repeated three times at one-week intervals.

was delayed due to late rains. But rainfed rice farmers who grow only one crop of rice and obtain yields of less than 2 t/ha cannot afford the power investment needed to reduce the time required for puddling and transplanting the monsoon crops. The introduction of a short-season variety, such as IR36, clearly has a much greater impact on production and profits than the increase in farm power. The situation may differ in areas where double cropping is feasible.

The decision on method of tillage for rainfed rice is related to weed control, among many factors. Tillage is, in fact, a form of weed control. More thorough tillage reduces the weed control problem later, but rainfed farms tend to be underpowered. Furthermore, the intermittent rains allow drying of the rainfed paddy. That not only enccurages weed growth but also makes it impossible to recommend the inexpensive preemergence herbicide treatments that are used in irrigated areas. Despite these difficulties, weed control is less of a problem in transplanted rainfed than in broadcast rice. If weeds are not adequately controlled in the early stages of stand establishment, the yield losses in broadcast rice may be severe.

Changes in technology and factor-price relationship have resulted in significant changes in land preparation and crop establishment methods. Because of the heterogeneity in the rainfed environment there can be no typical set of recommended practices.

However, it appears that no major effort should be undertaken to subsidize the mechanization of land preparation in most countries, unless other barriers to increased production are removed first. Lack of farm power does not appear to be the dominant constraint to increased production and cropping intensity.

REFERENCES

Binswanger, H. P. 1978. *The economics of tractors in South Asia, an analytical review*. Agricultural Development Council and International Crops Research Institute for the Semi-arid Tropics. 96 pp.

Blake, G. R. 1963. Objectives of soil tillage related to field operations and soil management. *Neth. J. Agric. Sci.* (Spec. issue) 11:130–139.

Bodman, G. B., and J. Rubin. 1948. Soil puddling. *Soil Sci. Soc. Am. Proc.* 13:27–36.

Bolton, F. R., and S. K. De Datta. 1979. Dry soil mulching in tropical rice. *Soil Sci. Plant Nutr.* (Tokyo) 25:173–181.

Briones, A. A. 1977. Aggregation in drained and reclaimed paddies. Department of Soils, College of Agriculture, University of the Philippines at Los Baños. Los Baños, Philippines (unpubl. mimeo.)

Clough, R. A. 1974. New developments in sod seeding rice. *Farmers Newsl.* (N.S.W., Australia) 91:18–22.

Curfs, H. P. F. 1974. Soil preparation and weed control for upland and irrigated rice growing. Pages 79–86 *in* International Institute of Tropical Agriculture. *Report on the expert consultation meeting on the mechanization of rice production*. Ibadan, Nigeria.

De Datta, S. K., and R. Barker. 1978. Land preparation for rice soils. Pages 623–648 *in* International Rice Research Institute. *Soils and rice.* Los Baños, Philippines.

De Datta, S. K., and M. S. A. A. A. Kerim. 1974. Water and nitrogen economy of rainfed rice as affected by soil puddling. *Soil Sci. Soc. Am. Proc.* 38:515–518.

De Datta, S. K., and J. C. O'Toole. 1977. Screening deep-water rices for drought tolerance. Pages 83–92 *in* International Rice Research Institute. *Proceedings, 1976 deep-water rice workshop, 8–10 November, Bangkok, Thailand.* Los Baños, Philippines.

De Datta, S. K., F. R. Bolton, and W. L. Lin. 1979. Prospects for using minimum and zero tillage in tropical lowland rice. *Weed Res.* 19:9–15.

De Datta, S. K., R. A. Morris, and R. Barker. 1978. *Land preparation and crop establishment for rainfed lowland rice.* IRRI Res. Pap. Ser. 22. 24 pp.

De Datta, S. K., H. K. Krupp, E. I. Alvarez, and S. C. Modgal. 1973. Water management practices in flooded tropical rice. Pages 1–18 *in* International Rice Research Institute. *Water management in Philippine irrigation systems: research and operations.* Los Baños, Philippines.

Ghildyal, B. P. 1978. Effects of compaction and puddling on soil physical properties and rice growth. Pages 317–336 *in* International Rice Research Institute. *Soils and rice.* Los Baños, Philippines.

Harada, T., R. Hayashi, and A. Chikamoto. 1964. Effect of physical pretreatments of the soils on the mineralization of native organic nitrogen in paddy soils. *J. Sci. Soil Manure, Jpn.* 35:21–24.

Huey, B. A. 1977. *Rice production in Arkansas.* University of Arkansas, Division of Agriculture, and U.S. Department of Agriculture. Circ. 476. 51 pp.

Huynh, T. N., X. Vo-Tong, and S. A. Bowers. 1974. Water requirement of rice grown on clay soil as influenced by structure, water level and fertility. Pages 79–89 *in* Colorado State University Water Management Research Project. Annual report.

IRRI (International Rice Research Institute). 1974. *Annual report for 1973.* Los Baños, Philippines. 266 pp.

IRRI (International Rice Research Institute). 1978. *Rice research and production in China: an IRRI team's view.* Los Baños, Philippines. 119 pp.

Jacobi, B. 1974. Some aspects of soil tillage for lowland rice in Orissa (India). Pages 87–92 *in* International Institute of Tropical Agriculture. *Report on the expert consultation meeting on the mechanization of rice production.* Ibadan, Nigeria.

Jamison, V. C. 1953. Changes in air-water relationships due to structural improvement of soils. *Soil Sci.* 76:143–151.

Johnson, L. 1963. Power requirements in rice production. Paper presented at the Conference on Agricultural Engineering Aspects of Rice Production, August 1963, International Rice Research Institute, Los Baños, Philippines. 29 pp.

Johnston, T. H., and M. D. Miller. 1973. Culture. Pages 88–134 *in Rice in the United States: varieties and production.* USDA Agric. Handb. 289. Washington, D.C.

Kisu, M. 1978. Tillage properties of wet soils. Pages 307–316 *in* International Rice Research Institute. *Soils and rice.* Los Baños, Philippines.

Koenigs, F. F. R. 1961. *The mechanical stability of clay soils as influenced by moisture conditions and some other factors.* Wageningen, Verslagen van Landbouwkundige onderzockingen. 677 pp.

Koenigs, F. F. R. 1963. The puddling of clay soils. *Neth. J. Agric. Sci.* 11:145–156.

Kuipers, H. 1974. The objectives of soil tillage. Pages 61–70 *in* International Institute of Tropical Agriculture. *Report on the expert consultation meeting on the mechanization of rice production.* Ibadan, Nigeria.

Larson, W. E. 1963. Important soil parameters for evaluating tillage practices in the United States. *Neth. J. Agric. Sci.* (Spec. issue) 11:100–109.

Mabbayad, B. B., and I. A. Buencosa. 1967. Tests on "minimal tillage" of transplanted rice. *Philipp. Agric.* 51:541–551.

Mabbayad, B. B., B. N. Emerson, and E. L. Aragon. 1968. Further tests on "minimal tillage" and rates of nitrogen application on transplanted rice. *Philipp. Agric.* 52:200–210.

Mittra, M. K., and J. W. L. Pieris. 1968. Paraquat as an aide to paddy cultivation. Pages 668–674 *in Proceedings 9th British weed control conference.* Vol. 2

Moens, A. 1963. Soil tillage related to other farm practices. *Neth. J. Agric. Sci.* (Spec. issue) 11:128–129.

Moomaw, J. C., S. K. De Datta, D. E. Seaman, and P. Yogaratnam. 1968. New directions in weed control research for tropical rice. Pages 675–681 *in Proceedings 9th British weed control conference.* Vol. 2.

Moormann, F. R., and R. Dudal. 1964. *Characteristics of soils on which paddy is grown in relation to their capability classification.* International Rice Commission Working Party on Rice, Soils, Water and Fertilizer Practices. Manila, Philippines. 18 pp.

Nichols, F. E. 1974. Research and development of low-cost technology for rice production at IRRI. Pages 104–111 *in* International Institute of Tropical Agriculture. *Report on the expert consultation on the mechanization of rice production.* Ibadan, Nigeria.

Orcino, N., and B. Duff. 1974. Experimental results from alternative system of land preparation. Paper presented at a Saturday seminar, 2 July 1974 International Rice Research Institute, Los Baños, Philippines. 5 pp., tables. (unpubl. mimeo.)

Reynolds, E. B. 1954. *Research on rice production in Texas.* Texas Agric. Exp. Stn. Bull. 775. 29 pp.

Rice, C. M. 1943. Page 327 *in* Edwards Bros. *Dictionary of geological terms.* Ann Arbor, Michigan. Litho.

Sanchez, P. A. 1973a. Puddling tropical soils. 1. Growth and nutritional aspects. *Soil Sci.* 115:149–158.

Sanchez, P. A. 1973b. Puddling tropical soils. 2. Effects of water losses. *Soil Sci.* 115:303–308.

Sattar, S. A. 1978. Land preparation for transplanted rice. Paper presented at a Saturday seminar, 8 July 1978, Bangladesh Rice Research Institute, Dacca, Bangladesh. 10 pp. (unpubl. mimeo.)

Seth, A. K., C. H. Khaw, and J. M. Fua. 1971. Minimal and zero tillage techniques and post-planting weed control in rice. Pages 188–200 *in Proceedings symposium on 3rd Asian-Pacific weed science conference, Kuala Lumpur, Malaysia.*

Valera, A., O. Giron, and T. Wickham. 1975. Land preparation: its speed in relation to rate of water supply, new instrumentation for irrigation flow measurements. Paper presented at a Saturday seminar, 15 February 1975, International Rice Research Institute, Los Baños, Philippines. 14 pp., tables, figs. (unpubl. mimeo.)

Varade, S. B., and B. P. Ghildyal. 1967. Effect of varying bulk densities on lowland rice growth. *Il Riso* 16:33–40.

Varade, S. B., and E. A. Patil. 1971. Influence of soil compaction and nitrogen fertilization on growth of rice. *Il Riso* 20:219–223.

Varley, J. E. 1970. *Gramoxone minimum tillage.* Central Research Institute of Agriculture (CRIA) Staff meeting Pap. 27. Bogor, Indonesia. 14 pp.

Wickham, T. H., and V. P. Singh. 1978. Water movement through wet soils. Pages 337–358 *in* International Rice Research Institute. *Soils and rice.* Los Baños, Philippines.

9

Water Use and Water Management Practices for Rice

Water is indispensable to plant life. A plant's water content varies by species and within various plant structures and also varies diurnally during the entire growth period. The formative water for the plant is obtained mainly from the soil through absorption by the plant roots. The plant uses less than 5% of the water absorbed. The rest is lost to the atmosphere through transpiration from the plant leaves.

Kramer (1969) summarized the functions of water for a plant:

- It is a vital constituent of cell protoplasm.
- It is a reactant or reagent in chemical reactions.
- It is a solvent for organic and inorganic solutes and gases facilitating their translocation within the plant.
- It gives mechanical strength to the plant by producing turgidity.

An adequate water supply is one of the most important factors in rice production. In many parts of tropical Asia, rice plants suffer from either too much or too little water because of irregular rainfall and landscape patterns.

Water management embraces the control of water for optimum crop yield and the best use of a limited supply of water. Proper management of water and irrigation systems, especially those that rely on stored water, enables a water supply during the dry season when yields generally are high due to high solar radiation and greater fertilizer nitrogen response.

EFFECTS OF FLOODING

A main reason for flooding a rice field is that most rice varieties maintain better growth and produce higher grain yields when grown in a flooded soil than when grown in a nonflooded soil. Water affects the physical character of the rice plant, the nutrient and physical status of the soils, and the nature and extent of weed growth.

Physical Characters of the Rice Plant

The height of the rice plant is directly related to the depth of water in the paddy; the plant height generally increases with increasing water depth. Tiller number of a rice plant, on the other hand, appears inversely related to water depth, at least over a relatively wide range of moisture conditions. With progressive drying of the soil, tiller number decreases, and much more sharply than under the influence of increased water depth.

Culm strength of the rice plant, and therefore lodging resistance, decreases as the plant height increases. Thus, culm strength decreases if culms elongate as water depth increases. There is no evidence that grain-straw ratio is affected by water management practice in the rice field.

Senewiratne and Mikkelsen (1961) compared growth of rice plants in submerged and upland soil. At the early growth stage of rice, plants grew larger in the nonsubmerged upland soil than on the submerged soil. But later there were more increases in tiller number, plant height, and leaf area in the submerged soil than in the nonsubmerged soil.

Nutrient Status and Physical Characteristics of Flooded Soils

One benefit of submerging rice soils is that it increases the availability of many nutrients, particularly phosphorus, potassium, calcium, silicon, and iron. But if the soil is highly permeable, nutrients will be leached downward from the root zone. The processes on nutrient transformations in flooded soil are described in Chapter 4. However, the question of whether internal drainage is desirable is often raised. There are advantages and disadvantages of internal drainage of rice soils.

The advantages of internal drainage are depression of the concentration of carbon dioxide, iron, and reducing substances, and prevention of the buildup of high concentrations of carbon dioxide, iron, and organic acids in cold soils. The disadvantages of internal drainage are loss of water, and loss of nutrients (Table 9.1).

Internal drainage is desirable on cold soils, saline and alkaline soils, and soils irrigated with saline and alkaline water. But in the tropics studies are critically needed to ascertain the benefits of internal drainage in normal soils.

Soil Toxicity

Flooding a soil causes chemical reduction of iron and manganese, as well as other elements in the soil. Various organic acids such as acetic and butyric, and gases

Table 9.1 Loss of Nutrients During One Season from Three Flooded Soils, in 210-liter Drums, Provided with Drainage at 1 cm/day at 20 cm Below the Soil Surface (From Ponnamperuma 1971)

Soil	Nutrient Loss (kg/ha)					
	NH_4^+–N	P	Ca	Mg	Fe	Mn
Luisiana clay	247	0.5	678	1072	1187	247
Maahas clay	56	0.7	1262	1678	55	289
Pila clay loam	61	10.8	1680	1990	22	118

such as carbon dioxide, methane, and hydrogen sulfide are produced. All except methane, when present in large amounts, may retard root development, inhibit nutrient absorption, and cause root rot (usually between the seedling and panicle initiation stages). These toxic effects are variously identified as physiological diseases, such as *Akiochi* (degraded paddy soils) in Japan and *bronzing* in Sri Lanka (see Chapter 10). Toxicity is most often noticed when oxygen in the soil is depleted due to the rapid decomposition of large quantities of organic matter.

Oxygen may be brought into the soil by allowing drainage with moderate drying. The reduced substances are then oxidized and the toxic gases may escape through the soil surface. Percolating water can bring oxygen into the soil and leach toxic substances beyond the rooting zone. Percolation rates of 2–3 mm/day may correct toxicity problems.

Water Temperature

Both high and low water temperatures have adverse effects on growth and grain yield of rice. In Japan, optimum water temperatures have been reported to be between 25 and 30° C. Water temperatures greater than 30° C have not shown adverse effects on IR8 and similar indica varieties in Pakistan. Where adverse effects of high water temperatures have been reported to reduce rice yield, reduced uptake of silicon, potassium (sometimes resulting in brown spot caused by *Helminthosporium oryzae*), reduced tiller number, and increased percentage of unfilled spikelet have also been reported.

At low water temperatures some indica varieties grow poorly, usually with reduced tillering. At later stages of rice growth, low water temperature delays panicle initiation, decreases panicle size, and increases sterility. Nutrient uptake is adversely affected by low (15° C) water temperature (Bhattacharyya and De Datta 1971).

Water Depth and Weed Population

Emergence of weeds and the types of weeds in a weed population are closely related to the moisture content of the soil and the water depth in the rice field.

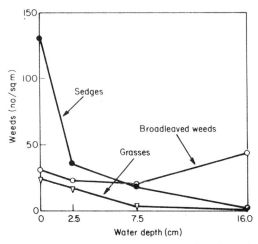

Figure 9.1 Effect of water depth on weed population in a field of IR8 at 28 days after transplanting. IRRI, 1968 wet season. (From De Datta et al. 1973b)

Conditions that favor weed growth also make weed control difficult. Moist, but unflooded soil, warm temperatures, and adequate light favor growth of grass. In addition, the lack of standing water hinders the effective distribution of granular herbicides, and higher temperatures and light may stimulate rapid decomposition of herbicidal components of some compounds.

Water control during the early stages of crop growth has a major effect on weed control. As weeds become established, it is much more difficult to control them through water management. For transplanted rice, proper management can substantially substitute for weeding. Grasses can be completely eliminated if continuous flooding to a 16-cm depth is maintained throughout crop growth. Even with 5 cm of continuous standing water, grasses are substantially controlled. Infestations of other weeds vary greatly with depth of standing water (Fig. 9.1). Sedges are almost completely controlled by continuous flooding at depths of 15 cm and greater. Broadleaf weed infestation at various water depths is difficult to predict.

Use of a 10–25-cm water depth has long been practiced in California where water seeding is a common method of stand establishment. However, considerable savings of water are possible if a combination of herbicides and good water management practices are followed (see Chapter 12).

TYPES OF WATER LOSS FROM RICE FIELDS

Water requirement varies among crops. Because rice is a semiaquatic plant, it requires more water than most other crops. Water to produce optimum yields of rice must satisfy the evapotranspiration needs of the crop and losses from the paddy areas through percolation and seepage. Timely water supply is equally important for optimum growth and high grain yields of rice. The amount of water required varies with the growth duration of the rice variety, type of soil, topography of land, and so on. Water consumption also varies with the stage of the crop.

The principal moisture losses from the rice paddy may be grouped into vapor losses and losses in liquid form. Both can be determined by field measurements. The vapor losses can be further grouped into loss by transpiration from the leaf surface and by evaporation at the water surface. The two types of liquid losses are:

- The downward movement or vertical percolation of free water, which relieves the surface soil and upper subsoil of superfluous moisture.
- The runoff of excess water over the field levees.

Evapotranspiration or Consumptive Water Use

When the soil is maintained in a saturated or flooded condition, evapotranspiration is primarily a function of the energy available for evaporation of the water. The combined losses of water resulting from plant transpiration and surface evaporation are called evapotranspiration (ET). It is also commonly referred to as consumptive water use. ET is generally measured in tanks, lysimeters, and plots; consumptive water use is measured for a larger area. Both are expressed in millimeters of water depth over the area considered.

Transpiration

Transpiration is the process by which plants release water vapor to the atmosphere through surface pores (stomatal openings) in the plant foliage in response to the atmospheric demand. The water of transpiration usually reaches a maximum value in the afternoon and the minimum value just before the sun rises. Soil moisture content and plant characters such as location and distribution of stomata, the reduction of transpiration surface (leaf rolling), and plant age affect transpiration rate.

Evaporation

Evaporation is the moisture lost in vapor form from the free water surface where rice is grown. Evaporation is one of the important factors that determine the effectivity of rainfall, particularly in arid or semiarid areas. Because shading of the water surface reduces evaporation, daily evaporation losses from the paddy surface are less for rice planted at close spacings and decrease as a crop approaches maturity.

Rate of Evapotranspiration

The ET rate is affected by:

- SOLAR ENERGY   The higher the solar energy, the higher the ET. The solar energy incident on water and plants varies greatly from area to area and from day to day within a given area.
- TEMPERATURE   Higher temperature increases evaporation of water.
- WIND OR AIR MOVEMENT   A dry wind continually sweeps away moisture vapor from a wet surface.
- RELATIVE HUMIDITY   ET is higher at lower relative humidity (e.g., 40%) than at higher relative humidity (e.g., 80%).
- PLANT CHARACTERISTICS   ET is influenced by plant characteristics including leaf morphology, depth of rooting, and growth duration.
- SOIL WATER REGIME   If the soil is saturated or submerged, ET will be at maximum rate. However, ET will be reduced with decrease in soil moisture content.

In most of the tropics, the evapotranspiration requirement during the wet season is 4–5 mm/day. During the dry season, 6–7 mm/day may be required for large irrigated areas. For small irrigated areas, the dry season ET may be high because of advective energy (energy brought in by wind from drier unirrigated areas).

Determination of Evapotranspiration

The amount of water available to the roots of rice plants depends on the balance between rainfall and evaporation and the relationship between soil moisture content, water potential and conductivity, effective rooting depth, and water table (Yoshida 1979).

Tomar and O'Toole (1979) suggest the following benefits of accurate measurement, or estimation, of evapotranspiration:

- Better engineering design, and management of irrigation facilities.
- Development of sound practices by irrigation agronomists.
- Determination of the water balance of rainfed rice and estimation of supplemental irrigation as water conservation goals to meet the crop's water requirement.
- Evaluation of cropping pattern's suitability based on the water balance estimate for a particular area.
- Classification of rice environment where genetic and agronomic technology may be transferrable.

The determination of actual or potential evapotranspiration has wide utility and has been studied in detail from lowland rice fields. Three approaches have been used to determine the evapotranspiration.

1 THE HYDROLOGICAL OR WATER BALANCE APPROACH This includes methods such as catchment hydrology, soil moisture sampling, and lysimetry. The most accurate is lysimetry but it is expensive.

For upland crops, weighing lysimeters are necessary for daily or short time interval measurements. For lowland rice however, a different principle is used. The change in water level in square or circular tank lysimeters is measured to refer to evapotranspiration. Recently, Tomar and O'Toole (1980) proposed a simple and sensitive microlysimeter to measure daily or hourly evapotranspiration from lowland rice fields. That microlysimeter consists of two principal parts: a polyvinyl chloride (PVC) cylinder with a closed bottom and a Mariotte system to maintain a constant water level in the PVC cylinder and to serve as a reservoir-cum-manometer. ET is calculated by measuring the decrease in water column height in the reservoir-cum-manometer. Before field installation, the unit is calibrated in the laboratory. The microlysimeter is installed 1 or 2 days before transplanting by excavating the soil. The excavated soil is replaced layerwise to the field level in the lysimeter cylinder. As shown in Fig. 9.2, the Mariotte system is connected to the PVC cylinder. After installation, the microlysimeter is ready for measurements.

2 MICROMETEOROLOGICAL APPROACH This includes such diverse methods as aerodynamic or mass transport (profile method; eddy correlation method), energy balance (Bowen ratio method), and combination of aerodynamic and energy balance methods. The energy balance method is a favored micrometeorological method.

The energy balance equation can be used to derive precise estimates of ET. The energy balance equation is

$$Rn = LE + H + G + P + M \qquad (1)$$

where Rn is the net radiation flux, LE is the flux of latent heat (L is the latent heat of vaporization and E is the quantity of water evaporated) and represents ET losses, H is the sensible heat flux above the surface, and G is the ground heat flux. P represents photosynthesis and M represents miscellaneous exchanges. Because the sum of P and M is small it is normally neglected in the estimation. Thus, equation (1) becomes,

$$RN = LE + H + G \qquad (2)$$

Using the Bowen ratio $\beta = H/LE$, equation (2) can be solved for LE as follows:

$$LE = \frac{Rn - G}{1 + \beta} \qquad (3)$$

Thus, by measuring net radiation and ground heat flux and knowing the value of β, ET can be estimated.

In the rice-growing areas of monsoonal Asia, monthly average solar radiation ranges from 300 to 650 cal/cm^2 per day. The corresponding range of potential evapotransporation is 3–7 mm/day. Yoshida (1979) observed that computed

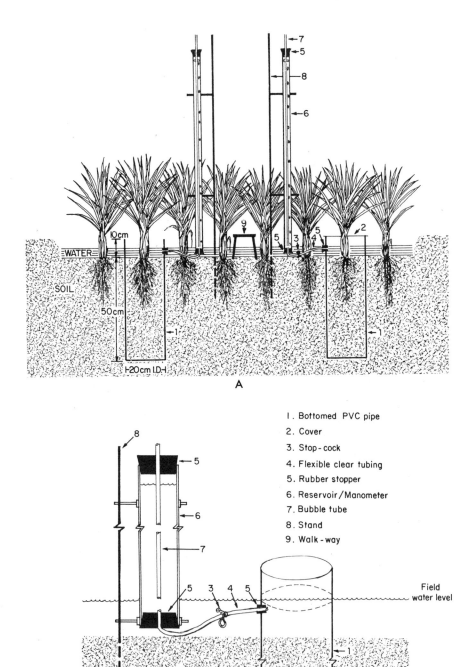

1. Bottomed PVC pipe
2. Cover
3. Stop-cock
4. Flexible clear tubing
5. Rubber stopper
6. Reservoir/Manometer
7. Bubble tube
8. Stand
9. Walk-way

Figure 9.2 Cross section of field-installed microlysimeters for measuring evapotranspiration and transpiration (A) and details of Mariotte System (B). (From Tomar and O'Toole 1980)

304

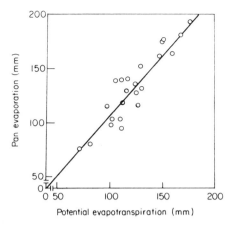

Figure 9.3   Relationship between potential evapotranspiration and pan evaporation for 1972 and 1973, IRRI. (From Yoshida 1979)

potential evapotranspiration was closely correlated with the observed United States Class A pan evaporation rate at IRRI (Fig. 9.3). On an average, the potential evapotranspiration is related to the pan evaporation as:

$$E = 0.93 \times \text{pan evaporation} \qquad (4)$$

Thus, the estimated potential evapotranspiration of a lowland rice field is slightly lower than the observed pan evaporation.

De Datta et al. (1973b) examined the relationship between solar radiation (Rs) and pan evaporation at IRRI and obtained two regression lines for 1971 and 1972. The two regression lines were basically the same and the combined regression line is given as:

$$\text{pan evaporation} = 0.7546 + 0.0096\ Rs \qquad (5)$$

where Rs is the incident solar radiation (cal/cm^2 per day).

Yoshida (1979) suggested further simplicity and the equation can be expressed by:

$$\text{pan evaporation} = 0.0113\ Rs \qquad (6)$$

Pan evaporation rates calculated from equations (5) and (6) generally agree for the range of solar radiation from 300 to 700 cal/cm^2 per day.

Tomar and O'Toole (1979) estimated evapotranspiration by using incident solar radiation as

$$\text{ET (mm/day)} = 0.9 + 0.0115\ Rs \qquad (7)$$

3   EMPIRICAL METHODS   Many empirical climatological estimates have been used to estimate potential evapotranspiration.

The Food and Agriculture Organization (Doorenbos and Pruitt 1977) suggested Blaney-Criddle, radiation, Penman, and pan evaporation methods to calculate reference crop evapotranspiration (ET_0) from available climatological data. ET_0 is defined as the rate of evapotranspiration from an extensive surface of 8–15 cm tall, green grass cover of uniform height, actively growing, completely shading the ground, and not short of water. The effect of crop characteristics on crop water requirements is given by the crop coefficient (kc), which presents the relationship between reference crop evapotranspiration (ET_0) and actual crop evapotranspiration, or $ET = kc \times ET_0$.

In several countries U.S. Weather Bureau Class A pan or sunken pan (installed in the rice field) data have been correlated with actually measured ET. Evaporation pans provide measurement that integrates the effect of solar radiation, wind, temperature, and humidity on evaporation from a specific open-water surface. To eliminate the effect of these meteorological factors, the values of E, transpiration, and ET have been divided by the value of pan evaporation (EP). These ratios can then be used to compare the difference between variety, cropping seasons, and latitudes (Tomar and O'Toole 1979).

The ET-EP ratio in rice starts around 1.0 after transplanting, shows a first peak (1.1 and 1.3) at the maximum tillering stage, and a maximum value at about heading stage. The value at heading time became 1.4–1.5 in Malaysia and 1.7 in the Philippines (Tomar and O'Toole 1979). On the average, this value is 1.43 for tropical countries. It is 1.33 for Japan and 1.37 for Australia (Evans 1971) (Fig. 9.4). Thus, on the average, this ratio is 1.0 at the time of transplanting, reaches 1.15 after 15–20 days, and attains a maximum of 1.3–1.4 at about heading.

Tomar and O'Toole concluded that over the whole rice crop growth period, ET from lowland rice fields is more than open pan evaporation during that period. They suggest the relationship:

$$ET = 1.2 \; EP \qquad \text{(Fig. 9.5)}$$

Reviewing the various methods of estimating evapotranspiration, Wickham and Sen (1978) contended that although Penman's equation is accurate, the input data required to use it is difficult to collect in many countries.

It is true that installation and operation of an evaporation pan is much less expensive than installation and operation of solarimeters. There are, however, cases when solar radiation data are available but pan evaporation data are not easily available. Consequently, there is a need for a simple model that relates solar radiation data to evapotranspiration.

Yoshida (1979) attempted to simplify energy balance of a lowland rice field with a model that uses solar energy data. The model agrees well with measurements and:

· Helps understand mechanisms of evapotranspiration.
· Provides means by which evapotranspiration can be estimated from solar energy data.
· Gives quantitative estimates of water deficit for drought stress.

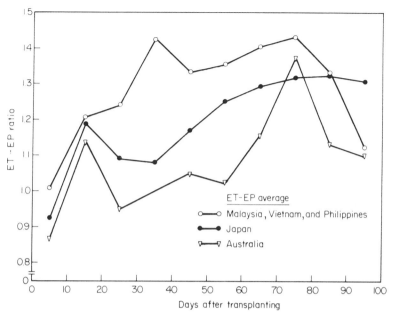

Figure 9.4 Growth stage relationships of evapotranspiration (ET)–pan evaporation (EP) ratio for lowland rice in different countries. (From Tomar and O'Toole 1979)

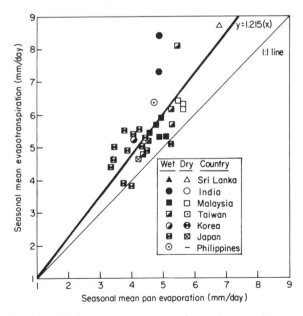

Figure 9.5 Relationship between pan evaporation and seasonal evapotranspiration at different locations. (From Tomar and O'Toole 1979)

• Provides means with which measured evapotranspiration can be examined if serious error in measurement is suspected.

Percolation Losses

Percolation losses are a function of the local soil and topographic conditions. Therefore, at any time the amount of rainfall or irrigation water entering a soil becomes greater than its water-holding capacity, loss by the downward movement of free water (vertical percolation) will occur. Percolation is often defined as the movement of moisture through saturated soils due to gravity, hydrostatic pressure, or both.

Where the soil is heavy and the water table is close to the soil surface, percolation losses are low—about 1 mm/day or less. Where soil is light and the water table is deep, percolation losses may be high—10 mm/day or more. If percolation rates are high, it is difficult to maintain a saturated or flooded soil.

Other factors that affect percolation losses are presence of a crop, the amount and distribution of rainfall, soil shrinkage and cracking, soil compaction, flooding and depth of water, and soil puddling.

Research in Japan has indicated that a percolation rate of 10–15 mm/day was favorable for supply of dissolved oxygen, the removal of harmful substances, and the maintenance of root activity. However, there was little benefit on yield under good soil conditions. In fact, with some situations, the loss of plant nutrients may be serious if the percolation rate is high.

Sugimoto (1969) evaluated percolation rate in relation to grain yield in an experiment in tanks in Malaysia. The enameled iron tanks had a drainage pipe of 1.25 cm diameter at the side near the bottom of the tank. The tanks had a 5-cm gravel layer and an 18.8-cm layer of soil over the gravel. The percolation rate was regulated by a pinchcock. Compared with the no-percolation, stagnant-water treatment, grain yield of Bahagia rice was increased by a 10 mm/day percolation rate. A percolation rate of 20 mm/day gave lower yield than percolation at 10 mm/day (Table 9.2). Circulated percolation decreased grain yield, probably due to accumulation of toxic substances.

Various studies suggest that the range of percolation varies between wide limits from less than 1 mm/day in compact soil up to several hundred millimeters per day in loose soil (Wann 1978). Field studies in the Philippines in the dry season have shown mean percolation rates to be 1.3 mm/day on alluvial and elastic soils with water tables between 0.5 and 2 m, and 2.6 mm/day when the water table is deeper than 2 m (Kampen 1970).

Seepage Losses

Seepage losses are most important where the lowland rice field borders a natural or artificial drainage channel. In that situation, the water lost usually is not

Table 9.2 Rice Yield and Its Components as Affected by Various Percolation Rates (Adapted from Sugimoto 1969)

Percolation Rate[a] (mm/day)	Heading Date	Angle of Lodging (°)	Weight of Straw (g)	No. of Spikelets per Panicle	Ripened Grains (%)	Weight of 100 Grains (g)	Relative Yield
None	9 Nov	50	45	148	71	25	100
10	10 Nov	65	50	146	73	25	108
20	10 Nov	45	58	141	68	25	103
20[b]	10 Nov	45	52	145	67	25	93

[a]Treatments were imposed after initial stage of spikelet differentiation.
[b]Circulated percolation regulated by a pinchcock.

available for crop production in the immediate area. But seepage losses in upper areas of a sloping landscape can provide substantial extra water to the lower area.

Seepage losses can be reduced to small proportions with adequate management of water and maintenance of paddy bunds. There are two kinds of seepage losses:

- *Perimeter seepage,* which is that water moving from a rice-growing area into a creek or areas not planted to rice, and is therefore considered a water loss.
- *Levee seepage,* which is the lateral subsurface movement of water within a rice-growing area, and is not a loss except where it occurs through the last levee (bund) separating the field from a drain.

Seepage is difficult to estimate because of wide variation in pore spaces and alternate wetting and drying of bunds near the water surface. Excessive seepage losses can often be traced to animal or insect burrows through bunds. Because perimeter seepage usually collects in drains, greater losses can be expected where the drainage density is high.

In a series of experiments, Thongtawee (1965) studied subsurface lateral seepage through a nonplastic loam soil near a creek. Lateral seepage to a dryland area was 25 liters/hour per meter of levee during the first weeks of the season, and increased to a maximum of 58 liters/hour per meter 8 weeks later. Average losses from the plot were 49 mm per day, which was about 10 times the daily rate of ET.

Along a 590-m length of creek draining fine silty clay soils of an irrigation system in Nueva Ecija province, Philippines, Wickham and Sen (1978) found a seepage loss of 15.6 liters/hour per meter of creek. The data were collected while rice fields on both sides of the creek had standing water.

Other causes of high seepage losses from rice-growing areas are:

- High perimeter-area ratios, which are usually found when farmers irrigate small areas separately in the dry season, or where rice-growing areas are separated by upland (dryland) or drains.
- Drying and cracking of soil, which results in high seepage losses when water is resupplied.
- Soil texture's effect on both seepage and percolation.

Most large creeks and rivers have light-textured soils along their banks; seepage losses are high when those areas are planted to rice because the soils have a high rate of internal drainage (Wickham and Sen 1978).

Combined Seepage and Percolation Losses

Seepage normally flows onto the soil surface or into streams, rivers, or drainage waterways, while percolation flow usually moves to the water table. Because they occur simultaneously and are difficult to separate in the field, seepage and percolation are usually considered together. Wickham (1978) reported combined losses of 0–2 mm/day in the wet season and 2–6 mm/day in the dry season on representative alluvial elastic soils in the Philippines. These rates were about half the rates of ET for each season. Seepage and percolation from soils with unfavorable conditions can be as high as 20 mm/day for both seasons.

Measurement of Seepage and Percolation

Seepage and percolation is strongly conditioned by the depth to water table, proximity to drains, sloping fields, poorly maintained bunds, and greater depths of standing water on the paddy. Three techniques of measuring seepage and percolation are suitable for field use (Wickham 1978):

1   A *cylinder method* uses a system of two cylinders buried in the soil with rice growing inside them. The tops of the cylinders are open above the soil and water surface. One cylinder is open at the bottom and the other closed (Fig. 9.6). Both cylinders are filled with water to a given reference level daily. The water required to refill the cylinder with the closed bottom is the ET rate, and the water required to refill the open bottom cylinder is ET plus seepage and percolation. Subtracting the first from the second gives the seepage and percolation rate. The prevention of lateral seepage, the unrepresentative soil placed inside the cylinders, and enhanced percolation at the interface between the cylinder wall and the soil are shortcomings of this measurement technique.

2   A *manometer method* measures the fall in water level with a simple meter. The fall is assumed to be the sum of ET and seepage and percolation,

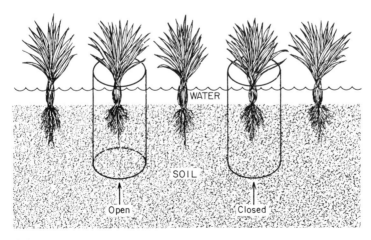

Figure 9.6   Seepage and percolation measurement with open and closed cylinders. (From Wickham 1978)

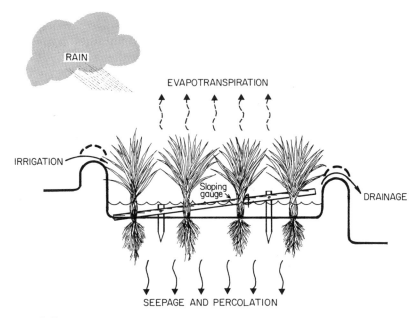

Figure 9.7   Seepage and percolation measurement with a sloping gauge. (Adapted from Wickham 1978)

provided there is no irrigation or rainfall contributing water and no surface drainage from the paddy. Seepage and percolation are determined by subtracting rough estimates of ET. The technique uses soils *in situ* and is quite accurate provided there is no surface water flow. Rainfall can be included in the calculations. For added precision, the meter can be placed in the water at an angle to magnify the difference in changes of water level (Fig. 9.7).

3   A *water balance technique* can measure the amount of added water and its loss components. This technique is probably the most useful because it is representative of a broad area rather than just one paddy or small area; at least 20 ha should usually be included if possible. However, the necessary flow measurements of irrigation and surface drainage are difficult and expensive, and frequently cannot be made with accuracy. In general, accuracy of water balance analysis increases when applied to larger areas.

Surface Drainage

Surface drainage represents a major water loss in almost all irrigated or rainfed sites and is most severe during the wet season. Drainage results from the supply of additional water to filled paddies. Although drainage water is sometimes available for later irrigation by use of pumps or a downstream check-dam, it is generally considered lost for a particular crop area and crop season.

WATER USE AND MOISTURE STRESS EFFECTS AT DIFFERENT GROWTH STAGES

Water use and moisture stress effects vary at different growth stages of rice. Sufficient moisture supply is more critical in some growth stages than others. Moisture stress reduces crop yield most when it occurs during the critical growth stages.

Water Use

It is generally believed that cereal crops show a marked sensitivity to moisture stress during the formation of the reproductive organs and during flowering. By and large, cereal crops can withstand and recover from mild or relatively brief periods of moisture stress if favorable conditions are quickly reestablished. With more severe stress, the preflowering stage is the least sensitive and the anthesis and spikelet-filling stages are the most sensitive. Matsushima (1962) reported that rice is most sensitive to moisture stress from 20 days before heading to 10 days after heading. Van de Goor (1950) had earlier reported that flooded rice used the maximum amount of water at that time. That suggests that the critical period for moisture stress coincides with the period in which plants use the most water.

Seasonal water requirement differs with different growth stages. For water management practices of rice, the growth stages of rice can be divided as the seedling, vegetative growth, reproductive, and ripening stages. In areas with low rainfall or a highly variable rainfall pattern, irrigation practices should be developed to assure needed water at the critical growth stages.

Seedling Stage

The water requirement of rice is low at the seedling stage. In fact, if seeds are submerged, the development of radicles is affected by lack of oxygen supply.

Vegetative Growth Stages

The production of an adequate number of tillers is an important factor in rice yields. Immediately after transplanting, sufficient water should be provided to facilitate early rooting. Following the early rooting stage, a shallow water depth facilitates tiller production and promotes firm root anchorage in the soil. Excessive water at this stage seriously hampers rooting and decreases tiller production. Leaf blades and leaf sheaths of the submerged plants become weak, turn light green, and break easily.

Reproductive Growth Stage

Reproductive growth starts when maximum tiller production is completed and includes the panicle primordia development, booting, heading, and flowering stages. A large amount of water is consumed in the major part of the reproductive growth period, which explains why rice is sensitive to moisture stress during reproductive growth.

Two factors should be considered for water management at this stage. Drought at this stage causes severe damage, particularly when it occurs from panicle initiation to flowering stages. Increased panicle sterility, caused by impeded panicle formation, heading, flowering, or fertilization, occurs if sufficient moisture is not provided.

The other factor is excessive water at the reproductive stage, particularly at the booting stage, which causes decrease in culm strength and increases lodging.

Ripening Stage

The last phase of the growing period includes the milk, dough, yellowish, and full ripening grain stages. Very little water is needed at this period and after the yellowish ripening stage no standing water is required. This allows draining of a field about 10 days before harvest and facilitates harvesting by machine. Rodent damage may increase on drained fields, however.

Moisture Stress Effects at Different Growth Stages

It is generally believed that the peak water demand of rice is between maximum tillering and the grain filling stage. In an IRRI experiment on Maahas clay (Aquic Tropudalf), moisture stress imposed at various growth stages was allowed to reach 50 centibars (cb), after which 5 cm of water was added to the plot.

Table 9.3   Grain Yield and Water Use of IR8 Grown in Tanks With and Without Bottoms; IRRI, 1969 Dry Season (From De Datta et al. 1973a)

		Tanks with Bottoms		Tanks without Bottoms	
Stress Period[a]	Time to Maturity (days)	Yield (g/m²)	Water Use (mm)	Yield (t/ha)	Water Use (mm)
None	123	910	618	7.16	1147
T–MT	131	770	653	5.84	1435
T–PI	133	730	632	4.68	1438
T–H	145	720	558	3.76	1121
MT–H	127	880	593	6.31	1178
PI–M	124	760	528	5.87	730
H–M	124	840	544	6.10	904
T–M	152	170	257	1.84	432

[a]T = transplanting; MT = maximum tillering; PI = panicle initiation; H = heading; M = maturity.

Table 9.3 shows the grain yield and water use of IR8. Moisture stress throughout the entire growth period reduced the grain yield to 20–25% of the yield of the continually flooded treatment. When moisture stress was allowed to develop only between the maximum tillering and the heading stages, the grain yields remained relatively high. The reduction in grain yield of IR8 grown on the Maahas clay was more related to the duration of moisture stress than to the stages of plant growth at which the stress occurred.

In subsequent greenhouse studies with IR8, IR5, and a traditional variety H-4, moisture stress early in the growth of the rice plant reduced tillering (Table 9.4), thereby reducing grain yield. If the stress was relieved before the reproductive phase began, some recovery in grain yield occurred through an increase in

Table 9.4   Yield Components of the Three Rice Varieties Subjected to Moisture Stress at Different Growth Stages in the Greenhouse (From De Datta et al. 1973a)

Stress Period[a]	Yield Component				
	Tillers (no./hill)	Panicles (no./hill)	Grains per Panicle (no.)	Weight of 100 Grains (g)	Unfilled Spikelets (%)
IR8					
None	7.2	7.2	89	2.55	22
T–MT	4.8	4.4	114	2.57	24
T–PI	4.0	3.9	116	2.52	17
T–H	4.1	3.5	91	2.77	17
MT–H	8.2	7.7	76	2.50	18
PI–M	7.3	6.5	67	2.50	20
H–M	7.7	7.4	79	2.40	34
T–M	7.8	6.2	—	—	—
IR5					
None	8.6	8.5	112	2.67	10
T–MT	5.5	5.4	128	2.63	10
T–PI	4.7	4.7	101	2.63	16
T–H	5.4	5.4	89	2.61	15
MT–H	9.2	9.2	84	2.53	8
PI–M	9.3	9.0	92	2.55	14
H–M	9.5	8.7	109	2.43	16
T–M	4.7	4.2	—	—	—
H-4					
None	8.2	7.6	150	2.57	21
T–MT	4.2	4.1	170	2.73	25
T–PI	4.0	4.0	148	2.60	15
T–H	3.5	3.2	118	2.52	27
MT–H	6.8	6.5	95	2.75	28
PI–M	8.0	7.3	75	2.53	46
H–M	10.9	9.8	69	2.32	53
T–M	3.5	1.4	—	—	—

[a]T = transplanting; MT = maximum tillering; PI = panicle initiation; H = heading; M = maturity.

number of spikelets per panicle. But if the stress period extended into the reproductive phase, a reduction in number of filled spikelets decreased grain yield further.

Moisture stress in the late vegetative and reproductive phases (MT-H and H-M) resulted in a decrease in grain yield through a reduction in number of spikelets per panicle, percentage of filled spikelets, and the 100-grain weight. Thus, the contribution of panicle number increases with later moisture stresses.

The traditional variety (H-4) was particularly sensitive to moisture stress in the spikelet filling stage; stress-caused filled spikelet percentage and the 100-grain weight decreased markedly. The results with H-4 are similar to the results reported by Matsushima (1962). In field and greenhouse experiments, there was a negative linear relationship between the relative grain yield of IR8 and the duration of the stress period, although the lines had different slopes for each experiment. The relative yield of IR8 after stress from maximum tillering to the heading stages fell significantly above the line of best fit to the other data points (Fig. 9.8). These data suggest that grain yield is less affected by moisture stress at this growth stage than at the other stages.

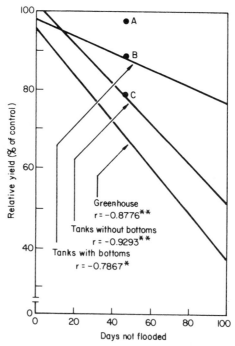

Figure 9.8   Relative grain yield as a function of the number of days soil was not flooded. Yield of IR8 from treatments with water stress from maximum tillering to heading in tanks with bottoms is shown by point A, in tanks without bottoms by point B, and in the greenhouse by point C. (From De Datta et al. 1973a)

Those results further suggest that varietal differences exist in response to moisture stress at different growth stages.

Total Water Requirements

Total water requirement includes water needed to raise seedlings, prepare land, and to grow a crop of rice from transplanting to harvest. The amount is determined by many factors. Those include soil type, topography, proximity to drains, depth of water table, area of contiguous rice fields, maintenance of levees, fertility of both top and subsoil, field duration of the crop, land preparation method, and, most of all, evaporative demand of the growing season.

Raising Seedlings

It is estimated that 150–200 mm of water is needed for nursery preparation and 250–400 mm is needed for irrigation to raise seedlings for 30–40 days, a period generally used by many farmers in South and Southeast Asia. For modern varieties, 21-day-old seedlings are generally considered optimum for transplanting. However, 16–35-day-old seedlings of modern varieties are not uncommon at the time of transplanting. Only 667 m^2 (for wet seedbed) is needed to raise enough seedlings for a hectare of the main field.

Land Preparation

The amount of water needed to prepare land depends mainly on soil type and water holding capacity, but most importantly on type of land preparation. Water is used 1–2 days before plowing to moisten the soil for easy working. After a prolonged dry period, large quantities of water are needed to fully moisten the soil for tillage by draft animals.

Kung and Atthayodhin (1968) suggested that 200 mm of water is commonly used for land preparation and initial flooding. However, the amount varies from country to country (Table 9.5).

In a comprehensive study on total water requirement for land preparation of irrigated lowland rice, Wickham and Sen (1978) reported that 656 mm was used for the 48-day period ending when 50% of the area was transplanted (Fig. 9.9). Land soaking, computed from the date water was first turned in to the date of primary tillage, used 110 mm of water. Seepage and percolation losses, plus water incorporated in the soil after it was plowed were 396 mm. Another 150 mm was lost through surface evaporation.

Field Irrigation

Rice plants require a large amount of water after transplanting. Field duration from transplanting to crop maturity is generally 90–120 days but with early-

Table 9.5   Water Use for Land Preparation in Various Asian Countries (Adapted from Kung and Atthayodhin 1968)

Place	Days Taken for Land Preparation (no.)	Depth of Water Required (mm)	Remarks
China, Szechwan		200–400	
China, Kwangtung		130–150	
Japan		120 (90–150)	
Rep. of Korea, Suweon		122 (36–164)	
Taiwan		150–200	
Taiwan	15	180	Heavy soil
Taiwan	2	150–180	Light soil
India, West Bengal		298[a]	
Bangladesh		180	
Thailand, Supanburi		300–400	
Malaysia		180	
West Malaysia	30	350	
Philippines	30	220–290	Clay soil
Philippines	52	723[b]	From first application of water to 50% of the area transplanted
Philippines	48	650[b]	

[a]Including land soaking.
[b]For large irrigation system blocks.

maturing varieties the field duration is reduced by 10–20 days. The amount of water required in the field depends on the water depth maintained, water management practices, soil types, and evaporative demand.

Water requirement in the field from transplanting to harvest is between 800 and 1200 mm, with a daily consumption of 6–10 mm (Kung and Atthayodhin 1968). In Japan, rice with an irrigation period of 90 days requires 1000 mm of water; rice with an irrigation period of 140 days requires 1400 mm of water.

Puddling the main field, as opposed to a nonpuddled system increases the amount of water use by a rice crop (see Chapter 8).

WATER MANAGEMENT SYSTEMS: CHARACTERISTICS AND LIMITATIONS

The effort required to implement a specific water management practice increases as the amount of water available decreases and as the desired degree of water control increases. Minimum efforts are required for continuous flooding

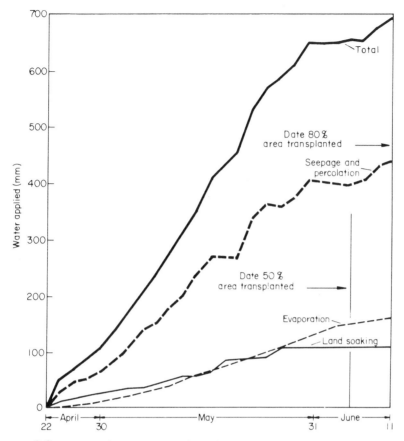

Figure 9.9 Cumulative water required (in mm) for land soaking, seepage and percolation, evaporation, and their sum during land preparation of a 145-ha area with clay loam soil. Talavera, N. E., Philippines, 1977 wet season. (From Wickham and Sen 1978)

practices, with adequate water supply. These relationships do not hold true in areas planted with deepwater rice.

The influence of water management practices on grain yield, growth characteristics of the rice plant, water requirements, and other management efforts are discussed below.

Continuous Flooding, Static—Shallow to Medium (2.5–7.5 cm)

Continuous flooding at a static 2.5–7.5 cm depth provides the potential to produce optimum rice yields. Experiments at IRRI show no differences in grain yield between 2.5 and 7.5 cm (Table 9.6).

Table 9.6   Effect of Water Management Practices on the Grain Yield and Efficiency of Water Use for IR8; IRRI, 1968 Dry Season[a] (Adapted from De Datta and Williams 1968)

Water Management Practice Treatment	Plots with Drainage			
	Total Water Use (mm)	Index (%)	Efficiency (g/liter)	Grain Yield (t/ha)[b]
Intermediate continuous flooding (7.5) cm	850	60	1.1	9.7 a
Shallow continuous flooding (2.5 cm)	805	57	1.2	9.5 ab
Intermediate continuous flooding + continuous soil saturation (7.5 cm + 1.0 cm)	800	56	1.2	9.4 ab
Continuous soil saturation + flooding at panicle initiation (1.0 cm + 7.5 cm)	780	55	1.2	9.1 abc
Deep continuous flooding (15 cm) + drainage at maximum tillering	1344	95	0.7	9.1 abc
Continuous soil saturation (1.0 cm)	647	46	1.4	9.0 abc
Deep continuous flooding (15 cm)	1418	100	0.6	8.9 bc
Deep continuous flooding (15 cm) + drainage at maximum tillering + drainage at panicle initiation	1240	87	0.7	8.5 c

[a] Evaporation = 378 mm (91 days). Evapotranspiration = 589 mm (91 days). Rainfall = 29.5 mm (91 days). Crop duration (days) = 126. Irrigation started, 28 January 1968. Irrigation stopped, 27 April 1968 (91 days).
[b] Means followed by the same letter are not significantly different at the 5% level. CV = 3.4%

Growth Characteristics

Changes in plant height, tiller number, and lodging resistance are minimal for a given variety.

Water Requirements

Between 600 and 800 mm of water is consumed during 85–91 days of field duration (from transplanting to crop maturity). Percolation losses vary with soil and water table condition.

Water Use Efficiency

Water use efficiency, measured by weight of grains per liter of water, was 1.1–1.2 g/liter depending on water depth (Table 9.6).

Water Management Requirements

With adequate water supply, management needed is minimal. With limited irrigation water, the paddies should be maintained to trap as much natural rainfall as possible.

Other Management Requirements

With a progressive increase in water depth from 2.5 to 7.5 cm, management practices such as fertilizer application and weed control are easily employed.

Water Temperature

Daytime water temperature fluctuation is moderate. The range is less than for the fluctuation of air temperatures.

Continuous Flooding, Static—Deep (15 cm or more)

Generally continuous flooding at 15 cm or more has the potential to produce yields similar to those at 2.5 cm water depth. However, in some dry seasons a 15-cm depth or more may reduce grain yield. Table 9.6 shows yield reduction was significant between 2.5 and 15 cm of water. Results will vary with variety. At IRRI, the traditional variety H-4 had a big reduction in yield due to deep flooding while the modern variety IR8 was not affected (Table 9.7).

Growth Characteristics

Plant height will increase substantially and tiller number will decrease at a water depth of 15 cm or more. The magnitude of these effects is related to the specific

Table 9.7 Effect of Water Management Practices on the Grain Yields of IR8 and H-4 under Natural Paddy Conditions; IRRI, 1966 Wet Season (Adapted from De Datta and Williams 1968)

Treatment	Grain Yield (t/ha)	
	IR8	H-4
Deep continuous flooding (10 cm)	5.1	1.6
Midseason drainage	5.3	2.4
Shallow continuous flooding (2.5 cm)	5.1	2.1
Rainfed	4.6	3.2

variety. For example, the traditional variety H-4 lodged because of weakened straw and additional height as a result of deep flooding.

Water Use Efficiency

Water use efficiency is lower (0.6 g/liter) at 15 cm or more than under shallow flooding (1.2 g/liter) (Table 9.6).

Water Requirements

Optimum water requirements are about 600 mm during a 90–100-day period. Seepage and percolation losses are generally higher because of greater water head. The total water requirement (excluding water for land preparation) is about 1400 mm at 15 cm of continuous flooding (Table 9.6).

Other Management Requirements

Management requirements are at a minimum, but care should be taken to provide water control structures and drainage facilities to minimize over-flooding.

Water Temperature

Daytime water temperature fluctuation is slight. Generally, the mean temperature of water is about the same as the mean air temperature.

Continuous Flowing Irrigation

The practice of continuous flowing irrigation may be useful if irrigation water temperature is high. Such is the case in the Kyushu and Shikoku districts in Japan, where field water temperature is often as high as 40°C in July and August. Therefore, the practice is desirable in high temperature areas because it will lower the water and soil temperatures, check abnormal soil reduction, and reduce sterility in rice grains. However, it is desirable to reduce the water temperature before it is introduced in the main field.

On the other hand in Hokkaido, Tohoku, or in intermountain areas in central Japan and in parts of the Republic of Korea it is not uncommon to register water temperatures of 25°C or lower. Under those conditions, continuous flowing irrigation may increase the water and soil temperature.

For both hot and cold water, continuous flowing irrigation may increase the availability of soil nutrients. Generally, flowing irrigation will keep the surface soil oxidized, a highly desirable effect in soils that are strongly reduced.

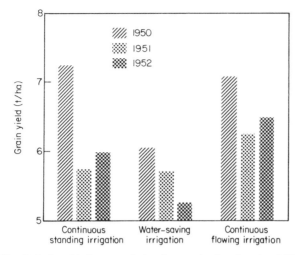

Figure 9.10   Relationship between irrigation method and grain yield of rice in a warm region in Japan. (Adapted from Matsubayashi et al. 1963)

Grain Yield

Continuous flowing irrigation has the potential to produce optimum rice yields. For example, in an experiment in Kyushu district in Japan, rice yields with continuous flowing irrigation (3–5 cm depth) were generally higher than the continuous, static submerged plot (5 cm depth) or a water-saving irrigation plot (Fig. 9.10). The water temperature was lowest in the continuous flowing irrigation plot and highest in water-saving irrigation plot.

In an experiment at Japan's Niigata Agricultural Experiment Station, flowing irrigation practiced from 35 days before heading to maturity gave about 7% increase in yield over that from continuous static submerged plots. There was, however, 8% reduction in grain yield if flowing irrigation was practiced throughout crop growth, indicating the period of flowing irrigation may be critical (Matsubayashi et al. 1963). A study in the Philippines showed that continuous flowing irrigation plots give yields similar to those from continuous static submerged plots at various depths (Table 9.8).

Growth Characteristics

Growth characteristics for rice in continuous flowing irrigation are essentially the same as in a similar flooding depth with static water conditions, except in those limited areas where temperature effects are noted. In one study in Japan the percentage of ripened grains decreased as the water temperature dropped to 25° C or lower for an early-maturing variety and dropped to 26° C and lower in the case of a late-maturing variety.

Table 9.8 Effect of Water Management Practices[a] on Rice Variety IR8; IRRI, 1967 Dry Season (From De Datta and Williams 1968)

Water Management Practice	Grain Yield		Straw Yield		Grain-Straw Ratio	Water Use		Efficiency		Percent Unfilled Spikelets	Weight of 100 Grains (g)
	(t/ha)	Index (%)	(t/ha)	Index (%)		Total (mm)	Index (%)	Grains (g/liter)	Grains + Straw (g/liter)		
Midseason drainage	8.5	101	8.1	100	1.0	867	95	0.98	1.94	18.7	3.0
Deep continuous flooding	8.4	100	8.2	100	1.0	910	100	0.93	1.82	16.1	3.0
Continuous flowing irrigation	8.3	98	8.2	100	1.0	4581	503	0.18	0.36	11.4	3.0
Shallow continuous flooding	8.2	97	8.5	104	1.0	804	88	1.02	2.07	21.4	3.0
Intermittent drainage + flooding	7.7	91	5.9	72	1.3	850	93	0.90	1.60	15.1	3.0
Alternate irrigation + flooding	7.5	88	6.2	76	1.2	832	91	0.90	1.64	22.7	3.0
Alternate irrigation	7.2	85	5.4	66	1.3	695	76	1.03	1.80	28.8	2.9
Irrigation at moisture stress	3.2	38	1.6	20	2.0	197	22	—	—	37.5	3.8

[a]CV = 17%. Number of irrigation days = 31 Jan–30 April 1967 = 90. Evapotranspiration = 441 mm. Evaporation = 285 mm.

Water Use Efficiency

Continuous flowing irrigation is the least efficient irrigation practice. Production was 0.18 g of grain per liter in continuously flowing irrigated plots (Table 9.8).

Water Requirements

With continuous flowing irrigation the environmental requirements of ET, and seepage and percolation losses, are similar to static flooding at the same depth. These total between 600 and 1000 mm during a 90–100-day irrigation period. The flow requirement will vary depending upon management practice. A flow requirement of 1200–1500 mm has been commonly found when deliberate drainage is practiced to provide a form of flow flooding in the Philippines and Malaysia. When designed as a part of controlled system, the flow requirement can be reduced considerably.

In IRRI experimental fields, as much as 4581 mm of water were used from transplanting to maturity for a continuous flowing treatment (Table 9.8).

Water Management Requirements

Where water supply is available in large quantities, the management requirement for continuous flowing irrigation is limited to maintaining the spillway height of the lowland rice field to control outflows consistent with irrigation inflow rates and the drainage system's capability. Where the water supply is limited, the water distribution must be managed to prevent excess water from leaving the productive area. For efficient water use, the water flowing from the paddies must be recaptured for use on lower paddies.

Other Management Requirements

A continuous flowing irrigation system causes some loss of nitrogen (5–10 kg N/ha) from the soils. However, if the adjacent paddy also belongs to the same farmer, nitrogen will be added there. Flowing irrigation may not be desirable during the time of herbicide and insecticide applications.

Water Temperature

With continuous flowing irrigation, water temperature tends to be lower in the hot areas and warmer in cold areas. Daytime water temperature fluctuations are relatively low.

Rotational Irrigation

Rotational irrigation is the application of required amounts of water to fields at regular intervals. The field may often be without standing water between irrigations, but ideally the soil does not dry enough for moisture stress to develop. Rotational irrigation is often recommended to irrigate a large area with a limited water supply to ensure better equity among water users. A major advantage of rotational irrigation is possibly the more effective use of rainfall in the rice field.

In Taiwan, for example, rotational irrigation is as effective as continuous static flooding if not better. In the Philippines, the method did not receive much attention until the Upper Pampanga River Irrigation Project was initiated. That system was designed to irrigate thousands of hectares at 5-day intervals, with an equivalent irrigation water flow of 13 mm/day at the offtake of a 40–50-ha block.

Grain Yield

Even though grain yield increases sharply with increase in irrigation intensity between 4 and 8 mm/day (Fig. 9.11), the rotational irrigation practice has the potential for optimum yield. In an experiment in the Philippines, grain yields did not vary greatly with 4–8-day intervals between irrigation. Grain yield, however, dropped 1 t/ha, or more, when the irrigation interval was increased to 10 days (Table 9.9).

Growth Characteristics

Plant height, tiller number, leaf area index, and dry matter production generally decreased as the irrigation interval increased from 4 to 10 days (Table 9.9). Lodging resistance is generally greater in the rotational irrigation system.

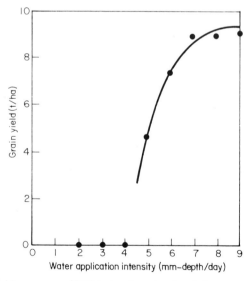

Figure 9.11   Yield response of IR8 to treatments of varied water application intensity. IRRI, 1969 dry season. (Adapted from Wickham and Sen 1978)

Table 9.9   Effects of Rotational Irrigation Practices on Two Rice Varieties; IRRI, 1971 Dry Season (From De Datta et al. 1973b)

Irrigation Interval (days)	Plant Height[a] (cm)	Tillers[a] (no./hill)	Leaf Area Index[b]	Total Dry Matter (g/hill)	Yield (t/ha)
IR20 (116)[c]					
4	102	18	6.5	53	7.2
6	97	18	5.4	47	7.1
8	98	17	5.2	49	6.8
10	92	12	4.2	37	5.6
IR480-5-9 (121)[c]					
4	93	17	6.5	55	7.0
6	87	15	5.5	46	6.6
8	86	15	5.1	52	6.4
10	79	13	—	47	5.2

[a]At harvest.
[b]At flowering.
[c]Growth duration in days.

Water Use Efficiency

Water use efficiency decreases as the growth duration and irrigation rate increase regardless of irrigation interval (Fig. 9.12).

Water Requirements

The water requirement for rotational irrigation is moderate to low. ET requirements are about 600 mm during a 90–100-day irrigation period. Seepage losses are relatively low. Percolation losses are slightly less than in continuous flooding methods. With good water management, the total requirement may be as low as 600–700 mm.

Management Practice Requirements

Rotational irrigation has not been widely adopted because it requires highly trained irrigation personnel as well as good farmer cooperation. Conveyance systems must be equipped with additional structures such as division boxes and flow measuring devices, and weed growth is greater when the plots lack standing water for a time. Water requirements for typical design in Taiwan vary between 1000 and 1300 mm with a moderate level of irrigation management within the system and on the farms. With effective water management, satisfactory yields were obtained with 650 mm delivered to the farms.

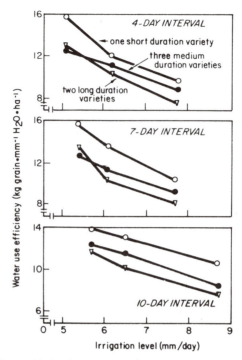

Figure 9.12 Effects of irrigation level and interval on the water use efficiency of rice varieties with different growth durations. IRRI, 1972 dry season. (From De Datta et al. 1973b)

Other Management Requirements

Generally, other management requirements are increased with rotational irrigation. For example, weed control and fertilizer nitrogen applications have to be coordinated with irrigation rates and intervals. Distribution and stability of herbicides are likely to be poor unless specific procedures for water management during the first 7–10 days after transplanting are developed for effective herbicide use. Under usual procedures, manual or mechanical weeding may be necessary.

Because of alternate moist (or dry) and fully flooded conditions some nitrogen loss is possible, which calls for nitrogen application in split doses.

Water Temperature

Temperatures are variable with rotational irrigation depending on the level of water in the field. Immediately after irrigation, conditions approximate those in the medium static flooding condition. Daytime water temperature fluctuations increase as the water level decreases, and remain high until the next irrigation.

Midseason Soil Drying

Japanese farmers practice midsummer soil drying or midsummer drainage (*nakaboshi*) of the rice paddy. The reason for the practice is not very clear. The primary benefit in Japan is to change the root zone temporarily to an oxidized state, which in some cases prevents root-rot disease. Removal of anaerobic toxins and carbon dioxide is a distinct advantage of midseason drying of some soils (Table 9.10). Another advantage is regulation of nutrient supply to the crop, particularly nitrogen supply at the later stages, to suppress growth of late tillers. Some irrigation water may also be saved, but where water supply and control are poor, as is the case in the Asian tropics, midseason drying may subject the rice crop to undue water stress. Other disadvantages include possible root pruning caused by soil shrinkage, the reversal of beneficial pH changes, and increased loss of nitrogen.

Midsummer drainage is done late in the tillering stage, prior to the early panicle formation stage. At that stage the number of panicles is fixed and the requirement of water by the rice crop is minimal. However, when physiological disease symptoms such as reddish brown sheath rot (*akagare*) occur earlier than the late tillering stage, water should be immediately removed from the paddies or at least reduced.

In Japan, midsummer drainage is advocated in soils where rapid soil reduction takes place and where excessive nitrogen applied as a fertilizer (which delays ripening) should be removed by rapid soil oxidation at the late stages of rice growth.

Water Management Between Crops

The choice of water management practice between crops is largely determined by the farmer's objectives. The choice is mainly between dry and flooded fallow.

Dry Fallow

The dry fallow condition is a common practice in most rice-growing areas. Allowing the soil to dry between crops saves water and hastens ammonification on reflooding. The disadvantages are reversal of reduction and loss of about 40–80 kg N/ha per season.

Flooded Fallow

With flooded fallow conditions, standing water is maintained between crops. The advantages are favorable chemical environment, reduction of nitrogen loss, and accretion of about 150 kg N/ha per year. The disadvantages are high water use, accumulation of organic substances, salts, and alkali, and an increase in zinc deficiency.

Table 9.10 Influence of Midseason Soil Drying on the Kinetics of CO_2, Iron, Organic Acids, and Reducing Substances (From Ponnamperuma 1971)

		Weeks Submerged						
		4	6	8	10	12	14
CO_2 (atm)	Submerged	0.43	0.50	0.49	0.42	0.40	0.26
	Submerged with MSD[a]	0.44	0.52	0.28	0.30	0.27	0.24
Fe^{2+} (ppm)	Submerged	51	120	245	289	385	235
	Submerged with MSD	57	138	94	144	210	112
Organic acids (mmole/liter)	Submerged	3.7	6.3	8.4	10.8	11.3	10.7
	Submerged with MSD	3.1	5.9	4.7	6.2	7.8	7.8
Reducing substances (me/liter)	Submerged	25	27	26	28	26	20
	Submerged with MSD	18	20	17	14	14	11

[a]7 weeks after submergence. MSD = midseason drainage.

Rainfed

More than 50% of the world's rice-growing area depends on rainfall for water, with rice grown mostly in rainfed paddies. Wherever possible, good water management is highly important for raising the production of rainfed lowland rice in South and Southeast Asia. If water supply to the crop is good, grain yield will be similar to irrigated rice.

Growth Characteristics

Plant growth for rainfed rice depends on the amount of rainfall and distribution throughout the season. Plant height is generally reduced and semidwarf rices often do not perform as well as intermediate-statured rices (about 130 cm tall). With dry soil conditions, the intermediate-statured rices will be as tall as the semidwarfs are with ideal irrigation. If there is stagnant water 30–50 cm deep in rainfed fields, the intermediate-statured rice has the best chance to produce 2–3 t/ha.

Water Requirement

Rainfall of 900–1000 mm during the growing season is adequate to produce optimum yields with moderate level of water management. More rainfall is necessary when management is less satisfactory. In an experiment during the 1966 wet season, total water use for rainfed rice was about half (457 mm) of that with 10 cm of continuous flooding (803 mm). Percolation losses were considerably lower for rainfed condition (Table 9.11).

Water Use Efficiency

Water use efficiency in rainfed fields is poor unless the rainfall is well distributed over the growing season and is near the total amount required by the crop.

Water Management Requirements

Where normal rainfall exceeds 1500 mm, which is fairly common in most Southeast Asia, and is reasonably well distributed throughout the growing season, relatively little management is required unless surplus water accumulates in a field due to low topographic position. Sometimes drainage may be needed to eliminate excess water. Nonuniform rainfall distribution is common in all of monsoonal Asia, however.

Where the normal rainfall is about 1000 mm or less, and distribution is often uneven, as is the case for most of South Asia, careful water management must be practiced. Bunds must be carefully maintained to minimize seepage and surface drainage losses. Higher bunds, carefully maintained, allow greater depth of water to be retained in the lowland rice fields.

Table 9.11 Effect of Irrigation Treatments on Total Water Use by IR8, Losses Due to Evapotranspiration, Evaporation, and Other Losses; IRRI, 1966 Wet Season (From De Datta and Williams 1968)

Treatment	Total Water Use[a]		Evapotranspiration (ET) Loss[b] (mm)	Evaporation Loss[c] (mm)	Other Losses Including Percolation	
	Depth (mm)	Index (%)			Depth (mm)	Index (%)
Deep continuous flooding (10 cm)	803	100	396	308	407	51
Midseason drainage (5–10 cm)	635	79	396	308	239	38
Shallow continuous flooding (2.5 cm)	568	71	396	308	172	30
Rainfed	457	57	396	308	62	13

[a]Total water use for 85 irrigation days. Recorded rainfall during 23 July and 16 September was 457 mm.
[b]Measured from the ET tank with rice plants growing inside and outside the tank.
[c]Measured from the evaporation tank with rice plants growing surrounding the tank.

Other Management Requirements

Except in those circumstances where continuous flooding can be maintained under natural rainfall, conditions for germination and growth of weeds tend to approach the ideal. At the same time, conditions for weed control with herbicides are poor. Hand or mechanical weeding provides the most effective weed control. Herbicides can be used if water is available to keep the paddies flooded during 7–10 days after transplanting.

Fertilizer nitrogen management is equally difficult in rainfed rice because of large losses of nitrogen. Because of alternate wetting and drying, nitrogen is lost through denitrification (see Chapter 4). If leaching losses are high due to high water head, nitrogen losses through leaching may also be substantial.

Water Temperature

Water temperature will vary depending upon the depth of water in the rainfed paddy and the climatic conditions at a given time.

WATER MANAGEMENT PRACTICES FOR CONTINUOUS CROPPING

With increased demand for all food crops, and particularly rice, efforts have been increased to raise total food production. One way to achieve that is to increase cropping intensity. Good water management is an important prerequisite to increasing cropping intensity.

Continuous Rice Cropping

When adequate and assured water supply is available, continuous rice cropping with at least three crops a year is possible where year-round favorable temperature provides that opportunity. Experiments suggest that continual flooding is not essential for high grain yield but modern rice varieties can tolerate at least 15 cm of water depth without adversely affecting grain yield. In an IRRI experiment, yields on continually shallow flooded plots were similar to those on plots continually flooded at 10-cm depth (Table 9.7). Deeper submergence has other advantages, however, such as suppression of weed growth, higher efficiency of fertilizer, and better insect and weed control with granular chemicals. Considering all factors, continuous submergence with 5–7.5 cm of water is probably best for continuous cropping of irrigated rice.

Rice-Based Cropping Systems

It is becoming increasingly evident that continuous year-round rice cropping is often undesirable even where there is an adequate water supply. The growing of other crops in rotation with rice means a large increase in both nutritional and economic benefits. In addition, rice pest pressure is less under a good rotational system.

In heavy textured soils, however, it is difficult to switch from puddled soil to the nonpuddled soil, which is essential to grow an upland crop. Strong arguments have been presented for the practice of furrow irrigating rice on nonpuddled soil as a desirable alternative to soil puddling for rice cultivation. Maintaining the soil in a nonpuddled or dryland condition allows the ready insertion of other crops into a rotation centered around rice.

Research data (IRRI 1971) show that more than one-third of the water involved in ET is from the surface of the standing water in a rice field. Thus, the practice of furrow irrigation in a nonpuddled soil might lower the requirement of rice for irrigation water. Evaporation from the field would be retarded by the mulching effect of the dry surface soil. Deep percolation losses could also be reduced. The importance and benefits to be derived by reducing the irrigation requirement of rice have been discussed by Young (1970).

WATER MANAGEMENT IN
DIRECT-SEEDED FLOODED RICE

In the Asian tropics rice is primarily transplanted for stand establishment. However, in most of Sri Lanka, and in parts of India, and Bangladesh, rice is broadcast-seeded either in the dry soil or a wet or moist soil. Seeding into standing water is not common in the tropics because lack of proper water control and low oxygen concentration in water under the high temperatures that occur in the tropics lead to poor stand establishment. There is experimental evidence, however, which suggests that broadcast seeding into water is a distinct possibility if proper water management can be provided.

In an experiment by De Datta et al. (1973b), as the water depth was increased for direct-seeded flooded rice, crop establishment or number of plants per square meter was decreased. A brief drainage period at the maximum tillering and panicle initiation stages reduced lodging but increased weed populations (Table 9.12).

In the United States, where rice is entirely direct-seeded, systems of water management for rice production vary widely depending on method of seeding, soil type, climate, crop rotation, diseases, and insects.

Essentially, there are two broad systems of water management revolving around seeding method. One is used in drill seeding or broadcast seeding in dry soil. The other involves water seeding, wherein fields are flooded just before aerial seeding and usually remain flooded until they are drained for harvest.

Dry-Seeded Flooded Rice

Dry seeding of flooded rice is generally practiced in Arkansas, Mississippi, Texas, and parts of Louisiana. Flood irrigation is used for all rice grown in the

Table 9.12 Effects of Water Depth and Management on Grain Yield and Growth Characteristics of Broadcast-seeded Flooded Rice (Average of IR8 and IR22 Rices); IRRI, 1970 Dry Season (From De Datta et al. 1973b)

Water Management Treatment	Yield (t/ha)	Plants[a] (no./m^2)	Weeds[b] (no./m^2)	Plant Height[c] (cm)	Tillers[c] (no./m^2)	Panicles[d] (no./m^2)	Lodging[d] (%)
2.5 cm continually flooded	6.5	262	311	64	768	463	48
5 cm continually flooded	6.6	237	254	66	709	466	52
10 cm continually flooded	6.4	206	158	68	649	413	54
20 cm continually flooded	5.8	197	80	76	535	360	81
5 cm drained at MT[e]	6.9	229	289	63	664	465	33
5 cm drained at PI[f]	7.0	248	215	66	664	465	31

[a]14 days after seeding.
[b]30 days after seeding.
[c]Maximum tillering.
[d]At harvest.
[e]MT = Maximum tillering.
[f]PI = Panicle initiation.

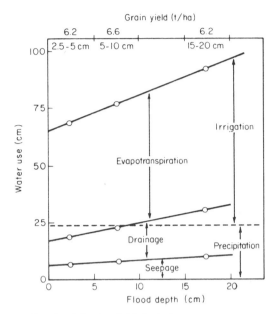

Figure 9.13 Flooding depth in rice fields in Arkansas versus water use of rice (average of 1968 and 1969 seasons). (Adapted from Huey 1976)

United States. In Arkansas, the first flooding may be as the rice is emerging or it may be delayed for 2 or 3 weeks depending on growing conditions and weed control methods used. Many Arkansas rice growers no longer drain their fields at midseason (Johnston and Miller 1973).

Huey (1976) studied total water requirements for maximum grain yield in Arkansas. On silt loam soils with an impervious subsoil, 450–600 mm of water was used per season. In areas of clay soils, with high transmission and seepage losses common, the water use range was 1120–1500 mm. A water depth of 5–10 cm produced maximum yields and minimum losses (Fig. 9.13).

Energy costs and a limited quantity of good quality water in some areas have caused increased interest in water conservation measures. Measures that will minimize water needs are:

- Selection of fields and soil types that will hold more water.
- Improved field topography.
- Accurate survey and control of levees.
- Use of underground pipes instead of canals for delivery.
- Drainage only when necessary.
- Flooding to only 5–10 cm depth.
- Use of short-season varieties.
- Reuse of water.

- Minimized pumping at peak power requirement periods.
- The shut-off of pumping before water reaches the last bund, allowing water in transit to fill the last bunds and avoiding excessive overflow (Huey 1977).

Where rice has been drilled or broadcast in a dryland soil and covered, the soil may need to be flushed if moisture is inadequate for germination or if a crust has formed as a result of drying following rains. A flush should be applied before the young seedlings lose penetrating power. Germination of seeds is not good if the seeds are covered by both water and soil.

A continuous flood is essential for high yield but during the vegetative stage, near-normal growth may be maintained without flooding, provided that moisture is sufficient for adequate growth during the reproductive stage. Losses due to evaporation, transpiration, seepage, and percolation may be as much as 12.5 mm/day for silt loam and 25 mm/day for clay soils when temperatures during that dry climate in July–August reach about 38°C (Huey 1977). The clay soils may be subjected to more cracking, which explains higher water losses due to percolation than in silt loam soils.

In Mississippi, the first flood is applied as early as 2 weeks after seeding but if germination of rice is low, it may be as late as 3 weeks after seeding. Most fields are drained 3–4 weeks after flooding and soil is allowed to dry for several days.

In Texas when the rice is drill-seeded on heavy soils the field is usually flushed for germination if rainwater is insufficient. Fields are flooded after the seedlings are established. The time and number of drainage periods vary with the maturity of the variety, availability of irrigation water, and insect and disease conditions.

In Louisiana, drill seeding and seeding into water each covers about half of the total rice-growing areas. Normally the rice is not flooded until plants are about 16–20 cm tall. Fields are drained as necessary for topdressing with fertilizer and for pest control.

Water-Seeded Rice

Most of California's and about half of Louisiana's rice farmers practice seeding into water. In California, presoaked rice is seeded by aircraft into 5–10 cm of water.

In water-seeded rice in Louisiana, water is drained after the rice seedlings have grown to 1.25 cm above the water surface. Reflooding follows after the seedlings are fully established.

Reasons for Drainage in Direct-Seeded Rice

Reasons cited for draining direct-seeded rice fields are grouped into two categories. First, during the growing season, drainage provides:

- Control of algae (scum) and aquatic weeds.

• Control of rice water weevil.
• Prevention of blight.
• Removal of salts in saline or alkaline affected areas.

Second, draining at the proper time before harvest is necessary to dry the soil enough to support harvesting equipment. Usually water is drained when the rice heads are turned down and are ripening in the upper parts (Johnston and Miller 1973).

IRRIGATION SYSTEM MANAGEMENT IN RICE

Providing adequate irrigation water is the most important single factor controlling the production of food crops in the tropics. To achieve this it is necessary to have an adequate water source, a conveyance system to carry the water to the area to be irrigated, and a distribution method to spread the water over the land. Irrigation infrastructure is generally more fully developed in the main stystem than at the terminal level. As a result, deficiencies in the conceptualization and management of irrigation systems often preclude the realization of full food-crop production potential within their command areas.

The greatest potential for improved water management in irrigation systems in South and Southeast Asia is often said to be at the farm level. But several research projects in the Philippines suggest that the problems of water distribution are greater in lateral and sublateral canals than at the farm level. Therefore, a prerequisite for further improvement in terminal level operations and management is the dependable but controlled flow of water in the main system, which requires adequate operations and management attention and equitable allocation and distribution of the water at the farm level.

A suitable irrigation system must consider:

• The consumptive water use by rice plants at different growth stages.
• Soil texture and its percolation and seepage rates.
• Seasonal distribution of rainfall.
• Topography of fields.
• The availability of irrigation water (Wann 1978).

The water supplied for irrigated agriculture faces many kinds of potential losses (Fig. 9.14).

Conveyance Loss of Water

The loss in conveyance of the water to the irrigated area varies. Losses are higher in:

• Coarse-textured soil.

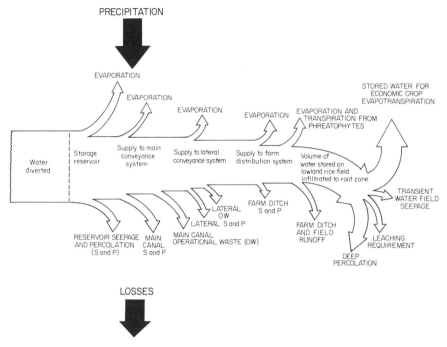

Figure 9.14 Disposition of water diverted for irrigation. (A. C. Early, IRRI, 1980, personal communication)

- Channels having a relatively higher elevation than the surrounding land.
- Channels that are relatively long.
- Channels with poor maintenance.

Water lost by conveyance may not be entirely lost to crop production because some seepage losses become available for crop use in adjoining areas. Percolation losses recharge the water table, which may naturally discharge water at other points in the landscape or be pumped for irrigation of new areas.

The magnitude of conveyance losses cannot be specified generally, but in the humid tropics they range between 15 and 40% of the diverted flow (Ongkingco and Levine 1970). With careful management, conveyance losses can be reduced to 5–10%. Generally, older canals lose less water because of progressive sealing of soil pores by sedimentation.

A suitable method to accurately determine the conveyance loss in a canal is to measure the difference between the quantity of inflow and that of outflow from a selected section of the canal (Wann 1978):

$$\text{rate of conveyance loss} = \frac{\text{inflow} - \text{outflow}}{\text{inflow}} \times 100$$

Distribution Loss of Water

The distribution losses of irrigation water are mainly due to:

- Use of an average pattern of minimal water measurement.
- Too few farm ditches.
- Independent control by the farm operator of water on his land.

Substantial portions of the water losses through distribution channels may be recovered downstream by return flow to creeks and rivers, which in turn may be dammed for irrigation diversion. The extent of this recycling of irrigation water in the total irrigation system has not been precisely estimated.

Water Distribution System

Water distribution for continuous application of water to rice fields at a certain depth is easiest to achieve if the supply of water is continuous and adequate.

Water distribution for intermittent application by a rotational irrigation system is difficult to accomplish. In a rotational irrigation system, water is distributed by:

- Rotation by section in the main canal, which requires bigger capacities for both conveyance and distribution systems than that of continuous irrigation.
- Rotation by the laterals or sublaterals, or their sections, which requires bigger capacity of laterals or sublaterals and farm ditches.
- Rotation by farm ditches which requires only bigger capacities of farm ditches.
- Selected combinations of the above.

Sources of Irrigation Water

The sources of irrigation are from surface water and groundwater.

Surface Water

Surface water is taken from rivers, creeks, ponds, or lakes. The amount of water available from these sources for irrigating rice depends largely on the season. Normally, the greatest quantity of surface water is available during the wet season. As a result, these surface sources serve as supplementary water sources to the rainwater received during the monsoonal season. In the dry season the amount of water available for irrigation is often low and demand is considerably higher. If irrigation water is available, rice yields can be at least 50% higher in the dry season than in the wet season because of higher solar radiation.

When the water level in a stream is lower than the area to be irrigated, the level may be raised by constructing a dam to divert water to the rice field. The dam may be a permanent concrete structure or a temporary dam, which is usually reconstructed or repaired yearly. Sometimes, pumps are installed if dam construction is not feasible. Storage reservoirs are means by which the extra water of the wet season can be stored for release during the dry season when rainfall is inadequate to sustain a crop.

Groundwater

Groundwater is usally obtained through a well. Wells are termed shallow or deep depending on the depth of the bore required to obtain the desired discharge from the water-bearing geologic formations. Pumps are used to lift water from the well to the ground level or above for its distribution to the service area.

Kinds of Irrigation Pumps

The types of pumps used for irrigation depend on the flow requirement, the lift from the water surface to the pump, and the height to which the water is to be raised (Ongkingco and Levine 1970). Types include:

*   *Propeller (axial flow) pumps,* used for low head (about 3 m) and high flow conditions.
*   *Centrifugal pumps,* used where the lift is less than about 7 m.
*   *Deep well turbine pumps,* generally used when pumping has to be done from a deep water level.

Quality of Irrigation Water

It is important to determine water quality before the water is used for irrigation. It is equally important to monitor water quality periodically against the potential hazards of crop damage by poor irrigation water. For example, some parts of IRRI's experiment station had the soil pH increased to 7.9, the electrical conductance to 12 mS/cm (12 mmho/cm), and the available boron content to 13 ppm as a result of irrigation with alkaline, slightly-saline, deep-well water high in bicarbonates of calcium and magnesium, boron, and silicon. The high pH, deposition of calcium and magnesium carbonate, and the accumulation of silicon depressed the availability of zinc (F. N. Ponnamperuma, IRRI 1980, personal communication).

In Arkansas, United States, many of the irrigation wells produce water containing high concentrations of calcium and magnesium carbonates. In parts of southwest and southeast Arkansas, the groundwater contains high levels of sodium. Continued application of such poor-quality water on a rice field may cause chlorosis and sometimes death of seedlings.

In California, Finfrock et al. (1960) described good rice irrigation water as that with:

- Specific conductance less than 0.75 mS/cm (0.75 mmho/cm).
- Boron contents of less than 1 ppm.
- Sodium adsorption ratio (SAR) index (tendency to form alkaline soil) less than 10.

When high-sodium water is used regularly each growing season, it may deflocculate the soil so that stickiness, compactness, and impermeability increase. The deflocculated soil makes tillage difficult and usually produces low yields.

Factors Affecting Irrigation Systems

Many irrigation projects have been less than successful, partly because they have been designed and managed almost exclusively from an engineering point of view. Because irrigation systems are usually built with large expenditures of public funds, their design should consider not only engineering aspects such as water storage, conveyance, and delivery, but also agricultural, economic, social, political, legal, and environmental conditions. Specialists from various disciplines, which include engineers, soil scientists, agronomists, economists, and other social scientists should work together in designing, monitoring, field investigating, and implementing any successful irrigation system (Thavaraj 1978).

The agricultural factors that are important in designing an irrigation system for lowland flooded rice include cropping schedule, the availability of labor and machines, and varietal difference in response to moisture levels.

Cropping Schedules

Cropping schedules are generally controlled by soils, rainfall distribution, and the availability of irrigation water. Figure 9.15 shows the amount of available water based on 80% probability of exceedance. Timing of irrigation is important for soaking fields for land preparation. For maximum water use efficiency it is best if vegetative and reproductive periods coincide with wet months and the ripening period coincides with dry months.

Availability of Labor and Machine

Labor requirements can be evenly distributed across the entire irrigation system, and peak water requirements for land preparation can be reduced by incorporating staggered planting of rice, introducing early maturing varieties,

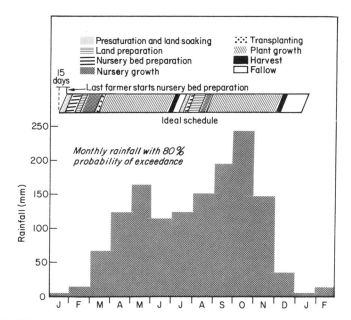

Figure 9.15 Calendar of proposed cropping schedule. Muda irrigation scheme, Malaysia. (From Thavaraj 1978)

communal nurseries, and some degree of mechanization at various levels of farm operations.

Varietal Differences

There are distinct varietal differences existing in tolerance for excess and shortage of moisture. Basic agronomic data are essential for design of systems for optimum management. These and other relevant information are discussed in details by Thavaraj (1978).

Water Control

Improved water control by better irrigation and drainage is perhaps the most important single factor in achieving the full yield potential of modern rice varieties.

Rice does not require continuous flooding but responds favorably to it. This is because many benefits are ascribed to continuous flooding such as better growth and yield, better nutrient supply, and better weed control.

Water requirements of lowland rice can be grouped as that:

- Required for land preparation

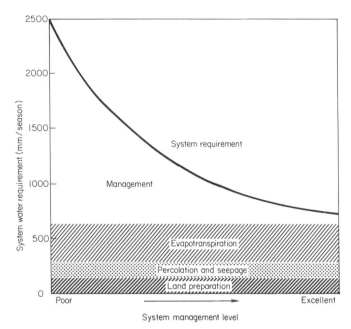

Figure 9.16 System water requirement as a function of management level. (From Levine 1969)

- Required for growth and transpiration
- Required for evaporation from the soil-water surface during crop growth
- Lost through percolation, seepage, and runoff.

As the management level to conserve moisture and maximize water use efficiency increases, water requirement from an irrigation system decreases (Fig. 9.16). Once demand for water for land preparation, percolation, seepage, and ET are satisfied, the better the water management practices, the less is the water requirement from an irrigation system.

An extended period of water shortage in an area of heavy elastic soils does not result in less total water use compared with continuous water supply if the area is resupplied with water following the stress period. This is because of increased seepage and percolation losses from fields that are dried and cracked.

Several factors are considered important in saving water in lowland rice:

- About 100–200 mm of water can be saved by preparing the land expeditiously.
- By growing an early-maturing variety that has a field duration of 75 days (100-day variety) instead of growing a variety that has field duration of 105 days (130-day variety), there can be a net savings of 200–350 mm of water.

- Rotational irrigation systems can be used where there are good irrigation water control and distribution systems and excellent farmer cooperation.

Whenever possible, continuous flooding with 5–7.5 cm of water depth is best for optimum grain yield, optimum nutrient supply, and excellent weed control with water depth and herbicides.

Because irrigation projects are expensive, it is important not only to learn from past experience in designing and managing new systems but also to draw judgment from specialists from various disciplines such as engineers, agronomists, soil scientists, economists, and farmers' cooperative specialists that will effectively implement new systems.

With increased emphasis on cropping intensity, dependable and controllable water supply through better irrigation facilities and better management is a must for total increase in production of rice and other crops.

REFERENCES

Bhattacharyya, A. K., and S. K. De Datta. 1971. Effects of soil temperature regimes on growth characteristics, nutrition and grain yield of IR22 rice. *Agron. J.* 63:443–449.

De Datta, S.K., and A. Williams. 1968. Rice cultural practices. B. Effects of water management practices on the growth characteristics and grain yield of rice. Pages 78–93 *in Proceedings and papers, fourth seminar on economic and social studies (rice production).* Committee for the coordination of investigations of the Lower Mekong Basin. Los Baños, Philippines.

De Datta, S. K., W. P. Abilay, and G. N. Kalwar. 1973a. Water stress effects in flooded tropical rice. Pages 19–36 *in* International Rice Research Institute. *Water management in Philippine irrigation systems: research and operations.* Los Baños, Philippines.

De Datta, S. K., H. K. Krupp, E. I. Alvarez, and S. C. Modgal. 1973b. Water management practices in flooded tropical rice. Pages 1–18 *in* International Rice Research Institute. *Water management in Philippine irrigation systems: research and operations.* Los Baños, Philippines.

Doorenbos, J., and W. O. Pruitt. 1977. *Guidelines for predicting crop water requirements.* FAO Irrigation and Drainage Pap. 24 (rev. 1977). FAO, Rome. 144 pp.

Evans, G. N. 1971. Evaporation from rice at Griffith, New South Wales. *Agric. Meteorol.* 8:117–127.

Finfrock, D. C., F. M. Raney, M. D. Miller, and L. J. Booher. 1960. Water management in rice production. University of California, Division of Agricultural Science, Leaflet 131. 2 pp.

Huey, B. A. 1976. Water management for rice production. Cooperative Extension Service, University of Arkansas, Division of Agriculture, and U.S. Department of Agriculture Leaflet EL566.

Huey, B. A. 1977. *Rice production in Arkansas.* University of Arkansas, Division of Agriculture, Circ. 476 (rev.). 51 pp.

IRRI (International Rice Research Institute). 1971. *Annual report for 1970.* Los Baños, Philippines. 265 pp.

Johnston, T. H., and M. D. Miller. 1973. Culture. Pages 88–134 *in* USDA Agric. Handb. 289. *Rice in the United States: varieties and production.* Washington, D.C.

Kampen, J. 1970. Water losses and water balance studies in lowland rice irrigation. PhD dissertation, Cornell University, Ithaca, New York. 416 pp. (unpubl.)

Kramer, P. J. 1969. *Plant and soil water relationships: a modern synthesis.* McGraw-Hill Book Co., New York. 482 pp.

Kung, P., and C. Atthayodhin. 1968. Water requirements in rice production. Pages 94–112 *in* Committee for the coordination of investigations of the Lower Mekong Basin. *Proceedings and papers, fourth seminar on economic and social studies (rice production).* Los Baños, Philippines.

Levine, G. 1969. Lowland irrigation requirements in the humid tropics with special reference to the Philippines. Paper presented at a Thursday seminar, 13 March 1969, International Rice Research Institute, Los Baños, Philippines. (unpubl. mimeo.)

Matsubayashi, M., R. Ito, T. Nomoto, T. Takase, and N. Yamada. 1963. *Theory and practice of growing rice.* Fuji Publishing Co. Tokyo, Japan. 502 pp.

Matsushima, S. 1962. *Some experiments on soil water plant relationship in rice.* Ministry of Agriculture and Cooperatives, Federation of Malaya, Kuala Lumpur. 35 pp.

Ongkingco, P. S., and G. Levine. 1970. II. Irrigation system requirements for rice. Pages 96–105 *in* University of the Philippines College of Agriculture in cooperation with the International Rice Research Institute. *Rice production manual.* Los Baños, Philippines.

Ponnamperuma, F. N. 1971. Critical examination of soil and water management practices in rice production. Second IRRI workshop on field experimentation. (unpubl. mimeo.)

Senewiratne, S. T., and D. S. Mikkelsen. 1961. Physiological factors limiting growth and yield of *Oryza sativa* under unflooded conditions. *Plant Soil* 14:127–146.

Sugimoto, K. 1969. *Studies on plant-water relationship of paddy in Muda river irrigation project area of West Malaysia.* Department of Agriculture, Malaysia. 145 pp.

Thavaraj, S. H. 1978. The importance of integrating nonengineering aspects in irrigation system design. Pages 15–24 *in* International Rice Research Institute. *Irrigation policy and management in Southeast Asia.* Los Baños, Philippines.

Thongtawee, N. 1965. Measurement of evapotranspiration and water losses in flooded rice field. Department of Agricultural Engineering, International Rice Research Institute, Los Baños, Philippines. (unpubl. mimeo.)

Tomar, V. S., and J. C. O'Toole. 1979. *Evapotranspiration from ricefields.* IRRI Res. Pap. Ser. 34. 15 pp.

Tomar, V. S., and J. C. O'Toole. 1980. Design and testing of a microlysimeter for wetland rice. *Agron. J.* 72:689–692.

Van de Goor, G. W. 1950. Research in irrigating rice [in Dutch, English summary]. *Landbouw* 22:195–222.

Wann, S. S. 1978. *Water management of soils for growing rice.* ASPAC Food Fert. Technol. Cent. Bull. 40. 12 pp.

Wickham, T. H. 1978. Water requirements for lowland rice. Pages 1–8 *in* International Rice Research Institute. Background papers on water requirements, agronomic, soils and yield response concepts in irrigation water management. Los Baños, Philippines. (unpubl. mimeo.)

Wickham, T. H., and C. N. Sen 1978. Water management for lowland rice: water requirements and yield response. Pages 649–669 *in* International Rice Research Institute. *Soils and rice*. Los Baños, Philippines.

Yoshida, S. 1979. A simple evapotranspiration model of a paddy field in tropical Asia. *Soil Sci. Plant Nutr.* (Tokyo) 25 (1):81–91.

Young, G. 1970. Dry lands and desalted water. *Science* 167:339–343.

10

Mineral Nutrition and Fertilizer Management of Rice

Development of a rational method of fertilizer application requires knowledge of the mineral nutrition of the rice plant at different growth stages. It is also essential to know the contributions of the nutrients absorbed to grain yield. A dynamic insight into the physiological condition of the plant at any stage of growth requires analytical studies of every part of the plant. The characteristics of various plant organs are greatly influenced by environmental factors of which one of the most important is mineral nutrition.

Aso studied the process of nutrient uptake of the rice plant at different growth stages as early as 1918. Subsequently, Gericke (1924) and Ishizuka (1932) systematically studied the mineral nutrition of the rice plant.

MINERAL NUTRITION

Nutrition is the supply and absorption of those nutrient chemical elements required by an organism. Crop nutrients are the elements, or simple inorganic compounds, indispensable for the growth of crops and not synthesized by the plant during the normal metabolic processes.

Essential Elements for Rice

For rice, 16 elements are essential—carbon, hydrogen, oxygen, nitrogen, phosphorus, potassium, sulfur, calcium, magnesium, zinc, iron, copper, molybdenum, boron, manganese, and chlorine. These are divided into major and minor elements. The major elements, C, H, O, N, P, K, Ca, Mg, and S, are needed by plants in relatively higher amounts than the minor elements, Fe, Mn, Cu, Zn, Mo, B, and Cl.

348

All essential elements must be present in optimum amounts and in forms usable by rice plants. Nitrogen, phosphorus, zinc, and potassium are nutrient elements most commonly applied by rice farmers. Sulfur is occasionally applied to some soils but sulfur is usually supplied as an ingredient of $(NH_4)_2SO_4$, K_2SO_4, and $CaSO_4$ (in ordinary superphosphate) even where it is not needed. Silicon, although not an essential element, is applied to degraded soils in Japan, Republic of Korea, and Taiwan.

All the other essential nutrients are provided by air, water, soil, and plant residues, or as contaminants in commercial fertilizers.

There are two ways to determine whether a specific element is essential for plant life:

- If plants are grown in a complete nutrient solution lacking only one specific element, they show abnormality compared with plants grown in a solution containing the missing element.
- If the specific element in question is added to the nutrient solution in which the abnormal plants are growing, the symptoms of abnormality disappear, or are reduced in severity.

Diagnosis of Nutrient Deficiencies and Toxicities

Nutrient deficiency symptoms in the rice plant are seen in color of the leaves, stems, and roots, in plant height and tillering habit, and in development of root systems.

Plant height can be normal or stunted (Yoshida 1975b). Deficiency symptoms are expressed by the tillers being normal, too few, or too many.

In leaves, deficiency symptoms will include:

- Yellow or dark green appearance.
- Presence or absence of interveinal chlorosis.
- Presence or absence of brown spots either on the lower or upper leaves.
- Brown spots at the tip or the marginal areas of the leaves, which vary in size with the severity of the symptoms.

The best time to observe deficiency or toxicity symptoms in rice is in the early stages of symptom development. For example, zinc deficiency in lowland rice usually appears within 2–3 weeks after transplanting, after which the crop apparently begins to recover when the deficiency is moderate (Yoshida 1975b). It is only when this deficiency is extremely severe that symptoms may persist until flowering or longer.

On the other hand, the symptoms of iron toxicity may appear in 1–2 weeks, or not until 1–2 months after transplanting.

Symptoms of some diseases are sometimes confused with those of nutritional disorders. For example, it is difficult to differentiate between zinc deficiency and grassy stunt virus in their early stages.

Functions and Deficiency Symptoms of Nutrients

Functions and deficiency symptoms of nutrients in rice are largely different for each element.

NITROGEN   Rice plants require a large amount of nitrogen at the early and mid-tillering stages to maximize the number of panicles. Nitrogen absorbed at the panicle initiation stage may increase spikelet number per panicle. Some nitrogen, however, is also required at the ripening stage.
   The functions of nitrogen in rice are:

· Gives dark green appearance to plant parts as a component of chlorophyll.
· Promotes rapid growth or increased height and tiller number.
· Increases size of leaves and grains.
· Increases number of spikelets per panicle.
· Increases filled spikelets percentage in panicles.
· Increases protein content in the grains.

Nitrogen deficiency symptoms are:

· Stunted plants with limited number of tillers.
· Narrow and short leaves which are erect and become yellowish green as they age (young leaves remain greener).
· Old leaves become light straw colored and die.

PHOSPHORUS   Phosphorus is involved in the supply and transfer of energy for all the biochemical processes in the rice plant. It:

· Stimulates root development.
· Promotes earlier flowering and ripening particularly under cool climate.
· Encourages more active tillering, which enables rice plants to recover more rapidly and more completely after any adverse situation.
· Promotes good grain development and gives higher food value to the rice because of phosphorus content of the grain.

Phosphorus deficiency symptoms are:

· Stunted plants with limited number of tillers.
· Narrow and short leaves that are erect and dirty dark green.
· Young leaves remain healthier than older leaves, which turn brown and die.

- Reddish or purplish color may develop on leaves of varieties that tend to produce anthocyanin pigment.

POTASSIUM   Potassium is not a constituent of any organic compound of the plant, but it is a cofactor for 40 or more enzymes. It:

- Favors tillering and increases the size and weight of the grains.
- Increases phosphorus response.
- Plays an important role in physiological processes in the plant including opening and closing of stomata, and tolerance to unfavorable climatic conditions.
- Renders resistance to diseases such as blast and *Helminthosporium*.

Potassium deficiency symptoms are:

- Stunted plants and tillering slightly reduced.
- Short, droopy, and dark green leaves.
- Yellowing at the interveins, on lower leaves, starting from the tip, and eventually drying to a light brown color.
- Brown spots sometimes develop on dark green leaves.
- Irregular necrotic spots may develop on the panicles.
- Long thin panicles form.
- Some symptoms of wilting when there is excessive imbalance with nitrogen (low K-N ratio in plant).

CALCIUM   The functions of calcium in the rice plant are as:

- A constituent of the cementing material of plant cells.
- An important constituent of calcium pectate, which strengthens the cell wall.
- Maintainer of turgidity of the cell walls.
- Promoter of normal root growth and development.

Calcium deficiency causes little change in general appearance of the plant, except in cases of acute deficiency. In those:

- The tip of the upper growing leaves becomes white, rolled, and curled.
- In an extreme case, the plant is stunted and the growing point dies.

MAGNESIUM   Magnesium functions in the rice plant as:

- A constitutent of chlorophyll molecule.
- A component of several essential enzymes.
- Functions similar to calcium.

Magnesium deficiency symptoms are:

- With moderate deficiency, height and tiller number are little affected.
- Wavy and droopy leaves due to expansion of the angle between the leaf blade and the leaf sheath.
- Interveinal chlorosis characterized by an orangish yellow color on lower leaves.

SULFUR   Sulfur functions in the rice plant as:

- A constituent of the amino acids cystine, cysteine, and methionine, and the plant hormones thiamin and biotin.
- An important factor in the functioning of many plant enzymes, enzyme activators, and oxidation-reduction reactions.

Sulfur deficiency symptoms are:

- Similar to those of nitrogen deficiency, which makes it impossible to distinguish between the two deficiencies by visual symptoms alone.
- Initially on leaf sheaths, which become yellowish, proceeding to leaf blades, with the whole plant chlorotic at the tillering stage.
- Reduced plant height and tiller number.
- Fewer panicles, shorter panicles, and reduced number of spikelets per panicle at maturity.

One point of distinct difference between nitrogen and sulfur deficiency is that sulfur deficiency produces general chlorosis in the whole plant but the older leaves do not dry quickly. Nitrogen deficiency induces more intense chlorosis in the older leaves. Sulfur deficiency delays growth and development whereas nitrogen deficiency hastens flowering and maturity. Sulfur does not move from older to younger leaves as nitrogen does. Data from Brazil suggest that sulfur deficiency reduces head rice yield and increases chalkiness in grains (Wang et al. 1976a).

ZINC   The functions of zinc in the rice plant are:

- Probable connection with the production of auxin.
- Activation of many enzymatic reactions.
- Close involvement in nitrogen metabolism.

Zinc deficiency symptoms are:

- The midribs of the younger leaves, especially the base, become chlorotic.

- Brown blotches and streaks in lower leaves appear, followed by stunted growth, although tillering may continue.
- Reduced size of the leaf blade but with the leaf sheath little affected.
- Uneven growth and delayed maturity in the field.

Yoshida (1968) made four observations on the zinc-deficient rice plants in Pakistan:

- The brown streaks and blotches of lower leaves appear 2–3 weeks after transplanting and severely affected plants are usually found in low-lying patches.
- Most affected plants recover but those severely affected die.
- The disorder usually appears in the first year of rice cultivation following reclamation.
- High rates of fertilizer aggravate the disorder and drainage alleviates it.

The symptoms observed by Yoshida were similar to those observed by Karim and Vlamis (1962). Total or available zinc in soil, measured chemically, has no relevance to incidence of zinc deficiency or plant zinc content (Yoshida 1968).

IRON   Iron functions in the rice plant as:

- Related to the formation of the chlorophyll, but not a constituent of it.
- A possible catalyst in an organic form or combined with organic compound as a component of redox enzymes.
- An inhibitor of the absorption of potassium by the rice plant (Yamasaki 1965).

Iron deficiency symptoms are:

- Entire leaves become chlorotic and then whitish.
- The newly emerging leaf becomes chlorotic if iron supply is cut suddenly.

MANGANESE   Manganese functions in the rice plant as:

- A factor in photosynthesis and in oxidation-reduction processes.
- An activator of several enzymes, such as oxidase, peroxidase, dehydrogenase, decarboxylase, and kinase.

Manganese deficiency symptoms are:

- Stunted plants with normal tiller number.
- Intraveinal chlorosis on the leaves.

- Chlorotic streaks spreading downward from the tip to the base of the leaves, which become dark brown and necrotic.
- Newly emerging leaves are short, narrow, and light green.

BORON   Boron functions in the rice plant as:

- A catalyst in the plant system.
- A regulator of physiological functions such as nitrogen metabolism and nutrient uptake, especially calcium metabolism.

Boron deficiency symptoms are:

- Reduced plant height.
- The tips of emerging leaves become white and rolled as in the case of calcium deficiency.
- The growing points may die in severe cases, but new tillers continue to be produced.

MOLYBDENUM   Molybdenum's function in the rice plant is related to reduction of nitrate to nitrite. No deficiency symptoms have been described.

COPPER   Copper functions in the rice plant as:

- A component of metalloenzymes.
- A regulator of enzymatic actions.

Copper deficiency symptoms are:

- Bluish green leaves, which become chlorotic near the tips.
- Development of chlorosis downward along both sides of the midrib, followed by dark brown necrosis of the tips.
- New leaves fail to unroll and maintain a needlelike appearance of the entire leaf or, occasionally of half the leaf, with the basal portion developing normally.

CHLORINE   Chlorine is essential in photosynthesis. Its deficiency symptoms have not been described in rice.

SILICON   On soils low in available silicon, the application of silicon will increase yields of a modern variety at high rates of nitrogen fertilizer application. The effects of silicon are classified into four categories:

1 Effects on normal growth of plants:
   - Promotes growth, strengthens culms and roots, and favors early panicle formation.
   - Increases number of spikelets per panicle and percentage of matured grains.
   - Helps maintain erect leaves, which is important for high rate of photosynthesis.
2 Effects on water economy:
   - Silicon-free plants suffer from internal water stress when they are placed in environments in which transpiration is greatly increased or water absorption is greatly impaired.
   - Silicon supply is critical during panicle initiation when root activity is somewhat reduced and transpiration loss of water is high.
3 Effects on disease and insect resistance:
   - A thick cuticle-silica layer provides an excellent barrier against attacks of fungi, insects, and mites because of its physical hardness (Fig. 10.1 shows the schematic representation of the rice epidermal cell with silica layer).
   - Silicon application diminishes the unfavorable action of nitrogen on the resistance of rice to diseases such as blast.

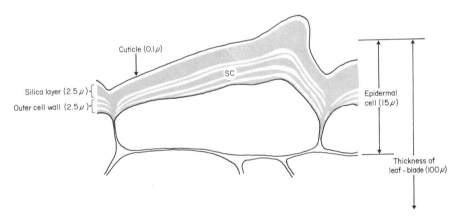

Figure 10.1 Schematic representation of the rice leaf epidermal cell. SC = silica cellulose. (From Yoshida et al. 1962, Yoshida 1975a)

4 Effects on other nutrients:
   - Silicon seems to promote translocation of phosphorus in the rice plant and retention of excessive phosphorus taken up.
   - Silicon makes soil phosphorus available to rice.

The leaves of silicon-deficient plants become soft and droopy.

Mineral Toxicity Symptoms

IRON   Iron toxicity symptoms are:

- Tiny brown spots on lower leaves, starting from the tips and spreading toward the bases (the spots are generally combined on interveins).
- Leaf usually remains green.
- In severe cases, the entire leaves become purplish brown.

ALUMINUM   Aluminum toxicity symptoms are orangish yellow interveinal chlorosis, which may become necrotic in serious cases.

MANGANESE   Manganese toxicity symptoms are:

- Stunted plant and limited tillering.
- Brown spots on the veins of the leaf blade and the leaf sheath, especially on the lower leaves (Tanaka and Yoshida 1970).

BORON   Boron toxicity symptoms include chlorosis at the tips of the older leaves, especially along the margins, followed by the appearance of large, dark brown elliptical spots in the affected parts, which ultimately turn brown and dry up. In the field, vegetative growth is not seriously depressed except in sensitive varieties. The first symptom of boron toxicity is a yellowish white discoloration of the tips of the older leaves, which appears about 6 weeks after transplanting. As the disease progresses, the tips and the leaf margins turn yellow. Two to four weeks later, depending on soil boron content and variety, elliptical dark brown blotches appear in the discolored areas in most rice varieties. Finally, the entire leaf blade turns light brown and withers (M. T. Cayton and F. N. Ponnamperuma, IRRI, 1979, personal communication).

SALT INJURY   Salt toxicity in rice is seen as stunted growth, reduced tillering, and whitish leaf tips. Frequently, some parts of the leaves become chlorotic (Tanaka and Yoshida 1970).

Critical Nutrient Levels in Plants and Soils

Determining the critical concentrations of elements in the rice plant, below which deficiency symptoms may develop or above which toxicity symptoms may become visible, allows diagnosis of abnormal rice plants growing in the field by chemical analyses of the plants.

Plant Analysis

In most cases, nutrient deficiency or toxicity symptoms appear when plants are young, which allows whole plants (excluding roots) to be sampled for chemical analysis. Plant samples are washed, dried, and ground. For analyses of metallic elements, Yoshida et al. (1972) suggest the use of an atomic absorption spectrophotometer, whereas for nonmetallic elements, such as chlorine and boron, colorimetric methods are normally used.

However, information on the critical concentrations of elements in the rice plant is required. The critical concentration determined from plants grown in the greenhouse is sometimes too high, and is not applicable to the field crop.

Tanaka and Yoshida (1970) provided a list of critical concentrations of various elements. These concentrations should be used only as a guide for diagnosis (Table 10.1). For example, the precise critical concentration for zinc deficiency is difficult to establish because of:

- Variation among sampled plants.
- Interactions between zinc and other elements (possibly with phosphorus).

Table 10.1 Deficiency and Toxicity Critical Concentrations of Various Elements in the Rice Plant[a] (Adapted from Tanaka and Yoshida 1970)

Element	Deficiency (D) or Toxicity (T)	Critical Concentration	Plant Part Analyzed	Growth Stage[b]
N	D	2.5%	Leaf blade	Til
P	D	0.1%	Leaf blade	Til
	T	1.0%	Straw	Mat
K	D	1.0%	Straw	Mat
	D	1.0%	Leaf blade	Til
Ca	D	0.15%	Straw	Mat
Mg	D	0.10%	Straw	Mat
S	D	0.10%	Straw	Mat
Si	D	5.0%	Straw	Mat
Fe	D	70 ppm	Leaf blade	Til
	T	300 ppm	Leaf blade	Til
Zn	D	10 ppm	Shoot	Til
	T	1500 ppm	Straw	Mat
Mn	D	20 ppm	Shoot	Til
	T	>2500 ppm	Shoot	Til
B	D	3.4 ppm	Straw	Mat
	T	100 ppm	Straw	Mat
Cu	D	<6 ppm	Straw	Mat
	T	30 ppm	Straw	Mat
Al	T	300 ppm	Shoot	Til

[a]Figures for critical concentrations collected from various references but adjusted to round figures.

[b]Mat = maturity; Til = tillering.

- Varying growth stages of sampled plants.
- Analytical errors.

Yoshida et al. (1973) developed criteria that, in addition to visible observations and soil analyses, would serve as a useful diagnostic tool for identifying zinc deficiency:

Zinc Concentration in the Whole Shoot	Diagnosis of Deficiency
< 10 ppm	Definite
10–15 ppm	Very likely
15–20 ppm	Likely
> 20 ppm	Unlikely

Reports on grain sulfur concentrations for rice vary between 0.034% in grain from a deficient plant, and 0.16% in grain from plants that had no response to sulfur application. Under those conditions, grain yields ranged from 0.75 t/ha to 8.0 t/ha and sulfur requirements varied from 0.26 to 12.8 kg/ha (Blair 1977). Yoshida and Chaudhry (1979) suggested that the critical sulfur content in leaf blades and straw varies with growth stages (Table 10.2). Wang (1976) concluded that the critical concentrations of sulfur in straw should be 0.05% for optimum grain yield (Fig. 10.2).

Soil Analysis

The measurement of soil pH is the simplest and most informative analytical technique in the diagnosis of nutrient deficiency or toxicity.

Chang (1978) reviewed various soil and plant analyses methods to evaluate the nitrogen, phosphorus, potassium, sulfur, zinc, and silicon available to lowland rice from submerged soils. The best correlation with the response of rice to the given elements was with determination of:

Table 10.2   Critical Sulfur Concentration of Leaf Blades and Straw at Different Growth Stages (From Yoshida and Chaudhry 1979)

| | Critical S Concentration[a] (%) | | | |
| | Leaf Blades | | Straw | |
Growth Stage	DC_{50}	DC_{100}	DC_{50}	DC_{100}
Tillering	0.06	0.16	0.06	0.16
Flowering	0.05	0.10	0.04	0.07
Harvest	0.04	0.07	0.03	0.06

[a]DC_{50}: S concentration required in plant tissue for 50% of maximum yield; DC_{100}: S concentration required for maximum yield.

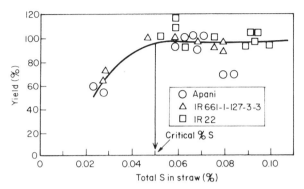

Figure 10.2   Relationship between total sulfur concentration in straw at harvest and percentage yield of three rice varieties under field conditions. (From Wang 1976)

- Available nitrogen by waterlogged incubation and alkaline permanganate methods.
- Available potassium by exchangeable potassium.
- Available phosphorus by Olsen and Bray P_1 methods.
- Available sulfur by extraction with $Ca(H_2PO_4)_2 \cdot H_2O$.
- Available zinc by extraction with buffered chelating agents or weak acids.
- Available silicon by extraction with sodium acetate.

Table 10.3 lists critical deficiency levels of some micronutrients in rice soils. With certain nutrients, total concentrations in soil do not reflect the availability of

Table 10.3   Critical Levels for Deficiency of Micronutrients in Rice Soils (From Randhawa et al. 1978)

Element	Method	Critical Level (ppm)
B	Hot H_2O	0.1–0.7
Cu	DTPA + $CaCl_2$ (pH 7.3)	0.2
Fe	DTPA + $CaCl_2$ (pH 7.3)	2.5–4.5
	$NH_4C_2H_3O_2$ (pH 4.8)	2.0
Mn	DTPA + $CaCl_2$ (pH 7.3)	1.0
	0.1 N H_3PO_4 and 3 N $NH_4H_2PO_4$	15–20
Mo	$(NH_4)_2(C_2O_4)$ (pH 3.3)	0.04–0.2
Zn	0.05 N HCl	1.5
	Dithizone + $NH_4C_2H_3O_2$	0.3–2.2
	EDTA + $(NH_4)_2CO_3$	1.5
	DTPA + $CaCl_2$ (pH 7.3)	0.5–0.8

nutrients to plants. For example, zinc status of rice plants was found to be unrelated to the total zinc content of the soil, but was closely related to the level of EDTA $(NH_4)_2CO_3$-extractable zinc. Zinc deficiencies were likely when the concentration in the extract fell below $23\mu M$ zinc (Forno et al. 1975).

Nutrient Absorption and Translocation

The rate of dry matter production of the rice plant, the proportionate weight of each plant part, and the mineral content of a given plant part change as the plant develops. The causes are changes in physiological status of the plant with the development of growth phases and changes in environmental conditions, such as temperature, availability of each nutrient in the soil, and so on, during growth and development.

The accumulation of nitrogen in the vegetative organs is high during the early growth stages and decreases with growth. After flowering, translocation of nitrogen from the vegetative organs to the grains becomes significant. There is some translocation of carbohydrates from the vegetative plant parts to the grains. Accumulation of carbohydrates in the vegetative organs is negligible during the vegetative phase, and after flowering a large amount of carbohydrates accumulates in the grains.

During the vegetative stages, protein synthesis is active; during the reproductive stage, synthesis of cell-wall substances, such as cellulose, lignin, and so on, becomes active although protein synthesis also continues; and during the ripening stage, starch synthesis in grains becomes active.

Nitrogen, phosphorus, and sulfur, which are components of proteins, are absorbed rapidly during vegetative growth and translocated from the vegetative organs to the grain after flowering. Potassium and calcium, which regulate various metabolic processes, are absorbed at a rate almost parallel to dry matter production but there is no marked translocation of these elements from vegetative organs to the grains during ripening. Magnesium is absorbed most actively during the reproductive growth, and its translocation after flowering occurs to some extent.

Nutrient mobility in the rice plant is in the sequence: $P > N > S > Mg > K > Ca$. The elements that are components of proteins have a high mobility, and the elements that are continuously absorbed until the end of growth have a low mobility. For example, silicon, which is continuously absorbed, has low mobility.

Nutrient Uptake at Different Growth Stages

The process of nutrient uptake at different growth stages is a function of climate, soil properties, amount and method of fertilizer application, and variety of the rice plant. Based on a number of experiments in a temperate climate, Ishizuka (1965) summarized nutrient uptake of elements at different growth stages:

- At the seedling stage, the percentage contents of nitrogen, phosphorus, and potassium increased as growth progressed and then decreased after reaching maximum value. The time of maximum percentage of nitrogen in the seedling differed with the amount of nutrient supplied.

- In the main field, the percentage of nitrogen in the plant decreased slightly after transplanting, and then increased until the initiation of flowering. After that, nitrogen percentage decreased continuously until the dough stage and remained nearly constant until complete ripening.

- The percentage of phosphorus decreased rapidly after transplanting, then increased gradually and reached a high percentage at the time of the start of flowering. This high percentage continued during flowering and then decreased until the dough stage. This coincided with the translocation and accumulation of starch in the grain, showing a close relationship between carbohydrate metabolism and phosphorus.

- The percentage of potassium decreased gradually according to the growth of the plant. It increased again from flowering until complete ripening.

- The changes in the percentage of calcium were similar to those of potassium.

- The percentage of magnesium was high from transplanting to the middle of tillering and then decreased gradually.

- The percentage of sulfur decreased with growth.

Figure 10.3 shows the amount of nutrients absorbed by one plant in the process of growth.

An example of nutrient uptake of the rice plant in countries at low latitudes is Tanaka's (1957, 1977) comparison of the pattern of nutrient uptake of Ptb 10, an early-maturing variety, and BAM-9, a late-maturing variety (Fig. 10.4). For both, nitrogen uptake had a distinct peak at about 37 days after transplanting but the late-maturing rice had an additional peak of nitrogen uptake about 95 days after transplanting.

Cultural practices, ecological conditions, and morphological differences in rice varieties account for differences in nutrient uptake between countries at different latitudes or different regions in a given country. Generally, the difference in the nutrient uptake patterns among different duration varieties in India is similar to the differences among latitudes in Japan from north to south (Ishizuka 1965).

Nitrogen Requirement and Yield-Determining Process

For high rice yield, optimum nitrogen nutrition is important at four growth stages. They are:

- Just after rooting.
- At the neck-node differentiation stage.

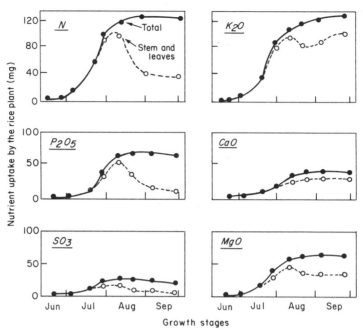

Figure 10.3   Nutrient uptake of the rice plant at different stages of growth in northern Japan. (Adapted from Ishizuka 1965, Tanaka 1977)

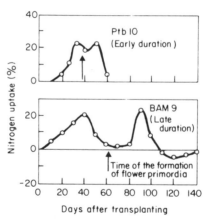

Figure 10.4   Rate of nitrogen uptake of rice varieties of different durations in India. (Adapted from Tanaka 1957, 1977, Ishizuka 1965)

• Just before the stage of reduction division of pollen mother cell.
• At full heading.

The relationship between the amount of nitrogen in the rice plant at the neck-node initiation stage and the number of panicles per square meter is highly

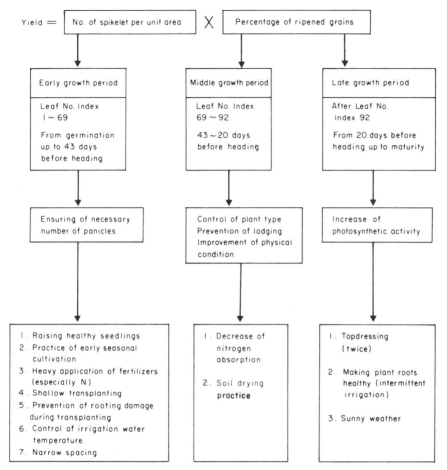

Figure 10.5 Schematic representation of points in practice in relation to nitrogen application in rice cultivation. (Adapted from Matsushima 1969)

significant. Based on nitrogen nutrition in relation to percentage of ripened grains as well as grain yield, Matsushima (1969) suggested draining and flooding the field at predetermined growth stages of rice and topdressings of nitrogen (Fig. 10.5). His suggestions were, however, developed in a temperate climate with low solar energy. If solar radiation is high, such as in California, in the United States, and southern Australia, or during the dry season in the tropics, high yields can be obtained without practicing the widely split nitrogen application.

Varietal Response to Nitrogen

Nitrogen nutrition plays a critical role in the rate and the duration of the dry-matter production after flowering (Tanaka 1969). The efficacy of dry-matter production depends mainly on the varietal differences in response of photosynthetic activity to nitrogen. These characteristics are reflected not only in the dry-matter production as a whole but also in the percentage of ripened grains and yields (Murata 1969).

In the field, the increment of yield increase caused by nitrogen application is generally more in number of panicles than in panicle size. There is a tendency for panicle size to decrease as panicle numbers increase. The larger the panicle size (with a high-panicle-weight variety), the taller the plant, making the high-panicle-weight type of rice more prone to lodging, especially at high nitrogen levels. Thus, the nitrogen response of the high-panicle-number type is higher than that of the high-panicle-weight type. With the former, an increase in yield can be accomplished by a heavy nitrogen application combined with dense planting.

Absorption of Phosphorus

Phosphorus uptake by the rice plant reaches its maximum by flowering time. The translocation of phosphorus from the leaves and culms to the panicle continues until the dough stage. This coincides with the translocation and accumulation of starch in the grain, showing a close relationship between carbohydrate metabolism and phosphorus (Ishizuka 1965).

Studies on phosphorus fertilization tend to get less attention because the effect of applied phosphorus on rice yield is seldom as dramatic as that of nitrogen. However, if phosphorus content in the soil is low, rice yield will be low.

Nutrient Removal by a Rice Crop

The analysis of various nutrient elements in rice grown in the tropics suggests that if the panicles and straw are removed from the field great amounts of silicon and significant amounts of potassium are removed. If only the grains are harvested, silicon removal is reduced greatly but the removal of nitrogen becomes relatively significant (Table 10.4).

Data from Japan show that the amount of nitrogen removed by a temperate climate crop of 12.68 t/ha is higher (Table 10.5) than that by a rice crop in a tropical country (Table 10.4). As expected, removal of silicon was also high in the data from Japan.

NUTRITIONAL DISORDERS

The visible toxicity symptoms of iron, aluminum, manganese, boron, and high salt content (salinity) were described earlier. The nutritional disorders of the rice plant that are caused by the toxic effects of iron, aluminum, manganese, boron, sulfide, organic acids, and salinity are described here.

Table 10.4 Nutrient Removal of a Rice Crop (Variety IR8) Yielding 7.9 t/ha Rough Rice[a] in Maligaya Rice Research and Training Center, Philippines, 1979 Dry Season

Nutrient Element	Mineral Concentration in Straw (%)	Mineral Concentration in Grain (%)	Amount of Mineral Removed by the Crop at Harvest (kg)			Amount of Mineral Removed per Ton of Rice Production (kg)	
			Straw	Grain	Total	Straw	Grain
N	0.53	1.09	37.4	86.1	123.5	5.3	10.9
P	0.08	0.20	5.6	15.8	21.4	0.8	2.0
K	1.36	0.31	95.9	24.5	120.4	13.6	3.10
Ca	0.39	0.05	27.5	4.0	31.5	3.9	0.51
Mg	0.26	0.11	18.3	8.7	27.0	2.6	1.1
S	0.07	0.10	4.94	7.90	12.84	0.7	1.0
Fe	0.02	0.0038	1.4	0.3	1.7	0.2	0.04
Mn	0.056	0.0048	4.0	0.38	4.38	0.6	0.05
Zn	0.003	0.0012	0.2	0.09	0.29	0.03	0.01
Cu	0.0003	0.0005	0.021	0.040	0.061	0.00298	0.00506
B	0.00089	0.000413	0.063	0.03	0.093	0.0089	0.0038
Si	7.40	1.68	521.70	132.72	654.42	74.00	16.80
Cl	0.18	0.16	12.69	12.64	25.33	1.8	1.6

[a]Straw yield = 7.0 t/ha.

Table 10.5   Nutrient Removal of a Rice Crop (Variety Kim-nampu) Yielding 12.68 t/ha Rough Rice in Toyama Prefecture in Japan in 1955 (Adapted from Shiroshita 1958)

Nutrient Element	Mineral Concentration in Straw (%)	Mineral Concentration in Panicle (%)	Total Amount of Mineral Removed by the Crop at Harvest (kg/ha)
N	0.64	1.06	206
P	0.051	0.220	32
K	1.70	0.32	251
Ca	0.296	0.028	40
Mg	0.12	0.07	23
Mn	0.109	0.004	14
Si	7.29	1.31	1051

Iron Toxicity

The iron concentration in the soil solution increases when the submerged soil is reduced. Iron toxicity occurs primarily on soils with low pH and high organic matter. Nutritional disorders of rice that are associated with iron toxicity are bronzing in Sri Lanka (Ponnamperuma et al. 1955), Akagare Type I in Japan (Baba and Tajima 1960), and Akiochi in the Republic of Korea (Park and Tanaka 1968).

The iron content is usually, but not always, high in plants showing bronzing, which makes diagnosis of iron toxicity by the iron content of the plant questionable. An interaction also exists between iron and potassium in the plant; plants exhibiting bronzing symptoms are usually low in potassium (Tanaka and Yoshida 1970).

The Akagare Type I disorder is remedied by the application of potassium and it is therefore considered to be associated with potassium deficiency. But plants affected by Akagare Type I are generally high in iron content, leading to the belief that bronzing and Akagare Type I disorders may be similar.

Iron toxicity is also a serious problem for rice grown in acid sulfate soils. Application of lime and phosphorus often minimizes the problem. Other suggestions for amelioration of iron toxicity are use of urea instead of ammonium sulfate, and application of phosphorus and potassium.

Aluminum Toxicity

Aluminum toxicity is sometimes a problem in upland rice grown on acid soils. It is also a problem on lowland soils, such as the acid sulfate soils in Thailand, where rice is grown as an upland crop for a few weeks before the field is flooded. Aluminum toxicity retards root growth, thereby reducing nutrient uptake and decreasing drought tolerance.

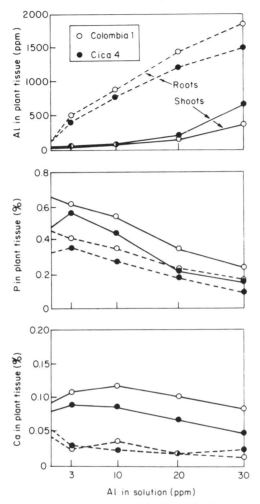

Figure 10.6 Effect of aluminum concentration in a solution culture on the aluminum, phosphorus, and calcium contents of shoots and roots of two rice varieties at 3 weeks after planting. (Adapted from Howeler and Cadavid 1976)

Howeler and Cadavid (1976) reported that distinct varietal differences exist in aluminum concentration in both shoots and roots when rice seedlings were grown in solution with varying aluminum concentrations. For example, the aluminum-tolerant line Colombia 1 had higher aluminum levels in the roots and lower levels in the shoots than the aluminum-sensitive CICA-4 (Fig. 10.6). This would indicate a better aluminum exclusion capacity of Colombia 1, resulting in aluminum precipitation in or outside of the root and less aluminum translocation to the shoot.

The phosphorus contents of shoots were higher than those of roots and generally decreased with increased aluminum concentration. Phosphorus con-

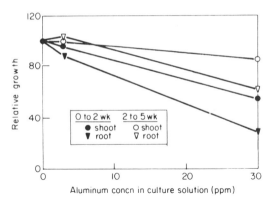

Figure 10.7   Effects of aluminum on growth of rice shoot and root length at two growth stages (av. of nine varieties). (From IRRI 1978a)

tents in both roots and shoots were consistently higher in Colombia 1 than in CICA-4. Similarly, the calcium contents in shoots were higher in Colombia 1 than in CICA-4 while those in the roots were not significantly different.

In 1977 IRRI initiated studies on varietal differences in tolerance for aluminum toxicity at 0–2 weeks after pregermination (IRRI 1978a). Because both tolerance for aluminum toxicity and deep root growth are desirable for acid upland soils, the studies used root length at 30 ppm aluminum to measure tolerance of rice for aluminum toxicity.

Aluminum retarded the growth of roots more than the growth of shoots (Fig. 10.7), with seedlings more susceptible to aluminum toxicity at 0–2 weeks old than at 2–5 weeks. Maximum and total root length were both affected. In general, lowland semidwarf varieties were more susceptible to aluminum toxicity than upland varieties, which confirmed earlier findings of Howeler and Cadavid (1976).

Recent study of aluminum toxicity at IRRI (Yoshida 1980, unpublished) suggests that aluminum reduces total root length but does not affect water uptake rate per unit root length. The total amount of phosphorus absorbed by the rice plant, however, is closely related to total root length which in turn is affected by aluminum.

Yoshida contends that aluminum toxicity under drought would increase plant water stress because maximum rooting depth is decreased. Thus, aluminum toxicity is considered an important soil factor affecting drought tolerance of rice. The mode of aluminum toxicity with rice growth is schematically represented in Fig. 10.8.

Manganese Toxicity

Rice has a high degree of tolerance for high manganese concentration in its tissues. With 3000 ppm manganese in the plant, rice usually yields well. The

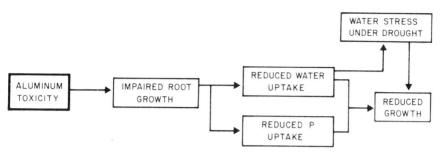

Figure 10.8 Schematic representation of mode of aluminum toxicity with rice growth. (From S. Yoshida, IRRI, 1980, personal communication)

critical tissue content for manganese toxicity was reported as 7000 ppm by Cheng and Ouellete (1971).

Tadano and Yoshida (1978) suggested that a high manganese content in rice tissues is frequently associated with high yields, possibly indicating that a high manganese content in the plant is associated with various favorable soil conditions.

Boron Toxicity

The first evidence of boron toxicity in rice was reported by Ponnamperuma and Yuan (1966) on a soil in the Silo region of Taiwan. The soil pH was 8.1. Soil solutions of the affected flooded soils contained 9.1 ppm boron compared with 1 ppm boron for two normal soils from the Philippines.

In an experiment with the Taiwan soil and two soils from the Philippines, Ponnamperuma and Yuan (1966) induced foliar symptoms of boron toxicity in rice at 20–60 ppm of applied boron as borax. These results suggested that the critical concentration of water-soluble boron for toxicity to rice may vary from soil to soil.

Irrigation water with boron content greater than 2 ppm causes accumulation of boron in soils (see Chapter 9).

Ponnamperuma (1979) summarized information on toxic limits of boron in a rice-growing soil:

- More than 3 ppm hot-water-soluble boron in soil is toxic (Hesse 1971).
- More than 2 ppm boron in irrigation water is harmful (FAO/UNESCO 1973).
- More than 2 ppm in culture solution is toxic (Lockard 1969).
- More than 1.5 ppm in saturated soil extract is toxic (USDA 1954).
- More than 9 ppm in the interstitial solution is toxic (Ponnamperuma and Yuan 1966).
- High potassium content aggravates boron toxicity (Houng 1975).

Amelioration of excess boron in soil is difficult. However, if irrigation water from a well has high (more than 2 ppm) boron content, it should be diluted with surface water with less or no boron content. Internal drainage should also minimize accumulation of boron (Ponnamperuma 1979).

Sulfide Toxicity

In a highly reduced flooded soil, sulfate is reduced to sulfide (Takai and Kamura 1966), which inhibits the respiration and oxidative power of the roots and retards the uptake of various elements. Mitsui et al. (1951) demonstrated that a hydrogen sulfide concentration as low as 0.07 ppm is toxic to rice.

On the other hand, sulfides can be oxidized by the oxidizing power of the roots and thus lose their toxic effects (Baba 1955, Takagi and Okajima 1956). Thus, the occurrence of sulfide toxicity in rice depends on the concentration of sulfide in the soil solution and the level of oxidizing power of the rice roots.

Hydrogen sulfide is considered the major cause of disorder of rice on degraded paddy soils. Sulfide toxicity occurs mainly on sandy, well drained, degraded paddy soils and on poorly drained organic soils (Tanaka and Yoshida 1970).

Toxicity of Organic Acids

Acetic acid is generally the major organic acid produced in a rice-growing soil. Other organic acids occurring in submerged soils are formic, propionic, and butyric (Motomura 1962, Gotoh and Onikura 1971). The production of organic acids is enhanced by incorporation of readily decomposable organic matter such as rice straw.

Organic acids are reported as toxic to rice at concentrations of 10^{-2}–10^{-3} M (Mitsui et al. 1959). Takijima et al. (1960) observed a distinct growth retardation of rice by acetic acids at a concentration of $1.5 \times 10^{-2} M$ in a submerged clay loam soil. Organic acids may aggravate iron toxicity in some soils (Tanaka and Navasero 1967).

Depending on pH of the soil solution, organic acids also appear to affect zinc uptake by rice (Forno et al. 1975).

Salinity

Salinity is generally associated with alkalinity in inland areas where evaporation is greater than precipitation. Salinity is also usually associated with acidity in coastal areas. The common problem in both cases is high salt concentration in the soil solution.

Under weakly saline conditions, and if sodium is the major cation and potassium is deficient, the sodium may benefit rice growth by partially replacing potassium. It appears that when potassium supply is high, sodium is antagonistic to potassium absorption, but when potassium supply is low, the beneficial effect of sodium by partial replacement of potassium is more pronounced than the antagonism between the two elements. As a result, the presence of moderately higher concentration of sodium chloride can promote rice growth (Yoshida and Castañeda 1969).

Salinity injury in rice varies with salt concentration, kind of salt, soil, temperature, variety, and development stage of the crop (Ponnamperuma 1979). Maas and Hoffman (1977) described salt injury in plants by the equation

$$Y = 100 - B(EC_e - A)$$

where

$$Y = \text{relative yield}$$
$$A = \text{salinity threshold value; equal to 3 for rice}$$
$$B = \text{yield decrease per unit increase in salinity; equal to 12 for rice}$$

Ponnamperuma (1978) suggested $EC_e < 2$ mS (2 mmho)/cm as best for rice.

Classification of Nutritional Disorders

Tanaka and Yoshida (1970) classified nutritional disorders of rice in Asia as shown in Table 10.6. At low pH, the main nutritional disorders are iron toxicity, phosphorus deficiency, manganese toxicity, and hydrogen sulfide toxicity. At high pH, the nutritional disorders are phosphorus, iron, and zinc deficiency, salinity, alkalinity, and boron toxicity. Deficiencies of phosphorus and zinc are discussed elsewhere in this chapter.

FERTILIZER MANAGEMENT FOR RICE

From time immemorial farmers have used common sense and experience to provide nutrition to their crops. They used raw organic matter, human and animal manure, ashes, fish bone, and other waste materials to make the crop plants more productive. Plant nutrition as a science began to develop in 1840 when the German physical chemist Justus von Liebig suggested that crop yields were directly related to the content of plant nutrients or mineral elements in the manure applied to the soil. Since Liebig's revolutionary theory on mineral nutrition there has been a complete modernization of science, technology, and resource consumption. The full benefits of that modernization are yet to be reaped in many parts of the world due to excessive population pressure on the land and lack of the necessary resources—fertilizers and other production inputs—required in successful agricultural development.

Table 10.6 Classification of Nutritional Disorders in Asia (From Tanaka and Yoshida 1970)

Soil		Soil Condition	Disorder	Local Name
Very low pH		(Acid sulfate soil)	Iron toxicity	Bronzing
Low pH	High in active iron	Low in organic matter	Phosphorus deficiency	
		High in organic matter	Phosphorus deficiency combined with iron toxicity	
		High in iodine	Iodine toxicity combined with phosphorus deficiency	Akagare Type III
		High in manganese	Manganese toxicity	
	Low in active iron and exchangeable cations	Low in potassium	Iron toxicity interacted with potassium deficiency	Bronzing / Akagare Type I
		Low in bases and silica, with sulfate application	Imbalance of nutrients associated with hydrogen sulfide toxicity	Akiochi
High pH		High in calcium	Phosphorus deficiency / Iron deficiency / Zinc deficiency	Khaira / Hadda / Taya-Taya
		High in calcium and low in potassium	Potassium deficiency associated with high calcium	Akagare Type II
		High in sodium	Salinity problem / Iron deficiency / Boron toxicity	

Fertilizer Efficiency

Fertilizer use efficiency is the output of any crop per unit of fertilizer nutrient applied under a specific set of soil and climatic conditions. Recently, Barber (1977) defined fertilizer efficiency as the increase in yield of the harvested portion of the crop per unit of fertilizer nutrient applied. For example, if changing agronomic practices increase yield without changing the rate of nutrient application, the efficiency of the fertilizer nutrient has increased. Barber emphasizes, however, the importance of obtaining high yields when measuring fertilizer efficiency. The highest fertilizer efficiency is obtained with the first increment of fertilizer and the size of the yield increase decreases successively with each additional increment of fertilizer added. Therefore, the lowest rate of fertilizer would give the highest efficiency but the yield obtained may be low and unprofitable.

The above contention is not entirely relevant to fertilization of rice in the developing world, particularly in tropical Asia where millions of farmers use low rates of fertilizer. While learning how to maximize grain yields at high rates of fertilizer application, maximization of grain yields at low rates of fertilizers should not be ignored. Any savings in fertilizer use, and hence cost, could help millions of small rice farmers.

Another approach to defining fertilizer efficiency is based on plant uptake with the least amount of fertilizer needed to match plant use with maximum grain yield as the most efficient rate of fertilizer application. Bartholomew (1972) classified nitrogen need according to plant use—good use efficiency, average use efficiency, poor use efficiency, and pollution range.

There are two possible reasons for not reaching expected yield levels:

- The fertilizer nutrients are not taken up by the plants because they are applied at the wrong time or in the wrong place, or transformation of nutrients has made them unavailable.
- Although taken up by the crop, the nutrients are not used for grain production due to other growth-limiting factors such as insufficient water or light or lack of other mineral elements.

Several factors determine the fertilizer efficiency in rice at the farm level. Among them are soil, crop and variety, season, time of planting, water management, weed control, insect and disease control, cropping sequence, time of fertilizer application, and fertilizer sources. Examples of chemical properties of common fertilizers for rice are given in Table 10.7.

For India, Mahapatra et al. (1974) summarized fertilizer efficiency in cereals as dry-season rice > wet-season rice > wheat > maize > millet > sorghum. In all those crops, a well planned and well executed program of field research would help identify nutrient deficiencies and determine yield responses from low, as well as high, rates of fertilizer application.

Table 10.7 Some Chemical Properties of Common Rice Fertilizers

Fertilizer	Available (%)			Chemical Formula	Solubility in Cold H_2O (g/100 ml)	Specific Gravity (g/cc)	Equivalent Acidity per kg of N (kg $CaCO_3$)
	N	P_2O_5	K_2O				
Urea	46	0	0	$CO(NH_2)_2$	119	1.3	1.8
Ammonium sulfate (24% S)	21	0	0	$(NH_4)_2SO_4$	71	1.2	5.3
Ammonium chloride (68% Cl)	26	0	0	NH_4Cl	37	1.2	1.8
Ammonium nitrate	33	0	0	NH_4NO_3	118	1.3	1.8
Ordinary superphosphate (12% S)	0	20	0	1-3 $Ca(H_2PO_4)_2$ 2-3 $CaSO_4$	2^b	1.2	—
Triple superphosphate[a]	0	46	0	$Ca(H_2PO_4)_2$	2	1.2	—
Ammonium phosphate sulfate (15% S)	16	20	0	1-2 $NH_4H_2PO_4$ 1-2 $(NH_4)_2SO_4$	23^b	1.8	5.3
Diammonium phosphate[a]	18	46	0	$(NH_4)_2HPO_4$	58	1.6	5.3
Muriate of potash (46% Cl)	0	0	62	KCl	65	1.6	—
Compound fertilizers[a]	14	14	14	Mixture of above	varies	1.5	5.3

[a] TSP, DAP, and compound fertilizers manufactured with wet process H_3PO_4 contain 2% or more S.
[b] Solubility of phosphorus ingredient.

NITROGEN

Nitrogen is generally needed in most rice soils, particularly in places where nitrogen-responsive modern rice varieties are grown with improved cultural practices. Low nitrogen use efficiency, the widespread need for nitrogen for food production, the anticipated increase in fertilizer nitrogen costs, and the anticipated world shortage of petroleum products all call for answers to the questions:

- What happens to the fertilizer nitrogen not utilized by the rice crop?
- How can fertilizer use and crop management practices be improved to conserve fertilizer nitrogen?

Nitrogen Fertilizer Effectiveness

Numerous nitrogen-response experiments have shown that the recovery of fertilizer nitrogen applied to the rice crop is seldom more than 30–40%. Even with the best agronomic practices and strictly controlled conditions the recovery seldom exceeds 60–65% (De Datta et al. 1968). Ironically, the soil and climatic conditions that favor rice growth adversely affect the recovery of nitrogen from the soil and are responsible for its rapid loss. Nitrogen may be lost from plants through root exudation, the flushing action of dew or rain, and natural or mechanical loss of plant parts.

Even considering all these possibilities, a portion of the applied nitrogen is not accounted for. Recent research in the United States indicates that there is a relationship between gaseous nitrogen loss and efficiency of nitrogen utilization by rice plants. These losses may need to be considered to determine nitrogen fertilizer effectiveness in rice (da Silva and Stutte 1979).

Management of Nitrogen Fertilizers

Detailed studies of fertilizer use and crop management to minimize nitrogen losses and to increase the efficiency of nitrogen fertilizer are needed. It is imperative to examine each fertilizer material and each management practice critically in order to increase efficiency of nitrogen utilization for rice.

Inorganic Sources of Nitrogen

The choice of nitrogen source for flooded rice depends on the method and the time of application. Many farmers apply fertilizer in two and three doses, with part at planting (during land preparation or shortly after planting) and the remainder as a topdressing during the various stages of growth. The nitrogen applied at planting should be in the ammonium (NH_4^+) form. The source of nitrogen used as topdressings is less critical; NH_4^+ and nitrate (NO_3^-) forms appear to be equally effective. This is because when the crop is fully established the NO_3^- form of nitrogen is rapidly taken up by the rice crop before it can be leached down to the reduced soil layer where it might be lost through denitrification. This may account for the equal performance of NO_3^--containing fertilizers such as NH_4NO_3 as compared to any other NH_4^+-containing or forming N sources such as ammonium sulfate or urea.

Numerous research reports and review articles have compared sources of fertilizer nitrogen for flooded rice. In most situations, ammonium-containing or ammonium-producing (urea) fertilizers are similar in effectiveness based on grain yield (De Datta and Magnaye 1969).

In some instances, however, different responses are obtained from urea and ammonium sulfate. For example, in certain problem soils low in iron, such as degraded paddy soils of Japan, urea, ammonium chloride, and other nonsulfur-containing fertilizer nitrogen are preferred over sulfur-containing nitrogen fertilizers such as ammonium sulfate. Without adequate iron, H_2S develops in a reduced soil and adversely affects rice plants. In the presence of adequate quantities of active iron, sulfur is precipitated as ferrous sulfide (FeS), which is not toxic to rice. On a sandy soil low in iron, Jayasekara and Ariyanayagam (1962) reported a rice yield increase of 73.4% with urea and only 33.7% with ammonium sulfate.

Experiments to evaluate six nitrogen sources for flooded rice in the United States, Sri Lanka, India, Japan, and Taiwan indicated that ammonium sulfate is as effective as ammonium chloride, followed by urea, and that nitrate-containing fertilizers, such as ammonium nitrate, are least effective (Engelstad 1967). Averages of data for 10 experiments in Thailand during 1966–1969 ranked the effectiveness of nitrogen sources as ammonium sulfate > ammonium chloride > ammonium sulfate/nitrate > urea > calcium ammonium nitrate. The reason for the apparent poor performance of urea in some of the Southeast Asian experiments was not clear. In many experiments yield levels and responses to nitrogen were rather low. When those data were expressed in percentage form the differences among sources appeared large.

Tests conducted in Arkansas, United States, during 1959–1960 indicated no differences among urea, ammonium nitrate, ammonium sulfate, and ammonium chloride topdressed in a two-split application. These results further confirm the contention that nitrogen source is of minor importance for topdressings during the growing season.

Anhydrous ammonia is agronomically equal in effectiveness to ammonium sulfate but its use in small paddies is prohibited by lack of suitable machines for its application.

It is apparent that urea is the principal nitrogen fertilizer for rice and will remain so in tropical Asia. Urea is the highest nitrogen-containing solid fertilizer material commercially available, and its high analysis permits considerable savings in shipping and distribution costs. These factors have made urea attractive to a large number of farmers and fertilizer manufacturers. Figure 10.9

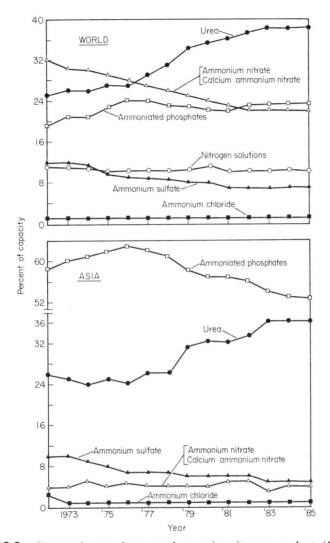

Figure 10.9 Changes in capacity to produce major nitrogen products (1972–1985). (From Tennessee Valley Authority and International Fertilizer Development Center, unpublished data)

shows the trend in production of nitrogen fertilizers in continental Asia and the world.

SLOW-RELEASE NITROGEN SOURCES   Fertilizer shortages, plus the heavy losses of fertilizer nitrogen in rice fields when application rates are high, as in the case of modern rice varieties, have led fertilizer technologists to develop fertilizer materials for more efficient use.

There are a number of proved or potential advantages of fertilizers that release their plant nutrients slowly throughout the growing season or longer. These slow-release fertilizers offer potential for increasing fertilizer efficiency, decreasing cost of application due to fewer applications, minimizing losses by leaching, fixation, or decomposition, eliminating or at least minimizing luxury consumption, and avoiding damage to seedlings.

Slow-release fertilizers to provide nitrogen that roughly coincides with that required by growing plants are made by:

- Synthesis of chemical compounds with inherently slow rates of dissolution.
- Coating of conventional soluble fertilizers to reduce their dissolution rates.

Most of the slow-release fertilizers are condensation products of urea and aldehydes. Horn (1979) reviewed various slow-release nitrogen sources that are relevant in lowland rice.

Five slow-release urea fertilizers tested for rice are:

1 *Ureaform (urea formaldehyde)*, which was first developed in the United States for turf grasses and ornamental plants. Ureaform is a generic term that covers a host of products that are mixtures of methylene ureas. They range from short-chain, water-soluble molecules to long-chain, water-insoluble molecules. The completely insoluble ureaforms are of no agronomic significance but those of higher solubility are of great importance.

2 *IBDU (isobutylidene diurea)*, which is a condensation product of urea and isobutyl aldehyde developed in Japan. It is now commercially manufactured and available in the Japanese markets. Nitrogen release from IBDU is generally controlled by granule size.

3 *CDU (crotonylidene diurea)*, which was prepared by the acid catalysis of urea and acetaldehyde. It was first manufactured in Germany and is now a commercial product in Japan.

4 *SCU (sulfur-coated urea)*, which was first developed by the Tennessee Valley Authority (TVA) in the United States by using sulfur, a microcrystalline soft sealant, and a microbicide. The porous nature of the sulfur coating was avoided by using a sealant that prevents rapid transfer of moisture through the coating (Stangel 1970). The release rate of nitrogen is controlled by varying the thickness of sulfur and sealant coating. Urea granules must have about the same thickness of coating on them regardless of their size to give the same dissolution rates. The granules should be spherical and have a smooth surface. If they do not, they are much more difficult to coat completely and effectively with sulfur.

A mixture of polyethylene and bright stock oil is now used instead of wax to seal the sulfur coating in the conventional process. Diatomaceous earth has been used as a conditioner because of its oil-absorbing properties. It is applied at about 2% of the total product weight.

5 *Sulfur-coated supergranules*, which were developed by the International Fertilizer Development Center (IFDC). Both supergranules of urea and sulfur-coated supergranules were produced for testing in Asian rice culture. Sulfur-coated supergranules were developed to combine the concept of deep placement and slow-release nitrogen fertilizer in increasing fertilizer nitrogen efficiency in lowland rice.

AGRONOMIC SIGNIFICANCE OF CONTROLLED RELEASE FERTILIZERS Field experiments in Japan showed that a single dose of IBDU gave 20% more rice than ammonium sulfate applied in split application, and 25% more yield than ammonium sulfate applied in a single basal dressing (Hamamoto 1966). Results with IBDU in India and the Philippines were equally promising (Prasad and De Datta 1979). In field experiments during 1967–1972 in the United States, Wells and Shockley (1975) also observed that IBDU was a suitable nitrogen source for rice.

Of the coated fertilizers, sulfur-coated urea (SCU) has been most widely tested on many crops, including rice. In many countries the crop response to SCU was generally superior to the response to urea in a single (basal) application and often superior to split application at the same nitrogen rate. Experiments with SCU in India have given excellent results. For example, in experiments conducted by the All India Coordinated Rice Improvement Project (AICRIP), SCU gave 25–30 kg grain/kg nitrogen compared with 16–20 kg grain/kg nitrogen with urea applied as a split application (Prasad and De Datta 1979). Results from 41 experiments in seven countries showed that SCU produced 6 more kg grain/kg N than urea (Table 10.8).

During 1975, IRRI initiated an informal collaborative program on fertilizer efficiency in rice in 10 countries in Asia. Collaborative programs on soil fertility and fertilizer management were developed jointly by national scientists, IRRI, and International Fertilizer Development Center under the International Network on Fertilizer Efficiency in Rice (INFER).

Table 10.8 Yield of Rough Rice Grown under Poor Water Management as Affected by Conventional and Slow-Release (Sulfur-Coated Urea) Nitrogen Fertilizer; Data are Averaged over 41 Experiments (From Doll 1975)

Nitrogen Source[a]	Grain Yield (t/ha)	Increased Yield Due to Nitrogen (t/ha)	Nitrogen Fertilizer Efficiency (kg rice/kg N)
No N	3.2	—	—
Urea	4.9	1.7	16
SCU	5.6	2.3	22

[a]All nitrogen was applied as basal treatments at or before transplanting. Equal amounts of nitrogen were applied as SCU and urea at rates varying from 80 to 250 kg N/ha.

Table 10.9 Performance of Urea Fertilizer Treatments Compared with Split Urea-N Applications at Two Rates of Fertilizer Nitrogen; Data are from International Network on Fertilizer Efficiency in Rice (INFER) Trials where Yield Response to Applied Nitrogen was Significant in 1975-1977 Dry and Wet Seasons (From IRRI 1978a)

| | Tests (%) Where Yields, Compared with Those of Best Split Application, Were | | | | | |
| | Significantly Higher at | | Significantly Lower at | | Not Significantly Different at | |
Urea Fertilizer Treatment	Low N-Rate[a]	Medium N-Rate[b]	Low N-Rate[a]	Medium N-Rate[b]	Low N-Rate[a]	Medium N-Rate[b]
Dry Season (22 trials)						
Liquid band	5	10	9	19	86	71
Mudball	57	38	0	5	43	57
Briquette	18	27	9	0	73	73
SCU	45	36	0	0	55	64
Wet Season (55 trials)						
Liquid band	2	12	6	10	92	78
Mudball	28	30	2	6	70	64
Briquette	20	23	2	4	78	73
SCU	18	33	4	2	78	65
Average (77 trials)						
Liquid band	3	11	7	13	90	76
Mudball	37	33	1	5	62	62
Briquette	19	25	4	2	77	73
SCU	26	34	3	1	71	65

[a]Low N-rate = 56 kg N/ha during the dry season, 28 kg N/ha during the wet season.
[b]Medium N-rate = 84 kg N/ha during the dry season, 56 kg N/ha during the wet season.

When the INFER data (IRRI 1978b) were summarized for 24 dry-season and 60 wet-season trials during 1975-1977, SCU showed the highest average yield during the dry season. Averages of two nitrogen rates and two seasons show that in 30% of INFER trials, SCU gave significantly higher grain yield than the best split application. In 2% of the trials, SCU gave a significantly lower yield and in 68% of trials no significant differences in grain yield (Table 10.9).

In 1978, the network name was changed to International Network on Soil Fertility and Fertilizer Evaluation for Rice (INSFFER) to include management of soil fertility through the use of biological nitrogen fixation including use of azolla.

Continuous flooding with good water control is the exception rather than the rule for rice paddies in the tropics. Often flooding is either delayed or intermittent after fertilizer application. The advantages of SCU have been most obvious in soils with intermittent flooding and drying. Under such conditions in Peru it was possible to produce the same crop yield with only 40% as much nitrogen as SCU as with urea (Sanchez et al. 1973). In Indonesia, SCU was superior to three applications of urea under intermittent flooding, but under constant flooding yields with SCU and three applications of urea were more often equal (De Datta 1978). In the United States, the use of SCU did not result in increased rice production or higher nitrogen efficiency in comparison with split applications of regular fertilizer grade urea (Westfall 1972).

NITRIFICATION INHIBITORS Another approach used to achieve controlled availability of plant nutrients is through the use of nitrification inhibitors.

Because a major portion of the fertilizer nitrogen lost from rice fields may be lost after the nitrification of amide and ammonium nitrogen to nitrates, nitrification inhibitors have been tested to minimize such losses. In almost all cases, the inhibitor is intended to block the conversion of ammonium nitrogen to nitrate by inhibiting *Nitrosomonas* growth or activity (Fig. 10.10).

Four examples of nitrification inhibitors and their chemical characteristics are:

1   *Nitrapyrin (2-chloro-6-trichloromethylpyridine),* which was developed in the United States, is possibly the most widely tested of all nitrification inhibitors. Nitrapyrin persists and is effective in the soil for as long as 6 weeks. Effectiveness of nitrapyrin declines as soil temperature increases. Nitrapyrin is recommended for banding with a nitrogen fertilizer.

2   *Dicyan-diamide,* which is sold commercially in Japan. The nitrification inhibitory effects of dicyan-diamide have been reported at concentrations of 5–10% (of ammonium N). Like nitrapyrin, its effectiveness declines with the increase in soil temperature.

3   *AM (2-amino-4 chloro-6-methyl pyrimidine),* which was developed in Japan has been proven to be effective at a concentration of 2 ppm. Increase in temperature greatly reduces its effectiveness, which reduces its potential for tropical conditions.

4   *ST (2-sulfanilamidiothiazole),* which was developed by the same Japanese concern that produced AM. ST has been claimed to have overcome the problems of AM. ST can be mixed with acid fertilizers and can be readily formulated with compound fertilizers.

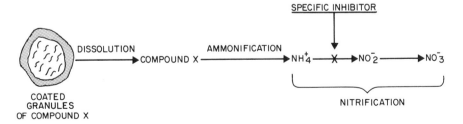

Figure 10.10   The coated-granule inhibitor approach. (Adapted from Stangel 1970)

AGRONOMIC SIGNIFICANCE OF NITRIFICATION INHIBITORS　In laboratory studies, several chemicals such as nitrapyrin, AM, dicyan-diamide, thiourea, ST, potassium azide, and some sulfur chemicals have effectively retarded nitrification. The effectiveness of inhibitors in the field has been less clear.

Field experiments in India and in Japan showed the benefits of the urea treatment over ammonium sulfate with a nitrification inhibitor, but research in the United States (Turner 1977) failed to obtain a significant increase in rice yield attributable to the same nitrification-inhibitor treatment. Recently, however, in 2 years of field testing in Arkansas, a nitrogen stabilizer containing nitrapyrin and urea fertilizer applied on Starbonnet rice growing on a Crowley silt loam soil showed increased grain yields, plant height, and nitrogen uptake when compared to equivalent rates of urea alone (Wells 1977).

In India, in tests of indigenous materials with nitrification-inhibiting properties, a 22% increase in rice yield was found due to treatment of urea with an acetone extract of neem (*Azadirachta indica* A. Juss) kernels. Because the acetone also extracts the oil, which can be utilized for medicinal purposes or for making soap, later studies were made with the cake after the extraction of the oil. An increase of 24% in rice yield was obtained by applying urea mixed with 20% by weight of neem cake. Neem cake appears to hold considerable promise for increasing the efficiency of nitrogen applied to rice (Prasad and De Datta 1979). The actual chemical compounds responsible for nitrification-inhibitory properties in neem kernels have yet to be identified.

Organic Sources of Nitrogen

Management of straw and other organic sources of nitrogen has received considerable attention. Their use has been influenced by two main factors:

- The fear that continuous application of certain inorganic nitrogen fertilizers, although they may give better yields in the first few years, can increase soil acidity and cause appreciable deterioration in soil productivity.
- The belief that compost and organic sources of nitrogen are essential for obtaining high yield.

The fear of soil acidity due to continuous application of inorganic nitrogen is particularly true if ammonium sulfate is used alone for a long time. Repeated applications of ammonium chloride in areas of high rainfall are believed to promote removal of calcium, magnesium, and phosphorus and gradually make the soil more acid. Digar (1958), however, did not observe any detrimental effect on rice yield or changes in soil characteristics even after 10 years of continuous application of ammonium sulfate.

In Japan, compost was a major factor for farmers in yield contests during 1948–1968. Contest-winning farmers obtained an average of 11.2 t/ha (9 t brown rice/ha), with the highest record of 13.1 t/ha (10.5 t brown rice/ha), by using compost as a supplement to inorganic fertilizers. The average amount of compost used by the contest winners was 18 t/ha but some farmers used as much as 30 t/ha.

Intensive use of compost in rice fields in Japan reached a peak about 1955 and then decreased steadily because of abundance of inexpensive nitrogen fertilizers (Tanaka 1978). The average use of compost was 6.5 t/ha in 1955 and 4.5 t/ha in 1970. The returns from chemical fertilizers became highly attractive while handling of rice straw for making rice straw compost became expensive because of the high cost of labor for making compost. Today, many farmers in Japan burn their straw in the fields; others incorporate it to minimize the cost of handling and avoid air-pollution problems caused by burning.

In China, compost and other organic nitrogen sources are used with inorganic fertilizers (primarily NH_4HCO_3). In general, compost and other organic sources are applied as a basal dressing. Swine manure is extensively used. Despite the fact that many large fertilizer plants are completed or scheduled to be completed soon, the recycling of organic waste will continue to play an important role in rice production in China.

In the Republic of Korea, use of compost has steadily declined over the years because of high labor cost. However, farmers are still encouraged to make compost for rice fields. With compost and inorganic use of fertilizer providing at least 150 kg N/ha, Korean farmers have achieved average national rice yields close to 7 t/ha (see Chapter 6).

The experiences of temperate East Asia indicate that dependence on compost and other organic sources of nitrogen declines with increased industrialization and associated high cost of labor rather than because of any decline of rice yields from continuous use of inorganic fertilizer in rice.

In tropical Asia, with increased availability of urea and other nitrogenous fertilizers and with greater use of short, stiff-strawed, nitrogen-responsive rice varieties, the use of organic sources of nitrogen provide a supplement rather than an alternative to inorganic fertilizers for lowland rice.

Nevertheless, because of fears of inorganic fertilizers' deleterious effects on soil productivity and farmers' inability to purchase inorganic fertilizers, organic sources of nitrogen for rice are being used in some tropical Asian countries. In India, 2200 million tons of compostable waste generated annually contribute about 22 million tons of plant nutrients.

With increased emphasis on cropping intensity, organic sources of nitrogen fertilizers, even in the developing countries of the Asian tropics, should remain as an important supplemental source of nutrients.

NATURE OF ORGANIC NITROGEN SOURCES Rice straw consists of organic substances such as cellulose, hemicellulose, pectin, lignin, proteins, and so on, and mineral elements such as nitrogen, phosphorus, potassium, silicon, and sulfur. When straw is burned, all organic substances and a large portion of nitrogen and sulfur in the straw are lost, and the other elements remain in the ash.

The composition of rice straw and compost prepared from it will vary greatly depending on the availability of plant nutrients in the soil on which rice and other

Table 10.10   Composition of Compost[a] (Adapted from Tanaka 1978)

	Maximum	Minimum	Mean	Coefficient of Deviation (%)
H_2O (%)	93.2	39.6	75.1	16.3
pH	9.4	5.9	7.9	10.1
C (%)	13.3	1.4	7.9	26.6
N (%)	1.07	0.07	0.39	43.5
C-N ratio	12:1	20:1	20:1	31.0
P_2O_5 (%)	0.57	0.03	0.19	47.3
K_2O (%)	2.22	0.09	0.70	64.2
CaO (%)	1.49	0.08	0.45	48.8
MgO (%)	0.49	0.02	0.13	38.5
Na_2O (%)	0.45	0.01	0.13	23.1
MnO (ppm)	600.00	41.00	248.00	45.0
B (ppm)	11.9	0.3	1.9	68.5
SiO_2 (%)	16.4	0.01	4.5	31.2

[a]105 samples were included in this survey collected in Ibaraki Prefecture in Japan. Materials used for preparation of composts included not only the rice straw but various other materials; animals include cattle, swine, poultry, and so on.

materials are grown. The composition of compost also depends on the conditions and the duration of decomposition. Table 10.10 gives an example of compost composition.

The Shanghai Academy of Sciences, China, reported nitrogen and phosphorus compositions of compost as:

- N from 0.3 to 0.4% and in some cases up to 0.6%
- P_2O_5 from 0.1 to 0.15% and in some cases up to 0.2%

COMPOST MAKING   There is no uniform way to make compost for rice production. In China, for example, the ingredients used to make compost vary from place to place but the basic steps are similar. In the Wu-shi area, droppings of swine, sheep, and rabbits are used to make compost. The farmyard manure is mixed with silt and the mixture is allowed to decompose. In the Hsu-han People's commune in Ja Din county near Shanghai, compost is made by mixing 70–80% silt from the bottom of fishponds, 15% straw, 9% night soil, and dead leaves. Most of the composting is done above ground (Fig. 10.11). The pit or pile is covered with mud. The mixture is stirred from time to time and the compost is ready for use after 3 months.

In most countries in South and Southeast Asia, water hyacinth, *Eichhornia crassipes* (Mart.) Solms, is a serious weed in canals, ponds, irrigation ditches, and

Figure 10.11   Compost piles covered with mud are seen near rice fields all over the People's Republic of China.

floodplains. In China, extensive use is made of water hyacinth for swine feed and as a material for compost.

GREEN MANURE   In many countries, green manure is regarded as an important nutrient source for rice. It is useful where organic sources such as compost and inorganic fertilizers are lacking. The adoption of the green manure practice is largely determined by the alternatives available to provide nitrogen and organic matter and the resources of rice farmers. In some instances green manure will be used because inorganic nitrogen is expensive; in other instances land may be better used to grow green manure rather than left idle. India is the country where research and use of green manures have received the greatest attention. Excellent results with green manure have been reported from India, Bangladesh, Indonesia, and China.

In many areas of China, green manure crops such as *Sesbania* sp. and *Crotalaria* sp. are grown to improve soil fertility. In areas where lowland rice is to be followed by an upland crop, the green manure crop is grown before the upland crop to improve soil structure. In Hua county, vetch is grown during winter, primarily for green manure.

The distinct benefits of green manure crops to soil fertility are that they provide organic matter and nitrogen. Green manuring, however, has the distinct disadvantage of using land and water that otherwise could be used for a crop of rice or other food crop. The requirement of adequate levels of phosphorus, and to some extent calcium and potassium, also raises serious questions regarding the use of green manure crops under most rice-growing situations in the world.

Nevertheless, green manuring will be practiced in India and some other countries until fertilizer supply and prices are within the reach of small farmers. In some countries, land is used for a green manure crop simply because other food or feed crops cannot be grown otherwise. In those situations green manure improves soil fertility and physical properties.

Green Manure Crops. Many plant species are grown as green manures. The most common green manure crops in India are *Crotalaria juncea* L., *Sesbania bispinosa* (Jacq.) W. F. Wight (Vachhani and Murty 1964). *Vigna unguiculata* subsp. *cylindrica* (L.) Verdc. is also used. Other examples of green manure crops in India with their green-matter production and nitrogen content are given in Table 10.11.

Joint vetch (*Aeschynomene americana* L.) is used as a green manure in Indonesia and Sri Lanka.

In California, the use of green manure crops has been a common commercial practice for 30 years. Estimates in 1955 indicated about 20% of the rice-growing area used a green manure crop of purple vetch (*Vicia benghalensis* L.) or common vetch (*V. sativa* L. subsp. *sativa*) (Williams et al. 1957).

Management of Green Manure Crops. Green manure crops are grown before and after a rice crop. The green manure crop can be broadcast into a standing rice crop shortly before harvesting. A green manure crop can also be broadcast into rice stubble without cultivation.

In India, for growing an early-maturing rice crop, a green manure crop is broadcast in April–May with the onset of the monsoon season. *Sesbania sesban* (L.) Merr. and *Vicia mungo* (L.) Hepper are most common, and produce 4–6 t green matter/ha within 6 weeks of seeding. In waterlogged areas, only *Sesbania* species are suitable. Where rice is transplanted late (August–September) slow growing species such as *Macroptilium lathyroides* (L.) Urb., joint vetch and *S.*

Table 10.11 Yield of Green Matter and Nitrogen Concentration of Different Green Manure Crops Evaluated in India (Adapted from Vachhani and Murty 1964)

Green Manure Crop	Green Matter at 8-week Stage (t/ha)	Nitrogen Concentration (%)
Sesbania aculeata	9.4	2.8
Crotalaria juncea	4.5	2.3
Phaseolus aureus	—	3.2
Aeschynomene americana	6.7	—

speciosa, which produce 6–8 t green matter/ha can also be grown in addition to species used for the early-maturing rice crop (Vachhani and Murty 1964).

In southern India, wild indigo [*Tephrosia purpurea* (L.) Pers.] is broadcast into the standing rice crop.

In many places in India, green manure crops are grown on the bunds of rice fields, or in fallow lands during the off-season, and their leaves incorporated into the rice fields. *Gliricidia sepium* (Jacq.) Kunth ex Walp. and *Leucaena leucocephala* (Lam.) de Wit are excellent species for production of green leaves.

Two factors are critical to the success of green manure production. One is the stage of the crop when it is ready for incorporation. Incorporation of the crop too early or too late gives less than the full potential of the green manure crop. The best time to incorporate a crop is when it is about to flower—about 8 weeks after seeding for most crops.

The time of incorporation of a green manure crop depends on the growth stage of the crop. If harvested at the optimum time of about 8 weeks after seeding, a green manure crop can be incorporated just before planting of a rice crop. But an older, woody, green manure crop should be turned under earlier because it requires more time for decomposition.

In California, vetch containing 3–4% nitrogen is a highly effective and inexpensive source of nitrogen if the crop is properly incorporated and the soil promptly flooded. Purple vetch capitalizes on the mild, moist winters of California and does well in areas without irrigation where the annual rainfall is 375 mm or more. Vetch is broadcast in the fully prepared field in October and harrowed in. The green manure crop is turned into the soil by disking or plowing in the spring during land preparation for rice.

In Texas, a suitable rotation with *Sesbania* species has been beneficial in rotation with rice; rice yields increased 20% or more. *Lespedeza cuneata* (Dum.) G. Don could be established by broadcasting the seed in rice stubble in late February or early March.

It is apparent that the short-term benefit of a green manure crop is its contribution as a source of nitrogen. Deep incorporation of green manure is desirable to minimize nitrogen losses in gaseous forms. In California, significant nitrogen losses occurred when vetch was incorporated less than 10 cm deep.

Timing of Nitrogen Application

If the crop absorbs a high proportion of the nutrients added as a fertilizer, fertilizer efficiency automatically increases. This means fertilizer efficiency can be increased by getting higher yields with the same amount of nutrient absorbed by the plant. One way to achieve this is to apply the fertilizer at a time to best meet the demand of the rice plant.

Rice plants of medium growth duration (145 days), when grown at low nitrogen levels (about 20 kg/ha), use fertilizer most efficiently for grain production during the maximum tillering stage and around the flowering stage (between the booting and milk stages). A high nitrogen supply tends to decrease the number of filled spikelets and the weight of 1000 grains. Therefore, split application of nitrogen, with one dose at transplanting and another at panicle initiation, is best for obtaining high grain yields, particularly in the case of medium- and long-season varieties. Nitrogen absorbed by the plant from tillering to panicle initiation tends to increase the number of tillers and panicles, and that absorbed during panicle development (from panicle initiation to flowering) increases the number of filled spikelets per panicle. Nitrogen absorbed after flowering tends to increase the 1000-grain weight (De Datta 1978).

In a direct-sowing experiment, Yanagisawa et al. (1967) observed that nitrogen applied at panicle initiation or at flowering reduced the rate of senescence of the lower leaves, probably as a result of a reduction in the amount of nitrogen that was translocated from the lower to the upper leaves.

Nitrogen uptake in relation to time of nitrogen fertilization has been studied extensively in Japan. If various rates of nitrogen are applied as a basal dressing, nitrogen uptake is progressively increased at different growth stages (Fig. 10.12). It was further reported that in the Hokkaido environment, which is cool during the rice-growing season, the desired high grain yields could not be obtained by increasing nitrogen rates as a basal dressing (Shiga et al. 1977).

The effect of nitrogen topdressing at the early panicle initiation stage on the patterns of nitrogen uptake by rice plants almost resembled that of a high level of basal dressing of nitrogen. The spikelet number increased but the percentage of ripened grain decreased. Lodging of rice plants sometimes occurred even when topdressing was 30 kg N/ha. The authors (Shiga et al. 1977) concluded that the topdressing of nitrogen at an early panicle initiation stage is inadequate for obtaining high grain yields in Hokkaido.

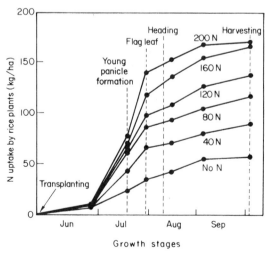

Figure 10.12 Nitrogen uptake by rice plants at successive growth stages from different levels of basal dressing of nitrogen (1972). (Adapted from Shiga et al. 1977)

The efficiency of nitrogen fertilizer uptake at different times of application can be determined by tracer techniques. In an experiment by Yanagisawa et al. (1967) in Japan, 55% of the nitrogen applied 15 days before heading was used by the plant. At flowering, use was 49%; at tillering, 34%; and at transplanting as a basal dressing, 7%. In another study, the application of two-thirds of the nitrogen at transplanting and the remaining one-third at the booting stage gave higher nitrogen recovery (51%) than when the entire amount was applied at transplanting (38%) or when topdressed at the tillering stage (32%). Studies in California showed that 18% of tracer ^{15}N was absorbed when the nitrogen was topdressed at maximum tillering, and 45% when topdressed at the booting stage (Patnaik and Broadbent 1967).

Studies at IRRI using ^{15}N-labeled fertilizer demonstrated that there is no advantage to a split application of nitrogen if the variety is lodging-resistant and is grown on a fertile heavy clay soil. On the other hand, split applications of nitrogen may result in significantly higher yields if the soil is not highly fertile and if the variety has a long growth duration. In such instances, split applications may prevent excessive vegetative growth and early lodging (De Datta et al. 1969).

Since the introduction of modern varieties in the Asian tropics in the mid-1960s, interest in proper timing of nitrogen fertilizers has greatly increased. Recent concern about high fertilizer prices has increased interest in efficient use of fertilizer by proper management, including optimum timing of nitrogen application. With decreased rates of fertilizer nitrogen, time of nitrogen applications becomes critical. For example, in a dry-season trial at IRRI, it was shown that the highest efficiency of nitrogen use for two modern rice varieties with medium maturity requirement (124–127 days) was from fertilizer nitrogen at half of the optimum rate (60 kg N/ha) applied just before panicle initiation. A shorter-season rice showed highest efficiency after basal application but responded well to the nitrogen applied just before panicle initiation (Fig. 10.13). When properly timed, split applications of nitrogen fertilizer gave higher yields than when all the fertilizer was applied as basal treatment.

More recent studies in India and in the Philippines, suggest that two to three applications of nitrogen per crop gives highest nitrogen efficiency and that more split applications are needed for long-duration varieties and for lighter soils (Prasad and De Datta 1979).

Coarse-textured soils usually have high percolation rates and also high nitrogen losses by leaching and volatilization. On such soils, split applications of nitrogen (usually two or three) are usually recommended. Finer-textured soils require fewer nitrogen applications provided good water management practices are followed.

Among the 84 trials conducted within the INFER program, 77 showed significant response to nitrogen fertilizer. In many of those trials yield responses from split application of urea nitrogen were similar to those from SCU and other methods of application (liquid band placement, placement of mudballs, and briquettes) (see Table 10.9). That indicates that if urea fertilizer nitrogen is

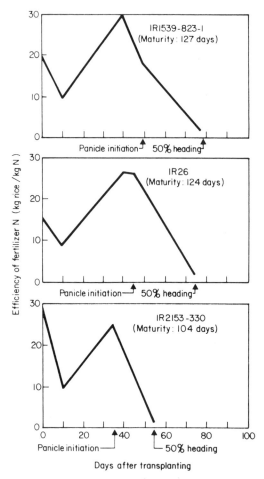

Figure 10.13 Efficiency (kg rice/ kg N) of 60 kg N/ ha as urea applied in a single dose at different times to three rice varietites. IRRI, 1974 dry season. (From De Datta et al. 1974)

applied at the proper time there is considerable room for improving its efficiency (IRRI 1978b).

It is clear that more research is needed to determine precisely the optimum timing of nitrogen application.

TIME OF NITROGEN APPLICATION IN MECHANIZED RICE CULTURE   In the United States, nitrogen for rice is used at somewhat higher rates in Arkansas than in the other southern states (Mikkelsen and Evatt 1973). Rates of more than 100 kg/ha are fairly common. Some differences in the fertilizer requirements of rice varieties are recognized with the stiffer strawed varieties being capable of using higher rates of nitrogen than lodging-susceptible varieties. Most rice farmers in

Louisiana and Texas prefer somewhat earlier application of the total amount of fertilizer than farmers in Arkansas and Mississippi (Thompson et al. 1962).

Results on time of nitrogen application in the United States vary with the rice-growing states. In Arkansas, the typical United States varieties such as Zenith, Nato, and Bluebonnet 50 produced high yields when fertilized 45–60 days before heading, indicating that if properly timed with the reproductive cycle of the plant, nitrogen application could increase the number of florets and eventually produce high grain yields (Hall 1965). Sims (1965) found that low rates of nitrogen are effective when applied in a single dose just before panicle initiation. He observed that under conditions of high soil fertility, nitrogen should be topdressed either during the early stages of crop growth, or not at all in the case of varieties similar to Bluebonnet 50.

A system of timing midseason nitrogen application based on plant development has been used in mechanized rice culture in the United States to minimize lodging and disease. The proper time to make the midseason nitrogen application is determined by cutting the main stems of representative rice plants and measuring the length of the elongated internodes. Optimum internode lengths for nitrogen application vary according to variety. Normally, if the midseason nitrogen rate is 45 kg/ha, one-half is applied at the proper internode length for the variety and the remainder 10–14 days later.

Timing of midseason nitrogen application is highly critical for some varieties (Huey 1977). Nitrogen applied too early will increase lodging; applying nitrogen too late will be less beneficial to yield and may delay maturity. Results in Arkansas suggest that it is possible to estimate the optimum time for applying midseason nitrogen by accumulating growing degree days (DD-50) from time of emergence to time of nitrogen application (Downey et al. 1976).

For California, Mikkelsen et al. (1967) reported that although topdressing with nitrogen supplements the nitrogen supply, it cannot be used as a substitute for an application before planting. They reported that topdressing should be done at early growth stages (30–40 days) and not 60 days after seeding, although the latter seemed to improve the appearance of the crop.

Where application of nitrogen to the seedbed was not sufficient to maintain normal color and growth of rice, supplemental applications have been profitable. For effective use of topdressed nitrogen on California rice varieties, the application should be made no later than the jointing stage (Mikkelsen and Evatt 1973).

Recent experiments in California with the new varieties ESD 7-1, M9, and M7 confirm earlier results with old varieties—preplant nitrogen applications are most efficient in mechanized rice cultivation. However, nitrogen application in split doses, with two-thirds preplant and one-third topdressed near panicle initiation, provides similar efficiency when near-optimum or excessive nitrogen rates are applied. When suboptimum nitrogen rates are applied, the preplant application provides the greatest yield and the maximum efficiency (University of California and United States Department of Agriculture 1978).

Results from Texas were somewhat different than those from California. In

Table 10.12   The Effect of Timing of Nitrogen Application in Relation to Flooding on Growth, Yield, and Quality of Rice (Adapted from Mengel and Leonards 1978)

N Timing	Plant Height (cm)	Lodging (%)	Rough Rice Yield at 12% Moisture (t/ha)	Whole Grain Mill Yield (%)
7 days preflood	127	81	7.8	67.8
5 days preflood	121	89	7.3	67.0
3 days preflood	124	88	7.2	67.2
1 day preflood	122	25	8.2	68.1
1 hour preflood	117	10	8.3	67.3
1 day postflood	105	0	7.1	69.2
3 days postflood	107	0	7.2	68.2
LSD[a] (5%)	6	9	0.7	1.2

[a]LSD = least significant difference at level given.

Texas, topdressing rice with nitrogen 82–84 days after seeding gave somewhat higher yields than applying the entire amount of fertilizer at seeding time (Reynolds 1954). Working with a crop grown under ideal conditions, Evatt (1969) indicated that nitrogen may be effectively applied in a single application. Evatt and Hodges (1975), however, suggested that if nitrogen is applied in a single dose, considerable blending with other nutrients is necessary before it is applied to the soil.

Application of significant amounts of nitrogen at the time of first permanent flood establishment (2–4 weeks after seedling emergence) is a common practice in Louisiana. Because it takes 5–10 days to flood a large field, a loss in effectiveness occurs due to leaving urea on the soil surface for several days. Recent results suggest that applying urea fertilizer to a dry soil surface up to 7 days prior to establishing a permanent flood is superior to broadcasting urea into standing water. Increased plant height, lodging, and in some cases, grain yields, indicate increased plant uptake of fertilizer nitrogen (Table 10.12).

Methods of Nitrogen Application

Since the classical work of Shioiri (1941) on denitrification losses of nitrogen from flooded rice soils, placement of nitrogen in the reduced soil layer has been considered the best method to decrease nitrogen losses and increase fertilizer nitrogen efficiency in lowland rice. The early history of the deep placement concept developed in Japan was recently reviewed by Mitsui (1977).

Based on the Shioiri concept of deep placement of fertilizer nitrogen, subsequent research in India (Vachhani 1952) indicated higher efficiency by deep placement of nitrogen fertilizers. A field experiment at IRRI using ^{15}N-labeled

Table 10.13   Effect of Placement of Nitrogen and Other Methods of Application on the Grain Yield and Efficiency of Fertilizer Nitrogen of Milfor 6(2) Rice (Adapted from De Datta et al. 1968)

Method of Application	Grain Yield[a] (t/ha)			Efficiency (kg rough rice/kg N)
	Nitrogen Sources (^{15}N-labeled)			
	Urea	Ammonium Sulfate	Mean	
Fertilizer incorporated at planting	6.8	6.8	6.8	14
Placed at 10 cm soil depth	8.6	8.3	8.4	43
Split application[b]	7.1	7.6	7.3	23
Mean (sources)	7.3	7.6	—	—

[a]Check (without fertilizer nitrogen) yield was 5.9 t/ha. LSD (5%) = 0.67 t/ha.
[b]30 kg N/ha at planting and 30 kg N/ha at panicle initiation.

fertilizer on a heavy clay demonstrated that fertilizer utilization and grain yield were highest where the fertilizer was placed at a 10-cm soil depth (Table 10.13).

Rice was fertilized in California before 1953 by broadcasting either on the dry seedbed before flooding and planting or, more commonly, on the floodwater by airplane after seeding. Fertilizer placement work of Mikkelsen and Finfrock (1957) demonstrated that nitrogen broadcast on the soil surface or applied to the flooded fields was not used efficiently by rice. However, ammonia nitrogen drilled 5–10 cm into the soil, where reducing conditions developed 3–5 days after flooding, remained in the soil and was continuously available to the rice plants.

Efforts to increase nitrogen fertilizer efficiency by deep placement were continued using labeled N. Results from experiments from different countries (IAEA 1970) were similar to the early work in Japan (Shioiri 1941) and IRRI (De Datta et al. 1968).

Using nonlabeled N fertilizer, Prasad and De Datta (1979) reported that deep placement was superior to broadcast application for both grain yield and nitrogen uptake. Use of pellets and nitrogen placement gave similar grain yields but more of the applied nitrogen was recovered from the pellet application.

Application of nitrogen in mudballs is a technique developed by the Japanese that has received considerable attention in recent years. In tests at IRRI, yields obtained with 60 kg N/ha applied as mudballs in the reduced soil layer and those obtained with 100 kg N/ha applied in split application were comparable (De Datta 1978).

In INFER trials (IRRI 1978b) in 11 countries during the dry and wet seasons of 1975–1977, 7 of 84 trials showed no significant response to nitrogen. In the remaining 67 trials, where positive nitrogen response was observed, mudball placement gave the best performance in the dry season. With a low nitrogen rate (50 kg N/ha), the mudballed urea was superior to the best split application in 57% of the trials and was comparable to the split application in 43% (see Table

10.9). In no case did the mudballs give a poorer performance than the best split application. With a medium nitrogen rate (84 kg N/ha), the percentage of trials where the mudballed urea was superior to the best split application was reduced from 57% to 38%; in 5% of the trials mudballed urea had poorer performance than the best split application. However, the labor required to make mudballs and apply them is a deterrent to the extension of this technique. It is practical mainly for small farmers with abundant family labor.

The concept of placement of fertilizer in the soil is becoming more relevant since the introduction of modified urea materials such as urea briquettes and urea supergranules. In the INFER trials (IRRI 1978b) placement of urea briquettes gave significantly higher grain yield over the best split application in the range of 18–27% of the trials (Table 10.9). Briquettes are clearly economically superior to mudballs because they do not require farm labor for manufacturing. Economic analysis of the INFER data suggested that the extra yield associated with placement of briquettes would give a return of $7 or more per dollar extra cost in countries with wage rates of $0.70/day or lower (IRRI 1978b).

Foliar application of nitrogen fertilizer is another method that has received attention. However, a large number of trials by AICRIP in India indicated that foliar application of nitrogen did not compete favorably with soil application. At IRRI, foliar application of urea in split doses did not give favorable results over conventional methods of application (De Datta 1978).

Plow-sole application of fertilizer, which combines primary tillage and deep placement of nitrogen fertilizer, showed promise in recent IRRI experiments (IRRI 1978a).

Complementary Practices for Increasing Fertilizer Nitrogen Efficiency

Removal of weeds and insects, manipulation of plant density, and management of water are examples of management practices that help improve the efficiency of fertilizer nitrogen in rice.

WEED CONTROL The response of rice to nitrogen fertilizer is markedly influenced by the adequacy of weed control. Experiments in farmers' fields in the Philippines indicate that if weeds are not controlled adequately fertilizer should not be applied. Likewise without the control of insects and diseases little response from nitrogen can be expected (De Datta 1978).

PLANT DENSITY Nitrogen increases tiller numbers in rice. Thus, applied nitrogen fertilizer influences the effect of plant density on rice yield. IRRI experiments demonstrated that with no added fertilizer nitrogen, rice yield increased progressively with increased plant density because high plant density tends to compensate for the adverse effects on yield of low tiller number with low soil fertility (Fig. 10.14). The results indicate that changes in plant geometry have potential for increasing fertilizer nitrogen efficiency, particularly at low levels of applied nitrogen fertilizer.

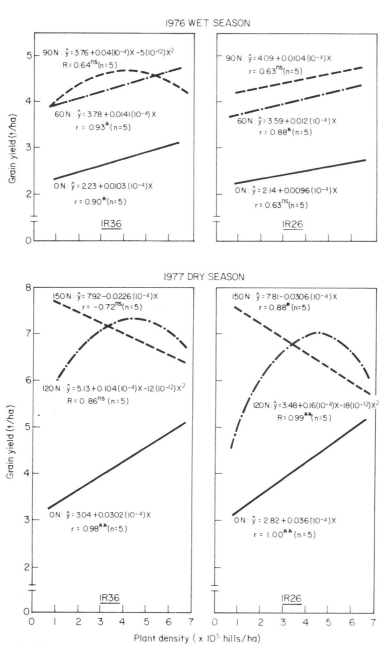

Figure 10.14 Effects of rates of nitrogen fertilizer application on yield response to plant density of IR36 and IR26 rices. ns = not significant at 5%; * = significant at 5%; ** = significant at 1% level by *t*-test. IRRI, 1976 wet and 1977 dry seasons. (From Nguu and De Datta 1979)

In the Asian tropics, it is important to exploit natural soil fertility and to maximize grain yield at a low solar energy level during the wet season when most of the rice is grown. Increasing the plant density of early-maturing rices may improve grain yield and the efficiency of rice plants to use fertilizer nitrogen. Hence, the level of applied fertilizer nitrogen could be decreased without affecting the yield of lowland tropical rice.

WATER MANAGEMENT   Water management may directly affect the efficient use of nitrogen fertilizer by lowland rice. This is because water regime of a soil has a profound effect on nitrogen transformations in the soil. Furthermore, the water regime determines the oxidation-reduction potential of the soil, which in turn regulates nitrification and denitrification.

In the rainfed rice crop, the moisture regime is often not ideal for efficient nitrogen use. However, if rain water provides adequate moisture, the efficiency of nitrogen fertilizer should be identical to that in an irrigated crop, provided no other factors such as insects, diseases, and weeds limit grain yield.

Experiments in a farmer's field in Nueva Ecija province, Philippines, demonstrated that nitrogen response in irrigated rice is greater than rainfed rice. Similarly, a recent survey made by IRRI economists indicated that nitrogen and water levels were the two major factors constraining yields of rice in some tropical Asian countries. It is, therefore, important that water control should be improved as much as possible in order to increase fertilizer nitrogen efficiency in lowland rice.

Drainage may directly affect the efficiency of nitrogen utilization by lowland rice. It is generally believed that temporary drainage is necessary when topdressing with nitrogen to bring fertilizer material as close to the soil particles as possible. Experimental results, however, suggest that temporary drainage for topdressing of nitrogen is not essential (De Datta et al. 1969).

The above data emphasize the interaction of fertilizer nitrogen response with good agronomic practices. Adequate weed, insect, and disease control coupled with proper management of water will increase yield response to fertilizer nitrogen and even make soil nitrogen more available to plants.

MANAGEMENT OF NITROGEN AND LEVELS OF SOLAR RADIATION   It is known that nitrogen response in rice varies considerably with the level of solar energy, which varies widely over the year and to a lesser extent from year to year. During the ripening period of the crop, the intensity of solar energy during an average day in the monsoonal tropics is about 350 cal/cm^2 per day, which is similar to the values reported during the rice-growing season in Japan. These year-to-year variations within the same date-of-harvest range have profound effects on the nitrogen response of rice varieties.

Using data from 12 crops grown on experiments during 1968–1969, De Datta and Zarate (1970) found correlation between grain yield and irradiance ranging from .50 to .77 during the 45 days before maturity, which includes about 15 days before flowering. In 12 harvests of four varieties tested, the highest nitrogen

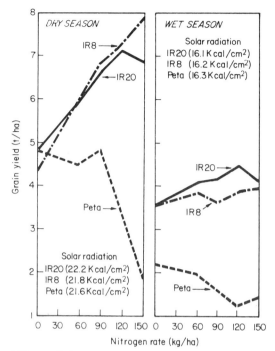

Figure 10.15 Effects of nitrogen levels on grain yield of IR8, IR20, and Peta. IRRI, 1968–1973. 1968 and 1969 dry seasons: 0, 30, 60, 90, and 120 kg N/ha; 1968 wet season: 0, 20, 40, 60, and 80 kg N/ha; 1969 wet season: 0, 25, 50, 75, and 100 kg N/ha. (From De Datta and Malabuyoc 1976)

response and grain yield (9.3 t/ha) were obtained when harvest was in May when the solar radiation was 20 kcal/cm². Subsequent analyses of the data from the same experiment of 1968–1970 show that nitrogen response was linear throughout up to 120 kg N/ha when the solar energy level for the ripening period was 22 kcal/cm² per month or higher. These data suggest that at that solar energy level the maximum grain yield was not reached at 120 kg N/ha.

Figure 10.15 shows the effects of solar radiation on yield response of rice varieties at different levels of nitrogen.

PHOSPHORUS

Rice, like any other cereal, requires a considerable quantity of phosphorus for vigorous growth and high grain yield. Phosphorus deficiency occurs on millions of hectares of Ultisols, Oxisols, acid sulfate soils, Ando soils (Inceptisols), Vertisols, and certain Inceptisols. Not only are these soils low in available phosphorus but they also fix fertilizer phosphorus in highly insoluble forms.

Besides, the increase in availability of phosphorus brought about by soil submergence is slight in these soils.

Management of Phosphorus Fertilizers

Management of phosphorus is largely dependent on soil characteristics such as soil reaction, degree of weathering, kind of clay minerals, water regime, cropping intensity, and cropping patterns. Previously flooded soils showed greater phosphorus sorption capacity than their nonflooded soil counterparts. Data from Australia suggest that soils that had previously grown flooded rice required more phosphorus for upland cropping than soil that had previously grown a crop such as maize (Willet et al. 1978).

Sources of Phosphorus

Except in extremely acid or alkaline soils, few significant differences have been found among the effects of various phosphorus sources in lowland rice. India, Pakistan, the Philippines, the Republic of Korea, and Burma use super-phosphate as their primary source of phosphorus on most soils, but Sri Lanka, Malaysia, Thailand, Khmer Republic, and Vietnam use phosphate rock. On acid soils of South and Southeast Asia, phosphate rock has been applied directly for rice. Phosphate rock is priced lower than acidulated phosphates, and can reduce fertilizer cost where transport costs are not too high.

In experiments at the Tennessee Valley Authority (TVA) and in fields of Thailand, several phosphate rocks with varying citrate solubility were evaluated. A procedure was developed that enabled a choice of phosphorus source on the basis of both agronomic effectiveness and prices of phosphorus in triple superphosphate (TSP) vs. phosphate rocks (Fig. 10.16).

Experiments by IRRI on Buenavista clay (pH 6.0, organic matter 0.8–1.3%, CEC 20–30 meq/100 g soil) demonstrated that urea-ammonium phosphate with various N-P ratios was as effective as superphosphate in lowland rice whereas Florida ground phosphate rock was a poor source of phosphorus. A method-ology has been developed, through the use of chemicals and physical character-ization of a representative sample for specific property, with which the agronomic effectiveness of a particular rock source can be predicted without going through extensive agronomic field trials. TVA researchers have extended this work to classify several rocks as to their probable effectiveness (Fig. 10.17).

The major sources of phosphorus for rice in the United States are ordinary superphosphate, concentrated superphosphate, ammoniated superphosphate, and the ammonium phosphates (Mikkelsen and Evatt 1973).

MODIFICATION OF PHOSPHORUS FERTILIZERS   Considering that the produc-tion of finished phosphate requires seven to eight times as much energy as phosphate rock for direct application, it seems certain that some changes in

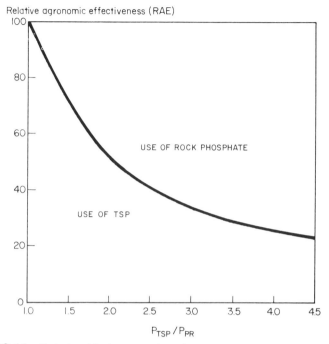

Relative agronomic effectiveness (RAE)

USE OF ROCK PHOSPHATE

USE OF TSP

P_{TSP} / P_{PR}

Figure 10.16   Relationship between the price ratios of triple superphosphate (TSP) and phosphate rock and relative agronomic effectiveness values (RAE). RAE can be used in making a choice between these sources. (Adapted from Engelstad et al. 1974)

phosphate processing are likely to occur. These will be aimed specifically at supplying the farmer a suitable phosphate product with minimal investments in energy, capital for processing, and contribution to the pollution problem.

There are a number of disadvantages of making direct application of phosphate rock. A major one is that to be effective, phosphate rocks must pass a 200-mm sieve (53 microns), which presents real problems in handling at all levels.

Several modifications of phosphate rocks have been suggested to increase its effectiveness and ease of handling. In these efforts, the International Fertilizer Development Center (IFDC) and TVA have cooperated to develop appropriate fertilizers for all crops including rice. Small phosphate rock particles (60–140 mesh) called minigranules, can be applied without the dust problem. Phosphate slurry is another possibility for direct application.

Iron-aluminum phosphate ores, such as those from Senegal and Christmas Island, can be calcined (heated) to produce useful phosphorus fertilizers. Such products, while limited in supply, have proven fairly effective on crops grown on high pH soils.

Such fertilizer materials are being agronomically evaluated for rice on high pH soils. Other possible sources are fused magnesium phosphate (produced mostly

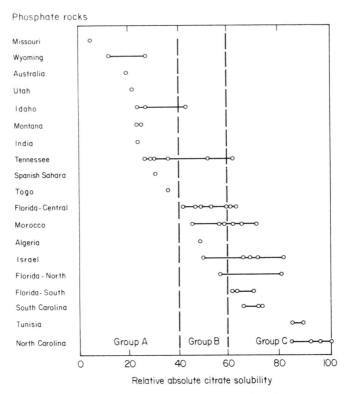

Figure 10.17 Comparison of absolute citrate solubilities of phosphate rocks (ACS value of MR-467 from North Carolina taken as 100). (From Stangel 1975)

in Japan, Brazil, and the Republic of Korea) and Rhenania phosphate ($CaNaKPO_4$) (De Datta 1978).

Time of Phosphorus Application

In general, phosporus is applied to rice at planting, but later application can be made, provided it is not later than the time of active tillering. Early application of phosphorus is essential for root elongation. Phosphorus applied during the tillering stage is not efficiently used for grain production.

Split application of phosphorus has not been proven of value because there is a great mobility of phosphorus from old leaves to young ones, because the availability of soil phosphorus increases with time during submergence, and because leaching losses are low (Chang 1976).

Method of Application

Despite the greater availability of phosphorus in a lowland soil, the recovery of fertilizer phosphorus by rice is generally less than 10%. Where extensive fertilization is practiced in the rice-growing areas, phosphorus is usually applied as a basal dressing. It is usually broadcast, with or without mixing with the surface mud, either during final land preparation or puddling, or immed'₊tely before sowing or transplanting (Table 10.14). Research on methods of fertilizer phosphorus application is limited. In India (International Rice Commission 1961) field experiments during 1954–1957 compared various methods of application of ammonium phosphate and triple superphosphate at rates of 10–20 kg/ha. The methods tested were:

- Broadcasting at puddling immediately before planting.
- Drilling at puddling before planting.
- Dipping the seedlings in a slush of mud with phosphorus.
- Application in a pellet form (prepared by mixing soil with five to ten times the fertilizer and making into small balls; applied 5–8 cm deep in the soil and 30 cm apart between rows 3–4 weeks after transplanting).

Pellet application was consistently superior to other methods of application in two stations, but other stations found no significant difference among different methods.

In the United States where rice is drilled in dry soil, fertilizer is applied by drilling bands at seeding or broadcasting immediately after seeding. The method and time of application of phosphorus fertilizers to rice in the United States are important because phosphorus stimulates grassy weeds. Because shallow water does not suppress germination and growth of grass seedlings, it is recommended that phosphorus fertilizers should not be broadcast before seeding on dry soil. If the rice is drilled, phosphorus fertilizers may be applied through the soil under the drill rows.

The effects of placement of phosphorus fertilizer in the soil are not clear. Muirhead et al. (1975), however, showed that maize and sunflower grown in

Table 10.14   Relative Efficiency of Various Phosphorus Placement Methods in Rice Based on the Percentage of Phosphorus Derived from the Fertilizer from the Surface Placement at 20 Days (Average of Seven Experiments Conducted by International Atomic Energy Agency in Five Countries) (From Davide 1965)

Placement Method	Days after Transplanting			
	20	40	60	80
Surface broadcast	100	96	89	114[a]
Surface broadcast and hoed in	138	124	97	111
10 cm deep at planting point	148	71	45	58
20 cm deep at planting point	109	61	34	48
10 cm deep between rows	72	72	45	46
20 cm deep between rows	50	48	37	40

[a]No significant difference between 60 and 80 days.

previous rice fields responded markedly to phosphorus fertilizer banded 10 cm below the seeds, but the response was not to the same extent as for fertilizer mixed with the surface 10 cm. This contrast in responses is characteristic of soils with high capacity to immobilize phosphorus. Later studies by Willet et al. (1978) suggested that phosphorus sorption was significantly positively correlated with percentage of phosphorus in the plant derived from the fertilizer but was not correlated with yield response to phosphorus with different methods of application.

POTASSIUM

Generally, the response of rice to potassium added to the soil is not as marked as for nitrogen or phosphorus. There are reports that most rice soils in Asia do not need potassium as much as nitrogen or phosphorus and that only a small and variable increase in rice yield is obtained with addition of potassium fertilizer. Although a large percentage of 566 experiments in the Republic of Korea showed positive response to potassium fertilizers the actual increase in yield was about 4% (Oh and Kim 1964).

Three trials in India yielded no significant difference in grain yield obtained with potassium additions. The rice requirement of potassium is sometimes supplied from plant residues turned under and from potassium in irrigation water. But India has rice-growing soils in the high rainfall areas that are deficient in potassium. Those soils are, in decreasing order of deficiency, the yellow, coastal alluvium, mixed red, and black laterite groups (Goswami et al. 1976).

Data from 665 agronomic trials in farmers' fields throughout India demonstrate that the highest potassium response (about 1.4 t/ha) was obtained with 60 kg K_2O/ha on red sandy or loamy soils. Because of those soils' lighter texture, losses of potassium were substantial.

Favorable response of rice to potassium has been obtained on sandy and coarse-textured soils in Vietnam, Sri Lanka, and Malaysia. Thenabadu (1973) reported that more than 75% of the rice soils in the wet zone of Sri Lanka have a potassium content of less than 0.15 meq/100 g of soil, and soils in the intermediate zone, as well as the boggy and peat soils, are deficient in potassium.

In Thailand, most rice-growing soils have moderate levels of exchangeable potassium except in the Khorat regions where the levels of exchangeable potassium are very low.

Most rice-growing soils in the Philippines are geologically young and release considerable potassium from weathering of primary minerals. Nevertheless, under continuous cropping with modern rice varieties, potassium response is becoming increasingly apparent in many lowland soils (De Datta and Gomez 1975).

The average increase in yield in Taiwan due to potassium is about 50%. Responses are most significant on red and yellow earths, and the effect of potassium is generally greater in the second crop than in the first.

Management of Potassium Fertilizer

Compared with nitrogen, phosphorus, and zinc, response to potassium in rice has been limited. Furthermore, management of potassium fertilizer is simpler than the management of nitrogen.

Sources of Potassium

Potassium fertilizers are obtained by mining and purifying natural deposits like sylvinite (halite and sylvite), carnallite, kainite, and langbeinite. Of these, more than 90% of the estimated world potassium resources occur as bedded potassium deposits principally sylvinite and carnallite.

Potassium chloride is the principal fertilizer source for potassium application to rice because of its low cost of production and high analysis (62% K_2O). It contains between 45 and 47% chloride, depending on the original rock composition.

Potassium sulfate (52% K_2O) may be used in areas of sulfur deficiency. It should not be used in degraded paddy soils such as those in Japan.

In several countries, some complete fertilizers such as 14-14-14, 15-15-15, 21-10-10, and 10-15-15 are applied basally. In Japan, N-K compounds such as 16-0-20 or 17-0-17 are widely used for topdressing rice.

Time of Potassium Application

Potassium should be applied during the final land preparation. However, based on experiments in Japan, Su (1976) summarized the benefits of adequate potassium status at different growth stages:

- Potassium absorbed at the maximum tillering stage increases the number of panicles and spikelets.
- Potassium absorbed at the panicle formation stage increases number of panicles and spikelets as well as weight of grains.
- Potassium absorbed after panicle formation mainly helps increase grain weight.

A large number of Taiwan experiments also summarized by Su (1976) showed a positive response of rice to split application of potassium. Research covering 25 years in Japan indicates that the grain yield response to potassium topdressing on degraded and poorly drained rice fields is remarkable (Noguchi and Sugawara 1966).

Sometimes, response to split application of potassium is related to an optimum nitrogen-potassium ratio. A basal application of potassium should not be applied when:

- Nitrogen supply from the soil and from the basal application is low.

- CEC of soil is low and drainage in soil is excessive.
- Plant spacing is wide.
- Low-tillering type varieties are used.
- Conditions are not favorable for tillering.

When a low rate of potassium (30 kg K_2O/ha) is applied, it is better to topdress once at the rapid tillering stage. In southern Japan, it is customary to apply one-half to two-thirds of the total potassium before transplanting and the remainder as one or several topdressings. Field trials in Orissa, India, showed a modest (about 8%) but significant benefit of split application compared to basal application of potassium.

Five years of data on clayey soils with high CEC in the Philippines did not show any beneficial effects of split application of potassium over single basal application (De Datta and Gomez 1975).

ZINC

As long as 50 years ago, zinc was recognized as an essential micronutrient (Sommer and Lipman 1926). Since then numerous studies have indicated that zinc deficiency is a serious nutritional problem for upland crops. Nene (1966) in India first reported zinc deficiency in lowland rice. Since then it has become recognized as a widespread and important nutritional problem throughout the rice-growing world.

Next to nitrogen and phosphorus deficiency, zinc deficiency now ranks first among the nutritional disorders that limit grain yield of rice. It occurs on Histosols, sodic, calcareous, and sandy soils, and on soils wet for prolonged periods (Ponnamperuma 1977).

Research on nutrition of zinc in rice was stimulated by early work of Yoshida (1968) who reported that a disorder of rice plant locally known as *hadda* in Pakistan was caused by zinc deficiency. Since then zinc deficiency has been reported in many areas in India, Japan, United States, Brazil, and the Philippines. In the Philippines zinc deficiency is the second most serious nutritional disorder limiting the yield of lowland rice. At least 13 provinces in the Philippines have large areas of zinc-deficient soils (Ponnamperuma 1977).

The incidence of zinc deficiency has increased recently because of:

- Replacement of traditional varieties by modern varieties, which have less tolerance for zinc deficiency or greater need for zinc because of higher production of rice.
- The replacement of ammonium sulfate by urea thereby temporarily increasing soil pH.
- Increased use of phosphorus fertilizer resulting in phosphorus-induced zinc deficiency.
- Increased use of concentrated fertilizers that contain no zinc or less zinc.

A number of physiological diseases in rice, such as *khaira*, *hadda*, *Akagare* Type II, *mentek*, *taya-taya*, and alkali diseases, have been identified as the result of a zinc deficiency (Table 10.6).

Management of Zinc-Deficient Soils

It is apparent that zinc deficiency is likely to occur on high pH or calcareous soils, and must be corrected if high yields are to be realized.

A possible solution to zinc deficiency is to drain the field and thereby increase zinc solubility, but that is often not possible or desirable because of other benefits that are expected from continuous flooding. In zinc-deficient soils, complete (NPK) fertilizers alone will provide limited yield advantage unless zinc deficiency is corrected first. Experiments in Pakistan showed that without zinc the grain yield was 4.3 t/ha; but by dipping seedlings in 1% ZnO suspension in water, the grain yield was 5.8 t/ha (Table 10.15) (Yoshida et al. 1970).

Table 10.15 A Comparison Between Methods of Zinc Application Used for Overcoming Zinc Deficiency at the Kala Shah Kaku Rice Farm, West Pakistan (Adapted from Yoshida 1975b)

Treatment No.[a]	Zn Application (kg/ha) Nursery Bed	Field Before Puddling	Grain Yield (t/ha)	Duncan's Multiple Range Test[b]
1	0	0	4.3	a
2	0	10	6.0	bcd
3	0	100	6.5	cd
4	100	0	5.2	abc
5	100	10	6.2	bcd
6	Broadcast Zn at 10 kg/ha after puddling		6.0	bcd
7	Broadcast Zn at 100 kg/ha after puddling		6.9	d
8	Broadcast Zn at 10 kg/ha when symptoms first appeared		5.7	bcd
9	Broadcast Zn at 100 kg/ha when symptoms first appeared		6.2	bcd
10	Spraying Zn at 10 kg/ha when symptoms first appeared		5.7	bcd
11	Dipping seedlings in 1% ZnO suspension		5.8	bcd

[a]All plots fertilized with 150 kg N/ha, 60 kg P_2O_5/ha, and 60 kg K_2O/ha.
[b]In a column, mean values with the same letter are not significantly different at 5% probability level.

Table 10.16 Effects of Zinc Application on Growth, Yield and Zinc Concentration of Rice Plants Growing on a Calcareous Soil, Cebu, Philippines (Adapted from Yoshida 1975b)

Treatment	Effect at 35 Days after Transplanting					Grain Yield (t/ha)	Relative Yield
	Plant Height (cm)	Tiller Number per Hill	Dry Weight per Hill (g)	Nutrient Concentration of Shoots			
				K (%)	Zn (ppm)		
NPK	31	8	1.0	2.0	15	1.8	100
NPK + dipping in 1% ZnO	52	16	5.4	3.1	22	5.3	301
NPK + ZnCl₂ (100 kg Zn/ha)	62	17	11.5	2.1	55	5.6	316
NP + ZnCl₂ (100 kg Zn/ha)	58	18	10.8	1.7	60	5.1	288

In Cebu, Philippines, with NPK alone the grain yield was 1.8 t/ha on a calcareous soil. With NPK plus a dipping of seedlings in 1% ZnO suspension in water the yield was 5.3 t/ha (Table 10.16). In Agusan del Norte, Philippines, Katyal and Ponnamperuma (1974) reported that NPK without zinc decreased yield at three of eight sites, whereas those with zinc oxide alone gave grain yield of 4.0 t/ha. NPK plus zinc oxide gave a further increase to 5.3 t/ha.

Transplanted Rice

Experiments on a highly zinc-deficient soil in Tiaong, Quezon province, Philippines, showed that without zinc application grain yield in two cropping seasons ranged from 0.1 to 0.5 t/ha in transplanted rice. The method of zinc application to rice is important for efficient use of zinc fertilizer. With a zinc oxide seedling dip the grain yields ranged from 3.3 to 4.1 t/ha (Table 10.17). The most practical means of correcting zinc deficiency was to dip the seedling roots in a 2% ZnO suspension in water prior to transplanting. Other treatments such as foliar spray with zinc sulfate and other zinc carriers (zinc-coated urea or zinc-coated urea-ammonium sulfate, zinc-lignosulfonate, and so on) were inferior.

Direct-Seeded Flooded Rice

For direct-seeded flooded rice, the most promising method of correcting zinc deficiency is to coat the pregerminated seeds with ZnO prior to seeding followed by a foliar spray with 0.5% ZnSO₄ · 7H₂O at 5–7 days before panicle initiation (Abilay and De Datta 1978). However, sprays are usually employed only when the growing crop shows deficiency symptoms.

Table 10.17 Zinc Response of Lowland Transplanted Rice on a Calcareous Soil; Tiaong, Quezon, Philippines, 1976–1977 Wet Seasons (Adapted from Abilay and De Datta 1978)

Zinc Source	Rate of Application (kg Zn/ha)	Method of Application	Grain Yield[a] (t/ha)		
			1976	1977	Average
Zinc oxide + $ZnSO_4 \cdot 7\,H_2O$	3.2 + 0.7 (0.5% solution)	Seedling dip + foliar spray	3.5 a	4.3 a	3.9
Zinc oxide	3.2	Seedling dip	3.3 a	4.1 a	3.7
$ZnSO_4 \cdot 7\,H_2O$	20.0	Broadcast and incorporated	3.2 ab	3.4 a	3.3
Zinc-lignosulfonate	4.0	Broadcast and incorporated	2.4 bc	3.4 ab	2.9
Zinc-coated ammonium sulfate	4.0	Broadcast and incorporated	2.0 cd	2.3 cd	2.2
Zinc-lignosulfonate	2.0	Broadcast and incorporated	1.2 def	2.6 bc	1.9
$ZnSO_4 \cdot 7\,H_2O$	2.1 (0.5% solution)	Three foliar sprays	1.4 de	1.4 de	1.4
Zinc-coated ammonium sulfate	2.0	Broadcast and incorporated	0.7 efg	1.1 e	0.9
Zinc-coated urea-ammonium sulfate	2.0	Broadcast and incorporated	0.6 efg	1.1 e	0.8
Zinc-coated urea	2.0	Broadcast and incorporated	0.4 fg	0.6 e	0.5
$ZnSO_4 \cdot 7\,H_2O$	20.0	Broadcast (seedbed)	0.2 g	0.6 e	0.4
No zinc	—	—	0.1 g	0.5 e	0.3

[a]In a column, mean values with the same letter are not significantly different at 5% probability level.

Table 10.18   Effect of Zinc Sources and Rates on Rice Yields on Three Soils in California (Adapted from Mikkelsen and Brandon 1975)

Source[a]	Zn Rate (kg/ha)	Yield of Rough Rice[b] (t/ha) Willows Clay	Plaza Silty Clay	Mormon Clay	Mean Zn Concentration in Seedlings[c] (ppm)
Control	0	1.2 d	0	2.1 d	13.5
ZnSO$_4$	2.2	3.8 b	3.8 b	7.5 c	21.0
ZnSO$_4$	4.4	4.5 a	4.0 ab	8.3 b	23.0
ZnSO$_4$	8.8	4.2 ab	4.5 a	9.0 a	25.0
ZnO	2.2	3.3 c	2.6 c	7.1 c	19.0
ZnO	4.4	3.7 b	3.0 bc	8.7 ab	22.0
ZnO	8.8	4.2 ab	3.2 b	8.4 b	25.0
Zn LS	2.2	4.1 ab	3.5 b	8.8 ab	43.0
Zn LS	4.4	4.5 a	4.3 a	9.5 a	47.0
Zn LS	8.8	4.6 a	4.6 a	—	51.0
Zn EDTA	2.2	3.3 c	2.4 c	—	15.0
Zn EDTA	4.4	3.7 b	3.1 bc	—	17.0
Zn EDTA	8.8	4.2 ab	3.5 b	—	20.0

[a]LS = lignosulfonate; EDTA = ethylenediaminetetra-acetic acid.
[b]In a column, mean values with the same letter are not significantly different at 5% probability level.
[c]Seedlings were 30 days old.

In California 4.4–8.8 kg/ha of Zn as ZnSO$_4$ corrected zinc deficiency in rice and increased grain yields from 0–2.1 t/ha (without zinc) to 4.0–9.0 t/ha with 8.8 kg Zn/ha. Average zinc concentration in seedling rice plants was increased with 13.5 ppm to 25 ppm (Table 10.18). Among the inorganic zinc sources, sulfate, chloride, and nitrate forms have been found equally effective in correcting zinc deficiency. Zinc oxide has been found to be slightly less effective than zinc sulfate. However, per unit of applied zinc, its cost is generally lower, thereby offsetting any economic disadvantage (Mikkelsen and Brandon 1975).

According to Giodarno and Mortvedt (1973), zinc sulfate was equally effective for rice when applied to the seed, soil, or water, and surface applications were as effective as soil-incorporated zinc. Mikkelsen and Brandon (1975) showed surface application of zinc was more effective for water-sown rice than was incorporation of zinc in calcareous soil. Surface application increased zinc availability at the soil-water interface and, thereby, promoted a higher rate of seedling survival. Without adequate zinc, water-sown seedlings died in 3–5 weeks (Mikkelsen and Kuo 1976, 1977).

In the United States, seed treatment with zinc materials has been successfully used in drill-seeded and water-seeded rice. Experiments in California and in Arkansas have shown that various zinc materials ($ZnSO_4$, ZnO, and Zn-lignosulfonate) applied to rice seed prior to planting correct zinc deficiency.

SULFUR

The requirement of sulfur for plant growth has been known for 130 years. However, for rice the earliest report on sulfur deficiency was reported in 1938 by Sen in Burma. Subsequently, Aiyar (1945) reported sulfur deficiency in rice from India. Recently, chlorosis of rice seedlings in India was corrected by sulfur application.

In Indonesia, *mentek* disease, which is associated with sulfur deficiency, is common in many parts of the country such as in Ngale (on Grumusols or Vertisols), East Java, and South Sulawesi (Ismunadji and Zulkarnaini 1977). In Brazil, sulfur deficiency is serious on soils high in organic matter (5–8%) in newly reclaimed Amazon swamps (Wang et al. 1976a, b). Recently, sulfur deficiency has been observed in many rice-growing areas in Bangladesh and Thailand. From Nigeria, sulfur deficiency in rice was reported by Osiname and Kang (1975).

Several reasons have been cited for the recent recording of sulfur deficiency in rice (Wang 1976). They are:

- Replacement of single superphosphate (13.9% S) with concentrated super-phosphate and of ammonium sulfate (24.2% S) with urea.
- Reduction in atmospheric pollution thereby minimizing sulfur enrichment of soil through rain.
- Less use of farmyard manure.
- Increased use of modern varieties, which need higher rates of nitrogen, phosphorus, potassium, and sulfur.

Blair (1977) suggested that low initial sulfur status may be a primary cause of deficiency, particularly in the tropics where some soils have low sulfur content because of low sulfur content in the parent material or extreme weathering and leaching losses, or both. Organic matter losses, leaching and erosion losses, and fertilizer use and management also often cause sulfur deficiency in the tropics.

Management of Sulfur-Containing Fertilizers

To produce between 4 and 9 t of grains/ha, a rice crop must remove between 8 and 17 kg S/ha (Wang 1976).

Sources of Sulfur

Sulfur additions are generally made as a component of ammonium sulfate (24.2% S), single superphosphate (13.9% S), potassium sulfate (16–18% S), and agricultural gypsum (15–18% S). Compound NPK fertilizers containing sulfur are also an excellent source of sulfur. Sulfur-coated urea (SCU) developed by TVA has also been suggested as an excellent sulfur-containing fertilizer for rice.

Single superphosphate is more difficult to store and use than ammonium sulfate. The former tends to be more hygroscopic and cakes badly, which makes ammonium sulfate a more convenient source of sulfur for rice production in humid tropical areas. In sulfur-deficient lowland soils in Sulawesi, Indonesia, elemental sulfur was as effective as ammonium sulfate and gypsum (Blair et al. 1978). However, because urea is the principal source of nitrogen in the Asian tropics, gypsum should be considered as an excellent source of sulfur for rice in that area.

Time of Sulfur Application

Sulfur, like nitrogen, is required for protein synthesis and should be applied early. It is important to apply sulfur before severe deficiency symptoms are observed, otherwise yields will be reduced. Reports from Brazil (Wang et al. 1976a) indicate that application of sulfur from 2 to 3 weeks before panicle emergence corrected sulfur deficiency on severely sulfur-deficient soils.

SILICON

For reasons that are still obscure, silicon is beneficial for normal rice growth. The first evidence that silicon was beneficial for rice was given by Sommer (1926).

Management of Silicon Fertilizers

Nearly one-third of the area in Japan (2.7 million hectares) that grows rice is deficient in silicon. The average dosage of silicate material (calcium silicate, a by-product of iron works) for these rice fields is 1.5–2.0 t/ha (Takahashi and Miyake 1977).

It is now generally believed that rice requires large amounts of silicon. In a rice crop producing 10 t/ha of grains, Si may be taken up to as much as 1 t/ha. From Japan, Imaizumi and Yoshida (1958) reported that the average uptake of Si by one crop of rice is 443 kg/ha.

Source of Silicon

The most common source of silicon fertilizer is slag containing calcium silicate, such as that used in Japan. In Korea, the most common slags are derived from the

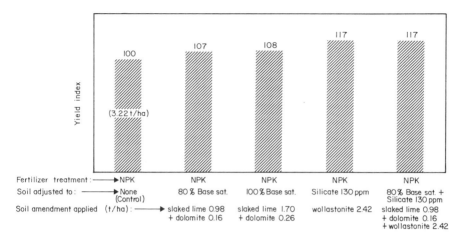

Figure 10.18 Average results of lime and wollastonite application in the Republic of Korea (8 provinces, 21 soil series, 4 replications), 1968, Office of Rural Development. (Adapted from Park 1979)

refuse of iron, ferro nickel, and manganese ore smelters. However, part of the silicon source used in Korea is from wollastonite (calcium metasilicate). Figure 10.18 shows response of rice to added silicon as wollastonite in Korean trials. The response recorded to silicon was in addition to the yield obtained with NPK fertilizers.

In Taiwan, silicate slag is effective in a wide range of soils. In general, slags of basic constituents all have higher silicon solubility, hence, of higher nutritional value.

Rate of Application

In most instances, 1.5–2.0 t/ha of silicate slag is adequate for rice.

Method of Application

Results from Taiwan indicate that a surface basal application of silicate slag is better than basal incorporation, although topdressing during early stages of crop growth is as good as basal application. Results from Japan show that both basal dressing and topdressing of silicon give high yield response.

REFERENCES

Abilay, W. P., Jr., and S. K. De Datta. 1978. Management practices for correcting zinc deficiency in transplanted and direct seeded wetland rice. *Philipp. J. Crop Sci.* 3:190–194.

Aiyar, S. P. 1945. The significance of sulphur in the manuring of rice. *Indian J. Agric. Sci.* 15:283–297.

Aso, K. 1918. *Graphical explanation of nutrient absorption of the rice plant* [in Japanese]. Shokobo Publ. Co., Japan.

Baba, I. 1955. Varietal differences of the rice plant in relation to the resisting capacity to root-rot disease induced by hydrogen sulfide, and a convenient method to test them [in Japanese, English summary]. *Proc. Crop Sci. Soc. Jpn.* 23:167–168.

Baba, I., and K. Tajima. 1960. Studies on the nutrition of rice plant in relation to the occurrence of the so-called 'Akagare' disease. VI. Changes in the growth, nutrient-absorption, and metabolism in the plant as influenced by the excessive supply of ferrous iron. *Proc. Crop Sci. Soc. Jpn.* 29:47–50.

Barber, S. A. 1977. Efficient fertilizer use. Pages 13–29 *in* Agronomic research for food. *Am. Soc. Agron. Spec. Publ.* 26. Madison, Wisconsin.

Bartholomew, W. V. 1972. *Soil nitrogen supply; processes and crop requirements.* N. C. State Univ. Tech. Bull. 6. 78 pp.

Blair, G. J. 1977. Position paper on sulphur research and development. International Fertilizer Development Center. (unpubl. mimeo.)

Blair, G. J., C. P. Mamaril, and E. Momuat. 1978. *Sulfur nutrition of wetland rice.* IRRI Res. Pap. Ser. 21. 28 pp.

Chang, S. C. 1976. Phosphorus in submerged soils and phosphorus nutrition and fertilization of rice. Pages 93–116 *in* ASPAC Food and Fertilizer Technology Center. *The fertility of paddy soils and fertilizer applications for rice.* Taipei, Taiwan.

Chang, S. C. 1978. Evaluation of the fertility of rice soils. Pages 521–541 *in* International Rice Research Institute. *Soils and rice.* Los Baños, Philippines.

Cheng, B. T., and G. J. Ouellette. 1971. Manganese availability in soil. *Soils Fert.* 34:589–595.

da Silva, P. R. F., and C. A. Stutte. 1979. Loss of gaseous N from rice leaves with transpiration. *Arkansas Farm Res.* 28(4):3.

Davide, J. G. 1965. The time and methods of phosphate fertilizer applications. Pages 255–268 *in* International Rice Research Institute. *The mineral nutrition of the rice plant.* Proceedings of a symposium at the International Rice Research Institute, February, 1964. The Johns Hopkins Press, Baltimore, Maryland.

De Datta, S. K. 1978. Fertilizer management for efficient use in wetland rice soils. Pages 671–701 *in* International Rice Research Institute. *Soils and rice.* Los Baños, Philippines.

De Datta, S. K., and K. A. Gomez. 1975. Changes in soil fertility under intensive rice cropping with improved varieties. *Soil Sci.* 120:361–366.

De Datta, S. K., and C. P. Magnaye. 1969. A survey of the forms and sources of fertilizer nitrogen for flooded rice. *Soils Fert.* 32:103–109.

De Datta, S. K., and J. Malabuyoc. 1976. Nitrogen response of lowland and upland rice in relation to tropical environmental conditions. Pages 509–539 *in* International Rice Research Institute. *Climate and rice.* Los Baños, Philippines.

De Datta, S. K., and P. M. Zarate. 1970. Environmental conditions affecting the growth characteristics, nitrogen response and grain yield of tropical rice. *Biometeorology* 4:71–89.

De Datta, S. K., C. P. Magnaye, and J. C. Moomaw. 1968. Efficiency of fertilizer nitrogen (^{15}N-labelled) for flooded rice. Pages 67–76 *in* International Society of Soil Science. *Transactions of the 9th International Congress of Soil Science*. Vol. IV.

De Datta, S. K., C. P. Magnaye, and J. T. Magbanua. 1969. Response of rice varieties to time of nitrogen application in the tropics. Pages 73–87 *in* Symposium on optimization of fertilizer effect in rice cultivation. *Trop. Agric. Res. Ser.* 3.

De Datta, S. K., F. A. Saladaga, W. N. Obcemea, and T. Yoshida. 1974. Increasing efficiency of fertiliser nitrogen in flooded tropical rice. Pages 265–288 *in* Proceedings of the FAI-FAO seminar on optimizing agricultural production under limited availability of fertilisers, New Delhi, India.

Digar, S. 1958. Effect of ammonium sulphate and farmyard manure on the yield of paddy and on soil. *J. Agric. Sci.* 50:219–226.

Doll, E. C. 1975. Development of more efficient fertilizers for rice. Paper presented at the International Rice Research Conference, 21–24 April 1975, International Rice Research Institute, Los Baños, Philippines. 10 pp. (unpubl. mimeo.)

Downey, D. A., B. A. Huey, and B. R. Wells. 1976. Using growing degree days to determine timing of midseason N application to rice. *Arkansas Farm Res.* 25(2):2.

Engelstad, O. P. 1967. *Nitrogen sources for rice on flooded soils*. Tennessee Valley Authority, Muscle Shoals, Alabama.

Engelstad, O. P., A. Jugsujinda, and S. K. De Datta. 1974. Response by flooded rice to phosphate rocks varying in citrate solubility. *Soil Sci. Soc. Am. Proc.* 38:524–529.

Evatt, N. S. 1969. Nitrogen recommendations for Texas Rice production. *Rice J.* 72(4):40.

Evatt, N. S., and R. J. Hodges. 1975. Developing efficient systems of fertilization of rice. Pages 31–36 *in* Texas Agricultural Experiment Station, in cooperation with the U. S. Department of Agriculture. *Six decades of rice research in Texas*. Res. Monogr. 4.

FAO/UNESCO (Food and Agriculture Organization/United Nations Educational, Scientific, and Cultural Organization). 1973. *Irrigation, drainage and salinity; an international source book*. Hutchinson, London. 510 pp.

Forno, D. A., S. Yoshida, and C. J. Asher. 1975. Zinc deficiency in rice. I. Soil factors associated with the deficiency. *Plant Soil* 42:537–550.

Gericke, W. F. 1924. The beneficial effect to plant growth of the temporary depletion of some of the essential elements in the soil. *Science* 59:321.

Giodarno, P. M., and J. J. Mortvedt. 1973. Zinc sources and methods of application for rice. *Agron. J.* 65:51–53.

Goswami, N. N., S. R. Bapat, C. R. Leelavathi, and R. N. Singh. 1976. Potassium deficiency in rice and wheat in relation to soil type and fertility status. Pages 186–194 *in* Potassium in soils, crops and fertilizers. *Indian Soc. Soil Sci. Bull.* 10.

Gotoh, S., and Y. Onikura. 1971. Organic acids in flooded soil receiving added rice straw and their effect on the growth of rice. *Soil Sci. Plant Nutr. (Tokyo)* 17:1–8.

Hall, V. 1965. Morphological development of rice plant in relation to time of nitrogen fertilization. *Rice J.* 68(7):47–48.

Hamamoto, M. 1966. *Isobutylidene diurea as a slow acting nitrogen fertiliser and the studies in this field in Japan*. Proc. Fert. Soc., London, 90. 77 pp.

Hesse, P. R. 1971. *A textbook of soil chemical analysis*. John Murray (Publishers) Ltd., London. 520 pp.

Horn, R. C. 1979. Research and development on nitrogen fertilizer materials for rice. Paper presented at the International Rice Research Conference, 16–20 April 1979, International Rice Research Institute, Los Baños, Philippines. 17 pp. (unpubl. mimeo.)

Houng, K. H. 1975. The physiology of boron and molybdenum in plants. Pages 61–73 *in* H. O. Kajima, I. Uritami, and K. H. Houng, authors. *Significant minor elements*. ASPAC Food and Fertilizer Technology Center, Taipei, Taiwan.

Howeler, R. H., and L. F. Cadavid. 1976. Screening of rice cultivars for tolerance to Al-toxicity in nutrient solutions as compared with a field screening method. *Agron. J.* 68:551–555.

Huey, B. A. 1977. *Rice production in Arkansas*. University of Arkansas, Division of Agriculture, and U. S. Department of Agriculture. Circ. 476, (Rev.) 51 pp.

IAEA (International Atomic Energy Agency). 1970. *Rice fertilization; a six-year isotope study on nitrogen and phosphorus fertilizer utilization*. Tech. Rep. Ser. 108. 184 pp.

Imaizumi, K., and S. Yoshida. 1958. Edaphological studies on silicon supplying power of paddy fields. *Bull. Natl. Inst. Agric. Sci.*, Ser. B, 8:261–304.

IRC (International Rice Commission). 1961. Eighth meeting of the IRC working party on rice soils, water and fertilizer practices. *Int. Rice Comm. Newsl.* 11(2):17–19.

IRRI (International Rice Research Institute). 1978a. *Annual report for 1977*. Los Baños, Philippines. 548 pp.

IRRI (International Rice Research Institute). 1978b. *Summary report (Agronomic data and economic analysis) on the first and second international trials on nitrogen fertilizer efficiency in rice (1975–1977)*. Los Baños, Philippines. 30 pp.

Ishizuka, Y. 1932. Absorption and utilization of nutrients at different stages of rice plants by means of water culture. [in Japanese]. *Bull. Agric. Chem., Jpn.* 8.

Ishizuka, Y. 1965. Nutrient uptake at different stages of growth. Pages 199–217 *in* The International Rice Research Institute. *The mineral nutrition of the rice plant*. Proceedings of a symposium at the International Rice Research Institute, February, 1964. The Johns Hopkins Press, Baltimore, Maryland.

Ismunadji, M., and I. Zulkarnaini. 1977. Sulphur deficiency in lowland rice in Indonesia. Pages 647–652 *in* Society of the Science of Soil and Manure, Japan. *Proceedings of the international seminar on soil environment and fertility management in intensive agriculture (SEFMIA), Tokyo-Japan, 1977*. Tokyo.

Jayasekara, E. H. W., and R. P. Ariyanayagam. 1962. Efficiency of urea and sulphate of ammonia as a source of nitrogen for rice in the sandy soils of the Batticaloa District. *Trop. Agric.* 118:103–108.

Karim, A. Q. M. B., and J. Vlamis. 1962. Micronutrient deficiency symptoms of rice grown in nutrient culture solution. *Plant Soil* 16:347–360.

Katyal, J. C., and F. N. Ponnamperuma. 1974. Zinc deficiency: a widespread nutritional disorder of rice in Agusan del Norte. *Philipp. Agric.* 58:79–89.

Lockard, R. G. 1959. *Mineral nutrition of the rice plant in Malaya: with special reference to Penyakit Merah*. Department of Agriculture, Kuala Lumpur, Malaysia. 148 pp.

Maas, E. V., and G. J. Hoffman. 1977. Crop salt tolerance—current assessment. *J. Irrigation Drainage Div. ASCE*, 103(IR2). Proc. Pap. 12993.

Mahapatra, I. C., S. R. Bapat, and M. P. Singh. 1974. Economics of fertilizer use. *Fert. Marketing News* 5(12):1–17.

Matsushima, S. 1969. Nitrogen responses to the rice plant at different stages of growth. Pages 59–72 *in* Symposium on optimization of fertilizer effect in rice cultivation. *Trop. Agric. Res. Ser. 3.*

Mengel, D. B., and W. J. Leonards. 1978. Timing of early season N in relation to permanent flood. Pages 42–44 *in* Rice Experiment Station, Louisiana State University, and U. S. Department of Agriculture. *70th annual progress report.*

Mikkelsen, D. S., and D. M. Brandon. 1975. Zinc deficiency in California rice. *Calif. Agric.* 29(9):8–9.

Mikkelsen, D. S., and N. S. Evatt. 1973. Soils and fertilizers. Pages 76–87 *in* USDA Agric. Handb. 289. *Rice in the United States: varieties and production.* Washington, D. C.

Mikkelsen, D. S., and D. C. Finfrock. 1957. Availability of ammoniacal nitrogen to lowland rice as influenced by fertilizer placement. *Agron. J.* 49:296–300.

Mikkelsen, D. S., and S. Kuo. 1976. Zinc fertilization and behavior in flooded soils. Pages 170–196 *in* ASPAC Food and Fertilizer Technology Center. *The fertility of paddy soils and fertilizer application for rice.* Taipei, Taiwan.

Mikkelsen, D. S., and S. Kuo. 1977. *Zinc fertilization and behavior in flooded soils.* Commonwealth Bureau of Soils Spec. Bull. 5. 59 pp.

Mikkelsen, D. S., J. H. Lindt, Jr., and M. D. Miller. 1967. Rice fertilization. *Calif. Agric. Exp. Stn. Ser.* 96. Revised Leaflet.

Mitsui, S. 1977. Recognition of the importance of denitrification and its impact on various improved and mechanized applications of nitrogen to rice plant. Pages 259–268 *in* Society of the Science of Soil and Manure, Japan. *Proceedings of the international seminar on soil environment and fertility management in intensive agriculture (SEFMIA), Tokyo-Japan, 1977.* Tokyo.

Mitsui, S., S. Aso, and O. K. Kumazawa. 1951. Dynamic studies on the nutrient uptake by crop plants. I. The nutrient uptake of rice root as influenced by hydrogen sulfide [in Japanese, English summary]. *J. Sci. Soil Manure, Jpn.* 22:46–52.

Mitsui, S., K. Kumazawa, and T. Hishida. 1959. Dynamic studies on the nutrient uptake by crop plants. 23. The growth of rice plant in poorly drained soils as affected by the accumulation of volatile organic acids (2) [in Japanese]. *J. Sci. Soil Manure, Jpn.* 30:411–413.

Motomura, S. 1962. Effect of organic matters on the formation of ferrous iron in soils. *Soil Sci. Plant Nutr.* (Tokyo) 8:20–29.

Muirhead, W. A., F. M. Melhuish, M. L. Higgins, and A. Ceresa. 1975. Rice stubble disorder. C.S.I.R.O., Australia. Division of Irrigation Research Report, 1974–75. 17 pp.

Murata, Y. 1969. Physiological responses to nitrogen in plants. Pages 235–259 *in* American Society of Agronomy. *Physiological aspects of the crop yield.* Madison, Wisconsin.

Nene, Y. L. 1966. Symptoms, cause, and control of Khaira disease of paddy. *Bull. Indian Phytopathol. Soc.* 3:97–101.

Nguu, N. V., and S. K. De Datta. 1979. Increasing efficiency of fertilizer nitrogen in wetland rice by manipulation of plant density and plant geometry. *Field Crops Res.* 2:19–34.

Noguchi, Y., and T. Sugawara. 1966. *Potassium and japonica rice; summary of twenty-five years' research*. International Potash Institute, Berne, Switzerland. 102 pp.

Oh, W. K., and Y. S. Kim. 1964. Response of paddy rice to potassic fertilizer in Korea. IRC Working Party on Rice Soils, Water, and Fertilizer Practices Working Pap. 9.

Osiname, O. A., and B. T. Kang. 1975. Response of rice to sulphur application under upland conditions. *Commun. Soil Sci. Plant Anal.* 6:585–598.

Park, C. S. 1979. Fertility management of flooded rice soil: a proposal to minimize the biological production potential-performance gap of high yielding varieties. Paper presented at a Saturday seminar, 27 September 1979, International Rice Research Institute, Los Baños, Philippines. (unpubl. mimeo.)

Park, Y. D., and A. Tanaka. 1968. Studies of the rice plant on an "Akiochi" soil in Korea. *Soil Sci. Plant Nutr.* (Tokyo) 14:27–34.

Patnaik, S., and F. E. Broadbent. 1967. Utilization of tracer nitrogen by rice in relation to time of application. *Agron. J.* 59:287–288.

Ponnamperuma, F. N. 1977. *Screening rice for tolerance to mineral stresses*. IRRI Res. Pap. Ser. 6. 21 pp.

Ponnamperuma, F. N. 1978. Electrochemical changes in submerged soils and the growth of rice. Pages 421–441 *in* International Rice Research Institute. *Soils and rice*. Los Baños, Philippines.

Ponnamperuma, F. N. 1979. Soil problems in the IRRI farm. Paper presented at a Thursday seminar, 8 November 1979, International Rice Research Institute, Los Baños, Philippines. (unpubl. mimeo.)

Ponnamperuma, F. N., and W. L. Yuan. 1966. Toxicity of boron to rice. *Nature* 221:780–781.

Ponnamperuma, F. N., R. Bradfield, and M. Peech. 1955. Physiological disease of rice attributable to iron toxicity. *Nature* 175:265.

Prasad, R., and S. K. De Datta. 1979. Increasing fertilizer nitrogen efficiency in wetland rice. Pages 465–483 *in* International Rice Research Institute. *Nitrogen and rice*. Los Baños, Philippines.

Randhawa, N. S., M. K. Sinha, and P. N. Takkar. 1978. Micronutrients. Pages 581–603 *in* International Rice Research Institute. *Soils and rice*. Los Baños, Philippines.

Reynolds, E. B. 1954. *Research on rice production in Texas*. Texas Agric. Exp. Stn. Bull. 775. 29 pp.

Sanchez, P. A., A. Gavidia O., G. E. Ramirez, R. Vergara, and F. Minguillo. 1973. Performance of sulfur-coated urea under intermittently flooded rice culture in Peru. *Soil Sci. Soc. Am. Proc.* 37:789–792.

Sen, A. T. 1938. Report on the operations of the Department of Agriculture, Burma, for the year ended the 31st March 1938. Pages 27–37 *in* Burma Department of Agriculture Report, 1937–1938.

Shiga, H., N. Miyazaki, and S. Sekiya. 1977. Time of fertilizer application in relation to the nutrient requirement of rice plants at successive growth stages. Pages 223–229 *in* Society of the Science of Soil and Manure, Tokyo, Japan. *Proceedings of the international seminar on soil environment and fertility management in intensive agriculture (SEFMIA), Tokyo-Japan, 1977*. Tokyo.

Shioiri, M. 1941. Denitrification in paddy soils [in Japanese]. *Kagaku (Tokyo)* 11:1–24.

Shiroshita, T. 1958. High yielding paddy field. Pages 444–463 *in* Research Bureau, Ministry of Agriculture and Forestry. *Handbook of soils and fertilizers.* Yokendo Ltd., Tokyo.

Sims, J. L. 1965. Nitrogen fertilization of rice growing on clay soils. *Rice J.* 68(6):31.

Sommer, A. L. 1926. Studies concerning the essential nature of aluminum and silicon for plant growth. *Univ. Calif. Publ. Agri. Sci.* 5:57–81.

Sommer, A. L., and C. B. Lipman. 1926. Evidence on the indispensable nature of zinc and boron for higher green plants. *Plant Physiol.* 1:231–249.

Stangel, P. J. 1970. *Modern chemical fertilizers: their potential and method of application—Asia.* ASPAC Food Fert. Technol. Cent. Ext. Bull. 2. 82 pp.

Stangel, P. J. 1975. Market trends and agronomic suitability of key fertilisers commonly sold in world trade. Pages 101–130 *in* FAI-FAO Seminar on optimizing agricultural production under limited availability of fertilisers, 1974.

Su, N. R. 1976. Potassium fertilization of rice. Pages 117–148 *in* ASPAC Food and Fertilizer Technology Center. *The fertility of paddy soils and fertilizer application for rice.* Taipei, Taiwan.

Tadano, T., and S. Yoshida. 1978. Chemical changes in submerged soils and their effect on rice growth. Pages 399–420 *in* International Rice Research Institute. *Soils and rice.* Los Baños, Philippines.

Takagi, S., and H. Okajima. 1956. Physiological behaviour of hydrogen sulfide in the rice plant. 6. Influence of iron, pH and rice plant root on oxidation of H_2S in water culture solution [in Japanese, English summary]. *J. Sci. Soil Manure, Jpn.* 26:455–458.

Takahashi, E., and Y. Miyake. 1977. Silica and plant growth. Pages 603–611 *in* Society of the Science of Soil and Manure, Japan. *Proceedings of the international seminar on soil environment and fertility management in intensive agriculture (SEFMIA), Tokyo-Japan, 1977.* Tokyo.

Takai, Y., and T. Kamura. 1966. The mechanism of reduction in waterlogged paddy soil. *Foliar Microbiol.* 11:304–313.

Takijima, Y., M. Shiojima, and Y. Arita. 1960. Metabolism of organic acids in soils and their harmful effects on paddy rice growth. 2. Effect of organic acids on root elongation and nutrient absorption of rice plants [in Japanese]. *J. Sci. Soil Manure, Jpn.* 31:441–446.

Tanaka, A. 1957. Health diagnosis of crops. On the conception of ideal growth [in Japanese]. *Nogyo Gijutsu, Jpn.* 12:302–306.

Tanaka, A. 1969. Physiological basis for fertilizer response of rice varieties. Pages 37–43 *in* Symposium on optimization of fertilizer effect in rice cultivation. *Trop. Agric. Res. Ser.* 3.

Tanaka, A. 1977. Nutrient requirement of field crops during growth. Pages 75–99 *in* U. S. Gupta, ed. *Physiological aspects of crop nutrition and resistance.* Atma Ram and Sons, Delhi.

Tanaka, A. 1978. Role of organic matter. Pages 605–620 *in* International Rice Research Institute. *Soils and rice.* Los Baños, Philippines.

Tanaka, A., and S. A. Navasero. 1967. Carbon dioxide and organic acids in relation to the growth of rice. *Soil Sci. Plant Nutr.* (Tokyo) 13:25–30.

Tanaka, A., and S. Yoshida. 1970. *Nutritional disorders of the rice plant in Asia.* IRRI Tech. Bull. 10. 51 pp.

Thenabadu, M. W. 1973. Response of rice to potassium fertilization in the wet zone of Ceylon. *Potash Review.* Subj. 16. Potash fertilizer and manuring. 65th Suite 8–9:1–8.

Thompson, L., R. Maples, J. Wells, R. J. Miears, and N. S. Evatt. 1962. Recommendations for rice fertilization in southern States. *Rice J.* 65(1):5, 6, and 40.

Turner, F. T. 1977. Terrazole as a nitrification inhibitor. *Int. Rice Res. Newsl.* 2(1):11.

University of California and United States Department of Agriculture. 1978. *Annual report comprehensive rice research.* 91 pp.

USDA (United States Department of Agriculture). 1954. *Diagnosis and improvement of saline and alkaline soils.* USDA Agric. Handb. 60. Washington, D. C. 160 pp.

Vachhani, M. V. 1952. Fertilizer use for stepping up rice production. *Indian Farming* 2(1):28.

Vachhani, M. V., and K. S. Murty. 1964. *Green-manuring for rice: results of investigations conducted at the Central Rice Research Institute, Cuttack.* Cent. Rice Res. Inst. Bull. 4. 26 pp.

Wang, C. H. 1976. Sulphur fertilization of rice. Pages 149–169 *in* ASPAC Food and Fertilizer Technology Center. *The fertility of paddy soils and fertilizer application for rice.* Taipei, Taiwan.

Wang, C. H., T. H. Liem, and D. S. Mikkelsen. 1976a. *Sulfur deficiency—a limiting factor in rice production in the lower Amazon Basin.* I. Development of sulfur deficiency as a limiting factor for rice production. IRI Res. Inst. Inc. Bull. 47. 46 pp.

Wang, C. H., T. H. Liem, and D. S. Mikkelsen. 1976b. *Sulfur deficiency—a limiting factor in rice production in the lower Amazon Basin.* II. Sulfur requirement for rice production. IRI Res. Inst. Inc. Bull. 48. 38 pp.

Wells, B. R. 1977. Nitrapyrin [2-chloro-6-(trichloromethyl)-pyridine] as a nitrification inhibitor for paddy rice. *Down to Earth* 32(4):28–32.

Wells, B. R., and P. A. Shockley. 1975. Conventional and controlled-release nitrogen sources for rice. *Soil Sci. Soc. Am. Proc.* 39:549–551.

Westfall, D. C. 1972. Use of sulfur-coated urea, a slow release fertilizer, as a source of nitrogen for rice. Pages 8–11 *in* Texas Agricultural Experiment Station. *Rice research in Texas, 1971.* Texas A&M University.

Willett, I. R., W. A. Muirhead, and M. L. Higgins. 1978. The effects of rice growing on soil phosphorus immobilization. *Aust. J. Exp. Agric. Anim. Husb.* 18:270–275.

Williams, W. A., D. C. Finfrock, L. L. Davis, and D. S. Mikkelsen. 1957. Green manuring and crop residue management in rice production. *Soil Sci. Soc. Am. Proc.* 21:412–415.

Yamasaki, T. 1965. The role of micronutrients. Pages 107–122 *in* International Rice Research Institute. *The mineral nutrition of the rice plant.* Proceedings of a symposium at the International Rice Research Institute, February, 1964. The Johns Hopkins Press, Baltimore, Maryland.

Yanagisawa, M., A. Irobe, S. Iida, and K. Yamazaki. 1967. On the efficiency of nitrogen received by direct sowing paddy rice at different growth stages. *J. Sci. Soil Manure, Jpn.* 38:37–42. Abstract: *Soil Sci. Plant Nutr.* (Tokyo) 13:62.

Yoshida, S. 1968. Occurrence, causes, and cure of zinc deficiency of the rice plant in calcareous soils. Presented at the Eleventh Session of FAO/IRC Working Party on Rice Soils, Water and Fertilizer Practices. Kandy, Ceylon. 11 pp., tables.

Yoshida, S. 1975a. *The physiology of silicon in rice.* ASPAC Food Fert. Technol. Cent. Tech. Bull. 25. 27 pp.

Yoshida, S. 1975b. *Minor elements for rice.* ASPAC Food Fert. Technol. Cent. Ext. Bull. 52. 25 pp.

Yoshida, S., and L. Castañeda. 1969. Partial replacement of potassium by sodium in the rice plant under weakly saline conditions. *Soil Sci. Plant Nutr.* (Tokyo) 15:183–186.

Yoshida, S., and M. R. Chaudhry. 1979. Sulfur nutrition of rice. *Soil Sci. Plant Nutr.* (Tokyo) 25:121–134.

Yoshida, S., J. S. Ahn, and D. A. Forno. 1973. Occurrence, diagnosis and correction of zinc deficiency of lowland rice. *Soil Sci. Plant Nutr.* (Tokyo) 19:83–93.

Yoshida, S., Y. Ohnishi, and K. Kitagishi. 1962. Histochemistry of silicon in rice plant. I. A new method for determining the localization of silicon within plant tissues. *Soil Sci. Plant Nutr.* (Tokyo) 8:30–35.

Yoshida, S., D. A. Forno, J. H. Cock, and K. A. Gomez. 1972. *Laboratory manual for physiological studies of rice.* International Rice Research Institute, Los Baños, Philippines. 70 pp.

Yoshida, S., G. W. McLean, M. Shafi, and K. E. Mueller. 1970. Effects of different methods of zinc application on growth and yields of rice in a calcareous soil, West Pakistan. *Soil Sci. Plant Nutr. (Tokyo)* 16:147–149.

11

Insects, Diseases, and Other Pests of Rice and Their Control

Rice grows in diverse soils and climates but it is best adapted to a warm, humid environment. In that environment, insects are more prolific than in a cooler, dryer environment. In addition, where year-round continuous cropping is practiced, there are overlapping insect generations throughout the year.

INSECT PESTS OF RICE AND THEIR CONTROL

Insects substantially reduce rice yields in the tropics. For example, in 24 separate experiments with lowland rice during six cropping seasons at IRRI, plots protected from insects yielded an average of 5.3 t/ha whereas unprotected plots averaged 2.9 t/ha.

Seventy insects are considered rice pests but only 20 species are of major importance. Those species infest all parts of the rice plant at all growth stages and some are vectors of virus diseases (Pathak 1968a, 1970).

The Major Insect Pests of Rice

The major rice insect pests of the tropics are discussed briefly here. Other publications with details on major rice insect pests are: Hatai (1975) for South and Southeast Asia, Soto and Siddiqi (1978) for Africa, Gifford (1973) for the United States, and Bowling (1979) for various rice-growing countries.

Stem Borers

Among the insect pests of rice, the stem borers are considered serious pests in tropical rice production. They occur at all growth stages and are found in all

Table 11.1   Important Species of Rice Stem Borers by Region (Adapted from Pathak 1969a, Soto and Siddiqi 1978)

Species	Common Name
Asian Region	
Tryporyza incertulas (Walker)	Yellow rice borer
Chilo suppressalis (Walker)	Striped rice borer
Tryporyza innotata (Walker)	White rice borer
Sesamia inferens (Walker)	Pink borer
African Region	
Maliarpha separatella Rag.	African white borer
Chilo zacconius Blesz.	African striped borer
Sesamia calamistis (Hampsoro)	African pink borer
Latin American Region	
Rupella albinella Cramer	South American white borer
Diatraea saccharalis (Fabricius)	Small sugarcane borer

types of rice crop—lowland, deepwater, floating, and upland. Common rice stem borers are listed by region in Table 11.1. They cause extensive damage and exhibit two kinds of symptoms:

- DEAD HEART   During the vegetative growth stages, the stem borer larvae bore into and feed in the leaf sheath causing broad, longitudinal, whitish, discolored areas at feeding sites. The central leaf whorl does not unfold, turns brownish, and dries. The affected tillers dry without bearing panicles. This condition is called dead heart.
- WHITE HEAD   During the reproductive growth stage, particularly after panicle initiation, the growing plant parts are severed resulting in drying of panicles. Empty, whitish heads (panicles) become conspicuous in the field and are called white heads. At this stage, the grain yield reduction by stem borer is severe.

Leafhoppers and Planthoppers

Many species of leafhoppers and planthoppers cause severe damage to the rice crop. When they occur in large numbers they cause a complete drying of the crop called hopper burn. Even a few hoppers per plant can cause reduction of grain yield.

Leafhoppers and planthoppers are also vectors of virus diseases. The brown planthopper (*Nilaparvata lugens* Stål) has become a serious threat to rice production throughout Asia. Although an important rice crop pest in Japan for many years, the brown planthopper only recently became a major pest in tropical Asian countries. In the past 6 years, the brown planthopper population has greatly increased and caused severe yield losses in several countries. Figure 11.1

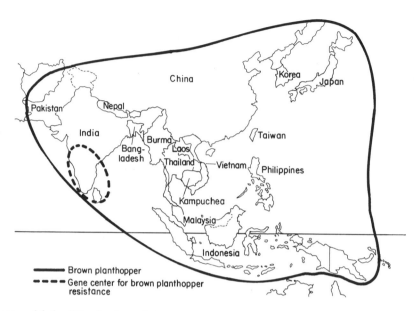

Figure 11.1 Distribution of the brown planthopper in Asia and the gene center for brown planthopper resistance. (From Khush 1979)

shows distribution of brown planthopper in Asia and the Pacific regions. Table 11.2 includes the common leafhopper and planthopper pests of rice. In addition, the leafhoppers and planthoppers recorded in Africa include *Nephotettix modulatus* Malichar and *Nilaparvata maender* Fennah (Soto and Siddiqi 1978).

Generally, the leafhoppers feed on the leaves and upper parts whereas the planthoppers are found at the basal region of the rice plant. Both damage the plants by sucking the sap and by plugging xylem and phloem tissues. Infestation by small populations during the early stages of plant growth can reduce the number of tillers, plant height, and general vigor. After panicle initiation, a similar small population can cause a high percentage of unfilled spikelets.

BIOTYPES OF BROWN PLANTHOPPERS With continuous rice cropping in irrigated areas, buildup of certain pests, such as the brown planthopper, causes heavy damage to the rice crop. The increase in severity of the insect appears to be associated with the technology used in modern rice culture. Figure 11.2 shows a farmer's rice crop completely damaged (hopper burn) by brown planthopper.

The recent discovery of several biotypes of the brown planthopper underscores the seriousness of this pest. The biotypes are not the basic cause of outbreaks but this development has caused the control of this pest to be more difficult (Heinrichs 1977). The biotypes vary not only in their ability to attack different varieties (Table 11.3) but in their susceptibility to control by pesticides. If a few brown planthopper-resistant rice varieties are planted intensively over wide

Table 11.2   Distribution of 10 Known Leafhopper and Planthopper Vectors of Rice Viruses (Adapted from Ling 1972, Pathak 1970, Heinrichs 1979a)

Common and Scientific Name	Distribution	Viral Diseases Transmitted
Cicadellidae (leafhoppers)		
Green leafhopper, *Nephotettix virescens* (Distant)	Malaysia, Korea, Indonesia, Japan, Taiwan, Philippines	Tungro, yellow dwarf, transitory yellowing
Green rice hopper, *Nephotettix cincticeps* (Uhler)	Korea, Japan, Taiwan	Transitory yellowing, yellow dwarf, dwarf
Green leafhopper, *Nephotettix nigropictus* (Stål)	South and Southeast Asia, Japan, Taiwan	Tungro, transitory yellowing, yellow dwarf
Zig-zag leafhopper, *Recilia dorsalis* (Motschulsky)	South and Southeast Asia, Japan, Taiwan	Tungro, dwarf, orange leaf
Delphacidae (planthoppers)		
Brown planthopper, *Nilaparvata lugens* (Stål)	South and Southeast Asia, Micronesia, China, Japan, Korea, Taiwan	Grassy stunt, ragged stunt
Small brown planthopper, *Laodelphax striatellus* (Fallen)	Japan, Korea, Taiwan	Stripe, black-streaked dwarf
Rice planthopper, *Sogatodes orizicola* (Muir)	Western Hemisphere	Hoja blanca
Rice delphacid, *Sogatodes cubanus* (Crawford)	Western Hemisphere	Hoja blanca
Rice delphacid, *Ribautodelphax albifascia* (Matsumura)	Japan	Stripe, black-streaked dwarf
Delphacid, *Unkanodes sapporonus* (Matsumura)	Japan	Stripe, black-streaked dwarf

Figure 11.2 A rice crop in a farmer's field in the Philippines completely damaged by brown planthopper—a damage known as hopper burn. (Courtesy of G. S. Khush, IRRI)

Table 11.3 Reaction of Varieties with Known Genes for Resistance to Three Brown Planthopper Biotypes at IRRI (From Heinrichs 1979a)

Variety	Gene for Resistance	Reaction[a] to		
		Biotype 1	Biotype 2	Biotype 3
TN1, IR8, IR20, IR22, IR24	None	S	S	S
Mudgo, IR26, IR28, IR29, IR30, IR34, Andaragahawewe, CO 10, MTU 9 Tibiriwewa	Bph 1	R	S	R
ASD 7, IR32, IR36, Ptb 19, Palasithari Murungakayan	bph 2	R	R	S
Rathu Heenati	Bph 3	R	R	R
Babawee	bph 4	R	R	R

[a]S = susceptible; R = resistant.

areas, new pest biotypes may develop through natural selection. Therefore, it is important to study races, strains, and biotypes of rice pests. Details on the brown planthopper are given in the IRRI publication *Brown Planthopper: Threat to Rice Production in Asia* (IRRI 1979).

Rice Gall Midge

The rice gall midge, *Orseolia oryzae* (Wood-Mason), is also a serious pest of rice in South and Southeast Asia. It has been reported from India, Pakistan, Bangladesh, Burma, Nepal, Sri Lanka, Indonesia, Thailand, Laos, Kampuchea, Vietnam, southern China, and Africa. It is, however, not recorded in the Philippines (Heinrichs and Pathak 1980).

The adult gall midge is about the size of a mosquito. Gall midge attack starts in the seedbed and continues until the booting stage, but most damage is limited to the vegetative growth stages. The gall midge damage changes rice tillers into tubular galls, which dry without bearing panicles. New tillers are initiated as the older ones are infested. As a result, early gall midge infestation causes profuse tillering. However, the new tillers are eventually attacked by the gall midge, thereby reducing vigor and stunting the rice plants.

Rice Whorl Maggot

A dull grey fly, called the whorl maggot, *Hydrellia* sp., is widespread in the Philippines. The maggots feed on the unopened whorl of the leaf by remaining in the center of the whorl and nibbling the innermost margin of the leaf (Pathak 1969a). The symptoms are manifested as small chewed discolored areas on the innermost margin of the central whorl. Severe whorl maggot infestation stunts the plant and reduces tillers in a given hill. The infestation and damage is limited to vegetative stages of the crop.

Rice Leaf Folder

The rice leaf folder, *Cnaphalocrocis medinalis* Guen., occurs most commonly in the Asian tropics, although it is known to occur in all rice-growing countries of Asia. The full grown larva is yellowish green with a dark brown head. The damage is caused by the larvae, which fold the leaf blades into a tubular shape and feed on green leaf tissue within the tubular structures.

Rice Bug

Several insects that suck the sap of the developing rice grains cause serious losses to the rice crop. The rice bug is an important example of the grain sucking group. The species of rice bugs of major economic significance are:

- *Leptocorisa acuta* (Thunb.), very common in India, Sarawak-Malaysia, and New Guinea.
- *L. oratorius* Fabricius, common in the Far East, Sri Lanka, China, India, Indonesia (Java), and Australia.
- *L. chinensis* Dallas, common in Asia.

The nymphs and adults of rice bugs are difficult to recognize in the rice field because of their color. The freshly hatched nymphs are tiny and green but become brownish as they grow. Both nymphs and adults feed on rice plants.

Armyworms and Cutworms

The armyworm, *Spodoptera frugiperda* B., and the cutworm, *Spodoptera litura* (F.), although occurring sporadically, may cause economic loss of rice crops (Pathak 1969a).

The armyworm has been recorded in rice fields in the southern United States and in Central American countries. The adult moths are of ash color with forewings that are mottled with irregular, white, or light grey spots near the extreme tip. Adults migrate from the grassy areas or upland crops to rice fields and deposit their eggs. The larvae may eat entire rice plants.

The common cutworm has been recorded on the Indian subcontinent, in Southeast and East Asia, China, Australia, and in several African countries. The cutworm is a problem mostly in upland rice. Lowland rice suffers only from larvae migrating from adjacent grassy areas. Seedlings may be cut at ground level; the larvae defoliate older plants.

Rice Water Weevil

The rice water weevil, *Lissorhoptrus oryzophilus* (Kuschel), is the most destructive rice insect in the United States, where it occurs in all rice-growing regions (Gifford 1973). The adult weevil is greyish and semiaquatic. It flies but can also swim just beneath the water surface. The adults feed on leaves of young rice plants resulting in longitudinal stripes on the leaf surface. In some fields of late rice in Texas, rice water weevils are so numerous and their feeding so heavy that occasionally some plants die as a result of shredding of the leaves (Bowling 1975).

Rice Stink Bug

In the United States, practically all rice fields in Arkansas, Louisiana, and Texas are infested with the rice stink bug, *Oebalus pugnax* (Fabricius). The adult stink bug is a straw-colored, shield-shaped insect about 12.5 mm long (Gifford 1973). The adult and nymphal stages of the stink bug feed on individual grains of rice as the panicle develops. When grain feeding is in the early milk stage the result is an empty glume or shriveled grain. As a result of stink bug damage, fungi may develop and cause black spots on grains; such grains are commonly called "pecky rice." Such damage may cause reduction in head rice yield or quality (Bowling 1975).

Other Insect Pests

There are a host of other insect pests that cause different degrees of damage both in the tropics and in the temperate regions. Some examples are rice leaf miner *Hydrellia griseola* var. *scapulasis* Loew in California, the planthopper *Sogatodes orizicola* Muir in Latin America, the chinch bug *Blissus leucopterus* Say in Arkansas, Louisiana, and Texas, and the green grasshopper *Conocephalus fasciatus* De Geer in Texas.

Control of Insect Pests of Rice

Control of insect pests of rice ranges, in practice, from chemical control through varietal resistance, biological control, and cultural control. Emphasis on varietal resistance increased greatly with the development of the modern rice for the tropics. A more recent development is attention to integrated pest control.

Chemical Control

At present, the control of rice insects in the tropics is largely dependent on insecticides, even though many traditional and modern varieties of rice have some degrees of resistance to one or more insect pests.

Two types of insecticides are used:

- CONTACT INSECTICIDES   A contact insecticide is applied on foliage and kills only those insects that come in contact with it. For the contact insecticide to be effective it must be applied when the insects are at their most vulnerable growth stage.
- SYSTEMIC INSECTICIDES   Systemic insecticides are chemicals that, when mixed in the paddy water or sprayed on the plants, are absorbed by the roots and other plant parts and distributed throughout the plant tissues. The advantages of systemic insecticides are that they can kill insects feeding inside the plant, and, because they are not easily washed away by rains, they have longer residual effects than contact insecticides. In granular form they can be easily applied with a minimum of equipment and experience.

CLASSIFICATION OF INSECTICIDES   Insecticides are classified by their chemical structure:

- Organochlorines (structures called chlorinated hydrocarbons). Examples: DDT, Perthane

- Organophosphates. Examples: Malathion, methyl parathion, diazinon
- Carbamates. Examples: Carbofuran, MIPC, carbaryl
- Formamidines
- Synthetic pyrethroids

INSECTICIDE FORMULATIONS Insecticides available commercially are sold as dust, wettable powder, emulsifiable concentrate, or granules.

Dust.   Dusts, which usually contain only 2–10% of the insecticide, are used in areas where it is difficult to obtain a large volume of water for spraying. Chemical dusts are blended with carriers such as talc. Dusts are applied as premix using hand dusters, ground-driven dusting machines, or aerial equipment. The dusts are powderlike formulations, which may consist of:

- Toxic agents only, such as sulfur.
- Toxic agents plus an active diluent that serves as a carrier.
- Toxic agents plus an inert diluent in the form of talc or clay.

Wettable Powder.   Wettable powder has an appearance similar to a dust but contains a wetting agent that disperses and suspends it when mixed with water and applied as a liquid spray. Spray solutions must be agitated to keep wettable powders in suspension.

Emulsifiable Concentrate.   An emulsifiable concentrate is an oil-based liquid compound containing a high concentration of insecticide. The compound is a mixture of insecticide, solvents, and emulsifiers to make it easily mixed in water, and wetting and sticking agents to make the material cover and adhere to the plants.

Granules.   Granular formulations consist of free-flowing grains of inert materials mixed or impregnated with an insecticide. They are the easiest formulation to apply.

METHODS OF INSECTICIDE APPLICATION Use of high-yielding modern rice varieties, which are resistant to many insects and diseases, does not guarantee adequate protection against all insect pests of rice. Use of insecticides is often the only practical way to effect immediate reduction in insect populations.
    Methods of insecticide application include:

- Foliar spray.
- Application to paddy water, the soil surface, or incorporation with top soil.
- Treatment of seeds and seedlings as seed treatment, seedbed treatment, or treatment of roots of rice seedlings prior to transplanting.
- Concentrated placement of insecticides in the soil.
- Application of dusts directly on the foliage.

Details on various methods of insecticide applications are discussed in a paper by Heinrichs (1979b).

Foliar Spray.   Foliar spray is the conventional method of rice insect control in tropical Asia. Contact insecticides are generally applied as foliar sprays from backpack-type sprayers.

One obvious advantage of foliar spray is quick kill of insects when they are visibly damaging the crop. The cost of an insecticide as one foliar spray is generally less than one application of a systemic granular insecticide. However, several foliar applications are often necessary to achieve the same degree of control as one application of a granular systemic insecticide.

Foliar spray is often unsatisfactory because the sprays do not adequately penetrate the thick foliage of the crop; rains easily wash off the insecticide, and spraying cannot be timed to coincide with the insect pest's most vulnerable stages, especially when generations overlap. In addition, foliar sprays may not reach such internal feeders as stem borers, the rice gall midge, and the rice whorl maggot (Pathak et al. 1974).

A recent trend in foliar spray application is use of ultralow volume (ULV). Ultralow volume concentrates are usually the insecticide in its original liquid form or its solid form dissolved in a minimum of solvent. With ULV, high concentrations of the insecticide are applied directly without adding water.

Application to Paddy Water, Soil Surface, or Soil Incorporation.   The disadvantages of using foliar sprays can be largely overcome by applying granular insecticides to the paddy water, or soil surface, or by incorporating them with the top soil (Koshihara and Okamoto 1957, Rao and Israel 1967, Pathak 1968b). Table 11.4 shows that weekly foliar sprays with 0.04% endrin were required to achieve the same degree of stem borer control as two applications of lindane into paddy water. By paddy water application, the insecticides are usually effective for 20–30 days as compared to 7–10 days for most foliar sprays (Pathak and Dyck 1973).

There is evidence that application of an insecticide on the soil surface without any standing water would be similar in effectiveness to application made into standing water varying in depth from 5 to 15 cm (Table 11.5). This is explained by the vaporizing effects of insecticides such as lindane, which produces a fumigating effect to kill insects. Experiments at IRRI demonstrated that lindane has a substantial fumigation effect on adult moths of the rice stem borer (IRRI 1966).

Similar results were obtained in the United States by Bowling (1970) with carbofuran, which caused nearly 100% mortality of a leafhopper, *Draeculacephala portola* Ball, caged on rice plants 2 hours after insecticidal application.

Treatment of Seeds and Seedlings.   Treatment of rice seeds and seedlings with insecticides has been used to reduce the amount of insecticide applied to the main fields. Treating the seeds is feasible but insecticide treatments in the seedbed, or of the seedlings after uprooting them for transplanting, are probably more convenient than treating seeds.

Table 11.4 Comparison of Two Applications of Lindane in Paddy Water with Weekly 0.04% Endrin Sprays for Rice Stem Borer Control[a]; IRRI, 1964 (Adapted from Pathak 1968b)

Variety	Treatment	White Heads (%)	Infested Tillers (%)	Larvae per 10 Hills (av.)	Productive Panicles per Hill (av.)	Yield (t/ha)
Chianung[b] 242	Lindane	2.1	6.5	1.5	10.4	5.2
	Endrin	3.3	12.8	4.3	10.4	4.4
Peta	Lindane	7.3	16.6	7.6	21.0	5.2
	Endrin	9.9	24.6	19.1	17.2	2.7
Chianung[b] 242	Lindane	3.0	20.6	9.3	11.1	3.6
	Endrin	4.9	18.3	11.3	8.7	2.9

[a]All observations, except yield, were made 10 days before harvest.
[b]Grown in different fields.

Table 11.5   Effectivity of Lindane (Gamma-BHC) Applied at 2 and 3 kg/ha at 50 and 80 Days after Transplanting in Rice Fields Containing Different Depths of Water; IRRI, 1965 (Adapted from IRRI 1966)

Treatment	Dead Hearts (50-day-old) (%)	No. larvae/ 10 hills (90-day-old)	White Heads (%)	Yield (t/ha)
No standing water + lindane	5.09	4.33	1.18	5.1
5 cm water + lindane	4.57	12.33	2.06	4.8
10 cm water + lindane	3.21	10.33	2.97	4.5
15 cm water + lindane	2.64	23.66	3.25	4.9
5 cm water without lindane (control)	8.49	26.66	2.18	3.6

Dipping uprooted seedlings in an insecticide solution just before transplanting is another method to provide crop protection against common insect pests for as long as a week after transplanting. Seedbed treatment or dipping seedlings is convenient and economical but insecticides with low mammalian (dermal) toxicity must be used to avoid health hazards for transplanters. The residual period of such treatments can be considerably increased if the insecticide is absorbed by the rice seedlings.

To evaluate the effectiveness of such treatments the roots of rice seedlings were dipped for 24 hours in 6 ppm solution of various insecticides prior to transplanting in a field experiment. The treatment was compared with that of applying the same compound at 2 kg/ha to the paddy water of plots planted with untreated seedlings. The seedling treatments were clearly inferior to application of insecticide in the paddy water as evidenced by a higher percentage of dead heart tillers from stem borer damage at an early stage (Fig. 11.3).

Later studies in the laboratory and greenhouse suggested that seedling treatment was effective against green leafhopper and the brown planthopper (IRRI 1971, Mitra et al. 1970). In field experiments, the root-soak treatment also provided some control of the rice whorl maggot and stem borers, but the results were inconsistent.

These data demonstrate that seedling treatment may be effective for brief periods but it needs to be supplemented with other treatments in most of Asia where the rice crop is subjected to infestation by many insect pest species. The insecticide adhering to the roots appeared more important in determining its residual period than the insecticide absorbed by the plants. Adding a sticker to the insecticide solution improved the initial effect of this treatment and prolonged the residual effect to about 30–40 days, which was 10–20 days longer than when no sticker was used (IRRI 1973).

Root-Zone Placement.   The success of the root-coat (insecticide solution plus a sticker) treatment led to the practice of placing insecticides in the root zone of the

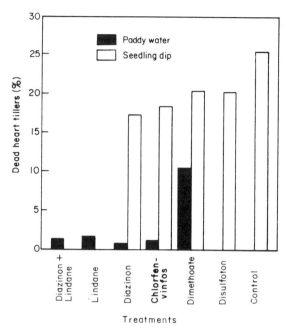

Figure 11.3   Effect on stem borer control of 6 ppm of various insecticide solutions as seedling root dip for 24 hours before transplanting or applied to paddy water at the rate of 2 kg/ha 5 days after transplanting. The data were recorded 56 days after transplanting. IRRI, 1966–1967. (Adapted from IRRI 1968)

rice plant. It was known that the anaerobic conditions that exist in the reduced soil layer make ammonium-containing or ammonium-producing (urea) fertilizers stable, which results in higher nitrogen efficiency in rice (see Chapter 10). Pathak and his associates at IRRI were the first to demonstrate that root-zone placement of insecticides was highly efficient in providing sustained control of insect pests in rice (Pathak et al. 1974). In one experiment at IRRI, root-zone placement gave significantly higher yield than an untreated control and gave yield comparable to four-time broadcast application to paddy water (IRRI 1977, Heinrichs et al. 1978). The root-zone placement also reduced toxicity to fish (Table 11.6).

When capsules containing insecticides are placed in the reduced soil layer of the root zone, insecticides are more readily available to the plants for a systemic effect. Inserting the insecticide in capsules below the soil surface protects it from heat, sunshine, volatilization, and loss with drainage water. Figure 11.4 data clearly demonstrate that root-zone application is more effective and lasts longer than paddy water application. The concentration of both chlordimeform and carbofuran insecticides was highest where each rice hill received the root-zone application (Aquino and Pathak 1976) because the roots are in direct contact with the insecticide.

Table 11.6   Effect of Root-Zone Application of Carbofuran on Yields of Rice (IR34) and Fish in an IRRI-Central Luzon State University Cooperative Study, March 1976 (Adapted from IRRI 1977, Heinrichs et al. 1978)

Application Rate (a.i./ha)[a]	Cost of Insecticide Application ($)	Rice Yield[b] (t/ha)	Fish[c] Yield (kg/ha)	Fish[c] Value ($/ha)	Income[d] ($)	Income Change[e] ($)
No insecticide	0	4.1 bc	155	127	701	—
Broadcast						
1 kg at 3 DT	30	4.3 bc	141	115	687	−14
1 kg at 3, 23, 43, and 63 DT	120	4.9 abc	0	0	566	−135
Root zone						
1 kg at 3 DT	46	5.1 ab	166	136	804	103
2 kg at 3 DT	92	5.6 a	150	123	815	114

[a] a.i. = active ingredient; DT = days after transplanting.
[b] Any two means followed by a common letter are not significantly different from each other at the 5% level.
[c] Fish seeded 7 days after first insecticide application at the rate of 3000 per ha.
[d] Income = value of rice plus value of fish minus insecticide and application costs. Based on price of rice at $0.14 and fish at $0.82 per kg.
[e] Income Change = income from insecticide treatments minus income from control (no insecticide).

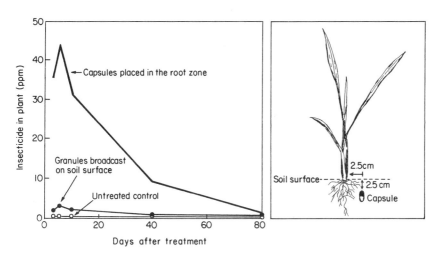

Figure 11.4   Putting insecticides in capsules and placing them in the rice root zones (right) protects the insecticides and makes them more readily available to the plants. Ten times as much insecticide was absorbed by plants when applied by the root zone method as when broadcast in the conventional manner. (From IRRI 1973)

The bottleneck of using the root-zone concept is development of a suitable machine to place the insecticides in the root zone.

RESISTANCE TO INSECTICIDE   At IRRI and in Japan, some rice insects, which were intensively controlled by insecticides, have developed insecticide resistance.

Carbofuran after about 7 years of use at IRRI, failed in 1977 to control the brown planthopper, which suggested resistance to that insecticide. The brown planthopper biotypes varied in their susceptibility to insecticide. Tests indicated that biotype 1 was least susceptible to carbofuran (Heinrichs and Valencia 1978).

In Japan, the brown planthopper and white-backed planthopper were successfully controlled for many years with lindane, DDT, malathion, and so on. The first case of resistance to lindane occurred in 1966 when a severe outbreak of the brown planthopper and the white-backed planthopper occurred (Nagata 1979).

Resurgence of the brown planthopper where certain insecticides are applied is common throughout tropical Asia. With the increase in number of insecticide applications, the resurgence ratio increases. Field studies to determine the effect of various methods of carbofuran application on resurgence suggested that brown planthopper resurgence was most common following foliar sprays (Fig. 11.5).

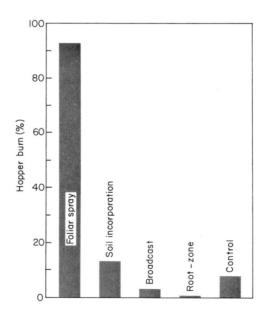

Figure 11.5   Hopper burn at 92 days after transplanting in plots treated with carbofuran by four application methods. Foliar spray was applied four times at 0.5 kg active ingredient/ha at 20-day intervals. IRRI, 1977. (From Aquino et al. 1979, unpublished)

Varietal Resistance

At the present level of technology, the use of insecticides is perhaps the only practical way to control most insect pests of rice. Insecticides, however, have limitations. They are expensive, require repeated applications, and sometimes have undesirable side effects.

Studies of the economics of rice production show that rice farmers generally use small amounts of insecticides, and then only when insect damage is visible. This, however, does not help if the insect transmits a virus disease to the rice crop. And, by the time a farmer observes insect damage, it is often too late to prevent crop damage and yield loss.

The use of insect-resistant varieties minimizes total farmer dependence on insecticides for controlling insect pests of rice. The first systematic study in insect resistance in rice was initiated when 1000 promising varieties were identified by screening 10,000 rices from the world germ plasm collection at IRRI (Pathak et al. 1969). That screening led to the identification of TKM-6, a variety from India that was resistant to stem borers and many other rice insects (Pathak 1969b). This led to the development of IR20, the first modern variety resistant to several insects and diseases. The success of IR20 showed that farmers would quickly adopt varieties with resistance to insect pests such as green leafhopper and stem borers.

For leafhopper resistance, the traditional variety from Bangladesh, Pankhari-203, and the first modern variety developed by IRRI, IR8, were excellent (Cheng and Pathak 1972). Subsequently, the variety Mudgo from India was identified for resistance to the brown planthopper (Pathak et al. 1969).

Because biotypes of planthoppers now pose a serious threat to rice production in the Asian tropics, systematic screening for resistance to existing biotypes has continued (IRRI 1978b). A detailed discussion of resistance to insects is presented in Chapter 6.

Biological Control

Biological control of rice insect pests depends on parasites, predators, and pathogens that kill the pests. Insect pests of rice have a well developed complex of natural enemies. However, systematic studies on importation, conservation, and augmentation of natural enemies for control of rice pests started only recently. One reason for this lag was that insecticides provided easy and effective insect pest control in improved rice farming. But even at the present level of insecticidal control, conservation of existing natural enemies, if not their manipulation, is called for in rice insect pest management.

The natural enemies of some major pests of rice have been identified, and their beneficial effects have been recorded in some cases (Heinrichs et al. 1979).

PARASITIC INSECTS   There are many records of parasitism of rice stem borers and other insect pests in South and Southeast Asia (Table 11.7). Mass rearing and release of *Trichogramma* species have been tested for control of rice stem

Table 11.7 Examples of High Parasitism by Parasites of Rice Stem Borers and Other Insect Pests Recorded in South and Southeast Asia (Adapted from Yasumatsu 1975)

Host	Stage Attacked	Parasitism (%)	Parasite	Area
Chilo auricillius Dudgeon	Egg	46.4	*Trichogramma* sp.	Malaysia
Chilo partellus (Swinhoe)	Egg	80.0	*Trichogramma* sp.	India
Chilo suppressalis Walker	Larva	50.0	*Apanteles chilonis*	Pakistan
Orseolia (*Pachydiplosis*) *oryzae* (Wood-Mason)	Larva	100.0	*Platygaster oryzae*	India
Orseolia (*Pachydiplosis*) *oryzae* (Wood-Mason)	Egg	76.0	*Polygnotus* sp., *Neanastatus* sp.	India
Pelopidas mathias F.	Larva	60.0	*Argyrophylax nigrotibialis,* *Halidaya luteicornis*	India
Pelopidas mathias F.	Larva	45.0	*Oncophanes hesperidis*	India
Tryporyza incertulas (Wlk)	Larva	55.5	*Tropobracon schoenobii*	India
Tryporyza incertulas (Wlk)	Egg	70.0	*Telenomus rowani,* *Trichogramma* sp.	Malaysia
Tryporyza incertulas (Wlk)	Egg	68.8	*Telenomus rowani*	Taiwan
Tryporyza incertulas (Wlk)	Egg	50.0	*Telenomus rowani*	Taiwan

borers. Results of such tests in Japan were not encouraging enough to consider the practice as economical (Yasumatsu 1975).

Recently, natural enemies of planthoppers and leafhoppers have been identified. Egg parasitism averaged about 30%. Parasites also attacked nymphs and adult hoppers, but average parasitism was lower (IRRI 1978b).

PREDATORS   Predators in this context are insects that attack a number of prey (other insects), devouring them or sucking their body fluids.

IRRI research has shown that spiders, *Microvelia* sp. and *Cyrtorhinus lividipennis* Reuter, are important natural predators of the green leafhopper and possibly the brown planthopper (IRRI 1978b). Records from sticky traps around several experimental fields indicated that adults of *C. lividipennis* flew primarily during the first half of the crop period, peaking about 25 days after rice was transplanted. Because that is a peak time when brown planthopper adults fly, biological control may be appropriate.

The use of parasites and predators will be effective, however, only if they are not killed by indiscriminate use of insecticides. In addition, rice should have resistance to those insect pests not controlled by parasites and predators.

Ducks herded in rice fields will reduce population size of the brown planthoppers.

PATHOGENS   In a few cases *Bacillus thuringiensis* Berliner is sprayed on rice to kill leaf-feeding larvae of the rice leaf folder.

Cultural Control

Many cultural practices, such as management of fertility, weeds, water, and crop residue, and plant spacings and cropping pattern greatly affect insect pests in rice.

One common example is the interaction of fertilizer application and insect control. If the variety is susceptible to a particular insect or a group of insects, heavy application of nitrogen fertilizer may aggravate the problem. Without insect control, it may be desirable to apply little or no fertilizer. Figure 11.6 shows that susceptible varieties (IR8 and IR20) with heavy brown planthopper infestation, produced hardly any grain without insect control. With insect control, IR20 fared better than IR8. The brown-planthopper-resistant IR1561-228-3 produced more than 2 t/ha without fertilizer and insecticides. With fertilizer and no insecticide, IR1561-228-3 produced at least 4 t/ha; this indicates the importance of growing varieties resistant to prevalent insects. Whenever possible, cultural practices should be used to minimize insect pests to a manageable level so that varietal resistance and insecticide applications produce effective control of insects.

Integrated Insect Pest Control

Integrated pest control uses a variety of technologies compatibly in a single pest management system (Smith and Apple 1978). In an integrated management

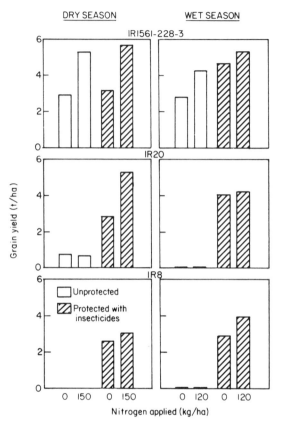

Figure 11.6 Nitrogen response of three rice varieties as affected by insect control with insecticides. IRRI, 1973. (From De Datta and Malabuyoc 1976)

system, realistic economic injury levels are used to determine the need for control actions. Insecticides are used when they can be justified based on economics and ecology. All precautions are taken to preserve natural enemies of insect pests. It is further recognized that the ultimate objective of integrated pest control is to produce optimum crop yield at minimum cost, taking into consideration ecological and socioeconomic constraints under a given agro-ecosystem.

The principles of integrated pest control are the same for insects, diseases, weeds, and other crop pests. However, in a restricted sense, it can be applied to a single pest group such as insects.

Smith and Apple (1978) described the principles evolved over the years. They are:

- IDENTIFYING PESTS   Proper identification of pests is critical to the development of integrated control measures.

- DEFINING THE AGRO-ECONOMIC SYSTEM   Knowledge of the interrelationships between pests and crop gives an understanding of the mobility of the key pest in the agro-ecosystem.
- DEVELOPING THE PEST MANAGEMENT STRATEGY   This includes coordinated use of multiple tactics in a single integrated system with an objective to hold pest numbers and damages to tolerable levels.
- ESTABLISHMENT OF ECONOMIC INJURY THRESHOLDS   The important objective is to determine the crop loss from the anticipated level of pest intensity in terms of quantity and quality of the crop and its economic value.
- DEVELOPING RELIABLE MONITORING AND PREDICTIVE TECHNIQUES   Monitoring of pest incidence and weather conditions can often reduce the cost of controlling insect pests.
- EVOLUTION OF A DESCRIPTIVE AND PREDICTIVE MODEL   This is the ultimate objective of any integrated pest management system but is not an absolute requirement for the development of integrated pest control.
- OVERCOMING THE SOCIOECONOMIC CONSTRAINTS TO THE ESTABLISHMENT OF THE INTEGRATED PEST CONTROL SYSTEM   It is critical to overcome social, economic, and political activities interfering with the development and establishment of integrated pest control.

Rice Insect Pest Management Programs

Research on management as a means of controlling rice insects is in early stages in the tropics. However, component technology is available for use in an integrated rice pest management system (Fig. 11.7).

Assuming that sufficient technology now exists to implement rice insect pest management programs, Heinrichs et al. (1979) suggested the following procedures in the formulation of pilot projects:

- Select a mechanism that assures *group planning and action* by all farmers within the pilot project area.
- Implement *training and extension programs* in pest management at levels from the research scientists to the farmer.
- Establish *monitoring systems* that provide reliable information on all fields throughout the project area and allow sound management decisions.
- Create an *input delivery system* to provide inputs when needed for the timely application of certain pest management practices.
- Set up an *evaluation system* to determine the degree of success of the pest management approach.

The complexity of the rice insect problem and the general socioeconomic conditions of rice farmers require that a pest management concept be adopted for practical, long-lasting insect control in some countries in Asia. Pilot programs are already operating in India, Indonesia, Philippines, China, and Thailand.

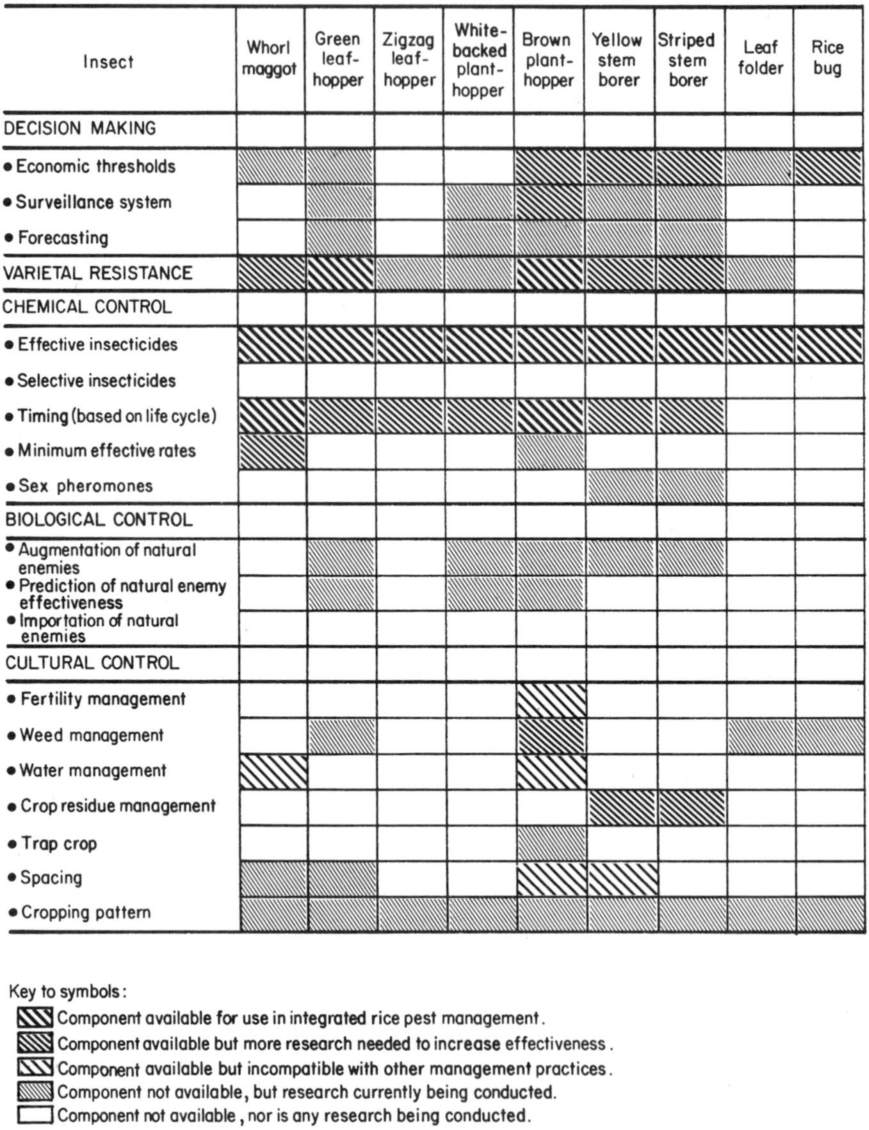

Insect	Whorl maggot	Green leaf-hopper	Zigzag leaf-hopper	White-backed plant-hopper	Brown plant-hopper	Yellow stem borer	Striped stem borer	Leaf folder	Rice bug
DECISION MAKING									
• Economic thresholds	▨	▨			▨	▨	▨	▨	▨
• Surveillance system		▨		▨	▨				
• Forecasting		▨		▨	▨				
VARIETAL RESISTANCE	▨	▨	▨		▨	▨	▨	▨	
CHEMICAL CONTROL									
• Effective insecticides	▨	▨	▨	▨	▨	▨	▨	▨	▨
• Selective insecticides									
• Timing (based on life cycle)	▨	▨	▨	▨	▨	▨	▨		
• Minimum effective rates	▨				▨				
• Sex pheromones						▨	▨		
BIOLOGICAL CONTROL									
• Augmentation of natural enemies		▨		▨	▨	▨	▨		
• Prediction of natural enemy effectiveness		▨		▨	▨				
• Importation of natural enemies									
CULTURAL CONTROL									
• Fertility management					▨				
• Weed management		▨			▨			▨	▨
• Water management	▨				▨				
• Crop residue management						▨	▨		
• Trap crop					▨				
• Spacing	▨	▨			▨	▨			
• Cropping pattern	▨	▨	▨	▨	▨	▨	▨	▨	▨

Key to symbols:
 ▨ Component available for use in integrated rice pest management.
 ▨ Component available but more research needed to increase effectiveness.
 ▨ Component available but incompatible with other management practices.
 ▨ Component not available, but research currently being conducted.
 ☐ Component not available, nor is any research being conducted.

Figure 11.7 The status of development of rice insect pest management systems at IRRI. (From Heinrichs et al. 1979)

DISEASES OF RICE AND THEIR CONTROL

The many diseases of rice are classified into four groups—fungus, bacteria, virus, and nematode—according to their disease-causing agents (Ou 1972, 1973, 1979). The distribution of rice diseases in the temperate and tropical regions is governed primarily by temperature and other weather factors. It is also influenced strongly by host varietal response and cultural practices. Although many diseases of rice extend through both tropical and temperate regions, some are specific to only one environmental regime.

Symptoms of diseases appear on the leaves, stems, leaf sheaths, inflorescence, and grains. The fungal diseases and one bacterial disease are usually exhibited by localized spots on the leaves, leaf sheaths, and stems. A major bacterial disease is systemic, causing either wilting of young plants or lesions on the margin of leaves. Viral diseases, being systemic, are generally characterized by abnormal growth (stunting of plant, depressed or excessive tillering, swollen veins, and so on) and change of leaf color to white, yellow, or orange (Ling 1972).

Fungal Diseases of Rice

The important fungal diseases are rice blast, sheath blight, leaf scald, sheath rot, bakanae, brown leaf spot, narrow brown leaf spot, and stem rot.

Rice Blast

Rice blast caused by *Pyricularia oryzae* (Cav.) occurs in all rice-growing countries. It occurs more severely in upland rice than in lowland rice. Its serious nature was recognized in Japan, Italy, the United States (Atkins 1973), and India early in the century (Parthasarathy and Ou 1963). With greater use of fertilizers for modern varieties, the threat of blast disease increased. Under regimes of low night temperatures (20–24° C) and high humidity, the severity of infection is high on susceptible varieties. The fungus that causes blast disease has many pathogenic races, which differ in their ability to infect rice varieties. Blast races may vary both within and among countries. Blast is particularly serious in Latin America and West Africa where upland rice is the major type of rice culture.

SYMPTOMS   The lesions of blast are found on the leaves, leaf sheaths, rachis, the joint of culms, neck (just below rice panicle), and the glumes. Figure 11.8 shows leaf symptoms of blast and other common diseases of rice. The most conspicuous symptoms of blast are on the leaves and neck. The center of the lesion has a pale green or dull greyish green, water-soaked appearance, with a dark brown outer rim; the center gradually becomes grey or almost straw-colored. Lesions on susceptible varieties tend to coalesce causing complete drying of infected leaves.

Near the joints of the rachillae or rachis of maturing panicles, brown to black spots or rings may be present. On the glumes, small brown to black spots are found on heavily infected panicles. The most striking symptoms are the lesions on the neck or on the nodes of the panicles near the base of the panicle, producing what is known as neck rot. This is the most destructive form of the disease and complete crop failure can result.

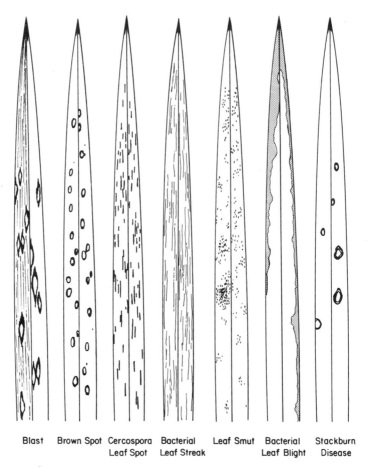

| Blast | Brown Spot | Cercospora Leaf Spot | Bacterial Leaf Streak | Leaf Smut | Bacterial Leaf Blight | Stackburn Disease |

Figure 11.8 Sketches showing typical lesions of some of the common leaf spot diseases of rice. (From Ou 1973)

Brown Spot

Brown spot, caused by *Helminthosporium oryzae* Breda de Haan, attacks rice plants at all growth stages. It occurs primarily in rice suffering from potassium imbalance or grown on soils low in fertility, particularly those deficient in nitrogen, or on saline soils. Excessive shading of the rice plants also aggravates this disease. Although the leaf spotting phase reportedly can cause heavy losses it seldom occurs in fields that are adequately fertilized.

SYMPTOMS The disease appears as small, circular, oval or elongated brown spots on the coleoptile, epicotyl, and seedling roots of germinating rice grains. It causes seedling and leaf blight and leaf spot of seedlings and mature plants. It also causes culm, sheath, and glume infection. At first, the lesions are dark reddish brown often with a yellow or gold halo. As the circular spots grow older, the center becomes greyish with distinct brown borders.

Sheath Blight

Sheath blight is caused by the fungus *Rhizoctonia solani* Kühn, which has the perfect stage *Thanatephorus cucumeris* (Frank) Donk. It is an important disease in rice in all regions. The fungus has many strains, which differ in morphological characters, rate of growth, and pathogenicity. The high temperature and humidity that prevail in the tropics favor the development of the disease. Modern rice varieties, which use higher rates of fertilizer, appear more susceptible to sheath blight than traditional varieties, which are often grown without fertilizer. The difference in susceptibility, however, is primarily due to intensive cultural practices followed with modern varieties.

SYMPTOMS Plants are usually attacked at about the tillering stage. Leaf sheaths become discolored at or above water level. The lesions are large, oblong or irregularly elongated, and appear on any part of the leaf sheath, some extending on the leaf blades. Eventually, the whole sheath parts and the leaf can easily be pulled off. Outer leaf sheaths are first affected and the fungus later extends to the inner sheaths.

   In the tropics, when sheath blight infection occurs it is not uncommon to find most leaves killed by the time the rice crop is mature. In general, the disease is more severe in a temperate climate where there are heavy and long dew periods due to the day-night temperature differential.

Leaf Scald

Leaf scald, caused by *Rhynchosporium oryzae* Hashioka & Yokogi, was first described in Japan. It has recently been reported in most tropical Asian countries, in the United States, and in many African and Latin American countries. The disease is becoming more serious as crop intensity increases or with increased rate of nitrogen application.

SYMPTOMS Leaf scald usually occurs near the tips of leaves, but sometimes starts at the leaf margin. The lesions are oblong or diamond-shaped, water-soaked blotches, which develop into large ellipsoid or oblong olive areas encircled by characteristic zonations (Ou 1972).

Stem Rot

The stem rot fungus, *Magnaporthe salvinii* Krause & Webster (1972) (*Sclerotium oryzae* Catt.), occurs in almost all rice-growing areas in the world. It lives in the soil as sclerotium for as long as 6 years (Atkins 1973). In the United States, it is less important in Texas than in Arkansas and Louisiana; it is most serious in California.

SYMPTOMS The first symptom of stem rot is the appearance of black, angular, water-soaked areas on the sheaths, near the water line usually at the maximum tillering or jointing stage. Generally, the lesions on outer sheaths become faded at maturity and are difficult to see. In time, the fungus infests the culms. When the culms are split, a cottony, greyish, mycelial growth is revealed inside the internodes. Later, when the rice approaches maturity, many black sclerotia can be found within the culms. Affected plants are damaged or killed. Stem rot makes plants susceptible to lodging (Atkins 1975).

Bakanae Disease

Bakanae disease is caused by *Gibberella fujikuroi* (Sawada) Ito (*Fusarium moniliforme* Sheld.). It has been known in Japan since 1928. The disease is widely distributed in all rice-growing areas of Asia.

SYMPTOMS The most conspicuous and common symptom is the bakanae symptom, an abnormal elongation of infected plants in the seedbed or the field. Infected seedlings are several centimeters taller than normal plants and are thin and yellowish green. The disease is seed-borne and infected plants are seen randomly scattered throughout the seedbed.

In the field, bakanae-infected plants are tall with lanky tillers bearing pale green flag leaves above the general level of the crop (Ou 1973). Most of the infected plants will die before producing a panicle or before maturity. Conidia formed on the lower portion of infected plants are blown to panicles of healthy plants infecting the flowers and causing further crop loss (M. C. Rush, IRRI, personal communication).

Sheath Rot

The causal organism of sheath rot is *Acrocylindrium oryzae* Sawada. It was first described in Taiwan in 1922, and is found in Japan, the United States, and all countries of tropical Asia.

SYMPTOMS The rot occurs on the uppermost leaf sheaths enclosing young panicles. The lesions start as oblong or somewhat irregular spots, with brown margins and grey centers, or they may be greyish brown throughout. They enlarge, often coalesce, and may cover most of the leaf sheath (Ou 1972). Severe sheath rot infection may prevent proper emergence of the panicles.

Rice Seedling Diseases

In water-sown rice in California, rice seedling diseases caused by *Achlya klebsiana* Pieters and *Pythium* species cause some problems of uniform stand establishment. The seed-rot and seedling disease of water-seeded rice, commonly called water-mold disease, was epidemic throughout the southern Louisiana rice area in 1971. Losses of 2–99% of water-seeded rice were observed in Louisiana fields during that year (Rush et al. 1972). These seed-rot and seedling diseases are generally more severe when temperatures during the planting season are cool and unfavorable for the growth and establishment of rice seedlings (Webster et al. 1973).

SYMPTOMS Seed-rot and seedling disease of rice become evident within few days of seeding into water. The most common symptoms are outgrowth of whitish hyphae from the surface of the seed or collar of the plumule. The hyphae grow from cracks in the glumes, and within a few days a radiating halo of mycelium from the infection point is visible on the seeds.

If the seedlings produce primary leaves and roots before infection by the fungi, they are usually stunted. The leaves and leaf sheaths become discolored and further development is retarded (Webster et al. 1970).

Control of Fungal Diseases

A successful disease control program for rice must be reflected in economic benefits through increased yield, increased quality, or both. Optimum management of rice diseases requires knowledge of the epidemiology of the disease. Rice diseases, irrespective of cause, are usually controlled by appropriate measures taken before the disease develops, not after an outbreak.

Chemical Control

The use of chemicals to control fungal, or any rice disease, must be economical. For blast, chemical control is possible but for some areas it is not economical. Under severe blast conditions chemical control is highly profitable. Recently, antibiotic preparations, often used in combination with organomercuric compounds, were found effective. Organomercuric compounds may cause phytotoxicity to some rices.

Spraying with systemic fungicides, such as benomyl, can control blast, but it is expensive. For sheath blight control, chemicals such as benomyl, validamycin A, neoasozin, thiophanate-methyl, and several coded compounds were found effective in reducing infection and increased grain yield over untreated control.

Varietal Resistance

In the tropics, breeding for resistant rice varieties is perhaps the only realistic solution to the fungus disease problem. However, a breeding program must take into consideration the variability of the causal organisms.

For example, breeding rice varieties resistant to all known blast races would be extremely difficult. However, by accumulating genes with a broad spectrum of resistance into a variety, more stable resistant varieties may be developed.

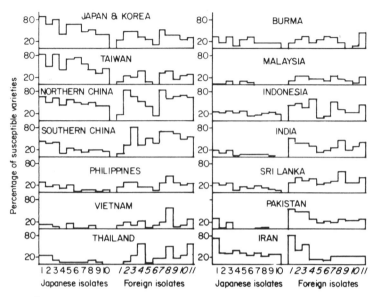

Figure 11.9 Percentage of varieties susceptible to each isolate. Names of Japanese and foreign isolates listed at bottom of the figure are Ken 53-33, Hiro 63-20, Ken 60-19, Naga 87, Hoku 373, Ken 62-39, Ai 62-22, Ken 64-38, Ina 168 and Naga 61-14, and Wp-06, Gu-03, In-28, Is-30, OE-10, Drass, IS-49, Vn-11, Th-10, Hk-02, and Is-56 in the listed order. (From Kozaka 1975)

The 1964 and 1965 results of the International Uniform Blast Nursery tests (Ou 1966) in 20 countries showed that the reaction of the tested rice varieties varied markedly among test areas, and three areas were clearly distinguishable:

· South Asia (India, Pakistan, and so on).
· Southeast Asia (Malaysia, Philippines, Indonesia, Thailand, and Vietnam).
· Temperate Asia (Republic of Korea and Japan).

Figure 11.9 shows the percentage of varieties in each country susceptible to each isolate; the diagrams indicate to which isolates most of the varieties are susceptible or resistant in each country. It appears that the diagrams are similar in shape in adjacent countries, and change gradually as the countries become farther apart (Kozaka 1975).

Rice varieties vary in their resistance to other fungal diseases such as narrow brown leaf spot, brown spot, sheath rot, leaf scald, stem rot, and bakanae. Breeding programs should take advantage of their general variability and select those varieties that are resistant and less susceptible. For brown spot disease, many modern varieties are resistant if optimum fertilizer is applied. No good source of resistance to sheath blight has been identified but several varieties have moderate resistance.

Bacterial Diseases of Rice

Two bacterial diseases are of importance in rice. Bacterial leaf blight, a systemic disease, is widespread throughout Asia and has recently been identified in Africa and Latin America (Ou 1979). It has been recognized as one of the most damaging diseases of rice in southern Japan for the past 60 years. Ishiyama (1922) first reported the cultural and physiological characteristics of the bacterial leaf blight bacteria. Considerable progress has been made on understanding the disease organisms and developing resistant varieties.

Bacterial leaf streak, which causes localized lesions, is also distributed widely in Asia.

Bacterial Leaf Blight

Bacterial leaf blight is caused by *Xanthomonas oryzae* (Uyeda and Ishiyama) Dowson. The bacterial isolates differ considerably in virulence. Studies in Japan show that many isolates collected from various countries of South and Southeast Asia showed high virulence as compared with those from Japan (Wakimoto 1975). Isolates from South Asia are more virulent than those from Southeast Asia (Buddenhagen and Reddy 1972). Even within some countries, such as India and the Philippines, the isolates of the bacterium vary in virulence.

SYMPTOMS In the tropics, the bacterial leaf blight has three types of symptoms—the leaf blight, the kresek, and the pale yellow plant. The relationship among them, based on inoculation test and observation, is seen in Fig. 11.10. Although *X. oryzae* is involved in all symptoms, kresek and leaf blight seem distinct and independent of each other, which means an individual rice plant may suffer from either.

Bacterial leaf blight at the seedling stage is important in the tropics (Ou 1973). Most strains are capable of causing kresek symptoms during seedling and early tillering stages of rice. Apparently, the bacterium enters the leaf through the cut surface of leaves or through the roots during transplanting, and systematically

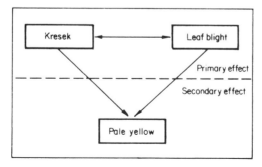

Figure 11.10   The relationship of the bacterial blight syndrome. (From IRRI 1978b)

reaches the crown or growing point. Leaves of the infected plants become water-soaked and begin to fold up and roll along the midrib. Finally, wilting of the whole leaves or plant occurs. The entire seedling ultimately dies. Bacterial leaf blight damage is often attributed to the stem borer.

The leaf stage of bacterial leaf blight is the most widespread and, therefore, causes the most damage. Early symptoms include yellow, undulate lesions along the margins of the upper portion of the leaf blades. The lesions develop rapidly parallel to the veins and extend laterally to the healthy regions. In extreme cases, a large portion of the entire leaf blade becomes infected, turns yellow or dirty white, and finally dies.

Bacterial Leaf Streak

Bacterial leaf streak, a disease limited to the tropics, is caused by *Xanthomonas translucens* (Jones, Johnson, and Reddy) Dowson f. sp. *oryzicola* (Fang et al.) Bradbury. The organism attacks chiefly the parenchymatous tissue between veins of the leaves and in the early stages remains confined to the interveinal spaces. It may enter the leaf through the stomata or through wounds, which are mainly caused during storms.

SYMPTOMS   Symptoms manifested are the appearance of fine, interveinal, long or short lines, which are water-soaked and greyish. The lesions extend and coalesce to form larger patches and become yellow from the death of cells. Eventually, a large portion of the entire leaf blade becomes yellow to dirty white. At this stage, the symptoms are difficult to distinguish from the bacterial leaf blight.

Control of Bacterial Diseases

Because rainfall has a close causal relationship with bacterial diseases, occurrence of these diseases is considered unavoidable in South and Southeast Asia where there is a heavy monsoonal season. This limits any successful chemical control of bacterial diseases. Varietal resistance is the most important single method for avoiding occurrence of bacterial diseases in the tropics.

Chemical Control

Chemicals such as copper or mercury compounds and antibiotics have been used to control bacterial diseases. Acetylene dicarboxamide has been used, as have sprays with streptomycin (3000 ppm) and cupric hydroxide.

Varietal Resistance

Sources of genetic resistance to bacterial diseases are available and have been used in developing modern rice varieties. These varieties are now grown in most rice-growing countries.

A simple inoculation method allows the rapid screening of rice varieties for resistance. In 1977 alone, 73,538 rices from the IRRI germ plasm bank and breeding program were screened for resistance to bacterial leaf blight. A high percentage (71%) of the breeding lines were resistant to pathotype 1 of *X. oryzae* in the Philippines (IRRI 1978a). However, because of the variation of the pathogen, varieties do not remain resistant indefinitely and different sources must be constantly used in breeding programs.

Viral Diseases of Rice

Viral diseases of rice are devastating. Many are transmitted by leafhoppers and planthoppers (Table 11.8). The important viral diseases are tungro, grassy stunt, ragged stunt, yellow dwarf, orange leaf, and hoja blanca. The viral diseases occur in almost all rice-growing countries in Asia. They also occur in Latin America (see Table 11.2) and in some countries in Africa.

Tungro

In tropical Asia, tungro virus is the most important rice disease in terms of rice-growing areas affected. From 1934 to 1936 in Indonesia, several crop failures were caused on 30,000–50,000 ha of rice land by a disease called *mentek*, which was later proven to be tungro virus (Ou and Rivera 1969). Since 1965, major tungro outbreaks have occurred in Thailand (1965–1967); in Bangladesh (1969); in Uttar Pradesh, Bihar, and West Bengal, India (1969); in South Kalimantan, Indonesia (1969); in North Krian, Malaysia (1969); in South Sulawesi Indonesia (1972); and in the Philippines (1970–1971) (Ling 1976).

Several environmental and biological factors influence the outbreaks of tungro virus. Tungro virus is transmitted by leafhoppers *Nephotettix malabanus, N. nigropictus* (Stål), *N. parvus, N. virescens* (Distant), and *Recilia dorsalis* (Motschulsky) (Ling 1976).

SYMPTOMS Stunting, yellowing of leaves, and slight reduction in tiller numbers are observed in tungro-susceptible varieties. Stunting is caused by shortening of the leaf sheath, leaf blade, and internode. Yellowing usually starts from the tips of the leaves; the color may vary from light yellow to orange-yellow or brownish yellow. On the yellow leaf and occasionally on the green leaf, irregularly spaced dark brown blotches frequently develop. Young leaves of infected plants are usually mottled with pale green to whitish spots, the pattern varying from mosaic to stripes of various lengths parallel to the veins. The grains are usually covered with dark brown blotches.

Grassy Stunt

Grassy stunt virus is transmitted by the brown planthopper *Nilaparvata lugens* (Stål). Grassy stunt was first observed in 1963 at IRRI. It became serious in

Table 11.8 Rice Virus Diseases in Tropical Asia (Adapted from Ling 1970, Ling et al. 1978a, b)

	Tungro	Grassy Stunt	Ragged Stunt	Yellow Dwarf	Orange Leaf	Hoja Blanca
Vector	*Nephotettix virescens*	*Nilaparvata lugens*	*Nilaparvata lugens*	*Nephotettix virescens* N. nigropictus N. cincticeps	*Recilia dorsalis*	*Sogatodes cubanus* S. orizicola
Symptoms	Tillering slightly reduced; stunting; yellowing of leaves; rusty blotches	Tillering excessive; severe stunting; chlorosis; narrow, erect leaves; rusty dots	Stunting of plants, appearance of ragged leaves, twisted leaves; vein swelling, delay in flowering; production of nodal branches; incomplete emergence of panicles; panicles bearing mostly unfilled spikelets	Tillering excessive; severe stunting; general yellowing; drooping leaves	Tillering reduced; inconspicuous stunting; orange and rolled leaves; rapid death	White and chlorotic stripes on the leaves, stunting of the whole plant, partial filling of spikelets at maturity

1973–1979 throughout large areas of the Philippines where the brown plant-hoppers were seen. The disease was later reported in most tropical Asian countries. Recently, grassy stunt became severe in Indonesia, but has caused little damage in other countries.

SYMPTOMS   Grassy stunt symptoms on diseased plants are severe stunting, excessive tillering, and an erect growth habit. The leaves are short, narrow, pale green or pale yellow, and often have numerous small, dark brown dots or spots of various shapes, which may form blotches. Young leaves of some varieties may be mottled or striped. The leaves may remain green when supplied with adequate fertilizer.

Ragged Stunt

Ragged stunt was first reported in North Cotabato in the Philippines in 1977 (Bergonia 1977). Because of the ragged leaf symptoms and stunting exhibited by infected plants, the name ragged stunt was adopted. The disease was characterized and identified as caused by a virus transmitted by a brown planthopper, *Nilaparvata lugens* (Stål) (Ling et al. 1978a,b).

After 1977, disease symptoms were observed in rice plants in India, Sri Lanka (Heinrichs and Khush 1978), in Indonesia (Hibino et al. 1977), in Thailand (Weerapat and Pongprasert 1978), and more recently in China.

SYMPTOMS   The symptoms of rice ragged stunt vary at different growth stages. They include stunting of plants, appearance of ragged leaves, twisted leaves, vein swelling, delay in flowering, production of nodal branches, incomplete emergence of panicles, and panicles bearing mostly unfilled spikelets. Other symptoms are not easily seen. For example, the color of the diseased plants and number of tillers often do not vary a great deal from those of healthy plants. However, stunting of plants often causes marked reduction in crop height.

Yellow Dwarf

Yellow dwarf virus is widespread in Japan and tropical Asian countries (Ou 1972, Ling 1972), but in tropical Asia the virus does not cause economic yield losses because of the long incubation period in the insect and the host plant. *Nephotettix cincticeps* (Uhler) is the vector that transmits yellow dwarf virus in Japan. *Nephotettix virescens* (Distant) and *N. nigropictus* (Stål) are vectors in other countries.

SYMPTOMS   General chlorosis, especially on the newly emerged and young leaves, is the first symptom of yellow dwarf. The color varies from yellowish to green. As the disease progresses, the infected plants become severely stunted, tillering increases markedly, and leaves become soft and droopy. The infected plants produce either no panicles or a few small panicles, which bear mostly unfilled spikelets.

Orange Leaf

Orange leaf disease was first found in Thailand in 1960, then in the Philippines in 1962 (Ou 1972, Ling 1972). The virus is transmitted by *Recilia dorsalis*. The disease does not cause serious reductions of rice yield.

SYMPTOMS In the field, orange-leaf-infected plants have golden yellow to deep bright orange leaves when the plants are about one month old or older. The infected plants die before flowering.

Hoja Blanca

Hoja blanca (white leaf) was recognized as a new viral disease of rice in Latin America in 1957 (Adair and Ingram 1957). Hoja blanca is limited to the western hemisphere. It is transmitted by *Sogatodes cubanus* (Crawford) and *S. orizicola* (Muir).

SYMPTOMS Plants with hoja blanca show stunting and reduced tillering with new leaves unfolding normally (Ling 1972). The major field symptoms are the appearance of white or chlorotic stripes on the leaves or the occurrence of completely white leaves, stunting of the white plant, and poor or partial filling of spikelets at maturity (Ou 1973).

Control of Virus Diseases

Control measures for viral diseases include practice of clean crop culture, such as removal of weeds and volunteer rice, rouging of diseased plants, use of insecticides to control the insect vectors, and planting resistant rice varieties.

Chemical Control of Vectors

Although there is no meaningful chemical control of viral diseases, controlling vectors with pesticides can be effective for some viruses (Pathak et al. 1967). However, once disease symptoms appear, chemical control of vectors does not effectively reduce incidence of viral diseases.

Varietal Resistance

Excellent sources of resistance are available for all tropical viral diseases of rice except the ragged stunt virus. Resistant genes for tungro, grassy stunt, yellow dwarf, and hoja blanca have been incorporated in many modern rice varieties.
There are three major steps in producing the resistant varieties (Ling 1972):

- Develop a method of testing and screening and of standardizing scales for measuring resistance.
- Identify a source of genes conferring resistance.

• Incorporate those genes with other desirable qualities of the crop.

The resistance of a rice plant to a viral disease can only be determined after the plant is exposed to the virus. The varietal reaction to a viral disease by a natural infection can be tested in the field but, because disease pressure in the field is not constant, the field reaction of a rice to viral diseases fluctuates from time to time and from place to place, regardless of variety. However, a rice that has a consistently low field reaction should be more resistant than a rice that does not show such a reaction.

Nematodes of Rice

Several rice diseases are caused by nematodes, also known as eelworms or roundworms. Nematode-caused diseases have been reported to occur in rice fields in the Indian subcontinent. The two major nematode-caused diseases are white-tip disease and ufra.

White-Tip Disease

White-tip disease is caused by the nematode *Aphelenchoides besseyi* Christie. It is present in many tropical countries, particularly in India and Bangladesh. In 1935–1945, the disease caused severe losses of rice in the United States in Arkansas, Louisiana, and Texas.

SYMPTOMS The most distinguishing symptom of white-tip is the presence of leaves with white tips of 2.5–5 cm long. The tips of the developing leaves may be twisted and wrinkled and the flag leaf may be twisted near the panicle. The infected plants are generally stunted (Davide 1970).

Ufra Disease

Ufra disease is caused by the nematode *Ditylenchus angustus* (Butler) Filipjev, and has been reported from India, Bangladesh, Burma, Thailand, Malaysia, and Egypt. The disease is also known as *dak pora* in Bangladesh, *akhet-pet* in Burma, and *yad-ngo* (twisted pathogen) in Thailand. In India, it is considered one of the important diseases of rice.

SYMPTOMS Infected plants are severely stunted, often with withered leaves, which sometimes show brown areas near the node. At times, the heads develop partially but are twisted and deformed. Developing heads may contain many nematodes in all stages from eggs to adults (Davide 1970).

Control of Nematodes

White-tip disease is controlled by dipping seeds in hot water (55–61°C) for 15 minutes. Chemicals such as parathion can also be used to treat the seeds.

Ufra disease can be controlled by field sanitation practices such as use of clean seeds, avoiding use of irrigation water from an infected field, and burning of stubble.

Multiple Resistance to Various Insects and Diseases

Building resistance to disease and insect attacks into the modern rice varieties remains a key element in rice breeding. Such resistant varieties mean higher rice yields and lower production costs—factors important to the rice farmer and, in the long run, to the rice consumer.

Details on multiple resistance to insects and diseases are given in Chapter 6.

OTHER PESTS OF RICE AND THEIR CONTROL

Besides insects and diseases, a variety of other pests affect tropical rice production. They are rodents, birds, snails, crabs, and some large animals. However, rodents and birds are the most serious pests among those mentioned.

Rodents in Rice Fields

Rodents, particularly rats, cause serious damage to the rice crop at all growth stages. They eat seeds (in direct-seeded rice) and seedlings, they gnaw off tillers, damage plants, and feed on rice grains at various stages. They also destroy stacked rice; stored, threshed rice; and hulled rice in retailers' bins.

There are many species of rats that destroy rice but *Rattus rattus* L. as a group is most prevalent in rice fields (Alfonso and Sumangil 1970). Rats move and feed at night. They burrow holes in levees, thereby causing water leakage, and decreasing water use efficiency in rice.

Control of Field Rats

Among rat control practices, keeping the levees (bunds) free from weeds; making fields, farm house, and storage areas ratproof; and baiting with poisons are common. However, none of these practices is foolproof. To protect experimental fields, electric rat fences are sometimes used at IRRI and at other experiment stations in Southeast Asia.

There are two kinds of poisons to kill rats:

- Acute poisons cause death shortly after ingestion. Sodium fluoroacetate (compound 1080) and zinc phosphide are most common.
- Slow-acting poisons are anticoagulants that induce death after a rat ingests poisons for several days. Examples are warfarin and coumarin.

BAITING Results from Taiwan suggest that the levels of bait acceptability for field rodents are variable and are influenced by the surrounding environment, such as the type and growth stages of the crops. For example, when bait is offered at the ripening stage of rice, acceptability by rodents is relatively low. After harvest, however, acceptability increases steeply as good food sources for the rodents are reduced to low levels. Therefore, a rodent control program should begin before crops are abundant and close to harvest, and provide a regular food source. The optimum control of rodents in rice paddies will occur when poison baits are offered before the rice flowering stage, and then continued after harvest (Ku 1979).

Birds in Rice Fields

Birds cause considerable damage shortly after seeding direct-seeded rice and from flowering time onward in any rice culture. Birds cause more damage to a shattering rice variety than to a nonshattering type. In tropical Asia, the weaver birds of the genus *Ploeus,* parakeets, Munia, and sparrows are most serious. Birds such as the migratory African species *Quelea quelea* (Linn.) cause serious damage to the rice crop in Africa.

Wild ducks and geese damage the rice crop in many countries including Australia, United States, and in some countries of Latin America.

For control of birds, various sound-making devices are often used. In some areas, various improved devices—gas operated cannons—are used.

Snails in Rice Fields

Snails such as *Lanistes ovum* (Trosch), *Ampullaria lineata,* and *A. glauca* destroy newly emerged rice seeds and seedlings. Snails are controlled by the insecticide lindane, which is generally used for rice stem borer control.

BEFORE USING ANY CHEMICALS, SEE THE MANUFACTURER'S LABEL AND FOLLOW THE DIRECTIONS. ALSO SEE THE PREFACE.

REFERENCES

Adair, C. R., and J. W. Ingram. 1957. Plans for the study of *hoja blanca,* a new rice disease. *Rice J.* 60(4):12.

Alfonso, P. J., and J. P. Sumangil. 1970. Control of rice-field rats. Pages 204–211 *in* University of the Philippines College of Agriculture in cooperation with the International Rice Research Institute. *Rice production manual.* Los Baños, Philippines.

Aquino, G. B., and M. D. Pathak. 1976. Enhanced absorption and persistence of carbofuran and chlordimeform in rice plant on root zone application under flooded conditions. *J. Econ. Entomol.* 69:686–690.

Aquino, G. B., E. A. Heinrichs, S. Chelliah, M. Arceo, S. Valencia, and L. Fabellar. 1979. Recent developments in the chemical control of the brown planthopper *Nilaparvata lugens* (Stål). Paper presented at a Saturday seminar, 10 February 1979, International Rice Research Institute, Los Baños, Philippines. (unpubl. mimeo.)

Atkins, J. G. 1973. Rice diseases. Pages 141–150 *in* USDA Agric. Handb. 289. *Rice in the United States: varieties and production.* Washington, D.C.

Atkins, J. G. 1975. Controlling diseases of rice. Pages 58–68 *in* Texas Agricultural Experiment Station in cooperation with the U.S. Department of Agriculture. *Six decades of rice research in Texas.* Res. Monogr. 4.

Bergonia, H. T. 1977. The infectious gall disease, a new rice malady in the Philippines: its distribution and possible cause. *Philipp. Phytopathol.* 13:2–3.

Bowling, C. C. 1970. Lateral movement, uptake, and retention of carbofuran applied to flooded rice plants. *J. Econ. Entomol.* 63:239–242.

Bowling, C. C. 1975. Insect pests in rice fields. Pages 69–75 *in* Texas Agricultural Experiment Station in cooperation with the U.S. Department of Agriculture. *Six decades of rice research in Texas.* Res. Monogr. 4.

Bowling, C. C. 1979. Insect pests of the rice plant. Pages 260–288 *in* AVI Publishing Co. *Rice: production and utilization.* Westport, Connecticut.

Buddenhagen, I. W., and A. P. K. Reddy. 1972. The host, the environment, *Xanthomonas oryzae,* and the researcher. Pages 289–295 *in* International Rice Research Institute. *Rice breeding.* Los Baños, Philippines.

Cheng, C. H., and M. D. Pathak. 1972. Resistance to *Nephotettix virescens* in rice varieties. *J. Econ. Entomol.* 65:1148–1153.

Davide, R. 1970. Common diseases of rice. Part III. Diseases caused by nematodes. Pages 235–236 *in* University of the Philippines College of Agriculture in cooperation with the International Rice Research Institute. *Rice production manual.* Los Baños, Philippines.

De Datta, S. K., and J. Malabuyoc. 1976. Nitrogen response of lowland and upland rice in relation to tropical environmental conditions. Pages 509–539 *in* International Rice Research Institute. *Climate and rice.* Los Baños, Philippines.

Gifford, J. R. 1973. Insects and their control. Pages 151–154 *in* USDA Agric. Handb. 289. *Rice in the United States: varieties and production.* Washington, D.C.

Hatai, N. 1975. Insect pests of rice and suitable insecticides for South-East Asian countries. Pages 393–401 *in* University of Tokyo Press. *Rice in Asia.* Tokyo, Japan.

Heinrichs, E. A. 1977. The brown planthopper threat to rice production in Asia. Pages 45–64 *in* Proceedings of the symposium on brown planthopper, third international congress of the Pacific Science Association. *The brown planthopper.* Bali, Indonesia, 22–23 July 1977.

Heinrichs, E. A. 1979a. Control of leafhopper and planthopper vectors of rice viruses. Pages 529–560 *in* K. Maramorosch and K. F. Harris, eds. Academic Press. *Leafhopper vectors and plant diseases agents.* New York.

Heinrichs, E. A. 1979b. Chemical control of the brown planthopper. Pages 145–167 *in* International Rice Research Institute. *Brown planthopper: threat to rice production in Asia.* Los Baños, Philippines.

Heinrichs, E. A., and G. S. Khush. 1978. Ragged stunt virus disease in India and Sri Lanka. *Int. Rice Res. Newsl.* 3(2):13.

Heinrichs, E. A., and P. K. Pathak. 1980. Resistance to the gall midge, *Orseolia oryzae* (Wood-Mason) in rice. Insect science and its application. (in press)

Heinrichs, E. A., and S. L. Valencia. 1978. Contact toxicity of insecticides to the three biotypes of brown planthopper. *Int. Rice Res. Newsl.* 3(3):19-20.

Heinrichs, E. A., G. B. Aquino, and R. Arce. 1978. Root-zone application of carbofuran in rice-fish culture. *Int. Rice Res. Newsl.* 3(5):16-17.

Heinrichs, E. A., R. C. Saxena, and S. Chelliah. 1979. *Development and implementation of insect pest management systems for rice in tropical Asia.* ASPAC Food Fert. Technol. Cent. Ext. Bull. 127. 38 pp.

Hibino, H., M. Roechan, S. Sudarisman, and D. M. Tantera. 1977. A virus disease of rice (Kerdil hampa) transmitted by brown planthopper, *Nilaparvata lugens* Stål. *Indonesia Contrib. Cent. Res. Inst. Agric. Bogor* 35:1-15.

IRRI (International Rice Research Institute). 1966. *Annual report 1965.* Los Baños, Philippines. 357 pp.

IRRI (International Rice Research Institute). 1968. *Annual report 1967.* Los Baños, Philippines. 308 pp.

IRRI (International Rice Research Institute). 1971. *Annual report for 1970.* Los Baños, Philippines. 265 pp.

IRRI (International Rice Research Institute). 1973. *Annual report for 1972.* Los Baños, Philippines. 246 pp.

IRRI (International Rice Research Institute). 1977. *Research highlights 1976.* Los Baños, Philippines. 108 pp.

IRRI (International Rice Research Institute). 1978a. *Research highlights 1977.* Los Baños, Philippines. 122 pp.

IRRI (International Rice Research Institute). 1978b. *Annual report for 1977.* Los Baños, Philippines. 548 pp.

IRRI (International Rice Research Institute). 1979. *Brown planthopper: threat to rice production in Asia.* Los Baños, Philippines. 369 pp.

Ishiyama, S. 1922. Studien über die weissfleckenkrankheit der reispflanzen [in Japanese]. *Mitt. K. Zentr. Landw. Versnchsst. Tokyo* 45:233-261. German abstract in Japan. *J. Bot.* 1(2):Abstr. 48. 1923.

Khush, G. S. 1979. Genetics of and breeding for resistance to the brown planthopper. Pages 321-322 *in* International Rice Research Institute. *Brown planthopper: threat to rice production in Asia.* Los Baños, Philippines.

Koshihara, T., and D. Okamoto. 1957. Control of rice stem borer by the application of BHC dust in the paddy field soil [in Japanese, English summary]. *Jpn. J. Appl. Entomol. Zool.* 1:32-35.

Kozaka, T. 1975. Reaction of rice varieties to major races of *Pyricularia oryzae* in Asian countries. Pages 318-326 *in* University of Tokyo Press. *Rice in Asia.* Tokyo, Japan.

Krause, R. A., and R. K. Webster. 1972. The morphology, taxonomy, and sexuality of the rice stem rot fungus, *Magnaporthe salvinii (Leptosphaeria salvinii). Mycologia* 64:103-114.

Ku, T. Y. 1979. *Field and laboratory tests on the acceptability of paraffin-rice mixed baits to rodents in Taiwan.* ASPAC Food Fert. Technol. Cent. Tech. Bull. 44. 14 pp.

Ling, K. C. 1970. Common diseases of rice. II. Virus diseases of rice. Pages 228–239 *in* University of the Philippines College of Agriculture in cooperation with the International Rice Research Institute. *Rice production manual.* Los Baños, Philippines.

Ling, K. C. 1972. Rice virus diseases. International Rice Research Institute, Los Baños, Philippines. 142 pp.

Ling, K. C. 1976. *Recent studies on rice tungro virus diseases at IRRI.* IRRI Res. Pap. Ser. 1. Los Baños, Philippines. 11 pp.

Ling, K. C., E. R. Tiongco, and V. M. Aguiero. 1978a. Rice ragged stunt, a new virus disease. *Plant Dis. Rep.* 62:701–705.

Ling, K. C., E. R. Tiongco, V. M. Aguiero, and P. Q. Cabauatan. 1978b. *Rice ragged stunt disease in the Philippines.* IRRI Res. Pap. Ser. 16. 25 pp.

Mitra, D. K., S. P. Raychaudhuri, T. R. Everett, A. Ghosh, and F. R. Niazi. 1970. Control of the rice green leafhopper with insecticidal seed treatment and pretransplant seedling soak. *J. Econ. Entomol.* 63:1958–1961.

Nagata, T. 1979. *Development of insecticide resistance in the brown planthopper and the white backed planthopper.* ASPAC Food Fert. Technol. Cent. Tech. Bull. 45. 16 pp.

Ou, S. H. 1966. International uniform blast nurseries, 1964–1965 results. *Int. Rice Comm. Newsl.* 15(3):1–13.

Ou, S. H. 1972. *Rice diseases.* Commonwealth Mycological Institute, Kew, Surrey, England. 368 pp.

Ou, S. H. 1973. *A handbook of rice diseases in the tropics.* International Rice Research Institute, Los Baños, Philippines. 58 pp.

Ou, S. H. 1979. Rice plant diseases. Pages 235–259 *in* AVI Publishing Co. *Rice: production and utilization.* Westport, Connecticut.

Ou, S. H., and C. T. Rivera. 1969. Virus diseases of rice in southeast Asia. Pages 23–34 *in* International Rice Research Institute. *The virus diseases of the rice plant.* Proceedings of a symposium at the International Rice Research Institute, April, 1967. The Johns Hopkins Press, Baltimore, Maryland.

Parthasarathy, N., and S. H. Ou. 1963. Opening address: International approach to the problem of blast. Pages 1–5 *in* International Rice Research Institute. *The rice blast disease.* Proceedings of a symposium at the International Rice Research Institute, April 1967. The Johns Hopkins Press, Baltimore, Maryland.

Pathak, M. D. 1968a. Ecology of common insect pests of rice. *Annu. Rev. Entomol.* 13:257–294.

Pathak, M. D. 1968b. Application of insecticides to paddy water for more effective rice pest control. *Int. Pest Control* 10:12–17.

Pathak, M. D. 1969a. *Insect pests of rice.* International Rice Research Institute, Los Baños, Philippines. 77 pp.

Pathak, M. D. 1969b. Stem borer and leafhopper resistance in rice varieties. *Entomol. Exp. Appl.* 12:789–800.

Pathak, M. D. 1970. Insect pests of rice and their control. Pages 171–198 *in* University of the Philippines College of Agriculture in cooperation with the International Rice Research Institute. *Rice production manual.* Los Baños, Philippines.

Pathak, M. D., and V. A. Dyck. 1973. Developing an integrated method of rice insect pest control. *PANS* 19:534–544.

Pathak, M. D., C. H. Cheng, and M. E. Fortuno. 1969. Resistance to *Nephotettix impicticeps* and *Nilaparvata lugens* in varieties of rice. *Nature* 223:502–504.

Pathak, M. D., D. Encarnacion, and H. Dupo. 1974. Application of insecticides in the root zone of rice plants. *Indian J. Plant Prot.* 1:1–16.

Pathak, M. D., E. Vea, and V. T. John. 1967. Control of insect vectors to prevent virus infection of rice plants. *J. Econ. Entomol.* 60:218–225.

Rao, Y. S., and P. Israel. 1967. Recent developments in and future prospects for the chemical control of the rice stem borer in India. Pages 317–324 *in* International Rice Research Institute. *The major insect pests of the rice plant.* Proceedings of a symposium at the International Rice Research Institute, September, 1964. The Johns Hopkins Press, Baltimore, Maryland.

Rush, M. C., M. A. Marchetti, and C. R. Adair. 1972. Stand establishment in early seeded rice in the south. *Rice J.* 75:32–34.

Smith, R. F., and J. L. Apple. 1978. Principles of integrated pest control. *Plant Prot. News* 4(4):4–7.

Soto, P. E. J., and Z. Siddiqi. 1978. Insect pests and rice production in Africa. Pages 175–179 *in* Academic Press. *Rice in Africa.* London.

Wakimoto, S. 1975. Bacterial leaf blight disease of rice in South-East Asian countries. Pages 327–339 *in* University of Tokyo Press. *Rice in Asia.* Tokyo.

Webster, R. K., D. H. Hall, J. Heeres, C. M. Wick, and D. M. Brandon. 1970. *Achlya klebsiana* and *Pythium* species as primary causes of seed rot and seedling disease of rice in California. *Phytopathology* 60:964–968.

Webster, R. K., D. H. Hall, J. Bostad, C. M. Wick, D. M. Brandon, R. Baskett, and J. M. Williams. 1973. Chemical seed treatment for the control of seedling disease of water-sown rice. *Hilgardia* 41:689–698.

Weerapat, P., and S. Pongprasert. 1978. Ragged stunt disease in Thailand. *Int. Rice Res. Newsl.* 3(1):11–12.

Yasumatsu, K. 1975. Insects injurious to rice cultivation and their natural enemies in South-East Asia. Pages 383–392 *in* University of Tokyo Press. *Rice in Asia.* Tokyo.

12

Weeds and Weed Control In Rice

Weeds are often called plants out of place. They are unwanted, useless, prolific, competitive, often harmful to the total environment—even poisonous—and they occur in every rice field in the world. As they interfere with agricultural operations, they reduce the potential production of rice. Weeds cause many direct damages to a crop.

- Weeds in cultivated fields reduce rice yield and quality by competing for nutrients, water, and light.
- Weeds intensify the problem of diseases, insects, and other pests by serving as their hosts.
- Weeds reduce the efficiency of harvesting.
- Aquatic weeds reduce the efficiency in irrigation systems.

It is estimated that annual crop losses due to weeds in Asia are 11.8% of potential production value; the corresponding figure for the world is 9.5%. One estimate at IRRI suggests weed growth in unweeded plots reduces yield by as much as 34% in transplanted rice, 45% in direct-seeded, rainfed lowland rice, and 67% in upland rice.

Effective weed control requires knowledge of the names, distribution, ecology, and biology of weeds in the rice-growing regions. Weeds are classified as grasses, broadleaf weeds, or sedges. Table 12.1 gives some examples of the common weeds in tropical rice fields. In most tropical Asian countries, moderate year-round warm temperature and high humidity encourage year-round growth of weeds. The number and diversity of weed species in the rice fields in subtropical and temperate areas are shown in Table 12.2.

In Japan, 43 families and 191 species of weeds have been reported from various rice-growing regions and 28 of these species can cause serious damage to rice yields.

In Taiwan, there are about 41 families and 145 weed species. Among them, 126 species occur in the first rice crop and 95 species appear in the second rice crop of the season. Most of them are annuals, a few are perennials (Chiu 1973).

460

Table 12.1 Some Common Weeds in Tropical Lowland and Upland Rice in South and Southeast Asia; Common Names are Given in Parentheses

I. Lowland rice

  a. Annual grasses (Gramineae family)

    Echinochloa glabrescens Munro ex Hook (formerly *E. crus-galli)* (Barnyard grass)

    Echinochloa crus-galli ssp. *hispidula* (Retz.) Honda (formerly *E. crus-pavonis)* (Barnyard grass)

  b. Annual broadleaf weeds (dicotyledonous)

    Monochoria vaginalis (Burm. f.) Presl. (Monochoria)

    Sphenoclea zeylanica Gaertn. (Goose weed)

  c. Annual sedges

    Cyperus difformis L. (Small-flowered umbrella plant)

    Cyperus iria L. (Umbrella sedge)

    Fimbristylis littoralis Gaudich. (Hoorahgrass)

  d. Perennial grass

    Paspalum distichum L. (Knot grass)

  e. Perennial sedge

    Scirpus maritimus L. (Bulrush)

II. Upland rice

  a. Annual grasses

    Echinochloa colona (L.) Link (Jungle rice)

    Digitaria sanguinalis (L.) Scop. (Large crab grass)

    Eleusine indica (L.) Gaertn. (Goose grass)

    Dactyloctenium aegyptium (L.) Beauv. (Crowfoot grass)

    Paspalum dilatatum Poir. (Dallis grass)

    Rottboellia exaltata L. f. (Itchgrass)

  b. Annual broadleaf weeds

    Ageratum conyzoides L. (Tropic ageratum)

    Amaranthus spinosus L. (Spiny amaranth)

    Calopogonium mucunoides Desv. (Calopo)

    Celosia argentea L. (Celosia)

    Commelina benghalensis L. (Hairy wandering jew)

    Commelina diffusa Burm. f. (Spreading dayflower)

    Eclipta alba (L.) Hassk. (Eclipta)

    Ipomoea triloba L. (Threelobe morning glory)

    Portulaca oleracea L. (Common purslane)

    Trianthema portulacastrum L. (Horse purslane)

  c. Annual sedge

    Cyperus iria L. (Umbrella sedge)

  d. Perennial grass

    Imperata cylindrica (L.) Beauv. (Cogon grass)

  e. Perennial sedge

    Cyperus rotundus L. (Purple nutsedge)

Table 12.2 Major Weeds of Rice Fields in Taiwan, Japan, and the United States; Common Names are Given in Parentheses (Adapted from Chiu 1973)

Family	Species	Taiwan	Japan	United States
Alismataceae	Alisma triviale Pursh (Common water plantain)			X
	Echinodorus cordifolius (L.) Griseb. (Burhead)			X
	Sagittaria cuneata Sheldon			X
	Sagittaria pygmaea Miq. (Arrowhead)		X	
	Sagittaria trifolia L.	X	X	
Amaranthaceae	Alternanthera nodiflora R. Br. (Common joyweed)	X		
	Alternanthera sessilis (L.) Dc. (Joyweed)	X		
Campanulaceae	Lobelia affinis Wall.	X		
Compositae	Ageratum conyzoides L. (Tropic ageratum)	X		
	Eclipta alba (L.) Hassk. (Eclipta)	X		
Cruciferae	Cardamine parviflora L. (Bittercress)	X		
Cyperaceae	Cyperus difformis L. (Small-flowered umbrella plant)	X	X	X
	Cyperus erythrorhizos Muhl. (Redroot flatsedge)			X
	Cyperus globosus All.		X	
	Cyperus iria L. (Umbrella sedge)		X	X
	Cyperus microiria Steud.		X	
	Cyperus rotundus L. (Purple nutsedge)	X		

Family	Species			
	Cyperus serotinus Rottb.		X	
	Eleocharis acicularis (L.) R. & S. (Slender spikerush)	X	X	
	Eleocharis kuroguwai Ohwi (Water chestnut)		X	
	Eleocharis palustris (L.) R. Br. (Creeping spikerush)			X
	Fimbristylis littoralis Gaudich. (Hoorahgrass)		X	
	Kyllinga brevifolia Rottb. (Green kyllinga)	X	X	
	Rhynchospora corniculata (Lam.) Gray			X
	Scirpus mucronatus L. (Roughseed bulrush)			X
	Scirpus wallichii Nees	X		
Euphorbiaceae	*Caperonia castanaefolia* (L.) St. Hil. (Mexicanweed)	X		
Elatinaceae	*Elatine triandra* Schk. (Elatine)			X
Gramineae	*Arundinella hirta* (Thunb.) Tanaka		X	
	Echinochloa colona (L.) Link (Jungle rice)		X	
	Echinochloa crus-galli (L.) Beauv. (Barnyard grass)		X	X
	Echinochloa crus-pavonis (H. B. K.) Schult. (Barnyard grass)	X		X
	Isachne globosa (Thunb.) O. Kuntze (Swamp millet)		X	X
	Leptochloa chinensis (L.) Nees (Sprangletop)			X

Table 12.2 Continued

Family	Species	Taiwan	Japan	United States
	Leptochloa fascicularis (Lam.) Gray (Scale grass)			X
	Oryza sativa L. (Red rice)			X
	Panicum repens L. (Torpedo grass)	X		
	Paspalum conjugatum Berg. (Sour grass)	X		
	Paspalum thunbergii Kunth ex Steud.		X	
Juncaceae	*Juncus prismatocarpus* R. Br.	X		
Lemnaceae	*Spirodela polyrhiza* (L.) Schleid. (Giant duckweed)		X	
Leguminosae	*Aeschynomene virginica* (L.) B. S. P. (Northern jointvetch)			X
	Sesbania exaltata (Raf.) Cory (Hemp sesbania)			X
	Ammannia coccinea Rottb. (Purple ammania)			X
Lythraceae	*Rotala indica* (Willd.) Koehne	X	X	X
	Rotala rotundifolia Koehne	X		
Marsiliaceae	*Marsilea quadrifolia* L. (Pepperwort)	X	X	

464

Family	Species			
Onagraceae	*Ludwigia prostrata* Roxb.	X		
Polygonaceae	*Polygonum lapathifolium* L. (Pale smartweed)	X		
Pontederiaceae	*Heteranthera limosa* (Sw.) Willd. (Duck salad)			X
	Heteranthera reniformis R. & P. (Mud plantain)			X
	Monochoria vaginalis (Burm f.) Presl. (Monochoria)	X	X	
Potamogetonaceae	*Potamogeton distinctus* A. Bennett (Largeleaf pondweed)		X	
	Potamogeton nodosus Poir. (American pondweed)			X
Scrophulariaceae	*Bacopa rotundifolia* Wettst. (Round-leaf water hyssop)			X
	Dopatrium junceum (Roxb.) Hamilt. (Dopatrium)	X	X	
	Lindernia cordifolia (Colsm.) Merr.	X		
	Lindernia pyxidaria All.	X	X	
Typhaceae	*Typha* spp. (Cattail)	X		X

In temperate East Asia, perennial weeds have developed because of continuous use of the same or similar herbicides. Ryang et al. (1976) reported that 22% of the Republic of Korea's total rice-growing area was infested with perennial weeds. Perennial weeds such as *Eleocharis kuroguwai* Ohwi cause serious problems in rice fields in Japan. Some examples of difficult-to-control perennial weeds in tropical lowland and upland rice are given in Table 12.3. The aquatic or semiaquatic algae and ferns such as *Salvinia molesta* Mitchell and *Marsilea minuta* L. also cause serious problems in some rice fields.

Table 12.3 Some Difficult Perennial Weeds in Lowland and Upland Rice Fields in Asia and Their Mode of Propagation; Common Names are Given in Parentheses (Adapted from De Datta 1974)

Difficult Weeds	Mode of Propagation
Lowland rice	
Broadleaf weeds	
Marsilea minuta L. (Fern) (Taiwan, Korea)	Rhizome and spore
Marsilea crenata Presl. (Fern) (Taiwan, Philippines, Indonesia, Malaysia)	Rhizome and spore
Sagittaria aginashi Makino (Korea)	Tuber
Potamogeton distinctus A. Bennett (Largeleaf pondweed) (Korea)	Tuber
Sagittaria pygmaea Miq. (Arrowhead) (Korea, Japan, Taiwan)	Tuber
Sedges	
Eleocharis acicularis (L.) R. & S. (Slender spikerush) (Taiwan, Korea, Japan)	Rhizome and seed
Eleocharis kuroguwai Ohwi (Water chestnut) (Korea)	Tuber
Scirpus maritimus L. and other species (Bulrush) (Philippines, Thailand, Vietnam, India)	Tuber
Cyperus serotinus Rottb. (Korea, Japan)	Tuber
Cyperus imbricatus Retz. (Philippines, prevalent on swampy areas grown to rice—Bicol, Bulacan)	Seed
Grass	
Paspalum distichum L. (Knot grass) (Philippines)	Rhizome and seed
Upland rice	
Broadleaf	
Mimosa invisa Mart. (Giant sensitive plant)	Seed
Sedge	
Cyperus rotundus L. (Purple nutsedge)	Tuber
Grass	
Imperata cylindrica (L.) Beauv. (Cogon grass)	Rhizome

CROP-WEED COMPETITION

Crops and weeds compete for the same resources—nutrients, water, space, light. Competition begins when crops and weeds grow in close proximity and the supply of any necessary growth factor falls below the demands of both. The overall effect of crop-weed competition is a reduction in the biomass of rice and a reduction in grain yield.

Crop competition from weeds is dependent on such factors as the environment, the crop grown and its density, the stage of the crop, the weeds and their density, the stage of the growth of weeds, and measures taken by the farmer to control weeds. Likewise, the environment associated with the crop determines the weed community growing in association with the crop.

Competition is greatest between plants that are similar in growth habits, such as root growth and foliage characteristics, because they make nearly the same demands upon the environment. Those species that best utilize environmental growth factors will dominate when population levels are similar. Crops and weeds may also influence each other by metabolic products, which are secreted or remain in the soil as residues of decaying plant material.

Components of Competition

Competition for light occurs whenever plants are grown close together and one plant shades another to the point that the intensity and quality of the light received by the shaded plant are less than necessary for optimum growth. For example, a rice crop suffers little competition for light from *Monochoria vaginalis* (Burm. f.) Presl. because of the latter's short stature.

Plants vary greatly in their requirement for water. Competition for water and nutrients is usually more important because it begins before competition for light. Competition for water is greater when roots of crop and weeds are closely intermingled and obtaining their water from the same volume of soil. This means weed competition for water is more serious for upland and rainfed rice than for irrigated rice because moisture is often limiting.

In competing for nutrients, weeds will absorb as much or more than the crop plant. Generally, the factors that give a plant a competitive advantage in water uptake also give it a competitive advantage in nutrient uptake.

Competition between weeds and crop for space and carbon dioxide is generally not critical.

Allelopathy

Reduction in growth and yield of crops caused by weeds may be due to a factor other than competition. An allelopathic effect may also occur. Allelopathy is the deleterious effect of one plant on another through the production of toxic chemical compounds that enter the environment. Allelopathic effects of weeds on crop plants, and vice versa, are known to occur.

Management Practices and Weed Competition

In tropical agriculture, weeds grow vigorously and cause major crop production problems. In rice, the crop-weed competition varies with the type of culture (upland, lowland, or deepwater), methods of planting (transplanting or direct seeding), variety (tall or semidwarf, low- or high-tillering), and cultural practices (land preparation, spacing, fertilization, seed purity, and so on). Such variation in competition presents an opportunity to develop combinations of practices to minimize weed problems in rice.

The widespread replacement of traditional tall tropical varieties with modern varieties may have increased weed problems throughout tropical Asia. The traditional varieties have droopy leaves; the semidwarfs are shorter in stature and have erect leaves. Therefore, more light penetrates the crop canopy, and more weeds emerge and survive. Furthermore, the high fertilizer rates used on modern rice varieties aggravate weed problems.

Competition in Flooded Rice

Flooded or moist soils favor an abundant supply of viable weed seed in rice fields. Vega and Sierra (1968) reported more than 800 million viable seeds per hectare within a soil depth of about 15 cm. Such weed seeds germinate and severely compete with the rice crop.

Rice yield losses caused by weeds in flooded rice fields vary with the time of weed infestation, soil fertility, rice varietal type, and planting method (De Datta et al. 1969). The most serious competition is from grass alone or from a mixed weed population (grasses, broadleaf weeds, and sedges) (Table 12.4).

Competition from weeds during early growth stages of the rice crop is more serious than no competition during early growth stages followed by competition during later growth stages (Table 12.5). Therefore, early weed control is more important to the achievement of high rice yields (De Datta et al. 1969). The duration of weed control, however, affects rice varieties in different ways. For example, 20 days of weed-free growth appears best in short-statured plant types such as IR8 (Table 12.6). For C4-63, an intermediate-statured variety, the weed-free period should be extended to the first 30 days after transplanting (Vega et al. 1967). In a rice crop, weeds have a significant effect on crop height, number of panicles, straw weight, percent fertility, and grain yield. In Taiwan, yield reductions were 85% for *Echinochloa crus-galli* (L.) Beauv., 72% for *Cyperus difformis* L., 62% for *Marsilea quadrifolia* L. and *M. vaginalis,* and 9% for *Spirodela polyrhiza* (L.) Schleid. Weed infestations of 100–200 weeds per square meter reduced grain yields 51–64% compared with the weed-free conditions (Chang 1970). On the same soil, *E. crus-galli* and *C. difformis* reduced rice yields more with high soil fertility than with low soil fertility; *M. vaginalis* and *C. difformis* caused substantial damage to the second crop. In the first crop, weeds emerging at 15, 30, 45, and 60 days after rice was transplanted reduced the grain yields by 69, 47, 28, and 11% respectively. In the second crop, weeds emerging 10

Table 12.4 Effects of Weed Competition and Weed Densities on the Yield of Lowland IR8 for Two Planting Methods; IRRI, 1968 Wet Season (Adapted from De Datta et al. 1969)

Weed Population	Weed Weight[a] (g/m²)		Grain Yield (t/ha)		Yield Reduction from Weeds (%)	
	Broadcast	Transplant	Broadcast	Transplant	Broadcast	Transplant
Grasses only	325	285	0.7	1.2	86	75
Sedges + broadleaf weeds	250	110	3.8	4.8	24	0
Grasses + sedges + broadleaf weeds	540	330	0	1.4	100	67
Weed free	0	0	5.0	4.8	—	—

[a]Determined at the heading stage of grassy weeds.

Table 12.5   Effect of Weed Competition and Weed Densities on the Yield of
Transplanted IR8 (Adapted from De Datta et al. 1969)

Competing Weeds	Crop Growth Stage	Weed Weight (g/m^2)	Yield (t/ha)
Grasses	Early	574	3.2
	Late	275	6.8
Grasses and	Early	479	3.6
other weeds	Late	211	6.5

Table 12.6   Effect of Duration of Weed Control on
the Yield of Two Lowland Rice Varieties (Adapted from
Vega et al. 1967)

Duration of Weed Control (days after transplanting)	Grain Yield (t/ha)	
	IR8	C4-63
0	0.6	0.4
10	0.7	1.9
20	2.2	2.8
30	2.2	3.6
40	2.2	3.2
50	2.5	3.8
60	2.4	3.4

to 20 days after transplanting reduced the yields by 53 and 13%, respectively, but
weeds that emerged after 20 days did not affect yields.

In Japan, where rice yields have steadily increased over the years, a general rule
to maximize rice yields is that a field should be kept weed-free until a
transplanted rice crop reaches the 4–6 leaf stage. Table 12.7 shows the effect of
density of *E. crus-galli* on rice yields in the United States.

DIRECT-SEEDED VERSUS TRANSPLANTED RICE   Weed control is more critical—
and more difficult—in rice grown from pregerminated seeds broadcast directly
into the field (a practice followed in Sri Lanka, northeastern India, and parts of
Bangladesh and the Philippines) than in transplanted rice. Hand weeders moving
through broadcast rice destroy some rice plants. Also, they cannot distinguish
between young grassy weeds and young rice.

Studies at India's Central Rice Research Institute indicate that, in the absence
of weed control, there was a 46% yield loss due to weeds if rice was direct-seeded
on dry soil, but only a 20% loss if direct-seeded on puddled soil. There was only an

Table 12.7   Effect of *E. crus-galli* Density on Yield of Bluebonnet 50 Rice, Stuttgart, Ark., 1962–1963[a] (Adapted from Smith et al. 1977)

Plants/m^2		Rice Yield (t/ha)	Yield per Hectare Lost to Grass Competition		
Rice	Grass		(t)	(%)	Value[b] ($)
33	0	4.90	—	—	—
33	11	2.10	2.84	57	278
33	55	0.98	4.01	80	393
33	275	0.26	4.74	95	465
110	0	5.68	—	—	—
110	11	3.42	2.30	40	225
110	55	1.94	3.82	66	374
110	275	0.64	5.14	89	504
341	0	6.14	—	—	—
341	11	4.61	1.56	25	153
341	55	3.12	3.08	49	302
341	275	1.30	4.94	79	484
LSD (5%)[c]		0.82	—	—	—

[a]*E. crus-galli* competed with rice all season.
[b]Rough rice valued at $5 per 50.8 kg.
[c]LSD = least significant difference at level given.

11% loss due to weeds when rice was transplanted in puddled fields. IRRI experiments gave similar results (De Datta et al. 1969).

ANNUAL VERSUS PERENNIAL WEEDS   The relative competitiveness of annual and perennial weeds largely depends on the weed species and the growing conditions. But in areas where perennial weeds such as *Scirpus maritimus* L. predominate, annual weeds are not as competitive as perennials. In a 1973 IRRI experiment, *S. maritimus* reduced yields more than annual weeds but the reduction was most severe (60%) when both annuals and perennials were present (Table 12.8).

In Japan, Yamagishi et al. (1976) reported the perennial weed *Cyperus serotinus* Rottb. as highly competitive in lowland rice. The faster the appearance of *C. serotinus,* and the greater its number, the greater the decrease in rice yields. Two periods of rice growth were found vulnerable to *C. serotinus* competition. The first period was about 40 days after transplanting. At that stage, the number of panicles was decided. The second vulnerable stage was booting, the stage when the number of spikelets was decided. In the mixed population of rice and *C. serotinus,* the reduction in growth and yield of rice was due to decrease in its leaf area index (LAI) and nitrogen content.

Table 12.8   Comparative Reduction of Grain Yield by Annual and Perennial Weeds; IRRI, 1973 Dry Season (Adapted from De Datta 1974)

| Weed Community | Weed Weight (g/m^2) | | Grain Yield[a] | |
	Annuals	Perennials	(t/ha)	Reduction (%)
Annual weeds only	277	0	4.8	23
Perennial (*Scirpus maritimus*) only	0	414	3.2	48
Annual weeds + *S. maritimus*	540		2.5	60
Weed free (hand weeding twice)	7	2	6.2	0

[a]Data are averages for IR20 and IR442-2-58.

Competition in Upland Rice

Weeds have always been a major problem of upland rice farmers. Using traditional upland varieties, Vega (1954) and Vega et al. (1967) reported as much as 83% grain yield reductions caused by weeds. The traditional rice variety, Palawan, had grain yields as low as 0.05 t/ha with poor weed control versus 3.0 t/ha with good control in these experiments.

In a 3-year study in India with the upland variety Ratna, Sharma et al. (1977) reported progressive decrease in rice yield when weed competition was maintained from 10 days after seeding until crop maturity.

Okafor and De Datta (1974) compared four weed communities—weed-free, annual weeds, *Cyperus rotundus* L., and *C. rotundus* plus annual weeds—in upland rice culture broadcast and drilled. Unchecked competition between IR5 rice and different weed communities significantly reduced grain yields (Table 12.9). In the broadcast-seeded crop, rice yield was reduced 82% when all weeds were allowed to grow. Annual weeds reduced grain yield by 67%. *Cyperus rotundus* (with a few *C. iria* L.) reduced grain yields by 51%. Grain yield reduction was due to reduced tiller and panicle numbers, reduced LAI, and reduced light transmission.

In a 1972–1973 IRRI study, nitrogen application to upland rice fields benefited *C. rotundus* more than the rice (Table 12.10). *Cyperus rotundus'* competition for moisture increased with increased nitrogen (Okafor and De Datta 1976a).

PRINCIPLES OF WEED CONTROL

The primary objective of all weed-control practices is to reduce the undesirable effect of weeds to a minimum. The principle of weed control is to make conditions as unsuitable for weed growth as possible. Often, weed control is

Table 12.9 Effect of Weed Densities on the Grain Yield of Upland Rice (IR5) under Two Methods of Seeding (Adapted from Okafor and De Datta 1974)

	Drilled			Broadcast		
	Weed Population[a] (per m²)	Grain Yield		Weed Population[a] (per m²)	Grain Yield	
Weed Community		(t/ha)	Reduction (%)		(t/ha)	Reduction (%)
Annual weeds	301	1.2	74	412	1.6	67
Cyperus rotundus	439	2.7	42	435	2.4	51
Annual weeds + C. rotundus	642	0.8	83	748	0.9	82
Weed free	79	4.7	—	34	4.9	—
LSD (5%)[b]	67	0.4		67	0.4	

[a]Taken 30 days after crop emergence.
[b]LSD = least significant difference at level given.

Table 12.10   Effect of *Cyperus rotundus* L. Competition on the Grain Yield of Upland Rice Grown at Different Levels of Nitrogen Fertilization (Average of Three Seasons); IRRI, 1972–1973 (Adapted from Okafor and De Datta 1976a)

Nitrogen Level (kg/ha)	C. rotundus		Grain Yield	
	Population per m^2	Dry Weight (g/m^2)	(t/ha)	Reduction (%)
0	0	0.4	1.4	
				29
0	717	110.4	1.0	
60	0	0.5	4.0	
				35
60	717	217.1	2.6	
120	0	0.5	3.9	
				30
120	717	159.0	2.4	

based on a single method rather than on the objective of manipulating the weeds' environment. Manipulation of the weeds' environment can be accomplished physically (tillage, fire, flooding, smothering), culturally (crop rotation), biologically, or chemically.

An example of the principle involved in physical weed control would be that young weed leaves depend on roots for carbohydrate supply. Therefore, cultivation should be delayed until the carbohydrate flow from the roots ceases and the leaves begin to manufacture enough food to start replenishing the root reserve. Thus, with most perennial weeds, tillage 10–14 days after emergence is more effective for carbohydrate starvation than tillage immediately after germination (Appleby 1967).

Herbicidal manipulation of the weeds' environment is the result of a chemical that disrupts a vital function of one plant and not of another. Such selectivity may depend on many mechanical, physical, chemical, or metabolic aspects. For example:

· In directed placement of a spray, the weed plants are sprayed (contacted) but the crop plants are not.
· Many weeds are controlled by the herbicide propanil (3', 4'-dichloropropionanilide) because the rice detoxifies propanil whereas the weeds have limited capacity to break down propanil.
· Selectivity exists between crops and weeds and between various weed species themselves. This can cause important ecological shifts of weed species in a mixed population of weeds.

Many important principles involving the use of herbicides are concerned with interaction between the herbicide and soil. Adsorption of herbicides by soil is a process that needs to be understood for more efficient use of herbicides. Among the factors that influence soil adsorption of herbicides, the amount of organic matter, soil texture, and type of clay are important.

Other principles involved in successful weed control are primarily agronomic in nature. The choice of method will depend on the technology available, types of rice culture, and farmers' resources and preferences. The cost of the weed-control measures must be compared with the resulting value of the increase in yield. But the weed-control method that the farmer selects should be one that subjects the weed population to a variety of pressures so that no particular weed or weed community will become established and cause continuous damage (Moody and De Datta 1977).

Among the weed-control practices that are critical for success, timeliness of weeding is dominant. Because weeds provide the greatest competition and cause the most damage at early crop growth, early control is important.

METHODS OF WEED CONTROL

It is entirely possible that weeds were a problem and that some primitive form of weed control was practiced when man first began to cultivate crops about 8000 B.C. Since then, man has been engaged in a continuous struggle against weeds. Several new weed-control measures have been developed and adopted, and old ones reexamined and modified, but none of the original practices has been completely abandoned.

There are no weed-control measures that will give continuous and effective weed control when used in isolation. In fact, there is no weed-control method that is best for use in all situations because weeds vary in their growth habits and life cycles. The final choice of any weed-control measure by the farmer will depend largely on effectiveness and economics. The cost of the weed control must be compared with the resulting increased value of the crop.

Although no single weed-control measure will give continuous and effective weed control when used in isolation, each control measure is covered separately in this discussion. The discussion of individual factors is followed by examples of multiple factors (integrated management practices) that affect weed control and management for tropical rice. Factors for control and management of weeds are grouped as substitutive, preventive, complementary, and direct methods according to type of rice culture.

Substitutive and Preventive Factors in Weed Control

Land preparation and water management are field operations that can substitute for another operation for a certain degree of weed control.

- Land preparation varies depending on the type of rice culture (lowland or upland) and the method of planting (transplanting or direct seeding).

    Minimized weed growth is an obvious benefit of land preparation, particularly puddling (see Chapter 8).

- Good water management is an important factor in weed control in rice. Emergence of weeds and types of weeds are closely related to the moisture content of the soil and the depth of irrigation. Many weeds will not germinate under flooding. But roughly 80% of the rice grown in South and Southeast Asia is subjected to uncontrolled water supply, which exposes the rice crop to various degrees of weed pressure (see Chapter 9).

Preventive measures to check the introduction and spread of weed seed are perhaps no less important than any other weed-control measure. It is probably easiest and most economical to practice preventive measures to substantially minimize weed seed introduction.

- Planting weed-infested seed is a common way of introducing weeds into a clean field. The problem in direct-seeded rice is more serious than in transplanted rice because it is difficult to separate rice and weed seeds. Red rice, a serious weed in direct-seeded rice fields in the United States is largely introduced through seeding (Smith et al. 1977).
- Other preventive measures to minimize the introduction of weed seed include weed-free seedbeds, clean levees and irrigation canals, clean tools and machinery, livestock kept out of fallow fields, and prevention of weed seed formation or vegetative propagation of weed tubers.

Complementary Practices in Weed Control

Perhaps the most useful factors of weed control are those that provide favorable stand establishment and growth for the crop, and that are simultaneously unfavorable to the weeds, particularly if such practices are economically attractive to farmers. Manipulation of time and method of planting, variety, planting density, fertilizer management, and crop rotation should aim for maximum benefit to the crop and minimum benefit to the weeds.

Planting Method

A farmer chooses a particular method of planting rice for various reasons, but crop competition with weeds and use of weeding methods are important considerations for a specific choice. Straight-row planting, either by transplanting or by direct seeding, makes weeding by hand pulling or by mechanical tools easier.

    Thorough puddling before transplanting incorporates weeds, thus giving the rice seedlings a head start over weeds that germinate later. If land preparation and water management are adequate, this competitive advantage is retained throughout most of the transplanted crop's growth. Furthermore, transplanting encourages effective use of preemergence herbicides and fewer herbicide applications than needed to control weeds in direct-seeded rice. The transplanting operation, as the transplanters move backward during planting, helps incorporate weeds and weed seeds into the soil.

Seedling age is easily manipulated in transplanted rice. In tropical Asia, many farmers use tall, old seedlings of traditional tall varieties in the wet season because they give better weed competition than young seedlings. With the modern semidwarf varieties, manipulation of seedling age for weed control is not important if preemergence herbicides are used, but if weeds are not controlled young seedlings are less desirable than somewhat older (21–30 days) seedlings.

In Japan, for example, the introduction of shorter (14 cm tall) seedlings for mechanical transplanting caused as much as 89% yield losses by weeds versus 59% in rice of normal seedlings (23.4 cm tall) (Matsunaka 1976). These results are no longer relevant in Japan because intensive weed control is practiced. However, the results explain why many farmers in tropical Asia, where land preparation and water management in lowland rice are generally poor, still prefer to use older (30–40 days) and taller seedlings than the 21-day-old seedlings normally recommended by rice researchers.

More weeds grow in direct-seeded than in transplanted rice, with direct seeding into water or onto mud resulting in less weed competition than direct seeding into dry soil.

Variety

Because rice farmers in tropical Asia had depended on tall, vigorously tillering plants to provide weed competition, the introduction of modern semidwarf rice varieties increased weed problems and the need for weed control became more obvious. The modern rices allow more light to penetrate the crop canopy thus allowing more weed seed to germinate and weeds to grow better. Furthermore, the high fertilizer rates used with modern rice varieties aggravate weed problems. Nevertheless, the IR8 plant type, with its high tillering capacity, is fairly competitive with weeds despite its short stature. Low-tillering varieties, with the same yield potential as IR8 when grown under ideal conditions, have not fared well under farm conditions.

The advantage of relatively taller plants (120–130 cm) is less in irrigated rice than in rainfed rice because even though weed control is easier, taller plants lodge earlier and yield less than semidwarfs. On rainfed rice farms, however, heavy-tillering varieties of medium stature may be better suited than semidwarfs if management practices such as land preparation, water, and fertilizer applications are less than optimal. Even at experiment stations, a recent study indicated that an intermediate-statured rice (IR442-2-58) competed better than a semidwarf variety (IR20) with both annual and perennial weeds. Detailed results on IRRI studies on varietal difference in weed competition are given elsewhere (De Datta 1977b).

Bangladeshi farmers generally do more intensive hand weeding with modern than with traditional varieties, whether transplanted or direct-seeded (Hoque et al. 1976a,b). This can be interpreted as meaning that weed problems, or expectations on returns from weeding, are less with traditional than with modern varieties.

It is also believed that it is critical to weed early-maturing rices earlier than late-maturing rices. That is because barnyard grass-type weeds grow fast and damage early-maturing rice more than late-maturing rice. If preemergence herbicide is used, however, the growth duration of rices may not be critical in irrigated rice but may still pose problems in rainfed rice.

Plant Spacing and Plant Density

Close spacing is essential to minimize weed infestation and to obtain high yields. That is because the closer the rice plants are sown, the greater the competition against weeds that grow in association with rice. In a study by Estorninos and Moody (1976), more weeds grew at the 25×25 cm than at 15×15 cm spacing (Table 12.11). The grain yield decrease due to weeds averaged 18% at 15×15 cm, 30% at 20×20 cm, and 52% at 25×25 cm, compared with weed-free plots. However, the increased cost of close spacing should be compared with that of somewhat wider spacing combined with other weed-control practices.

In wet-seeded rice, Moody (1977a) reported less weed competition as seeding rates increased in the range of a 80–200 kg/ha. However, grain yield advantage was not recorded for different seed rates. Again, the increased cost of increased seeding rates should be compared with that of lower rates combined with other weed-control practices, particularly if the increased seeding rate does not increase grain yields.

Table 12.11   Effect of Plant Spacing on Weight of Weeds Growing in Association with Transplanted IR28 and IR30 at Harvest (Adapted from Estorninos and Moody 1976)

Plant Spacing (cm)	Weed Weight (kg/ha)		
	IR28	IR30	Average
15 × 15	1984	2458	2221
20 × 20	3296	3076	3186
25 × 25	3470	3818	3644

Fertilizer Application

It is critical to apply fertilizer when it will benefit the crop most and the weeds least. In a 1974 study in two farmers' fields in Laguna province, Philippines, applied nitrogen increased the grain yield by a maximum of 0.5 t/ha where weeds were not controlled. However, without nitrogen fertilizer, weed control

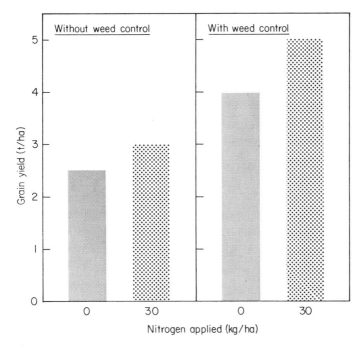

Figure 12.1   Grain yield response to low rates of nitrogen in IR26 rice in farmers' fields with and without weed control (av. of two experiments in 1974 wet season). (From De Datta and Barker 1977)

alone increased grain yield by about 1.5 t/ha (Fig. 12.1). Where weeds were controlled, the grain yield with 30 kg N/ha was 1.0 t/ha more than without fertilizer nitrogen, emphasizing the interaction between fertilizer response and good agronomic practices (De Datta and Barker 1977). Weeds will absorb as much, or more, nutrients than the crop plant. For example, Moody (1977a) reported that the crop plus the weeds on an untreated plot absorbed the same amount of nitrogen as the crop alone on a weed-free plot (Table 12.12).

Table 12.12   Nitrogen Uptake of Weeds and Rice Growing in Association (Adapted from Moody 1977a)

Plant Species	Nitrogen Uptake (kg/ha)	
	Weeds Present	Weeds Few or Absent
Rice	26	106
Echinochloa sp.	75	0.6
Total	101	107

In an earlier study (De Datta et al. 1969) it was found that increased application of nitrogen benefited the grass population more than the broadleaf weeds and sedges.

The application of nitrogen as a basal treatment in dry-seeded rice is unwise unless preemergence herbicides are used. Topdressing the dry-seeded crop with nitrogen after weeding is desirable to maximize nitrogen fertilizer efficiency and to minimize weed growth.

Cropping Systems

The possibility of a buildup of a certain weed species or group of species is greater if the same crop is grown year after year. Crop rotation, therefore, minimizes the undisturbed development of weeds (De Datta and Jereza 1976). In many instances, crop rotation can eliminate or at least reduce difficult weed problems. Recent results show that various crop and soil management practices cause the weed community to shift from weed species that are difficult to control to weeds that are easier to control (Table 12.13). Rotation of the herbicides used in

Table 12.13 Population of Annual and Perennial Weeds and Yield of Rice and Other Crops in Untreated Check under Three Cropping Systems; IRRI, 1976–1977 (Adapted from De Datta and Jereza 1976)

| | | | Weed Population (no./m^2) | | |
| | | | | Perennials | |
Season	Crop	Annuals	*Scirpus maritimus*	*Cyperus rotundus*	Grain Yield[a]
Cropping System I					
1976 dry	Lowland rice	990	205	0	0.2
1976 wet	Lowland rice	1168	300	0	0
1977 dry	Lowland rice	277	191	0	0
1977 wet	Lowland rice	88	421	0	0
Cropping System II					
1976 dry	Maize + mung bean	1571	60	30	0
1976 wet	Lowland rice	639	191	0	1.3
1977 dry	Maize + soybean	620	59	8	24,500 + 213
1977 wet	Lowland rice	176	209	0	1.0
Cropping System III					
1976 dry	Maize	2825	15	75	0
1976 wet	Dry-seeded rainfed rice	1422	78	115	0
1977 dry	Maize	521	26	38	27,500
1977 wet	Dry-seeded rainfed rice	3740	35	9	0

[a]For rice, t/ha; for maize, marketable ears/ha; for mung bean and soybean, kg/ha.

continuous cropping of rice is also essential. This puts pressure on various weed species so that no single species or group of species develops undisturbed.

In the future, more effective treatments need to be developed, for different weed compositions and relevant cropping system, with a view to developing several equally effective alternatives that make most efficient use of the farmers' available resources. The weed-control technique finally chosen will then be determined by the individual crop situation and the farmer's most limiting resources, because it is unlikely that any one technique could be expected to suit the resource base of all farmers.

Direct Methods of Weed Control

The substitutive, preventive, and complementary weed-control practices generally minimize weed populations to a manageable level but do not eliminate them. Farmers must still resort to the most direct methods of weed control, including hand weeding, mechanical weeding, and herbicides.

Often, the most important factor is when, not how, weeds are controlled. The direct methods of weed control used in Asia vary greatly and depend on the type of rice culture. Farmers' resources such as land, labor, and capital are important considerations in making the final choice of weeding method (De Datta and Barker 1977).

Hand Weeding

Hand weeding is common in almost all tropical rice-growing areas. But the degree of hand weeding and the problems associated with it largely depend on the type of rice culture.

Several hand tools—the hoe, narrow spade, Swiss hoe, knife, machete, and even pointed sticks—are used with various degrees of success to control weeds in dry-seeded rice in many countries. In West Malaysia, repeated cutting with a *tajak*, a kind of scythe, followed by flooding and soil puddling, is used to control weeds in transplanted rice. In Bangladesh, weeds in rice fields are controlled by a hand tool locally known as *niranee*. Several other tools are also used to control weeds. In Burma, implements called *dawset* and *gwinset* are drawn by animals to chop down weeds and incorporate them with the soil.

TRANSPLANTED RICE   Hand weeding is the most common weed control method in transplanted rice. One hand weeding, or two at the most, should be sufficient to control weeds adequately in transplanted rice. Studies in 1967 showed that delay of a single hand weeding beyond 42 days after transplanting significantly reduced the yield of IR8 and increased the labor needed for weeding (Table 12.14). The differences in grain yields between one hand weeding 21–42 days after transplanting and two hand weedings were not significant. This sugggests that one properly timed hand weeding may be adequate to reduce the weed populations enough to obtain high yields with modern varieties. However, early

Table 12.14   Effects of Time and Frequency of Hand Weeding on Weeding Time and Yield of Transplanted IR8 Rice; IRRI, 1967 Dry Season (From De Datta 1979)

Hand Weeding			
Number	Timing (DT[a])	Time Required to Weed (hours/ha)	Grain Yield (t/ha)
1	21	164	7.4
1	28	227	7.1
1	35	201	8.0
1	42	418	8.3
1	49	410	6.7
1	56	349	6.4
1	63	546	5.0
2	21 and 42	280	8.3
2	21 and 56	388	8.1
2	21 and 63	399	7.8
2	42 and 63	470	8.6
No weeding		—	2.4
LSD (5%)[b]			1.7

[a]DT = days after transplanting.
[b]LSD = least significant difference at level given.

hand weeding, such as at 21 days after transplanting, has the advantage of requiring less labor (Table 12.14) because the weed population at that stage is lower (De Datta 1979).

DIRECT-SEEDED FLOODED RICE   For direct-seeded flooded rice, two hand weedings are sufficient to remove most weeds. The time spent in weeding is considerably less and weed control can be more timely if rice is drilled rather than broadcast. When hand weeding broadcast rice the weeders destroy some rice plants (De Datta and Bernasor 1973).

DRY-SEEDED RICE   Dry seeding is done in either bunded fields or in nonbunded fields (upland rice). For both, rain provides all of the moisture for crop growth. Weeds grow profusely in dry-seeded rice and without weed control there is a total yield loss (De Datta 1978).

   In a study with a traditional upland variety, Palawan, the maximum yield for a single hand weeding was when weeding was 25 days after seeding. The yield (2.6 t/ha) was, however, more than a ton below that of rice weeded three times (3.7 t/ha). A delay in hand weeding beyond 15–25 days sharply reduced yields. Based on the differences between weeding at 25 and 45 days, each day's delay in hand weeding reduced yield by 43 kg/ha per day and sharply increased labor

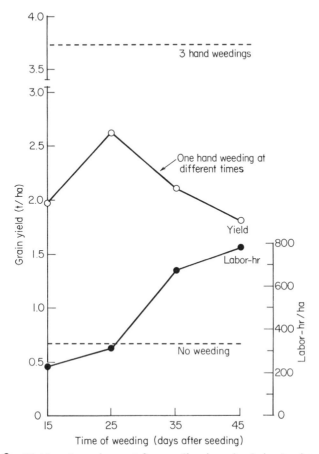

Figure 12.2   Yield and requirement for weeding in upland rice (variety: Palawan), 1965 wet season. (From De Datta 1979)

requirements. A single hand weeding made at 15 days after seeding required only 240 hours of labor/ha but when the hand weeding was delayed until 45 days after seeding, labor requirements were 780 hours/ha (Fig. 12.2). The yield from a nonweeded control was only 0.67 t/ha. In plots weeded when weeds first appeared to cause competition, yields were 3.7 t/ha and required only 530 hours of labor/ha (De Datta 1979).

Results from these and other experiments (Bhan 1978) demonstrate that hand weeding can be effective in controlling weeds in dry-seeded rice but it is extremely labor consuming.

Mechanical Weeding

Weeding by machine is feasible only where rice is planted in straight rows. Rotary weeders are pushed by hand or powered by a small gasoline engine.

Figure 12.3 The push-type rotary weeder mixes weeds with mud between rows. Timely use of a rotary weeder often provides weed control as good as hand weeding.

TRANSPLANTED RICE   In lowland rice, particularly transplanted rice, the push-type rotary weeder is by far the most successful tool (Fig. 12.3). In a 1966 study, although the labor for two rotary weedings (115 hours/ha) was half that for two hand weedings (202 hours/ha), the grain yield was lower (Table 12.15). The slightly lower yield with two rotary weedings was perhaps due to the inability of

Table 12.15   Efficacy of Weed Control Methods in Transplanted IR8 Rice; IRRI, 1966 Wet Season (From De Datta 1979)

Weed Control Method[a]	Time Required to Weed (hours/ha)	Grain Yield (t/ha)	Rice per Labor-hour (kg/ha)
Hand weeding	202	4.9	17
Rotary weeding	115	4.1	22
Rotary weeding fb hand weeding	153	4.6	20
MCPA spray fb hand weeding	212	4.3	13
MCPA spray fb rotary weeding	65	4.3	42
Unweeded control	—	1.6	—

[a]Weeding in all cases was 20 and 40 days (two weedings) after transplanting; fb = followed by.

rotary weeding to remove weeds within or close to rice hills. As a result, when the data were computed for rice per hour of weeding, they were nearly identical for two hand weedings, two rotary weedings, and one rotary weeding followed by one hand weeding (Table 12.15).

DIRECT-SEEDED FLOODED RICE   It is possible to use push-type rotary weeders in direct-seeded flooded rice if the seeding is onto mud in straight rows.

DRY-SEEDED RAINFED RICE   If rice is seeded in rows in dry soil, a number of hand tools or a tractor-driven rototiller can be used to smother and incorporate weeds. In the Philippines, upland rice farmers use an animal-drawn implement called a *kalmot*. It also smothers some rice plants, but the local rice varieties recover because of their deep root system. The *kalmot* is not an ideal method of weed control but considering the farmers' resources it effectively minimizes weeds.

Chemical Weed Control

Attention to study of weeds and their control increased following the discovery of the herbicidal property of 2,4-D (2,4-dichlorophenoxy acetic acid) in the United States in the early 1940s. In some instances, herbicides offer the most practical, effective, and economical means of reducing weed competition, crop losses, and production losses.

Herbicides are chemicals used for killing or inhibiting the growth of certain plants. There are nearly 200 herbicides—often chemically and functionally diverse and highly selective—available for use on various crops throughout the world. For most weed problems, there are herbicides available. The choice or use of any particular herbicide depends primarily on economics. In some situations returns from a herbicide do not justify its use, in others, socioeconomic conditions do not allow it to be used extensively.

TYPES OF HERBICIDES   Herbicides differ in how they kill plants, in the types of plants that they kill, and the amounts needed to kill plants. Herbicides may be grouped on the basis of one or more common characteristics such as their chemistry (molecular composition and configuration), biological effect (mode of action, selectivity, fate in plants), time of application (preplant, preemergence, postemergence), or use (control of annuals or perennials, grasses, broadleaf weeds or sedges, herbaceous or woody plants).

Herbicides are applied by hand or sprayed by hand sprayers. In the Asian tropics, granular herbicides are generally applied by hand. If liquid herbicides are used, they are applied by hand sprayers. Several types of sprayers are available. To have a uniform and accurate herbicide application, the constant-pressure knapsack sprayer is considered the best (Vega and Paller 1970).

It is impossible to make a rigid classification of herbicides. Frequently, a herbicide can fit into more than one place in a classification depending on the purpose for which it is needed and the rate at which it is used. In this discussion, three classification systems are used.

1   TIME OF APPLICATION   The terms pre- and postemergence are used to denote time of application of herbicides to both the crop and the weeds. Many people prefer to use such terms solely with respect to the crop but to avoid confusion the stage of development of both the crop and the weeds should be stated.

•   PREPLANT   A preplant treatment is any treatment made before the crop is planted. Examples of herbicides that can be applied preplant are dalapon (2,2-dichloropropionic acid), glyphosate [N-(phosphonomethyl) glycine], and paraquat (1,1'-dimethyl-4,4'-bipyridylium ion). A preplant incorporated treatment is one in which the herbicide is incorporated into the soil—to prevent volatilization or to place the chemical in the zone in which it is needed—prior to the planting of the crop. Examples of preplant incorporated herbicides are dinitramine ($N'N'$-diethyl-2,6-dinitro-4-trifluoromethyl-m-phenylenediamine), EPTC (S-ethyl dipropylthiocarbamate), and trifluralin (2,6-dinitro-NN-dipropyl-4-trifluoromethylaniline).

•   PREEMERGENCE   A preemergence treatment is made after the crop has been planted but prior to the emergence of the crop or weeds, or both. The herbicide is usually applied to the soil surface. Examples of preemergence herbicides are atrazine (2-chloro-4-ethylamino 2',6'-isopropylamino-1,3,5-triazine), butachlor (N-butoxymethyl-α-chloro-2',6'-diethylacetanilide), and propachlor (α-chloro-N-isopropylacetanilide).

•   POSTEMERGENCE   A postemergence herbicide is applied after the crop or weeds, or both, have emerged from the soil. Examples of postemergence herbicides are 2,4-D and dinoseb (2-sec-butyl-4,6-dinitrophenol).

2   MODE OF ACTION   Mode of action includes the entire series of events that occur within the plant from its first contact with the herbicide to its ultimate effect, which could be death of the plant.

•   CONTACT HERBICIDES   The contact herbicides are applied to the foliage and kill plant tissues at or close to the site of application. Contact herbicides are ineffective against perennial weeds except when they are in the seedling stage. Young grasses, which have their growing points below the soil surface, will frequently be more tolerant than broadleaf plants, which have their growing points exposed. Examples of this type of herbicide are paraquat and dinoseb.

•   SYSTEMIC HERBICIDES   The systemic (or translocated) herbicides can be applied to the foliage or to the soil and are capable of movement within the plant to exert effects away from the site of application. Examples are 2,4-D and atrazine.

3   SELECTIVITY   A herbicide may be selective to a particular crop at recommended rates but if higher rates are used the herbicide may damage or even kill the crop.

•   SELECTIVE HERBICIDES   A herbicide that kills or stunts some plant species with little or no injury to others is selective. Examples of selective herbicides are atrazine for maize, propazine [2-chloro-4,6-di (isopropylamine)-1,3,5-

triazine] for sorghum, alachlor (α-2′,6′-diethyl-N-methoxymethylacetanilide) for peanut, and butachlor for rice.
* NONSELECTIVE HERBICIDES   A herbicide that kills all plant species is non-selective. Examples are paraquat and glyphosate.

HERBICIDE FORMULATIONS   Herbicides are frequently insoluble in water and are usually applied at such low rates that they must be combined with a liquid or solid carrier to make their applications easier, to increase their effectiveness, and to be distributed evenly during application. In addition, the formulation can have an effect on herbicide solubility, volatility, biological activity, and other characteristics.

* WATER- AND OIL-SOLUBLE HERBICIDES   Inorganic herbicides and the salts of some organic herbicides are soluble in water. The solution thus formed is a stable, physically homogeneous mixture. These herbicides are sold either as solids or aqueous concentrates depending upon their solubility rate or on the physical properties of the present compound. Examples are the sodium and amine salts of 2,4-D and sodium salt of dalapon.

  Some herbicides, dinoseb and pentachlorophenol for example, are soluble in oil. They are often dissolved in oil and used to increase the toxicity of oil sprays (referred to as fortified oils). Ester formulations of 2,4-D and related products are also soluble in oil.

* EMULSIONS   An emulsion is formed when one liquid is dispersed in another liquid, each maintaining its original identity. Some herbicides that are insoluble in water may be dissolved in oil or another organic solvent together with a suitable emulsifying agent to form an emulsifiable concentrate. When added to water they form emulsions usually of the oil-in-water type but occasionally of the water-in-oil type (invert emulsions). Oil-in-water emulsions have the same viscosity as water and can be applied with conventional spraying equipment, but special equipment may be needed to apply invert emulsions, which may be too viscous to be sprayed by conventional equipment. These formulations require continuous agitation to keep them mixed uniformly. Examples are trifluralin, chlorpropham (isopropyl 3-chlorophenylcarbamate), ester formulations of 2,4-D and related compounds.

* WETTABLE POWDERS   Some herbicides are nearly insoluble in water, oil, or organic solvents. Therefore, concentrated solutions or emulsions cannot be prepared. These are made into wettable powders, which consist of a finely divided solid herbicide, an inert diluent such as clay, and dispersing agents (deflocculating agents and wetting agents) that form suspensions when mixed with water. To prevent settling of the solid particles, the suspension should be agitated. Wettable powders are highly effective as soil-applied herbicides but are of much lower efficiency when applied to the foliage.

* GRANULES   Granular formulations vary in shape and size and are usually prepared by spraying a solution of the herbicide onto a preformed granular

carrier or by agglomeration of a powdered mixture of herbicide and carrier. Some granules dissolve slowly in water and others are insoluble. The herbicide in insoluble granules is released slowly to the soil or soil solution.

Granular herbicides, even though they are about twice as costly as nongranular ones, can be applied by hand in association with fertilizer and may be more acceptable to the farmer. Nongranular herbicides require water and a sprayer for application. Furthermore, the proportion of the active ingredient in granules is constant whereas other herbicides have to be mixed and applied carefully to obtain the same precision. Imperfections in application of granules are overcome by diffusion and redistribution of the herbicide in the water. Numerous herbicides which are formulated as liquid or solids are also formulated in granules. Examples are thiobencarb (S-4-chlorobenzyl diethylthiocarbamate), butachlor, and 2,4-D.

- WATER-DISPERSIBLE SLURRY   The water-dispersible slurry usually consists of a finely divided powder suspended in an oil base, which is capable of suspension in water. It requires less water (only 10–20 liters/ha) for application than conventional herbicides. Examples are atrazine and fluorodifen (4-nitrophenyl 2-nitro-4-trifluoromethylphenyl ether).

- SLOW-RELEASE COMPOUNDS   Methods have been developed to achieve slow or controlled release of pesticides by trapping small amounts of the pesticide in tiny capsules, which are then suspended in a liquid. The liquid can then be mixed with water and applied through a regular sprayer. Herbicides are also being formulated as polyvinyl impregnated pellets and clay or casting plaster pellets.

Advantages of controlled slow-release herbicides are reduced residue levels, increased herbicidal activity, reduced loss by evaporation or leaching, and reduced cost of weed control. These formulations have been used primarily for control of aquatic weeds and for weeds in ornamentals.

HERBICIDES FOR RICE   Several herbicides are now available to Asian rice growers at reasonable cost, and many more are being developed to provide effective and economical weed control under the different conditions of various cultural practices by which rice is grown in Asia. Some examples of herbicides for rice are:

- PHENOXY ACID HERBICIDES   Herbicides belonging to this group are classified as translocated herbicides. Examples are 2,4-D and MCPA (4-chloro-2-methylphenoxy acetic acid).

- PROPANIL   Propanil is a contact herbicide that controls annual grass weeds and a few broadleaf weeds and sedges. Fields must be drained to expose the leaves to the herbicide. Complete coverage by the spray is essential to obtain good contact and effective weed control.

- MOLINATE   With early postemergence, molinate (S-ethyl NN-hexamethyl-enethiocarbamate) controls primarily annual grasses but does not control broadleaf weeds and sedges.

- NITROFEN   Nitrofen (2,4-dichloro-4'-nitrodiphenyl ether) is a selective pre-emergence herbicide.
- BUTACHLOR   Butachlor controls most annual grasses, broadleaf weeds, and sedges and can be applied pre- as well as postemergence. In tropical Asian countries it is available in granular and in liquid formulation.

    In the United States, tank or formulated mixtures of butachlor—and propanil—control many grass weeds and aquatic broadleaf and sedge weeds. Because of its residual activity it is highly effective on problem soils.
- THIOBENCARB   Similar to butachlor, thiobencarb is highly effective against most annual grasses, broadleaf weeds, and sedges. Remarkable selectivity against *Echinochloa* spp. and other weeds and good residual effects make thiobencarb an excellent herbicide for flooded rice.
- BENTAZON   Applied postemergence to weeds, bentazon [3-isopropyl-(1*H*)-benzo-2,1,3-thiadiazin-4-one 2,2-dioxide] spray controls many succulent perennial sedges in flooded rice. It is a very safe chemical to use in rice.

Herbicide use is limited to those areas where labor is not plentiful and the wage rates are somewhat high. Herbicides are extensively used in the United States (Baker 1978a, Huey 1977, Smith and Seaman 1973), Australia, and Europe. They are also commonly used in Taiwan, Republic of Korea, and Japan (De Datta and Barker 1977).

    The choice of herbicide and the timing of its application vary with the type of rice culture.

CHEMICAL CONTROL OF ANNUAL WEEDS

Transplanted Rice.   The effectiveness of phenoxy acid herbicides, such as 2,4-D, has been demonstrated repeatedly. 2,4-D and MCPA have been used on transplanted rice to control broadleaf weeds and sedges 2–3 weeks after those weeds have emerged (Vega 1954, Vacchani et al. 1963, Moomaw et al. 1966, Mukhopadhyay 1978). The time of application to avoid toxicity losses has been found to vary from 15 days to 50 days after transplanting or sowing. It has been observed frequently, however, that early applications of phenoxy acetic acids, cause no damage (Smith et al. 1977) and that with suitable cultural practices, especially tillage and water management, 2,4-D or MCPA may provide as good weed control as the best combination of selective herbicides and hand weeding. In fact, it was observed in the Asian tropics that all commonly marketed formulations and derivatives of 2,4-D and MCPA can control grasses as well as broadleaf weeds and sedges when applied 4–5 days after transplanting (De Datta et al. 1968, De Datta et al. 1971). Some weed species that are susceptible at preemergence of weeds (4 days after transplanting) and postemergence of weeds (25–30 days after transplanting) are listed in Table 12.16.

    2,4-D is effective against most annual grasses (Fig. 12.4) and many annual broadleaf weeds and sedges and helps produce yields as good as those from use of selective herbicides, such as thiobencarb in combination with 2,4-D or hand

Table 12.16   Effect[a] of Various Herbicides on Weeds of Lowland Rice
(Adapted from Madrid et al. 1972)

Weed Species	2,4-D 4–6 DT[b]	2,4-D 25–30 DT	Butachlor 4–6 DT
Cyperus difformis	S	S	S
Fimbristylis littoralis	S	S	S
Echinochloa colona	S	T	S
Echinochloa glabrescens	S	T	S
Monochoria vaginalis	S	S	S
Sphenoclea zeylanica	S	FT	S

[a]S = susceptible; T = tolerant; FT = fairly tolerant.
[b]DT = days after transplanting.

Figure 12.4   2,4-D is effective against most annual grasses and many broadleaf weeds
and sedges and under good water control gives similar yields as selective herbicides.

weeding two times (De Datta 1979). Most 2, 4-D users, however, still consider it
as a postemergence herbicide for broadleaf weeds and sedges in rice fields.

Some selective herbicides, such as butachlor and thiobencarb, can be applied
as preemergence treatments (4–5 days after transplanting) as well as postemergence treatments (6–8 days after transplanting).

Propanil is also used in transplanted rice but strictly after weeds have emerged,
particularly in combination with other weed-control practices (Mukhopadhyay
1978). In the Asian tropics, propanil application alone, however, gives incon-

sistent weed control in rice where irregular rainfall often washes away the chemical or induces a *burn* of the rice crop (De Datta 1977c).

Direct-Seeded Flooded Rice.   Direct-seeded flooded rice culture will become an acceptable alternative to transplanted rice as the cost of labor rises, as less expensive selective herbicides become available, and as water control becomes better (De Datta 1977b).

In the tropics, butachlor, thiobencarb, piperophos/dimethametryn (*S*-2-methyl-piperidinocarbonylmethyl *OO*-dipropyl phosphorodithioate)/[2-(1,2-diethylpropylamino)-4-ethylamino-6-methylthio-1,3,5-triazine], and butralin [4-(1,1-dimethylethyl)-*N*-(1-methylpropyl)-2,6-dinitrobenzene amine] effectively control weeds in direct-seeded flooded rice (De Datta and Bernasor 1973, De Datta 1979). Among these, butachlor and thiobencarb are perhaps the most widely tested in direct-seeded rice.

In the United States, propanil and molinate are used extensively in drill-seeded and water-seeded rice (Smith 1977). Propanil applied at 3.3–5.5 kg/ha controls *E. crus-galli* 2.5–7.5 cm tall (1–4 leaf stage) with little or no injury to rice. Propanil is ineffective as a preemergence application or when applied to *E. crus-galli* at its tillering stage or later (Smith et al. 1977). When *E. crus-galli* has reached the 1–4 leaf stage, rice usually has one or two leaves. Examples of weeds that are controlled by various herbicides in the United States are shown in Table 12.17.

In a recent test in drill-seeded rice in Louisiana, Baker (1978a) applied herbicides as preplant, preplant incorporated, preemergence, postemergence, or postflood treatments. Although all herbicide treatments gave a significant degree of weed control, only oxadiazon [5-*tert*-butyl-3(2,4-dichloro-5-isopropoxyphenyl)-1,3,4-oxadiazol-2-one] applied at emergence gave a significantly higher degree of *E. crus-galli* control than the propanil check (Table 12.18). Based on experience in Arkansas, Huey (1977) reported that propanil at 3.3–5.5 kg/ha controls *E. crus-galli* and many broadleaf weeds effectively when applied to rapidly growing grass in the 1–3 leaf stages (5–7.5 cm tall). Somewhat larger weeds may be killed with higher rates. Rice may show yellowing and tip burn soon after treatment, but plants recover quickly under good growing conditions. Healthy rice plants are tolerant of propanil at rates as high as 8.8 kg/ha. Timing of application of propanil should be based on the size of weeds, and not on the size of rice. Weed sizes for effective control with propanil are given in Table 12.19.

Temperatures affect weed control with propanil. For example, an extremely high temperature (>35°C) enhances the activity of propanil on rice and may cause excessive leaf burn and injury.

Because rapidly growing weeds are most susceptible to propanil, sufficient moisture is essential for effective weed control. Also, propanil is a contact herbicide and good coverage of the weeds is necessary. Flooding should be applied within 4–5 days to prevent reinfestation by weeds because propanil has no residual activity. Shallow flooding should be started 24 hours after treatment. Mixtures of propanil and several new herbicides with residual activity have been

Table 12.17 Common Rice Field Weeds and Their Susceptibility to Control by Herbicides (Adapted from Smith et al. 1977)

Common Name	Scientific Name	Herbicide and Control[a]					
		Propanil	Molinate	2,4-D	MCPA	2,4,5-T	Silvex
Barnyard grass	Echinochloa Beauv. spp.	Excellent	Excellent	Poor	Poor	Poor	Poor
Bulrush	Scirpus L. spp.	Poor	Poor	Poor	Poor	Poor	Poor
Crab grass	Digitaria Heist. spp.	Excellent	Excellent	Poor	Poor	Poor	Poor
Fimbristylis	Fimbristylis Vahl spp.	Excellent	Poor	Excellent	Excellent	Excellent	Excellent
Goose weed	Sphenoclea zeylanica Gaertn.	Poor	Poor	Fair	Fair	Fair	Fair
Knot grass	Paspalum L. spp.	Fair	Poor	Poor	Poor	Poor	Poor
Morning glory	Ipomoea L. spp.	Poor	Poor	Excellent	Excellent	Excellent	Excellent
Paragrass	Panicum purpurascens Raddi	Excellent	Good	Poor	Poor	Poor	Poor
Red rice	Oryza sativa L.	Poor	Poor	Poor	Poor	Poor	Poor
Spikerush	Eleocharis R. Br. spp.	Good	Good	Good	Good	Good	Good
Sprangletop	Leptochloa Beauv. spp.	Fair	Good	Poor	Poor	Poor	Poor
Umbrella sedge	Cyperus L. spp.	Fair	Poor	Good	Good	Good	Good

[a]Excellent—one application at normal rates kills 80–100% of the weeds. Normal rates for propanil, 3.3 kg/ha; molinate, 3.3 kg/ha; phenoxy herbicides, 1.1 kg/ha.

Good—one application at rates higher than normal kills 60–80% of the weeds.

Fair—weed is not controlled selectively in rice, but can be controlled with high rates or with repeated application on leves and in canals where rice injury is not important.

Poor—weed is not controlled even at high rates.

Table 12.18 Crop Tolerance, Weed Control, and Grain Yield of Drill-seeded Rice as Affected by Various Herbicides under Mechanized Culture[a] (Adapted from Baker 1978a)

Herbicide[b]	Rate (kg a.i./ha)	Time of Treatment[c]	Rice Tolerance[d]	Weed Control Rating[e]				Yield (t/ha)
				ES	BR	AS	AV/SB	
Untreated check	—	—	5.0	1.0	1.0	1.0	1.0	2.2
Propanil check	3.36	Post	4.5	3.0	5.0	5.0	5.0	3.9
Propanil fb 2,4-D	3.36/1.68	Post/MS	4.375	4.0	5.0	5.0	5.0	4.1
Propanil + bentazon	3.36 + 0.84	Post	4.33	3.67	5.0	5.0	5.0	4.1
Propanil + butachlor	3.36 + 2.8	Post	5.0	4.0	5.0	5.0	5.0	4.4
Propanil fb bifenox	3.36/2.24	Post	4.75	4.0	5.0	5.0	5.0	4.2
Propanil + oxadiazon	3.36 + 1.12	Post	4.5	3.75	5.0	5.0	5.0	4.1
Oxadiazon	1.12	AE	3.25	5.0	5.0	5.0	5.0	3.5
HSD[f]	—	—	1.2	1.3	<1	<1	<1	1.1

[a]Drill-seeded 1 May 1978; flushed 4 May 1978; 75% emerged 15 May 1978; flooded 29 May 1979.

[b]Tank mix treatments indicated by + and sequential treatments indicated by fb (followed by).

[c]At emergence treatment (AE) applied 16 May 1978. Postemergence treatments (Post) applied 25 May 1978. Midseason treatment (MS) applied 27 June 1978.

[d]Rice tolerance ratings: 1 = no tolerance; 5 = excellent tolerance.

[e]Weed control ratings: 1 = no control; 5 = excellent control. ES = *Echinochloa* sp. BR = signal grass (*Brachiaria platyphylla*); AS = annual sedge (*Cyperus iria*); AV/SB = complex of northern jointvetch (*Aeschynomene virginica*) and hemp sesbania (*Sesbania exaltata*).

[f]Honestly significant difference at 5% level.

493

Table 12.19   Weed Sizes for Most Effective Weed Control with
Propanil (Adapted from Huey 1977)

Weed	Weed Height (cm)	Stage (no. of leaves)
Barnyard grass	2.5–7.5	1–3
Sprangletop	1.25	1–2
Morning glory	7.5	2–3
Duck salad	<2.5	2
Red stem	2.5	2–4
Smartweed	<5	2–4
Volunteer rice	10	4–6
Northern jointvetch	<30	—
Hemp sesbania	90	—

used experimentally to improve weed control and increase yields of rice (Huey 1977).

Molinate is effectively used in California (Oelke and Morse 1968) and Australia in water-seeded rice.

Current research in the University of California suggests that thiobencarb and molinate plus bifenox [methyl 5-(2,4-dichlorophenoxy)-2-nitrobenzoate] give excellent *E. crus-galli* control (University of California and USDA 1978).

Upland Rice. About 10% of the world's rice-growing area is dry-seeded on unbunded fields. In West Africa and Latin America, upland rice is the major growing system, and in Asia it is also an important system. Recent experiments in many parts of the world strongly suggest that chemical weed control in upland rice is not only possible but also effective and economical (De Datta 1977b).

Much of the further expansion in rice-growing areas may be as upland culture because most of the potential lowland areas are already cultivated. In such new areas, suitable weed-control practices, including use of herbicides, will play an important role in increasing rice production. For example, Indonesia has 1.28 million hectares of upland rice and the area is gradually expanding in southern Sumatra and Kalimantan. If weeds such as *Imperata cylindrica* (L.) Beauv. and *Cyperus rotundus* can be controlled, further increases in area are possible.

In most Latin American countries, a combination of mechanical and chemical methods has brought large areas, especially in Brazil, into rice production.

In many experiments around the world, no chemical alone gives as good results as hand weeding, but in combination with other weed-control measures several chemicals give acceptable weed control.

Many researchers have suggested propanil for weed control in upland rice (Mukhopadhyay et al. 1971, Bhan 1978). In Ghana, propanil plus butachlor appeared promising (Aryeetey 1973). Recent results at IRRI suggest that butachlor, oxadiazon, and pendimethalin [*N*-(1-ethylpropyl)-3,4-dimethyl-2,6-dinitrobenzene amine] can be used in upland rice with limited success.

Dry-Seeded Rainfed Bunded Rice.   Most of the herbicides that have given good weed control in wet-seeded or upland rice do not provide sustained weed control. Many are also phytotoxic, particularly when rainfall causes sudden submergence of a field. Some chemicals such as butachlor, thiobencarb, butralin, and piperophos/dimethametryn appeared promising in recent trials despite some phytotoxicity that caused stand reduction (Table 12.20). One follow-up hand weeding 5–6 weeks after the herbicide application sometimes helps in minimizing grain yield reduction. Stand reduction of rice due to herbicide phytotoxicity can be minimized by increasing seeding rates from 80 to 100 kg/ha. With that, even if there is as much as 30% stand reduction, it has little effect on the final yield.

Herbicide phytotoxicity can be reduced by applying them after a germinating rain rather than applying them immediately after seeding. No loss in weed control, and in some instances better weed control, has been observed when herbicides are applied at that time (Moody 1977b). When the herbicide is applied within 3–4 days after seeding, several weeks may pass before there is sufficient rainfall to provide quick germination. During that dry spell, herbicides may break down and weed control will be less than desirable than if rain had fallen immediately after seeding.

Despite some problems in improving herbicide effectiveness, it seems that herbicides will play a major role in dry-seeded rice culture.

CHEMICAL CONTROL OF PERENNIAL WEEDS   Perennial weeds pose a serious threat to both lowland and upland rice culture.

Lowland Rice.   In tropical Asia, perennial weeds such as *Scirpus maritimus* pose serious weed problems in lowland rice (De Datta and Lacsina 1974). Problems with *S. maritimus* and *S. mucronatus* L. have also been reported in Europe and the Americas. In the Philippines and some other tropical Asian

Table 12.20   Effect of Different Herbicides on Weed Weight and Yield of Dry-seeded Rice (From Moody 1977b)

Treatment[a]	Rate of Application (kg/ha)	Stand Reduction (%)	Weed Weight (37 DS) (kg/ha)	Yield (t/ha)
Thiobencarb	3.0	37	76	5.3
Manual weeding (15 fb 30 DS)	—	0	0	5.2
Butralin	2.0	16	88	4.5
Butachlor	2.0	19	112	4.4
Piperophos/dimethametryn	1.2/0.3	22	182	4.2
Butachlor + propanil	1.5 + 2.0	14	288	3.3
Untreated	—	0	1946	2.7

[a]fb = followed by; / = proprietary mix; + = tank mix; DS = days after seeding.

countries, *Paspalum distichum* L. is a perennial weed that causes serious concern where there is a reduced tillage or poorly prepared land. *Paspalum distichum* is not a serious problem where there is good tillage and water management.

The direct methods of control for annual weeds are applicable in various degrees for perennials. In temperate rice-growing countries, chemical control of weeds including perennials is dominant.

In the Republic of Korea, chemicals that minimize perennial weed problem in rice fields have been identified. For example, for *Cyperus serotinus* control, glyphosate, perfluidone [1,1,1-trifluoro-*N*-2-methyl-4-(phenylsulphonyl) phenylmethane sulphonamide], and bentazon were effective. Control of *Potamogeton distinctus* A. Bennett, *Eleocharis acicularis* (L.) Roem. et Schult., and *S. polyrhiza* was excellent with piperophos + dimethametryn and molinate + simetryn [2-methylthio-4,6-bis(ethylamino)-s-triazine] and thiobencarb + simetryn (Ryang et al. 1976).

In Japan, paraquat applied before planting is reported to control *E. acicularis* in rice fields (Denize 1968).

In tropical Asia, several herbicides have been identified that look outstanding for controlling *S. maritimus* in lowland rice fields. For example, the selective herbicides bentazon and fenoprop [(±)-2-(2,4,5-trichlorophenoxy)propionic acid] gave excellent control of *S. maritimus* in both direct-seeded and transplanted rice. But with a mixed population of annual weeds and *S. maritimus*, no single chemical can control both groups of weeds (De Datta 1979).

Upland Rice.   In upland rice, *C. rotundus* is a serious weed for which suitable control measures are needed. For example, the upland rice-growing area in southern parts of Sumatra and Kalimantan, Indonesia may increase if *C. rotundus* and *Imperata cylindrica* can be effectively controlled (De Datta 1977a).

For *C. rotundus* control in upland rice, dymrone [*N*-(α,α-dimethylbenzyl)-*N*-*p*-tolylurea] incorporated before seeding rice gave excellent control with no visible crop damage and increased grain yield (Okafor and De Datta 1976b). Dymrone, however, does not control grassy weeds and has to be followed by a selective grass herbicide such as butachlor.

Recent IRRI results suggest that methyl dymrone [1-(α,α-dimethylbenzyl)-3-methyl-3-phenylurea] is a better herbicide than dymrone for *C. rotundus* control in upland rice. In fact, reports from Japan suggest that it is effective against both purple and yellow (*C. esculentus*) nutsedge (Takematsu et al. 1976).

For chemical control of the perennial grass *I. cylindrica*, Eussen and Soerjani (1976) made a detailed review.

Biological Control of Weeds

Biological control is the intentional use of living organism to control a pest. In the southern United States, *Aeschynomene virginica* (L.) B.S.P. weed in rice has been successfully controlled by an endemic fungus, *Colletotrichum gloeosporioides* (Penz.) Sacc. f. sp. *aeschynomene*. The weed is killed about 5 weeks after it is inoculated with conidia of the fungus.

Biological control, however, has the disadvantage of controlling one specific weed by a particular insect or disease. Nevertheless, biological control offers an important approach for the future with advantages such as lack of harmful residues in crops and in the environment.

FACTORS IN INTEGRATED WEED MANAGEMENT

Each weed-control method has its advantages and disadvantages. None is applicable in all cases because weeds vary so much in their growth habits and life cycle. There is a growing concern that continuous usage of the same method of weed control such as land preparation, interrow cultivation, or herbicides will lead to a buildup of weed species that are tolerant of the control method used. It is, therefore, essential to combine or change various methods of weed control if one wishes to avoid possible increased weed problems.

The concept of pest management or integrated weed control is not new. For many years farmers have been using it, although perhaps not exactly as it is practiced today. The best integrated weed management practices are those that take into account the crop grown, weeds infested, and the farmers' socio-economic environment. Smith and Reynolds (1966) defined *integrated weed control* as a weed population management system that uses all suitable techniques in a compatible manner to reduce weed populations and maintain them at levels below those causing economic injury.

Noda (1977) listed the following weed-control methods as components of integrated weed-control management: preventive land preparation, preventive manual weeding, mechanical weeding, water management, crop rotation, chemical and biological weed control, and other practices such as fertilizer management. The combinations of these measures used depend on the rice-production practices of each country.

Smith et al. (1977) suggested that weed management systems combine cultural, mechanical, chemical, and biological control programs. Management of cultural and mechanical weed-control practices may be used effectively to control specific weeds. Preventive methods such as use of weed-free crop seed, use of irrigation water free of weed seed, and use of clean equipment are required to avoid weed problems before they begin in rice fields.

Velasco et al. (1961) reported that the most feasible and economical schedule for weed control in upland rice was one that involved chemical and mechanical methods as well as cultural practices.

Herbicide selectivity for a given species may lead to infestation of more noxious weeds that were formerly of secondary importance. Therefore, combinations of several weeding methods are likely to prove more effective than a single factor to alleviate buildup of a difficult-to-control weed or group of weeds.

In the past, many new control measures were developed and it is likely that many new measures will be developed. Thus, numerous combinations of practices are possible to develop an integrated weed management system in most crop-weed situations.

Good water management is effective in controlling or suppressing grassy weeds in flooded rice and a combination of water management with other methods, such as a chemical application and soil preparation, would provide higher efficiency in weed control.

In flooded rice, the depth and time of flooding govern the infestation level of *E. crus-galli* and *Leptochloa* sp. as well as the water management practices that should be used in combination with herbicide applications. Effective control of those weeds can be achieved by use of clean rice seeds, water management, fertilizer management, and proper application of herbicides such as molinate and propanil (Smith 1977).

In Sri Lanka, rice is often seeded densely to prevent weed emergence during the early stages of rice growth. Once rice is ahead of weeds, harrowing is done to reduce seedling density. Considerable work is clearly required to develop suitable integrated weed-control technology for transplanted, direct-seeded flooded, and dry-seeded rainfed (upland and lowland) rice.

Most weed-control trials have dealt with a single control practice without considering other control methods. The following are examples of advantages of multiple-factor methods for integrated weed management in rice.

Transplanted Rice

In Japan, by using combinations of chemical, mechanical, and hand weeding methods of weed control, the chances of a complete weed control are vastly improved. Some practices used by Japanese farmers are given in Table 12.21. The farmers do not necessarily use all of the methods but various combinations of the methods can be used depending on weed species present, the degree of infestation, and stage of the crop (Noda 1977).

Often, herbicides have to be followed by one hand weeding or a mechanical weeding. For example, in a study in India, weed control was superior and grain yield was significantly higher when herbicides such as butachlor or propanil were followed by one hand weeding (Rangiah et al. 1974). Two hand weedings produced grain yield similar to treatment with chemical followed by one hand weeding.

Studies in the Philippines and Thailand suggest marked complementarity between fertilizer application and need for a higher weed control measure. It is apparent that the benefits from added fertilizer increase markedly with higher level of weed control (De Datta and Barker 1977). In a similar study in Bangladesh, when fertilizer and weed control were applied together, significantly higher yields were obtained than when they were applied singly.

Table 12.21 Standard Weed-Control Practices for Transplanted Rice in Japan (Adapted from Noda 1977)

Weed-Control Practice	Time of Application	Herbicides Recommended[a]
Land leveling and soil puddling	Before transplanting	—
Herbicide	0–3 days before transplanting	chlomethoxynil, chlornitrofen, nitrofen, oxadiazon, thiobencarb/chlornitrofen
Herbicide	1–10 days after transplanting	chlornitrofen, nitrofen, thiobencarb/chlornitrofen
Interrow cultivation	10–15 days after transplanting	—
Herbicide	2–4 days after interrow cultivation	chlomethoxynil, chlornitrofen, nitrofen, thiobencarb/chlornitrofen
Herbicide	10–15 days after transplanting	molinate/simetryn, molinate/simetryn/MCPB, swep/MCPA, thiobencarb/simetryn, thiobencarb/simetryn/MCPB
Herbicide	Early stages of Potamogeton sp.	ACN, prometryn, simetryn
Herbicide	20–40 days after transplanting	2,4-D, MCPA, MCPB
Hand weeding	Middle stages of rice growth	—
Hand weeding	After heading (especially for Echinochloa sp.)	—

[a]A slant bar (/) between chemical names indicates that the compounds are formulated as a proprietary mixture.

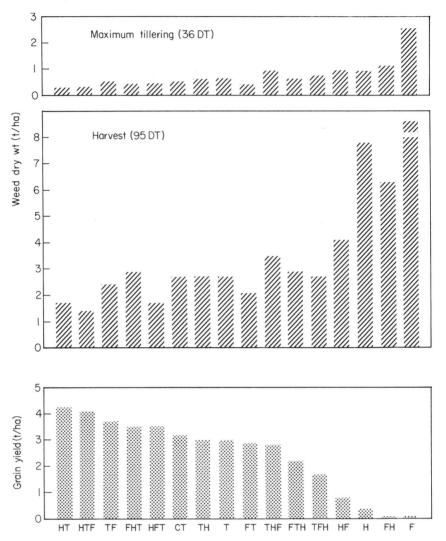

Figure 12.5 An illustration of the relationship between weed dry weight as influenced by land preparation treatments and rice yield. H = herbicide application; T = tillage operation; F = flooding; C = conventional. IRRI, 1976 dry season. (Adapted from De Datta et al. 1979)

In an IRRI experiment, three components of reduced tillage—preplant herbicide application, tillage operation, and flooding—were tested alone and in combination. Each component requires a different resource; the preplant herbicides require mainly capital, the flooding period requires irrigation water, and the tillage operation requires mainly power and labor. All require time. When preplant herbicide application was followed by a tillage operation, control of weeds was satisfactory and the rice yields were the same as those with conventional tillage (Fig. 12.5). Results confirm that preplant herbicide, tillage, and the period of flooding may be complementary and possibly compensatory (De Datta et al. 1979).

Direct-Seeded Flooded Rice

In a recent experiment at IRRI, herbicides were more effective under continuous flooding than when direct-seeded flooded plots were dry for 28 days then saturated. For example, the grain yields were significantly higher with thio-bencarb followed by 2,4-D than for the untreated control when continuous submergence was not maintained. With continuous submergence, the grain yields were similar between the untreated control and the herbicide combination of thiobencarb followed by 2,4-D (Navarez et al. 1978).

A number of cultural practices help minimize infestation by a specific weed species in direct-seeded flooded rice. Some examples of practices that minimize competition from a specific weed species are shown in Table 12.22.

Upland Rice

The effect of variety and tillage on weed control in upland rice was studied with three different varietal types (short-, intermediate-, and tall-statured) and three degrees of tillage (one plowing followed by one, two, or three rototillages). One plowing followed by two rototillages had significantly lower dry weed weight than one plowing followed by one rototillage. Further increase beyond two rototillages did not significantly affect the dry weed weight (Fig. 12.6).

Dry-Seeded Bunded Rainfed Rice

In dry-seeded bunded rainfed rice, the crop begins its life cycle as an upland crop. As the season progresses, water accumulates in the field and the crop finishes its life cycle as a lowland crop. In this system, weed control with a single method is extremely difficult. It is essential to combine several methods—direct and indirect—to control weeds in dry-seeded bunded rainfed rice.

TRENDS IN WEED CONTROL IN RICE

Hand weeding is commonly used in almost all areas in South and Southeast Asia where rice is transplanted. Rotary mechanical weeding is common in some Southeast Asian countries, but not in others. In most South and Southeast Asian countries, puddling operations are largely used for controlling weeds in rice fields (De Datta and Barker 1977).

Table 12.22 Control of Ricefield Weeds by Selected Cultural Practices (Adapted from Smith et al. 1977)

Weed	Hand Weeding	Seedbed Preparation	Dry Seeding	Timely Flooding[b]	Timely Draining[b]	Rice Stand[c]	Crop Rotation
					Cultural Weed-Control Practice[a]		
Algae	Poor	Good	Good	Poor	Good	Good	Good
Cyperus difformis	Poor	Fair	Fair	Poor	Good	Good	Good
Echinochloa crus-galli	Poor	Good	Poor	Fair	Poor	Fair	Fair
Eclipta alba	Poor	Poor	Poor	Fair	Poor	Good	Fair
Fimbristylis littoralis	Poor	Poor	Good	Poor	Good	Good	Poor
Jussiaea sp.	Poor	Good	Fair	Poor	Fair	Good	Good
Leptochloa filiformis	Poor	Fair	Poor	Poor	Poor	Good	Good
Oryza sativa (Red rice)	Good	Good	Poor	Poor	Poor	Fair	Good
Paspalum distichum	Poor	Good	Fair	Poor	Fair	Good	Good
Sphenoclea zeylanica	Poor	Good	Good	Poor	Good	Good	Fair

[a]Poor—practice cannot be used economically in commercial rice, or fails to control the weed.
[b]After crop emergence.
[c]A rice stand of 130–215 plants/m^2 reduces many weed problems.

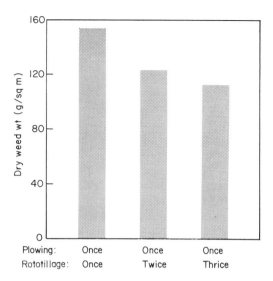

Figure 12.6   Dry weed weight as affected by degree of tillage in upland rice. IRRI, 1976 wet season. (From De Datta 1979)

Use of Herbicides

In Japan and the Republic of Korea, weeding of almost the entire rice-growing area is by herbicides because of high labor cost. The use of herbicides started with the use of phenoxy acids, particularly when granular formulations were developed. Then PCP (pentachlorophenol) with a mixture of phenoxy acids became most popular but it was abandoned because of PCP toxicity to fish. Subsequently, several herbicides that were less toxic to fish were used. They are chlornitrofen (4-nitrophenyl 2,4,6-trichlorophenyl ether), nitrofen, prometryn [2,4-di(isopropylamino)-6-methylthio-1,3,5-triazine], and thiobencarb. With a growing infestation of perennial weeds such as *Cyperus serotinus, Scirpus juncoides* Roxb. var. *hotarui* Ohwi, and *Sagittaria pygmaea* Miq., the control and management of weeds in Japan became extremely difficult and the use of herbicides continued to rise because of high labor cost of hand weeding. For example, in 1975, the saving in labor for weeding was $592/ha (Noda 1977).

Some modifications in herbicide use had to be made as more areas were transplanted by machine. Usually, seedlings are transplanted with machine 7–10 days earlier than hand transplanted fields. As a result, there is more time for weeds to germinate. Therefore, sequential weeding operations are followed with repeated applications of herbicides (Table 12.21).

To sustain mechanical transplanters, the soil has to be made firm by repeated puddling. During puddling and transplanting, more weeds grow as the time between the operations increases. In addition, rice seedlings for machine transplanting are shorter and are more subject to herbicide injury, particularly if the weather condition is unfavorable. Herbicides such as chlornitrofen, which are least toxic to ordinary seedlings, are somewhat toxic to seedlings for machine transplanting. And, because seedlings are irregularly planted by machine, rotary and hand weeding are more difficult.

In Taiwan, the cost of hand weeding increased steadily while the cost of herbicides was relatively stable until 1972. Weed control systems in Taiwan included soil preparation, water management, two to three weedings by hand, and chemicals. Farmers started to use herbicides in 1966 and use increased steadily: 1.4% of the total rice-growing area in 1969, 6.4% in 1971, and 32.8% in 1973. A recent estimate suggests almost 100% of Taiwan's rice farmers now use herbicides.

Despite their high cost, more selective herbicides, such as butachlor, are replacing poorer ones for rice in Taiwan and the Republic of Korea.

Korea's rice fields suffer from perennial-weed problems similar to those in Japan and Taiwan. As in both of those countries, labor cost in Korea has increased and chemical weed control has become the most important single method. All of these East Asian countries must consider integrated methods of weed control if perennial weeds need to be contained beyond the present level of infestation.

Among the tropical Asian countries butachlor and thiobencarb are most extensively used. The Philippines and Malaysia use herbicides on 10–30% of the rice-growing area. In Punjab, India, butachlor is extensively used in rice fields because of the high labor cost for hand weeding.

This raises the question of what factors mainly influence the adequacy of weed control. It appears to be scarcity of rice land. In Java, Indonesia, and in Taiwan, and Japan, farms are extremely small (less than 1 ha) and population pressure demands every effort to maximize the return to land by producing yields as high as possible on the scarce land. Weeds are simply not tolerated. By contrast, in the Philippines and Thailand farms are generally larger (2–3 ha) but are generally not so well irrigated. Thus, weeds are frequently a greater problem and less manpower is available to cope with them. In those countries, it is not uncommon to find farmers who essentially do not weed.

Weed-control recommendations of the future will be based on more scientific data than those of today. Basic knowledge of the biochemistry and physiology of the weeds to be controlled needs to be known, and the best method and the time to apply herbicides should be established more precisely for the most effective results (Flinchum 1975).

PROBLEMS OF WILD RICE AND RED RICE

Wild rice and red rice as weeds are a serious threat to rice production in many countries in Africa and Latin America. For example, in Mali 10,000 ha out of 30,000 ha in the Mopti area is infested with wild rice species. The problem is equally serious in Swaziland. Red rice has been a serious weed problem in rice fields in Latin America (Cheaney et al. 1976) and the United States (Smith et al. 1977, Sonnier 1978). It reduces yields and lowers the market value of rough and milled rice because seeds are difficult to remove when they are mixed with ordinary rice.

Wild Rice Species

There are a number of wild rice species that invade rice fields. Some examples are *Oryza longistaminata* Chev. and Roehrich, *O. breviligulata* Chev. and Roehrich, and *O. glabberima* Steud.

- *O. longistaminata* is a rhizomatous rice that thrives in medium- to deepwater areas. It propagates primarily through rhizomes, although reproduction through seeds is possible.
- *O. breviligulata* is easily recognizable at the flowering stage because of its long red awn. It grows in deep water and propagates by seeds.
- *O. glaberrima* is a cultivated species of rice. However, it becomes a difficult weed in areas where *O. sativa* rices are grown.

All three species are prevalent in Mali, but *O. longistaminata* is most difficult to control (Martin and Guegan 1973). In Swaziland, *O. punctata* Kotschy ex Steud. has become a serious threat in many rice fields.

Red Rice Species

Red rice (*Oryza sativa* L.) belongs to the same species as the cultivated rice (*O. sativa*). Another red rice species is *O. rufipogon* Griff.

Other names for red rice are Vermillion red rice, Italian rice, and common red rice. There are many types of red rice—those with short, medium, or long grains; those with straw-colored, red, or black hulls; and those with short or long awns on the spikelet (Smith et al. 1977).

Red rice is an annual that propagates by seed. Red rice tillers profusely and it is difficult to distinguish it from cultivated rice. Red rice shatters easily and therefore poses a continuous problem because the seeds remain dormant in the soil.

Control of Wild and Red Rice

It is difficult to control wild and red rice in cultivated rice fields because both are rice plants belonging to the same genus or species. Although there are many cultural practices that can be used to minimize red rice problem, there is a need for chemical weed-control program which would include chemical control of red rice as well as other weeds.

Nonselective Chemical Control

Chemical control is slightly more expensive than cutting the weeds but because control is better with herbicides, they compare favorably with manual cutting or with deep plowing (Martin and Guegan 1973).

Dalapon and diuron [3-(3,4-dichlorophenyl)-1,1-dimethylurea] have been suggested for controlling *O. longistaminata*. At 10–15 kg/ha dalapon can destroy up to 95% of all the rhizomes of *O. longistaminata* when it is 25–35 cm tall. At 15–35 cm height, it can be controlled with diuron at 5–6 kg/ha. Both dalapon and diuron are persistent chemicals. Therefore, fields treated with either herbicide cannot be used for one year for growing any crop.

Recently, glyphosate has been evaluated against wild rice species in Mali. Because of its nonpersistent nature, glyphosate offers a good opportunity for controlling wild rices.

For red rice control in Cali, Colombia, atrazine at 2 kg/ha gave excellent control and left no residue after the 30-day flooding period in the experiment. The combination of paraquat and linuron[3-(3,4-dichlorophenyl)-1-methoxy-1-methylurea] also gave excellent red rice control and left no residue problem after 30 days (Cheaney et al. 1976).

Herbicide Protectants and Antidotes

Herbicide protectants are used to allow increased herbicide doses without injury to the crop or the use of more potent herbicide to obtain a wider spectrum of weed control, longer period of weed control, or greater reliability. For example, rice has been successfully protected by pelleting pregerminated seed with activated charcoal, which decreased injury from butachlor and oxadiazon but only if the pregerminated rice seed was sown on nonflooded puddled soil (Nangju 1973). Perhaps the success with pregerminated rice is due to the emerged radicle being coated with activated charcoal as there is evidence that charcoal may migrate along the roots.

Selectivity may be governed by one or more factors including herbicide penetration, uptake, translocation, and metabolism. One approach to increasing selectivity and the period over which weeds are controlled, and to providing greater reliability under varying environmental conditions, is the use of an antidote, a protectant that works inside the plant to counteract the herbicide rather than merely preventing the entry of herbicide into the plant (Blair et al. 1976). It is also referred to as a chemical safening agent.

One example of an antidote is NA (1,8-naphthalic anhydride), which can be used as a seed dressing to protect a wide range of crops such as sorghum, rice, and oats. As a seed dressing, NA is used to protect rice and helps control wild rice (*O. punctata*). Rice sometimes is damaged slightly when treated with NA at 0.5–2% on soils with high nutrient contents (Parker and Dean 1976).

Smith showed in 1971 that coating with NA could be used to protect rice seeds against molinate and thereby make possible the selective control of red rice (*O.*

rufipogon), which is very closely related to *O. sativa*. This treatment has not been extensively used because of the need for incorporation of molinate into the soil before planting. Parker and Dean (1976), however, confirmed the effectiveness of NA against molinate and perfluidone injury to rice. NA can also be used against thiobencarb, a herbicide widely used in rice. This was shown in a recent experiment with drill-seeded rice in Louisiana by Baker (1978b) with seeds of four varieties treated with NA at the rate of 1 kg/100 kg seed. Molinate and thiobencarb, which have shown good selectivity in water-seeded rice in previous experiments, also showed good selectivity in the drill-seeded rice. The tolerance of the commercial varieties is marginal to poor when drill-seeded. Of the new herbicides tested, only butam [2,2-dimethyl-*N*-(1-methylethyl)-*N*-(phenylmethyl) propanamide] was promising (Baker 1978b).

Nonchemical Red Rice Control

Although herbicides provide an excellent opportunity for control of red rice, greater attention should be paid to suitable soil management and cultivation techniques such as plowing at the beginning of the dry season. For example, adequate red rice control can be achieved by good seedbed preparation and crop rotation. Wherever feasible, red rice should be hand pulled.

Continuous flooding also controls red rice, as shown by Sonnier (1978) who compared three water managements and two rice seeding-date managements on the red rice control in Louisiana and found that the behavior of red rice differed greatly at the different seeding dates. Continuous flooding resulted in the lowest number of red rice plants in both dates of seeding (March and April). An extremely dense population of red rice occurred under the prolonged drying treatments.

ECONOMICS OF ALTERNATIVE WEED CONTROL PRACTICES

There is a wide variety of direct and indirect methods of weed control for rice. It is clear that the optimum combination of weed-control practices may differ greatly, depending on the type of rice culture and the resources available to the farmer.

One of the best guides to the choice of alternative weed control methods is the relative cost of labor and chemicals. Knowing the daily wage rate and the approximate time required for one hand weeding in a given area, the cost per hectare of hand weeding can be determined (Table 12.23) and compared with the cost of herbicides. It is apparent that at existing wage rates in South and Southeast Asia, chemicals such as 2,4-D, which control weeds at about $10/ha, will be adopted in many areas, but the more expensive herbicides are likely to have very limited acceptance unless the wage rate relative to the chemical cost rises fairly substantially, such as in Japan, Republic of Korea, and Taiwan (De Datta and Barker 1977).

Herbicides are frequently looked upon as a threat to employment. But an analysis in the Philippines suggested that in areas where weed control was

Table 12.23   Cost per Hectare for Hand Weeding at Different
Weeding Intensities and Farm Wage Rates (From De Datta and
Barker 1977)

Daily Farm Wage Rate ($/ha)	Cost for Hand Weeding ($/ha)			
	75 hr	125 hr	175 hr	225 hr
0.50	4.70	7.80	10.90	14.10
0.75	7.30	11.70	16.40	21.10
1.00	9.40	15.60	21.90	28.10
1.25	11.70	19.50	27.30	35.20

traditionally poor, use of herbicides complemented increased use of labor to raise
farm production (De Datta and Barker 1977).

In much of tropical Asia, developing integrated methods of weed control,
using limited quantities of low-cost chemicals in combination with direct and
indirect weed-control techniques, may be the most attractive alternative from
agronomic, economic, and ecological points of view.

BEFORE USING ANY CHEMICALS, SEE THE MANUFACTURER'S LABEL AND FOLLOW THE
DIRECTIONS. ALSO SEE THE PREFACE.

REFERENCES

Appleby, A. P. 1967. Some general principles of weed control. Pages 20–21 *in*
Proceedings first Asian Pacific weed control interchange. East-West Center,
University of Hawaii, Honolulu, 12–22 June 1967.

Aryeetey, A. N. 1973. Chemical weed control in rice in Ghana. *Ghana J. Agric. Sci.*
6:199–204.

Baker, J. B. 1978a. Rice weed control studies (a preliminary report). Pages 106–121 *in*
Rice Experiment Station, Louisiana State University, and U. S. Department of
Agriculture. *70th annual progress report.*

Baker, J. B. 1978b. Chemical control of red rice (a preliminary report). Pages 122–130 *in*
Rice Experiment Station, Louisiana State University, and U. S. Department of
Agriculture. *70th annual progress report.*

Bhan, V. M. 1978. Weed control in drilled rice. *In* Indian Council of Agricultural
Research. *National symposium on increasing rice yield in kharif (8–11 February
1978),* Central Rice Research Institute, Cuttack, India. (unpubl. mimeo.)

Blair, A. M., C. Parker, and L. Kasasian. 1976. Herbicide protectants and antidotes–a
review. *PANS* 22:65–74.

Chang, W. L. 1970. The effect of weeds on rice in paddy field. 1. Weed species and population density. *J. Taiwan Agric. Res.* 19(4):18–25.

Cheaney, R. L., R. Lopez, and J. D. Doll. 1976. Algunos metodos promisorios para el control de arroz rojo y arroz espontaneo de variedades indeseables [in Spanish, English summary]. *COMALFI* 3:206–217.

Chiu, C. C. 1973. A comparison of weeds and weed control practices in the United States and Asian rice culture. A professional paper presented in partial fulfillment of the requirement for the Master of Agriculture. University of Idaho. (unpubl.)

De Datta, S. K. 1974. Weed control in rice: present status and future challenge. *Philipp. Weed Sci. Bull.* 1:1–16.

De Datta, S. K. 1977a. Approaches in the control and management of perennial weeds in rice. Proceedings 6th Asian-Pacific Weed Science Society Conference, Jakarta, Indonesia, 11–17 July 1977. I: 205–226.

De Datta, S. K. 1977b. Weed control in rice in Southeast Asia: methods and trends. *Philipp. Weed Sci. Bull.* 4:39–65.

De Datta, S. K. 1977c. Weed control and soil and crop management in rainfed rice at IRRI and other locations in tropical Asia. Pages 201–211 *in* I. W. Buddenhagen and G. J. Persley, eds. Academic Press. *Rice in Africa.* London.

De Datta, S. K. 1978. Land preparation for rice soils. Pages 623–648 *in* International Rice Research Institute. *Soils and rice.* Los Baños, Philippines.

De Datta, S. K. 1979. Weed problems and methods of control in tropical rice. Pages 9–44 *in* Weed Science Society of the Philippines, Inc., and Philippine Council for Agriculture and Resources Research. Symposium "Weed control in tropical crops." Los Baños, Philippines.

De Datta, S. K., and R. Barker. 1977. Economic evaluation of modern weed control techniques in rice. Pages 205–288 *in* J. D. Fryer and S. Matsunaka, eds. University of Tokyo Press. *Integrated control of weeds.* Tokyo.

De Datta, S. K., and P. C. Bernasor. 1973. Chemical weed control in broadcast-seeded flooded tropical rice. *Weed Res.* 13:351–354.

De Datta, S. K., and R. Q. Lacsina. 1974. Herbicides for the control of perennial sedge *Scirpus maritimus* L. in flooded tropical rice. *PANS* 20:68–75.

De Datta, S. K., and H. C. Jereza. 1976. The use of cropping systems and land and water management to shift weed species. *Philipp. J. Crop Sci.* 1:173–178.

De Datta, S. K., F. R. Bolton, and W. L. Lin. 1979. Prospects for using minimum and zero tillage in tropical lowland rice. *Weed Res.* 19:9–15.

De Datta, S. K., R. Q. Lacsina, and D. E. Seaman. 1971. Phenoxy acid herbicides for barnyardgrass control in transplanted rice. *Weed Sci.* 19:203–206.

De Datta, S. K., J. C. Moomaw, and R. T. Bantilan. 1969. Effects of varietal type, method of planting and nitrogen level on competition between rice and weeds. Pages 152–163 *in* Proceedings 2nd Asian-Pacific weed control interchange. Los Baños, Philippines.

De Datta, S. K., J. K. Park, and J. E. Hawes. 1968. Granular herbicides for controlling grasses and other weeds in transplanted rice. *Int. Rice Comm. Newsl.* 17(4):21–29.

Denize, J. R. 1968. New production techniques and chemical aids. *Weed Abstr.* 17:7–8.

Estorninos, L. E., Jr., and K. Moody. 1976. The effect of plant density on weed control in transplanted rice. Paper presented at the 7th Annual Conference, Pest Control Council of the Philippines, 5–7 May 1976, Cagayan de Oro City. (unpubl. mimeo.)

Eussen, J. H. H., and M. Soerjani. 1976. Problems and control of "Alang-alang" [*Imperata cylindrica* (L.) Beauv.] in Indonesia. Pages 58–65 *in* Proceedings 5th Asian-Pacific Weed Science Society Conference.

Flinchum, W. T. 1975. Controlling weeds in rice. Pages 51–57 *in* Texas Agricultural Experiment Station in cooperation with the U.S. Department of Agriculture. *Six decades of rice research in Texas*. Res. Monogr. 4.

Hoque, M. Z., P. R. Hobbs, and N. I. Miah. 1976a. Report on the 1975 T. *Aman* crop cut studies in the BRRI pilot project area. Bangladesh Rice Research Institute. *Rice cropping system Bull*. 4. 16 pp.

Hoque, M. Z., P. R. Hobbs, and N. I. Miah. 1976b. Report on the 1975 *Aus* crop cut studies in BRRI pilot project area. Bangladesh Rice Research Institute. *Rice cropping system Bull*. 3. 15 pp.

Huey, B. A. 1977. *Rice production in Arkansas*. University of Arkansas, Division of Agriculture, and U.S. Department of Agriculture. Circ. 476 (Rev.). 51 pp.

Madrid, M. T., Jr., F. L. Punzalan, and R. T. Lubigan. 1972. *Some common weeds and their control*. Weed Science Society of the Philippines, College, Laguna, Philippines. 62 pp.

Martin, P., and R. Guegan. 1973. The problem of wild rices in Mali: approach to the possible control measures, and results obtained by IRAT during three annual trials. West Africa Rice Development Association (WARDA) S/P/13/28. 43 pp. (unpubl. mimeo.)

Matsunaka, S. 1976. Main weeds in rice culture and their control. Pages 131–139 *in Plant protection in Japan*. Agriculture in Asia. Spec. issue 10.

Moody, K. 1977a. Weed control in rice. Lecture prepared for the participants of the 5th BIOTROP Weed Science Training Course, 14 Nov.–13 Dec. 1977, Kuala Lumpur, Malaysia.

Moody, K. 1977b. Weed control in sequential cropping in rainfed lowland rice growing areas in tropical Asia. Paper presented at the Workshop on Weed control in small scale farms during the 6th Asian-Pacific Weed Science Society Conference, 11–17 July 1977, Jakarta, Indonesia. 30 pp.

Moody, K., and S. K. De Datta. 1977. Integrated control of weeds in rice. Paper presented at the Seventh session of the FAO panel of experts on Integrated pest control and resistance breeding, 21–28 April 1977, Rome, Italy. 11 pp.

Moomaw, J. C., V. P. Novero, and A. C. Tauro. 1966. Rice weed control in tropical monsoon climates: problems and prospects. *Int. Rice Comm. Newsl*. 15(4):1–16.

Mukhopadhyay, S. K. 1978. Weed control in different rice culture systems. I. Weed control in lowland rice under submergence. *In* Indian Council of Agricultural Research. *National symposium on increasing rice yield in kharif (8–11 February 1978)*. Central Rice Research Institute, Cuttack, India.

Mukhopadhyay, S. K., B. C. Ghosh, and H. Maity. 1971. Weed problems in upland rice and approaches to solve the problem by use of new herbicides. *Oryza* 8:269–274.

Nangju, D. 1973. Seed pelleting as an approach to herbicide selectivity in direct-seeded rice. *Diss. Abstr. Int*. B 33:4079.

Navarez, D. C., L. L. Roa, and K. Moody. 1978. Weed control in wet-seeded rice grown under different moisture regimes. Paper presented at the 9th Annual Conference Pest Control Council of the Philippines, 3–6 May 1978, Manila, Philippines.

Noda, K. 1977. Integrated weed control in rice. Pages 17–44 *in* J. D. Fryer and S. Matsunaka, eds. University of Tokyo Press. *Integrated control of weeds*. Tokyo.

Oelke, E. A., and M. D. Morse. 1968. Propanil and molinate for control of barnyardgrass in water-seeded rice. *Weed Sci.* 16:235–239.

Okafor, L. I., and S. K. De Datta. 1974. Competition between weeds and upland rice in monsoon Asia. *Philipp. Weed Sci. Bull.* 1:39–45.

Okafor, L. I., and S. K. De Datta. 1976a. Competition between upland rice and purple nutsedge for nitrogen, moisture and light. *Weed Sci.* 24:43–46.

Okafor, L. I., and S. K. De Datta. 1976b. Chemical control of perennial nutsedge (*Cyperus rotundus* L.) in tropical upland rice. *Weed Res.* 16:1–5.

Parker, C., and M. L. Dean. 1976. Control of wild rice in rice. *Pestic. Sci.* 7:403–416.

Rangiah, P. K., A. Palchamy, and P. Pothiraj. 1974. Effect of chemical and cultural methods of weed control on transplanted rice. *Madras Agric. J.* 61:312–316.

Ryang, H. S., M. K. Kim, and J. C. Jeon. 1976. Control of perennial weeds in paddy rice in Korea. Pages 293–297 *in* Proceedings 5th Asian-Pacific Weed Science Society Conference.

Sharma, H. C., H. B. Singh, and G. H. Friesen. 1977. Competition from weeds and their control in direct-seeded rice. *Weed Res.* 17: 103–108.

Smith, R. F., and H. T. Reynolds. 1966. Principle definitions and scope of integrated pest control. 1. Pages 11–17 *in* Proceedings FAO symposium on integrated pest control, 11–15 October 1965. Rome.

Smith, R. J., Jr. 1971. Red rice control in rice (Abstr.). Page 163 *in* Southern Weed Science Society. Proceedings of the 24th annual meeting.

Smith, R. J., Jr. 1977. *Comparisons of herbicide treatments for weed control in rice.* Agriculture Experiment Station, University of Arkansas, and Agriculture Research Service, USDA, Rep. Ser. 233. 28 pp.

Smith, R. J., Jr., and D. E. Seaman. 1973. Weeds and their control. Pages 135–140 *in* USDA Agric. Handb. 289. *Rice in the United States: varieties and production.* Washington, D.C.

Smith, R. J., Jr., W. T. Flinchum, and D. E. Seaman. 1977. *Weed control in U. S. rice production.* USDA Agric. Handb. 497. Washington, D.C. 78 pp.

Sonnier, E. A. 1978. Red rice studies. Pages 131–142 *in* Rice Experiment Station, Louisiana State University, and U.S. Department of Agriculture. *70th annual progress report.*

Takematsu, T., H. Kubo, N. Seki, N. Nato, and Y. Omura. 1976. Control of nutsedge (*Cyperus rotundus* and *Cyperus esculentus*) and other weeds with K-223 and K-1441. Pages 121–124 *in* Proceedings 5th Asian-Pacific weed science society conference, Tokyo, Japan.

University of California and U.S. Department of Agriculture. 1978. Pages 1–4 *in* University of California, Davis. *Annual report comprehensive rice research, 1978.*

Vacchani, M. V., M. S. Chaudhri, and N. N. Mitra. 1963. Control of weeds in rice by selective herbicides. *Indian J. Agron.* 8:368–377.

Vega, M. R. 1954. The effect of herbicides on weeds in rice fields. *Philipp. Agric.* 38:13–47.

Vega, M. R., and E. C. Paller, Jr. 1970. Weeds and their control. Pages 147–170 *in* University of the Philippines College of Agriculture in cooperation with the

International Rice Research Institute. *Rice production manual.* Los Baños, Philippines.

Vega, M. R., and J. N. Sierra. 1968. Population of weed seeds in a lowland rice field. Pages 1-9 *in* Proceedings 1 Philippine weed science society conference, Makati, Rizal, Philippines.

Vega, M. R., J. D. Ona, and E. C. Paller, Jr. 1967. Evaluation of herbicides for weed control in upland rice. Pages 63-66 *in* Proceedings 1 Asian-Pacific weed control interchange. Honolulu, Hawaii.

Velasco, J. R., M. R. Vega, P. A.Llena, and S. R. Obien. 1961. Studies of weed control in upland rice. *Philipp. Agric.* 44:373-393.

Yamagishi, A., A. Hashizume, and Y. Takeichi. 1976. Studies on control of some perennial weeds in paddy field. VII. Competition between *Cyperus serotinus* Rottb. and rice [in Japanese, English summary]. *Bull. Chiba-Ken Agric. Exp. Stn.* 17:1-20.

13

Postproduction Technology of Rice

The importance of postproduction handling of rice has increased because of the increased yields possible from the modern rice varieties. Losses that had been small in absolute terms expanded proportionately with increased farm-level yields. And the emphasis on increasing rice production by increasing cropping intensity resulted in a crop to harvest at the height of the wet season, dictating substantial changes in postproduction operations.

The problem of determining the magnitude and source of losses in the post-production sequence of operations is important and complex. All postharvest operations such as handling, threshing, drying, and storage are interdependent. The technique employed, the timing of each operation, and the environment in which they are carried out all contribute to the quality and quantity of rice that farmers ultimately sell or retain for home consumption. In tropical Asia, the magnitude of postproduction losses of the rice crop makes consideration of their causes, pattern, and eventual control a matter of primary concern.

Duff and Toquero (1975) emphasized the close relationship between performance at the processing level and operations at the farm level. When rice is improperly threshed, dried, and cleaned at the farm level, it is impossible to achieve high milling efficiency at the processing plant. In most of tropical Asia, a proper range of rice processing technology, particularly for smaller farmers and millers, is not yet available at an acceptable cost. In addition, because markets are fragmented between rural household demand and urban areas, investment capital is limited and incentives for new technology are often lacking or distorted.

The attempts that have been made in most countries to improve rice processing have not focused on the total system, but have taken a piecemeal approach. That has upgraded some processing components but left others unchanged and has resulted in little or no significant impact on the quality or quantity of rice processed. It is important to consider the entire postproduction operation as a system, an approach taken in most temperate-climate rice-growing countries.

Low rice yields in farmers' fields may be further aggravated by later losses from antiquated, wasteful techniques in postproduction grain handling. In the tropics, these losses (both quantitative and qualitative) take place at various stages in the harvest and postharvest sequence of operations. It is estimated that losses from harvesting through milling amount to about 25% of all rice grain harvested (Samson and Duff 1973).

HARVESTING AND POSTHARVEST OPERATIONS

In the market place, the price of rice is often determined by the appearance of the rice grain. Rice breeders have included grain appearance and good grain milling quality in their objectives for developing improved varieties. But the milling and market quality of rice are greatly affected by harvesting and all other postharvest operations.

Harvesting

In the Asian tropics, rice harvesting is usually done by hand. The rice straw is usually cut with a sickle 15–25 cm above the ground and either carried away from the main field or dried in place. In some areas, particularly Indonesia and some upland rice-growing areas in the Philippines, only the panicles are removed.

In China, where intensive cropping is practiced, rice is completely removed from the main field shortly after harvesting (Fig. 13.1). In Malaysia, an estimated 200 combine harvesters (4.87-m width cut) are used in the Muda irrigation scheme. These harvesting machines are privately owned and appear to be highly profitable. The use of such machines is expanding to the other areas of Malaysia outside the Muda scheme (Ayob 1979, Rayarappan 1979). In Japan, small self-propelled combine harvesters are used (50–150-cm cutting width). Those can harvest 0.05 ha/hour. In the United States, Europe, and Australia, rice is harvested with self-propelled combine harvesters.

Time of Harvest

In the tropics, it is essential to harvest the crop on time, otherwise grain losses may result from feeding by rats, birds, and insects and from shattering and lodging. There is conclusive evidence that both early and late harvests are detrimental to the grain yield and to milling returns of rice (Faulkner and Wratten 1962, Langfield 1957, Morse et al. 1968). Timely harvesting ensures good grain quality, a high market value, and improved consumer acceptance. Bushel weight (bulk density) and germination percentage also have been found to be low at early harvest (Malabuyoc et al. 1966, Oelke et al. 1968).

Farmers judge their harvest timing by examining the percentage of ripened grains in the panicles. The crop should be ready to harvest when 80% of the panicles are straw-colored and the grains in the lower portions of the panicle are in the hard-dough stage. Farmers in some South and Southeast Asian countries

Figure 13.1   In Beijing area in People's Republic of China, rice is removed to leave a clean soil surface shortly after harvesting so that rice or other crops can be planted after the first rice crop.

usually harvest at maturity to minimize field losses resulting from shattering of overripe grains, unfavorable weather, and pilferage. Delays in harvest are caused primarily by lack of labor, especially during the peak months of harvest. In terms of optimum time of harvest for lowland rice, Nangju and De Datta (1970) reported it to be between 28 and 34 days after heading during the dry season and 34 and 38 days after heading during the wet season (Fig. 13.2). The earlier dry-season harvest is due to higher temperature and solar radiation during the ripening period. An upland rice crop matures sooner after heading than either a transplanted or direct-seeded lowland crop.

Nangju and De Datta (1970) showed that applying nitrogen fertilizers to varieties with chalky grains, particularly in the dry season, improved head rice (unbroken polished grains) recovery. In another study, Seetanun and De Datta (1973) found that a topdressing of nitrogen at heading increased the percentage of head rice, which was associated with the higher protein content resulting from the nitrogen treatment. Based on maximum grain yield, with highest milling recovery and seed viability, the best time for harvesting transplanted rice was between 30 and 42 days after heading in the wet season and between 28 and 34 days after heading in the dry season.

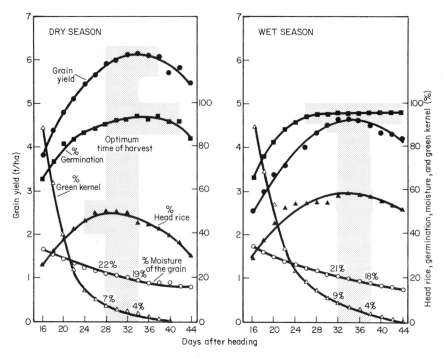

Figure 13.2 Optimum time of harvest on the basis of maximum grain yield and high percentages of head rice and germination as indicated by percent moisture of the grain at harvest and percent green kernels. IRRI, 1968 dry season and wet season. (Adapted from Nangju and De Datta 1970)

From India, Govindaswami and Ghosh (1968) reported that harvesting between 27 and 39 days after flowering at high moisture content (18–23%) gave maximum head rice recovery. Harvesting before or after that period resulted in an increase of broken grains.

Horiuchi et al. (1971) found that filled spikelet recovery in Malaysia was the highest 34 days after heading. During both seasons, the highest seed and volume weights were obtained from harvest 35 days after heading. In Japan, Eikichi (1954) found that the best time to harvest was 20–35 days after heading.

In California, some growers reported high head rice yields by harvesting at 22–26% moisture. In Arkansas, rice is harvested at 18–22% moisture content (Huey 1977). At that stage, the heads have turned downward and the grains in the lower parts of the head are in the hard-dough stage. Rice harvested before that stage may have a high percentage of light, chalky kernels, which reduce yields of head rice and total milled rice. If harvested later, there may be considerable loss from shattering. Also, there will be a reduction of head rice because of checking (stress cracking) of the kernels.

Cutting Height

An important consideration affecting the technique of harvesting is the method of threshing and the subsequent use for which the stalks will be used. In a Philippine study, Toquero and Duff (1974) reported that more than 50% of farmers used a sickle and harvested stalks 61–91 cm from the base of the panicle. About 30% used a hand-held knife (*ani-ani*) that cut the stalks 46 cm or less from the base of the panicle. The shorter stalk length facilitates the handling and stacking process. In many instances, rice straw is used as feed for animals.

In combine harvesting, standing rice is cut just below the heads to reduce the volume of stems passing through the threshing and separation units. Combine cutting height is set primarily to avoid excessive shattering and unharvested grain.

Losses in Harvesting

Wimberley (1972) reported that during the field drying process, overmature rice begins to shatter and fall and is lost in the puddled soil. Germination in the panicles prior to harvesting contributes to loss, and birds and rodents add to losses during harvesting and transporting.

In the Philippines, an estimated 10% of the rice crop is lost when grains are left in the field to dry to 14–16% moisture before harvest (de Padua 1970). Under normal harvesting conditions in Thailand, total losses run from 60 to 120 kg/ha (FAO 1968).

Recently, Djojomartono et al. (1979) measured harvesting losses from *ani-ani* and sickle harvesting of IR36 rice at different maturity levels; one level 5 days before and three levels with 3-day intervals beyond optimum maturity level based on farmers' judgment. Results showed losses are stable when the *ani-ani* is used irrespective of the harvesting date (Fig. 13.3).

A delay in harvesting beyond the stage of grain maturity affects the harvesting field losses as well as the quality of husked rice. Exogenous factors such as rainy days increase the percentage of broken grains at milling.

In Thailand and the Philippines, ducks are often used to glean the field for scattered grains after harvesting the rice crop. This reduces harvesting losses but unless they own the ducks, farmers may not derive any benefit from this practice.

The effect of combine adjustment and harvest losses of rice was studied by McNeal (1950). His study suggests four types of combine losses—by cutter bar, cylinder, rack, and shoe. Curley and Goss (1964) suggest that several loss checks should be made on a machine in a given area to determine the effect of adjustment or change in ground speed. Electronic harvester attachments are now available to monitor grain losses in the threshed straw.

Threshing

Threshing methods for rice vary greatly from country to country. The methods are generally classified as manual, animal, or mechanical.

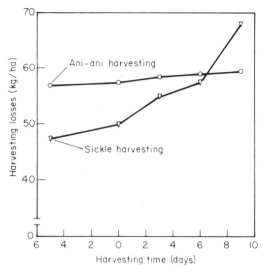

Figure 13.3   Harvesting losses at different harvesting dates of IR36 rice (dry season). (Adapted from Djojomartono et al. 1979)

Principle of Threshing

Panicle retention forces vary with the size, form, and structure of the plant tissue holding the kernels. Other critical threshing variables are grain weight, degree of ripeness, and kernel moisture content (Esmay et al. 1979).

Studies at IRRI determined the separation of rice grains from the panicle by centrifugal force. The separation force was determined to be

$$F = 3.9943 W_g\ R\ N_c^2$$

where

F = threshing force in kg
W_g = grain weight in kg
R = average radius of panicle in m
N_c = revolutions per second (Casem and Khan 1968)

The common method of separating grain from the panicle is by impact, which may be applied by hand beating, treading, or mechanically by a rotating drum with spikes or rasp bars. Hand threshing is often done by beating the panicles on a slotted bamboo platform. Figure 13.4 shows hand threshing in a farmer's field in the Philippines. Occasionally, grain is threshed by having animals, or humans, tread on the harvested crop. Mechanical threshers are still uncommon in most of tropical Asia. In some parts of the Philippines, mainly Central Luzon, some farmers use a McCormick-type thresher. These threshers are owned by individuals who provide custom threshing services to farmers. The large

Figure 13.4   Hand threshing is done in the Philippines primarily by beating the rice heads on a perforated platform made of bamboo.

machine, which cannot be efficiently run by an individual small farmer, has a threshing capacity of more than 1.5 t/hour (Duff and Toquero 1975). Small threshers, however, either individually owned or custom operated, are fast replacing the big threshers.

Types of Mechanical Threshers

There are various throw-in-type threshers in use in the Philippines; the Bicol-type, a popular small thresher with a single drum, has low efficiency because it has no separation system. Output per horsepower, however, is highest of the types tested (Policarpio and McMennamy 1978). That machine is rapidly being replaced by IRRI-designed portable threshers or some modified versions of them. Several portable power threshers have been designed at IRRI and are being evaluated in different rice-growing tropical countries.

The large McCormick-type thresher that is popular in Central Luzon, Philippines, has high capacity, but is a high horsepower type. It requires eight men and a 65–80-hp tractor to operate it. Labor output at 306 kg/hour is high and total grain loss is a low 4.12%. Purity of output is also high compared to that of other types of threshers. An IRRI-designed portable axial-flow thresher (Fig. 13.5) has a high labor output of 300–400 kg/hour and a total grain loss of 2%.

In Japan, in most of temperate Asia, and in other temperate rice-growing countries in the world, harvesting and threshing operations are combined by various sizes of rice combines. In general, the japonica varieties grown in

Figure 13.5 IRRI-designed, portable axial-flow threshers are now being used by farmers in India, Sri Lanka, Thailand, and the Philippines.

temperate Asia are harder to thresh than the indica varieties grown in most tropical Asian countries. The Japanese use a Kyowa-type double drum thresher that has a blower but no separator. An 8-hp engine and six men are required to operate the machine, and it has a low output per man-hour.

Losses in Threshing

Threshing losses vary with the thresher used. A study in Indonesia suggested that short straw posed problems in feeding the pedal and mechanical threshers and left unthreshed grains at the base of the panicles (Djojomartono et al. 1979). A Philippine study by Toquero et al. (1977) compared threshing losses of four types of machines. Grain losses were: large stationary thresher, 8.14% > manual thresher, 6.82% > IRRI-designed axial-flow thresher, 2.07% > IRRI-designed portable thresher, 1.97%.

Drying

The moisture content of rough rice must be below 14% before it can be safely stored. Rice is normally harvested at a moisture content of 20% or more. If the moisture content is not reduced to below 14% shortly after threshing, grain quality deteriorates because of microbial activities and insect damage. According to Huey (1977) drying should begin within 12 hours but not later than 24 hours after harvesting.

Principles of Drying

Rice, like other cereals, is a hygroscopic material and will change in moisture content in relation to the temperature and the relative humidity of the surrounding air. Details on the principles of drying are given by Esmay et al. (1979).

The drying process is basically the transfer of heat by converting the water in grain to a vapor and transferring it to the atmosphere. Heat is transferred to the kernel for the evaporation process by convection, radiation, or conduction. Convection is most commonly used. Convection drying requires the heating of the air to lower its relative humidity sufficiently to absorb moisture from the grain.

In most of Asia, rice grain is commonly dried in the sun, although this is difficult during the wet season. In general, 4–5 days of sun-drying are required to reduce the grain moisture content to an acceptable level. At IRRI, however, if weather conditions are good, one full day in the sun is enough to bring the moisture content from 24% to 14% (Duff et al. 1974).

Sun-drying operations vary greatly from country to country. In the Philippines, two methods of sun-drying are used. In the first, the harvested rice is placed in loose bundles and left to dry in the field for several days. The length of drying time depends on weather, practices in the area, and the availability of a thresher.

In the second method, immediately after threshing, the wet grains are spread on drying surfaces such as concrete pavement, mats, plastic sheets, canvas, and so on (Fig. 13.6). Repeated stirring is necessary to obtain uniform drying.

Sun-drying can produce grain fissures (sun checks), which results in low head-rice yields and high bran production in milling (Angladette 1963). Good sun-

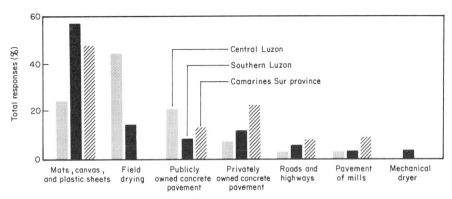

Figure 13.6   Methods of drying rough rice in three regions of the Philippines, 1974. (From IRRI 1976)

drying, with frequent stirring can, however, produce good quality milled rice. Sun-drying requires a large surface area and labor for spreading, stirring, and collecting the grain, and uncertain weather can result in poor control of the drying operations. The problem is that most operators do not stir often enough to minimize the differential heating that occurs and results in thermal stress in the grain. Another problem of sun-drying is that if it is not completed quickly it causes fissuring of the grain. This often results from the repeated wetting and drying of the grain that can occur as moisture condenses from the air at night (Kunze and Prasad 1978).

With increased crop intensification with photoperiod-insensitive, early-maturing rices, rice may often be harvested at the peak of the wet season. Therefore, an alternative to sun-drying must be developed if substantial losses in quality are to be avoided.

In Japan, a simple batch dryer has been used for many years. Recently, circulating batch dryers have gained rapid popularity for on-farm rice drying in Japan. These dryers mechanically load, unload, or circulate the grain to minimize labor. Both the simple flat-bed and the circulating batch dryer can dry 1.0–1.5 t in 8–10 hours (Khan 1973). Such dryers have good potential for adoption in tropical Asia provided they are locally produced and made available to farmers at reasonable prices.

In Texas, rice is bin-dried with unheated or slightly heated air (supplemental heat), and by multipass drying in continuous-flow hot air dryers. Bin-drying is particularly suited for on-farm installations because rice handling is reduced to a minimum. Drying is completed within 10–20 days by forcing a low volume of unheated or slightly heated air through a 2.5–3-m layer of rice.

In Arkansas, combine-harvested rice is usually dried by passing it through a multipass heated-air dryer at temperatures ranging from 38 to 54°C, or it may be put into bins where either heated or unheated air is passed through it until the moisture content is low enough for safe storage. Rice growers in California ordinarily maintain a rice drying temperature of 35°C. The 35°C temperature increased drying speed by 37% over ambient air drying and there was no increase in rice breakage. Rice temperatures increased with each pass (Willson 1979).

The continuous-flow dryers are usually commercial operations in which a large volume of heated air is forced through a thin layer of rice (10–25 cm thick) during a number of short duration passes. Between dryer passes, the grain is tempered in a storage bin for 6–24 hours. Drying is completed within 2–6 days, but the total time during which rice is exposed to heated air in the dryer generally is 2 hours or less (Calderwood et al. 1975).

Drying Rate and Temperature

Unlike most cereals, rice is consumed primarily as unbroken kernels, so that the market value for whole kernels is much greater than that for broken kernels. Therefore, more care is needed to avoid breakage in drying rice than in drying other cereals (Johnston and Miller 1973).

To minimize quality deterioration and maximize system efficiency, Esmay et al. (1979) suggested that careful consideration be given to the selection of the drying air temperature, the maximum grain mass temperature during drying, and the duration of the grain mass exposure to high temperatures at various grain moisture levels. The selection of optimal drying conditions also depends somewhat on the rice variety, the permitted extreme high drying temperature, and the initial grain moisture content. As the rice kernel is dried, the outer portion shrinks, setting up stresses and strains. When moisture is removed too rapidly, chalking or shattering of the kernel results.

Grain Losses in Drying

Mechanical drying produces significantly higher head-rice recoveries than sun-drying. Total rice recovery, however, is usually only marginally greater than that attained after sun-drying, although the recovery rate is a function not only of the method of drying but also the type of rice mill in which rice is subsequently processed. Figure 13.7 summarizes the trends and comparative efficiencies observed in comparing solar with mechanical drying methods at different moisture content at harvest resulting from different harvest dates.

Sun-drying of small amounts of rough rice is relatively easy to control and can give a good quality of dried rice. In an Indonesian study, differences in losses due to drying were negligible and difficult to detect (Djojomartono et al. 1979). Quality tests showed a head rice percentage of 53% for controlled sun-drying compared to 52% for mechanical drying using an IRRI batch-type dryer.

Cleaning

Cleaning of rice to remove foreign seed and trash is important because of their subsequent effects on the storability and milling quality. Rice with impurities is more likely to deteriorate in storage. Impurities also reduce the milling recovery rates, particularly if there are stones mixed with the rice. Unclean rice also increases the maintenance requirements on the milling machines.

Principles of Grain Cleaning

Grain cleaning is based on the function of air velocity, which is used to separate materials by weight, density, and wind resistance. As air velocity increases from 0 to 200 m/minute, the percentage of trash removed from grain increases sharply but grain losses are negligible. An increase in air velocity beyond 200 m/minute provides no improvement in trash removal but increases grain losses. Grain losses will attain about 1% at an air velocity of 400 m/minute (Harrington 1970).

Grain cleaning is based on three characteristics of the grain:

· *Air cleaning* takes advantage of differences in the weight and aerodynamic characteristics of the grains in separating them from other materials.

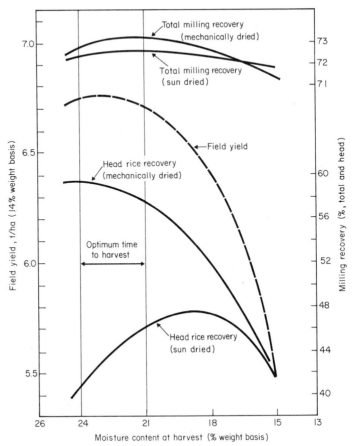

Figure 13.7 Advantages of mechanical versus solar dryer. (Adapted from Duff and Toquero 1975)

- *Mechanical cleaning* uses sieves for separation based on grain size and shape.
- *Gravity cleaning* with cleaners such as rough-rice separators uses differences in the specific gravity and bulk density of the grain to separate materials that have little difference in size and total weight.

Cleaning Operation

Cleaning is mostly done by a hand winnower, which takes advantage of wind. Other rough-rice cleaners include the winnowing basket, wooden or metal boxes with perforations, and a combination of the above (Duff and Toquero 1975). In some areas, a manually operated wooden winnower is used for cleaning paddy. These are not equipped with screens and high air velocities are used to blow away most of the impurities (Khan 1973).

For field cleaning, winnowing is undoubtedly most common, but mechanical separation, often in conjunction with air cleaning, is widely used in threshing machines and during rice milling.

Storage

Rice, like other cereal grains, is subject to deterioration because of changes in temperature and relative humidity. Insects and diseases are also more active after increases in temperature and relative humidity.

Principle of Storage

Rice grains will change moisture content until they are in equilibrium with the surrounding air temperature and relative humidity. It is generally considered that 13% (wet basis) moisture content is an acceptable value for rice. It is critical to have storage structures that can keep the moisture content and the relative humidity of the grains stable throughout the storage period.

Storage for Seeds

Low temperature and low moisture are necessary for the long-term storage of rice for seed (Adair 1966, Ito 1975, Ota 1971). Nishiyama (1977) evaluated the loss and recovery of germination activity of rice seed stored in controlled humidity. He found that rice seed of 10–14% moisture content can be stored in good condition at 18°C for more than 2 years. However, if the moisture content in seed was higher than 19%, germination decreased after about a year. If the moisture content of seed was between 5 and 6%, germination was much lower, but the germination rate did not change during 9 months of storage. Some of the decreased germination activity returned after rehumidification of the seed.

The use of heated-air systems to aerate it may prevent seed from absorbing moisture in storage in the tropics. The heated air should also prevent mold growth if the air is below 60% relative humidity.

In a United States study in Kansas, seed storage at 44°C and 55% relative humidity resulted in progressive loss of seed viability. However, storage at high temperatures should have little effect on the food value of rice (Pfost 1978).

In the United States, a maximum moisture content of 12% is recommended for seed rice, but up to 14% usually is safest in bulk storage.

Storage Facilities

In tropical Asia, rice is stored as rough or milled rice. Rough rice (unhusked) is stored for home consumption in sacks, metal or wooden boxes, bamboo baskets, cans, drums, and small granaries. If it is stored in large quantities, a special room within the farmer's house is used. Rough rice to be sold is stored either in the farmer's house or in a warehouse. Few farmers, however, store their rough rice in the mill's warehouse. It is desirable to store rice in gunny sacks on wooden platforms that allow free passage of air between grain sacks and the floor.

In the warehouse, gunny sacks are the most common container for storage but crude bulk storage in which rough rice is piled in one corner of the mill is used by some millers. The most important factors in warehouse storage are to store well-dried rough rice and to keep the building moisture-proof. In Japan and other temperate rice-growing countries, rough rice is stored in large storage facilities which are protected from major losses due to insects and rodents.

For storage bins, any construction material is satisfactory if it results in a storage structure that will keep the grain dry, cool, and free of insects and other pests. It should provide safety and convenience for those moving and inspecting the grain (Johnston and Miller 1973). If dried rough rice is to be stored longer than a few months, or if damp rice is to be held before drying, the storage structure should be equipped for aeration.

Losses in Storage

Losses due to pilferage, rats, birds, and insects are the major rice storage problems in tropical Asia. The moisture content of dried grains often increases during storage causing bacterial and fungal diseases. Few farmers in South and Southeast Asia treat their stored rough rice with chemicals, even though it is known that fumigation is the only practical method to control insect damage. Empty jute sacks often contain insects and should be treated with chemicals or fumigated before use for rice.

Storage insect control depends on:

- Proper bin construction and management.
- Proper sanitation.
- Proper drying of grain.
- Proper physical control.
- Proper chemical control.

Sanitation can be improved through good housekeeping practices.

In bulk storage, it is essential to provide continuous aeration by forcing a stream of air through grains. Rat-proofing a building consists of changing structural details to prevent the entry of rodents (Esmay et al. 1979).

RICE PROCESSING

Interest in rice processing activities is related to the possibility of increased yield of milled rice and the potential for low cost increases in output through the use of improved processing systems.

Parboiling

Parboiling is a hydrothermal treatment of rough rice prior to milling.

Parboiling affects the milling, storage, cooking, and ultimately consumer preference of rice. A comparison of raw and parboiled rice is given in Table 13.1.

In the parboiling process, rough rice is soaked, steamed, and redried before milling. In the United States and Europe, parboiling is done with modern equipment but various methods of parboiling are used in India, Bangladesh, Pakistan, Sri Lanka, Burma, and Malaysia. In many areas, crude methods are still used.

Parboiling gelatinizes the starch within the rice grain, thus, causing swelling and fusion of starch within the kernel. A preparboiling moisture content of 30–35% (wet basis) and 26 cal of heat/kg of rough rice are needed (Esmay et al. 1979).

The three important steps for parboiling are:

· Soaking (sometimes called steeping) of rough rice in water to increase its moisture content to about 30%.

· Heat treating the wet rough rice, usually by steam, to complete the physicochemical changes.

· Drying the rough rice to a moisture level safe for milling.

Table 13.1 Range and Mean of Some Properties of Seven Samples of Raw and Parboiled Rice (Adapted from Raghavendra Rao and Juliano 1970)

Property	Raw		Parboiled	
	Range	Mean	Range	Mean
Brown rice				
Hardness (Kiya tester)				
Breaking, kg	4.6–6.4	5.4	6.3–12.1	9.6
Crushing, kg	7.8–9.9	8.9	14.4–16.3	15.4
Protein, % dry basis	7.57–12.8	9.9	7.45–12.7	9.86
Total extractable protein	7.10–10.95	8.10	3.49–6.12	4.60
Milled rice				
Head rice, % of milled rice	42.4–92.6	75.3	99.4–100	99.8
Amylose, % dry basis	2.0–27.2	17.0	2.0–27.4	16.4
Alkali test values				
Spreading	2.2–7.0	4.2	1.1–6.0	4.7
Clearing	1.2–6.0	3.7	3.0–5.7	4.4
Cooking test (presoaked grain)				
Water-uptake ratio	2.52–3.53	3.09	2.63–3.12	2.88
Solids in cooking water, %	7.4–15.8	9.9	4.0–7.1	5.5

The advantages of parboiling are:

- Easier dehulling.
- Reduction in the number of broken grains during milling because the grains are strengthened during parboiling.
- Harder grains, which are less vulnerable to insect attack during storage.
- Parboiled milled rice is richer in the B vitamins than milled raw rice (Padua and Juliano 1974).
- Cooked parboiled rice is flakier (less sticky) than freshly harvested raw rice.

The disadvantages of parboiling are:

- Parboiled rice tends to become rancid during storage.
- Longer cooking time is required.
- Additional cost (and more energy) is involved in the drying operation following parboiling.
- Parboiled rice requires more energy to achieve the same degree of milling.

Parboiling Methods

Many traditional and modern parboiling processes have been used in different countries. Most parboiling of rough rice in Asian countries is by:

- Soaking in large concrete tanks and steaming in small kettles.
- Using a small metal tank for soaking and steaming without using a boiler.
- Using large metal tanks for soaking and steaming with a boiler.

These methods have proven economical over many years of operation. Each method, when operated properly, produces good quality parboiled rough rice at a minimum operating cost (Wimberley 1980).

The most conventional method of parboiling consists of soaking the rough rice and then steaming it. The parboiled rice is then dried in the sun. The process takes from 24 hours to several days depending on the weather. Care is taken to use clean water and rice free of foreign matter.

Experimentally, Khan et al. (1973) used ultrahigh-conduction drying of rough rice at temperatures of 100–200°C, which reduced the moisture content of the rice from about 28 to 18% in a few seconds. Rice was simultaneously parboiled and dried to about 18% by placing it in contact with heated sand in the laboratory. The high temperature drying process produced a high gelatinization of the starch granules which fused sun-checks and other fissures. The fusing process resulted in higher output of head rice after milling.

In the United States, the rough rice is dried after parboiling by putting it through a continuous-flow mechanical dryer. If mechanical equipment is used for drying, a sequence has been worked out for a more uniform operation. The sequence of operation is important for two reasons:

- The plant is designed for a given capacity with a certain sequence of operation. If not followed, daily capacity drops considerably.
- Maximum utilization of equipment gives the lowest operation cost.

The following is the sequence of operation for one batch of rough rice of one ton or more. The sequence can be paralleled with other batches using staggered times, thus giving a continuous operation, maximum use of equipment, and maximum capacity.

Step	Operation	Time Required (hour)
A	Loading water and rough rice into the parboil tank	1
B	Soaking and steaming	4
C	Moving rough rice to the dryer, drying, moving the rough rice to the tempering bins	3
D	Tempering	6
E	Second drying pass	3
	Total	17

The parboiling processes are described in detail by Wimberley (1980).

Milling

Rice milling involves the removal of hulls and bran from rough rice to produce polished rice. Milling ranges from hand pounding with the simple wooden mortar and pestle in many remote areas of South and Southeast Asia and West Africa to the modern equipment developed in Japan, Europe, and the United States.

A great quantity of rice is lost during the milling process. Therefore, milling requires careful planning and use of properly designed and operated equipment. A good rice milling operation should:

- Produce the maximum yield of edible rice.
- Obtain the best possible quality.
- Minimize losses.
- Minimize the processing cost (Esmay et al. 1979).

Milling Recovery

Time of harvest and season affect the milling yield of rice. In an IRRI study, the percentage of head rice decreased with time less abruptly in the wet season than in the dry season (Fig. 13.8). Earlier and more severe breakage of the grains due to sun-cracking occurred during the dry season (Seetanun and De Datta 1973).

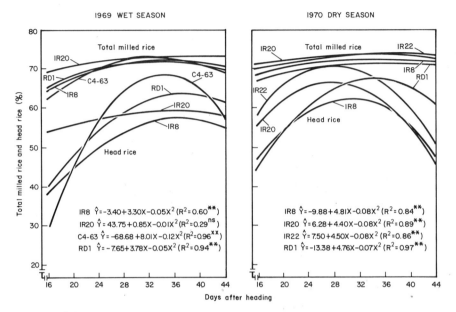

Figure 13.8   Effect of time of harvest on the percentage of total milled rice and head rice in IR8, IR20, IR22, C4-63, and RD1 (average of three nitrogen treatments). IRRI 1969 wet season and 1970 dry season. (Adapted from Seetanun and De Datta 1973)

The Milling Process

The output of milled rice and its quality are highly affected by the milling process and equipment used. In many rural areas, hand pounding is still practiced with various indigenous devices. The percentage yield of whole kernels from those hand pounding devices is low.

Three types of mills are generally used—the steel huller Engelberg type, the under-run disc sheller-cono system, and the modern mill:

- The steel huller (Engelberg type) removes the husk and bran in one operation.
- The under-run disc sheller mill (cono system) has a cleaner, disc shelling unit, a separator, and cone-type polishers.
- The modern rice mill has extensive provisions to remove foreign matter, dehusk the grain, remove and rehusk unhulled grains, remove the bran layer, grade rice by separating broken grains from head rice, and whiten to put a final polish on the milled grain. Some have automatic blending and bagging facilities.

ENGELBERG TYPE   The most popular rice milling equipment in many areas is the simple Engelberg type. About 53% of the rice produced in the Philippines is processed by this type of mill. The Engelberg mill has a high power requirement and subjects rough rice to excessive pressure and friction, resulting in heating, low milling recovery, and high grain breakage. The milling recovery of the steel huller mill is 60–63% compared with 68–70% in modern mills (Wimberley 1972).

To improve performance and reduce loss, some Engelberg operators recycle the grain two or three times or arrange two or three steelhullers in series. The inefficient performance of this rice mill causes enormous loss of grain. The by-products of milling are often used for animal feed.

The steel huller has the advantages of:

- Low initial cost.
- Ease of operation and maintenance.
- It can mill small-lot sizes conveniently and economically.

Most Engelberg mills are operated as a part-time family enterprise and payment for milling is usually in by-products or as a percentage of the milled rice. The Engelbergs appear to have considerable utility in most areas because of their wide dispersal (convenient to most farm households), production of by-products for livestock feeding, and low cash requirements for operation. Until transport systems and marketing practices are improved, or a comparable low cost alternative practice is available, it is unlikely the Engelberg will be replaced.

CONO SYSTEM   The cono system mill processes the greatest share of rice entering the commercial market. It usually consists of one or more under-run disc sheller units operated in conjunction with a number of vertical cono polishers. Materials are fed into a central opening and centrifugal action causes radial movement of the rough rice into the space between the discs.

In the Philippines, cono installations produce appreciably higher recovery rates (up to 68% by weight) than the Engelberg unit. In addition, head rice recovery is higher, bran and hulls are separated, and the degree of milling can be closely controlled (Duff and Estioko 1972). Some alternative rice milling systems are shown in Fig. 13.9.

MODERN MILLS   Modern milling equipment has the rubber roller system, which was developed in Japan. It consists of two closely spaced rubber rollers rotating in opposite directions at different speeds. Contact with the rubber rollers creates a shearing action on rough rice, which removes the husk from the grain. Operated in conjunction with an aspirator and paddy separator, it is possible to recycle unhulled paddy through the machine to achieve complete hulling.

When correctly adjusted for optimal operation, the rubber-roll method produces the highest total recovery and best head rice yield.

Despite its high technical efficiency, however, high investment costs, combined with a lack of demand for high-quality rice, have retarded diffusion of this technology in South and Southeast Asia. Furthermore, they are often run at a loss because of the high-volume requirement of modern rice milling systems

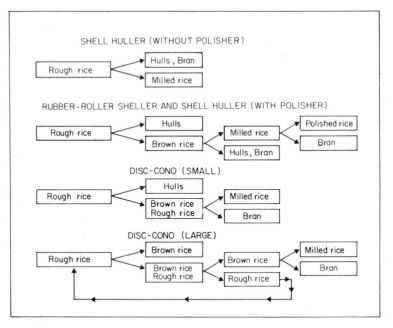

Figure 13.9   Alternative rice milling systems. (Adapted from Duff and Estioko 1976)

(1–10 t/hour), which is usually not guaranteed in big mills in South and Southeast Asia.

Milling Efficiency

Evaluation and comparison of milling efficiency should be based on the entire milling process rather than on the hulling unit. Many of the rice milling systems in the tropics do not, however, have a complete line of standard components such as rough rice cleaners, separators, and the like. Table 13.2 shows some generalized data on the efficiency of four milling systems (Esmay et al. 1979) and Fig. 13.10

Table 13.2   Milling Efficiency of Rice as Affected by Four Milling Systems (From Esmay et al. 1979)

Milling Process	Husk (%)	Bran (%)	Total Husk and Bran (%)	Head (%)	Broken (%)	Total Head and Broken (%)
Hand pounding	—	—	40.0	40.0	20.0	60.0
Steel hullers	—	—	36.6	46.5	16.9	63.4
Disc shellers	—	—	32.5	55.9	11.6	67.5
Rubber rollers	22	8	30.0	62.0	8.0	70.0

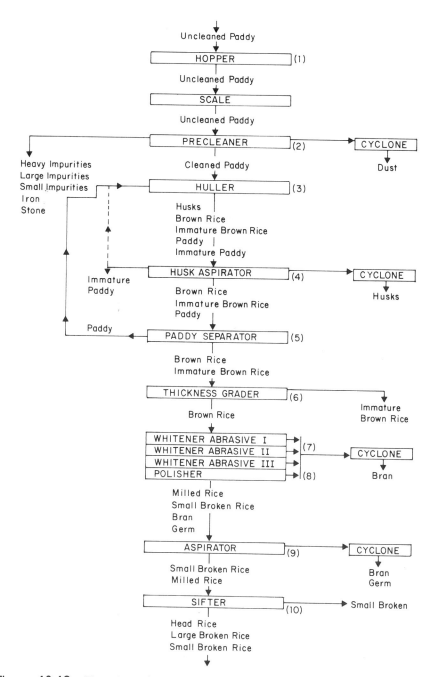

Figure 13.10 Flow chart of the complete modern milling process. (From Esmay et al. 1979)

533

shows a flow diagram of a 10-step milling process. The products of each operation are shown directly below and the by-products are shown on the side.

Milling Potential

The milling potential of rough rice can be ascertained as follows:

$$\text{dockage (\%)} = \frac{\text{wt of impurities}}{\text{wt of sample (500 g)}} \times 100$$

$$\text{weed and crop seeds (\%)} = \frac{\text{wt of weeds and crop seeds}}{\text{wt of sample}} \times 100$$

$$\text{other varieties (\%)} = \frac{\text{wt of other varieties}}{\text{wt of sample used (25 g)}} \times 100$$

$$\text{chalky and immature (\%)} = \frac{\text{wt of chalky and immature kernels}}{\text{wt of sample used}} \times 100$$

$$\text{damaged kernel (\%)} = \frac{\text{wt of damaged kernels}}{\text{wt of sample used}} \times 100$$

$$\text{yellow and fermented (\%)} = \frac{\text{wt of yellow fermented kernels}}{\text{wt of sample used}} \times 100$$

$$\text{red rice (\%)} = \frac{\text{wt of red rice}}{\text{wt of sample used}} \times 100$$

$$\text{milling recovery (\%)} = \frac{\text{wt of milled rice}}{\left(\begin{array}{c}\text{wt of rough}\\\text{rice sample}\end{array}\right) - \left(\begin{array}{c}\text{wt of unhulled rice}\\\text{used in hulling}\end{array}\right)} \times 100$$

$$\text{head rice (\%)} = \frac{\text{wt of whole milled rice}}{\text{wt of milled rice used (100 g)}} \times 100$$

$$\text{broken rice (\%)} = \frac{\text{wt of broken milled rice}}{\text{wt of milled rice sample used}} \times 100$$

To gauge the performance of a huller, the coefficient of hulling and wholeness must be determined (IRRI 1978). The coefficient of hulling indicates the quantity of rough rice actually dehusked during a single hulling pass; the coefficient of wholeness indicates the quantity of unbroken brown rice in the brown rice yield. This breakage can be the result of the huller performance or can be related to the quality of the rough rice—cracked grain, moisture content, and so on.

The efficiency of a huller can be determined using the equations from Camacho et al. (1978):

$$\text{Hulling efficiency (\%)} = \text{coefficient of hulling} \times \text{coefficient of wholeness} \times 100$$

where

$$\text{coefficient of hulling} = \frac{\text{wt of brown rice}}{\text{wt of sample (mixture of rough and brown rice)}}$$

$$\text{coefficient of wholeness} = \frac{\text{wt of whole brown rice}}{\text{wt of sample (mixture of whole and broken brown rice)}}$$

Glossary of Terminology

The terminology used in postharvest technology and rice quality are summarized in Table 13.3.

GRADES AND STANDARDS

Rough rice grades and standards are applied to rice with the hull on it. After the rice is milled, a set of milled rice standards is applied. The latter permits a more precise measure of value. Most pricing of rice is based on a milled sample or on the grade of the entire lot after milling.

Milled Rice Grades for Export Market in Thailand

Thailand is the leading exporter of rice in Asia and a leading rice exporter of the world. The quality standard for milled rice followed by the Thailand export market is extremely rigid. The Ministry of Commerce recently revised the standards with regards to production, milling, trading, and international quality standards.

Rice grain is divided into four classes:

1 *Extra long grain* is the whole grain having a length of more than 7.0 mm.
2 *Long grain* is the whole grain having a length of more than 6.6 mm but not more than 7.0 mm.
3 *Medium grain* is the whole grain having a length more than 6.2 mm but not more than 6.6 mm.
4 *Short grain* is the whole grain having a length of 6.2 mm or less.

Table 13.3 Glossary of Terminology Used in Postharvest Technology (From Duff and Estioko 1972)

Terminology	Meaning
Hulling, husking, or shelling	Process of removing hulls from the grain
Scouring or pearling	Process of bran removal (locally known as polishing)
Polishing	Removal of very fine bran clinging to the milled rice and smoothing of its surface to appear glossy and clean (often called whitening)
Milling degree	Extent or degree of removing the different bran layers which cover the rice grain as a result of polishing
Milling recovery	Weight of milled or commercial rice recovered from the original rough rice usually expressed in percent
Kernel	Edible portion of rice seed
Rough rice (paddy)	Unhulled rice
Milled rice	Whole or broken kernels where the hull and at least the outer bran layers and a part of the germ are removed; cannot contain more than 1% seeds or foreign matter
Head- or whole-grain recovery	Weight of whole grain recovered from the milled rice expressed in percent of the original rough rice (kernels not less than 3–4 of the whole grain)
Brokens	Milled rice with 25% or less than 3–4 to 1–2 of a whole grain
Brewer	Milled rice whose sizes range from 1–2 to 1–4 of a whole grain
Chalky kernels	Milled rice containing 25% or larger white portions
Flinty kernels	Kernels which are less than 2% white
Foreign matter	Impurities such as weed seeds, stones, sand, dust, etc., which are foreign to milled rice.
Red rice	Rice with any degree of redness

Whole grain means the full kernel without any broken part. The grades and grade requirements are shown in Table 13.4.

United States Grades and Standards for Rough and Milled Rice

In the United States, the primary factors relating to rough rice grade are seeds and head-damaged kernels, red rice, chalky kernels, and rice of other classes. For each of these factors, a tolerance has been established and limits are also stated for each grade (Table 13.5).

A SYSTEMS APPROACH TO POSTPRODUCTION OPERATIONS

In temperate countries, the entire sequence of postharvest operations is highly mechanized and has developed into an interdependent system. Harvesting, threshing, milling, storage, and marketing operations are closely interrelated. This has led to a highly developed postproduction system that results in minimal losses in quantity and marketable quality of rice. As a result of mechanization of postproduction operations, the risk and uncertainty to the rice farmer are reduced.

In tropical Asia, mechanization of postharvest operations is limited. However, several new machines for threshing, cleaning, and drying have been designed by IRRI and other national centers, and offer the opportunity to partially mechanize the postharvest operations. But the approach to improved post-harvest operations is often piecemeal. For example, one program deals with improvement of storage facilities, another with improving drying operations, and still another with better milling. Such independent programs may be beneficial, but savings—of time and costs—are often possible when all components are viewed as interdependent and development attempts integrate these components to produce more efficient operations. The interdependent nature of all postharvest operations means each operation should be treated with an awareness of its interaction with others. For example, if harvesting is carried out at the wrong time or using the wrong method, it may be impossible to dry, mill, or store the resulting crop properly. This can result in poor quality rice.

Figure 13.11 shows a simplified diagram of postharvest operations. Wimberley (1980) suggests using a systems approach to reduce costs and improve efficiency that includes matching the cleaning and drying facilities to purchasing programs, adjusting storage capacities to receiving and milling schedules, and matching milling capacities and facilities to the storage system and rice distribution requirements. This systems approach assures maximum utilization of existing facilities and minimum investment in new facilities, resulting in the lowest possible operation cost.

The size of storage and processing plants is also a major consideration. Large plants have higher investment costs but lower operation costs and use fewer

Table 13.4 Standards of White Rice Stipulated by Ministry of Commerce, Thailand (Adapted from Board of Trade, Thailand, 1978)[a,b]

Grain Classification columns are grouped under "Length of Grain" (Extra Long, Long, Medium, Short Grain) plus "Size of Brokens"; Grain Composition and Maximum Allowance of the Following Mixtures follow.

Grades of White Rice	Extra Long 7 mm (%)	Long 6.6–7.0 mm (%)	Medium 6.2–6.6 mm (%)	Short Grain less than 6.2 mm (%)	Size of Brokens	% Whole Grain	% Head Rice Big Brokens	% Brokens	% Small White Brokens C.1	% Red Streaked Kernels	% Red Kernals	% Chalky Kernels[a]	% Damaged Kernels	% Yellow Kernels	% Shrivelled Kernels	% Immature Kernels	% Split Kernels	% Foreign Matters	% Seeds	% Glutinous Rice	Paddy per 1 kg	Milling Degree	Moisture not higher Than
100% Class A (70–100)	+70 (70–100)	±25 (0–30)	−5 (0–5)	(0–5)	−8.0 / +5.0	+60	±36	−4	—	—	—	0.5	—	—	—	—	—	—	—	0.5	5	Extra well milled	14%
100% Class B (45–55)	+50 (45–55)	±35 (30–40)	±10 (0–25)	(0–5)	−8.0 / +5.0	+60	±35.5	−4.5	—	—	—	0.5	—	—	—	—	—	—	—	0.5	10	Extra well milled	14%
100% Class C (30–40)	+35 (30–40)	±45 (40–50)	±15 (5–30)	(0–5)	−8.0 / +5.0	+60	±35	−5	—	—	—	0.5	—	—	—	—	—	—	—	0.5	15	Extra well milled	14%
5% (20–25)	+20 (20–25)	±35 (30–40)	±35 (25–50)	(0–10)	−7.5 / +3.5	+60	±33	−7 (3–7)	—	2	—	2.5	0.25	0.5	—	—	0.5	0.1	—	0.5	15	Well milled	14%
10% (10–15)	+10 (10–15)	±30 (25–35)	±45 (35–55)	(10–15)	−7.0 / +3.5	+55	±33	−12 (8–12)	—	2	—	3	0.5	1	—	—	0.75	0.2	—	0.5	20	Well milled	14%
15% (5–10)	+5 (5–10)	−20 (0–20)	−40 (20–40)	(35–50)	−6.5 / +3.0	+55	±28	−17 (13–17)	—	4	1	3	1	1	—	—	0.75	0.2	—	0.5	25	Reasonably well milled	14%
20% (0–10)	+0 (0–10)	−15 (0–15)	−30 (10–30)	(55–65)	−6.0 / +3.0	+50	±27	±22 (18–23)	−1	5	2	5	2	1	0.5	0.5	0.75	0.25	—	0.5	25	Reasonably well milled	14%
25% (Super) (0–8)	+0 (0–8)	−35 (17–35)	—	(65–75)	−5.0 / +3.0	+40	±32	±27 (23–28)	−1	4	1	3	1	1	—	—	0.75	0.2	—	0.5	30	Reasonably well milled	14%
25% (0–8)	+0 (0–8)	−35 (17–35)	—	(65–75)	−5.0 / +3.0	+40	±32	±26 (23–28)	−2	6	4	8	2	1	1	1	0.75	0.5	0.5	0.5	30	Ordinarily milled	14%
35% (0–8)	+0 (0–8)	−35 (17–35)	—	(65–75)	−5.0 / +3.0	+32	±28	±38 (33–40)	−2	7	4	10	2	1	1	1	0.75	1	0.5	0.5	30	Ordinarily milled	14%
45% (0–8)	+0 (0–8)	−35 (17–35)	—	(65–75)	−5.0 / +3.0	+28	±22	±47 (42–50)	−3	8	4	10	2	1	1	1	0.75	1	0.5	0.5	30	Ordinarily milled	14%

[a] Except the grade 100% Class A, allowance of higher percentage than that specified above has been approved by the authority.

[b] + = not less than; − = not more or less than; ± = more or less than.

538

Table 13.5 Grades and Grade Requirements for the Classes of Long-Grain Milled Rice, Medium-Grain Milled Rice, Short-Grain Milled Rice, and Mixed Milled Rice (Adapted from Huey 1977)

	Maximum Limits of										
	Seeds, Heat-Damaged, and Paddy Kernels (singly or combined)	Heat-Damaged Kernels and Objectionable Seeds (no. in 500 g)	Red Rice and Damaged Kernels (singly or combined) (%)	Chalky Kernels[a]		Broken Kernels				Other Types[c] (%)	Color[a] and Milling Requirements[d]
Grade	Total (no. in 500 g)			Long-Grain Rice (%)	Medium- or Short-Grain Rice (%)	Total (%)	Removed by a 5 Plate[b] (%)	Removed by a 6 Plate[b] (%)	Through a 6 Sieve[b] (%)		
U.S. No. 1	2	1	0.5	1.0	2.0	4.0	0.04	0.1	0.1	1.0	Shall be white or creamy and shall be well milled
U.S. No. 2	4	2	1.5	2.0	4.0	7.0	0.06	0.2	0.2	2.0	May be slightly gray and shall be well milled
U.S. No. 3	7	5	2.5	4.0	6.0	15.0	0.10	0.8	0.5	3.0	May be light gray and shall be at least reasonably well milled
U.S. No. 4	20	15	4.0	6.0	8.0	25.0	0.40	2.0	0.7	5.0	May be gray or slightly rosy and shall be at least reasonably well milled

Table 13.5 Continued

		Maximum Limits of									
	Seeds, Heat-Damaged, and Paddy Kernels (singly or combined)		Red Rice and Damaged Kernels (singly or combined) (%)	Chalky Kernels[a]		Broken Kernels				Other Types[c] (%)	Color[a] and Milling Requirements[d]
	Heat-Damaged Kernels and Objectionable Seeds (no. in 500 g)	Total (no. in 500 g)		Long-Grain Rice (%)	Medium- or Short-Grain Rice (%)	Total (%)	Removed by a 5 Plate[b] (%)	Removed by a 6 Plate[b] (%)	Through a 6 Sieve[b] (%)		
Grade											
U.S. No. 5	25	30	6.0	10.0	10.0	35.0	0.70	3.0	1.0	10.0	May be dark gray or rosy and shall be at least lightly milled
U.S. No. 6	75	75	15.0[e]	15.0	15.0	50.0	1.00	4.0	2.0	10.0	May be dark gray or rosy and shall be at least lightly milled

U.S. sample grade

U.S. sample grade shall be milled rice of any of these classes which: (a) does not meet the requirements for any of the grades from U.S. No. 1 to U.S. No. 6 inclusive, (b) contains more than 15.0% moisture, (c) is musty, or sour, or heating, (d) has any commercially objectionable foreign odor, (e) contains more than 0.1% foreign material, (f) contains live or dead weevils or other insects, insect webbing, or insect refuse, or (g) is otherwise of distinctly low quality.

[a]These color requirements do not apply to parboiled milled rice.
[b]Plates should be used for southern production rice and sieves should be used for western production rice, but any device or method which gives equivalent results may be used.
[c]These limits do not apply to the Mixed Milled Rice Class.
[d]These color requirements do not apply to undermilled milled rice.
[e]Grade U.S. No. 6 shall contain not more than 6.0% of damaged kernels.

541

Figure 13.11   Sequence of postharvest operations. (Adapted from Wimberley 1980)

people per ton of rice handled. On the other hand, a smaller plant requires less capital investment, has higher operational costs, and uses more people per ton of rough rice handled.

Using Philippine data, Habito and Duff (1980) attempted to simulate a system formulated to study the rice postproduction system consisting of harvesting, threshing, cleaning, and grain drying. The bulk of the work was spent deriving the various interrelationships among factors within the system, which are in the nature of agronomic, technical, and economic relationships. Their results suggest that mechanized threshing will enjoy wider acceptance in the future. However, with the range of existing technologies, farmers will still favor sun-drying for some time to come.

It is obvious that a systems approach to all postharvest operations must be developed before rice production and postproduction yields and returns can increase.

REFERENCES

Adair, C. R. 1966. Effect of storage treatment on germination of rice. *Int. Rice Comm. Newsl.* 15.

Angladette, A. 1963. *Rice drying principles and techniques.* FAO Informal Working Bull. 23. 73 pp.

Ayob, A. M. 1979. The economics and adoption of the combine harvester in the Muda Region of Malaysia. Paper presented at the Workshop on the Consequences of Small Rice Farm Mechanization, 1–4 October, International Rice Research Institute, Los Baños, Philippines. 31 pp. (unpubl. mimeo.)

Board of Trade of Thailand. 1978. *Notification of the Ministry of Commerce re: standard of rice in effect as from January 30.* B. E. 2517. Bangkok, Thailand. 46 pp.

Calderwood, D. L., J. W. Sorenson, Jr., and H. W. Schroeder. 1975. Drying, storing and handling rice. Pages 81–91 *in* Texas Agricultural Experiment Station in coopera-tion with the United States Department of Agriculture. *Six decades of rice research in Texas.* Res. Monogr. 4.

Camacho, I., P. Hidalgo, B. Duff, and E. Lozada. 1978. Comparison of alternate rice milling systems in the Bicol region. Pages 101–198 *in* Department of Agriculture, Thailand, and the Southeast Asia Cooperative Post-harvest Research and Development Programme. *Proceedings of the workshop on grain-harvest technol-ogy.* Bangkok, Thailand.

Casem, E. O., and A. U. Khan. 1968. *Physical properties of the rice plant and paddy.* International Rice Research Institute. Los Baños, Philippines. 5 pp.

Curley, R. G., and J. R. Goss, 1964. Estimating combine losses in rice. Pages 15–17 *in* University of California, Department of Agricultural Engineering. *Agronomy notes.* Los Angeles, California. Feb. 1964.

de Padua, D. 1970. Basic principles in grain drying and milling. Pages 247–260 *in* University of the Philippines College of Agriculture in cooperation with the International Rice Research Institute. *Rice production manual.* Los Baños, Philippines.

Djojomartono, M., K. Abdulla, and R. Syarief ST. 1979. In field post rice production losses on farm in West Java. Pages 92–109 *in* National Logistic Agency and the Department of Agriculture, Indonesia, and Southeast Asia Cooperative Post-Harvest Research and Development Programme. *Proceedings of the workshop on grain post-harvest technology.* Los Baños, Philippines.

Duff, B., and I. Estioko. 1972. Establishing design criteria for improved rice milling technologies. Paper presented at a Saturday seminar, 26 August 1972, International Rice Research Institute, Los Baños, Philippines. 28 pp. (unpubl. mimeo.)

Duff, B., and I. Estioko. 1976. Design criteria. Pages 291–307 *in* E. V. Araullo, D. B. de Padua, and M. Graham, eds. International Development Research Center. *Rice postharvest technology.* Ottawa, Canada.

Duff, B., and Z. Toquero. 1975. Factors affecting the efficiency of mechanization in farm level rice post-production systems. IRRI Pap. 75-04 AE, Los Baños, Philippines. Paper presented at a workshop on rice post-production technology, University of the Philippines at Los Baños. 45 pp.

Duff, B., F. E. Nichols, J. K. Campbell, and C. C. Lee. 1974. Agricultural equipment development research for tropical rice cultivation. International Rice Research Institute, Los Baños, Philippines. Semiannual Prog. Rep. 18. Contract No. AID/CSD-2541. 52 pp. (unpubl. mimeo.)

Eikichi, I. 1954. *Rice crops in its rotations in sub-tropical zone.* Bunshodo Printing Co., Tokyo.

Esmay, M., Soemangat, Eriyatno, and A. Phillips. 1979. *Rice postproduction technology in the tropics.* The University Press of Hawaii, Honolulu. 140 pp.

FAO (Food and Agriculture Organization). 1968. *Pilot study of paddy losses in Thailand during harvesting, drying, and threshing.* IRC/AE/WP 29. FAO, Rome. 29 pp.

Faulkner, M. D., and F. T. Wratten. 1962. Rice drying and processing. Pages 109–141 *in* Louisiana Rice Experiment Station. *54th annual progress report.*

Govindaswami, S., and A. K. Ghosh. 1968. Assessment of losses of paddy and rice during harvesting, drying, threshing, cleaning, storage, and processing. Int. Rice Comm./AE/WP 13. 3 pp.

Habito, C. F., and B. Duff. 1980. A simulation model to evaluate mechanization of rice postharvest operations in the Philippines. International Rice Research Institute, Los Baños, Philippines. 49 pp. (unpubl. mimeo.)

Harrington, R. E. 1970. Thresher principles confirmed with a multicrop thresher. 8th Annual meeting, Indian Society of Agricultural Engineers, Ludhiana, Punjab, India.

Horiuchi, T., S. J. Samy, C. C. Phang. 1971. Grain loss during hand harvesting in the rice cultivation in Kedah, West Malaysia. *Tonan Ajia Kenkyu* (Southeast Asia Studies) 9:220–226.

Huey, B. A. 1977. *Rice production in Arkansas.* University of Arkansas, Division of Agriculture, and USDA Cooperating Circ. 476 (Rev.). 51 pp.

IRRI (International Rice Research Institute). 1976. *Annual report for 1975.* Los Baños, Philippines. 479 pp.

IRRI (International Rice Research Institute). 1978. The technical and economic characteristics of rice: postproduction systems in the Bicol River Basin. A report. International Rice Research Institute and the University of the Philippines at Los Baños to the Bicol River Basin Development Program. Los Baños, Philippines. 143 pp. + appendices.

Ito, H. 1975. Long term storage of crop seeds and its use for breeding (1) [in Japanese]. *Agric. Hortic.* 50:849–852.

Johnston, T. H., and M. D. Miller. 1973. Culture. Pages 88–134 *in* USDA Agric. Handb. 289. *Rice in the United States: varieties and production.* Washington, D.C.

Khan, A. U. 1973. Rice drying and processing equipment for Southeast Asia. *Trans. Am. Soc. Agric. Eng.* 16:1131–1135.

Khan, A. U., F. E. Nichols, and B. Duff. 1973. Agricultural equipment development semiannual research for tropical rice cultivation. Agricultural Engineering Department, International Rice Research Institute. Progress Report 16. Los Baños, Philippines. 66 pp.

Kunze, O. R., and S. Prasad. 1978. Grain fissuring potentials in harvesting and drying of rice. *Trans. Am. Soc. Agric. Eng.* 21:361–366.

Langfield, E. C. B. 1957. Time of harvest in relation to grain-breakage on milling in rice. *J. Aust. Inst. Agric. Sci.* 23:340–341.

Malabuyoc, J. A., N. G. Mamicpic, P. S. Castillo, R. M. Miranda, and H. P. Callao. 1966. Grain characters, yield and milling quality of rice in relation to dates from heading. *Philipp. Agric.* 49:696–710.

McNeal, X. 1950. *Effect of combine adjustment on harvest losses of rice.* Arkansas Agric. Exp. Stn. Bull. 500. 26 pp.

Morse, M. D., J. H. Lindt, E. A. Oelke, M. D. Brandon, and R. G. Curley. 1968. The effect of grain moisture at time of harvest on yield and milling quality. Pages 7–8 *in* California Cooperative Rice Research Foundation, Inc., University of California, and United States Department of Agriculture. *Rice research in California.*

Nangju, D., and S. K. De Datta. 1970. Effect of time of harvest and nitrogen level on yield and grain breakage in transplanted rice. *Agron. J.* 62:468–474.

Nishiyama, I. 1977. Decrease in germination activity of rice seeds due to excessive desiccation in storage. *Jpn. J. Crop Sci.* 46:111–118.

Oelke, E. A., R. B. Ball, C. M. Wick, and M. D. Miller. 1968. Seed quality and seedling vigor of rice harvested at different grain moisture content. Pages 9–10 *in* California Cooperative Rice Research Foundation, Inc., University of California, and United States Department of Agriculture. *Rice research in California.*

Ota, Y. 1971. Long term storage and dormancy in rice seeds [in Japanese]. *Beibaku Kairyo* 1971(1):17–25.

Padua, A. B., and B. O. Juliano. 1974. Effect of parboiling on thiamin, protein and fat of rice. *J. Sci. Food Agric.* 25: 697–701.

Pfost, H. B. 1978. High temperature, high humidity grain storage. Pages 462–467 *in* Department of Agriculture, Thailand, and Southeast Asia Cooperative Postharvest Research and Development Programme. *Proceedings of the workshop on grain post-harvest technology.* Bangkok, Thailand.

Policarpio, J. S., and J. A. McMennamy. 1978. The development of the IRRI portable thresher: a product of rational planning. *Agric. Mechanization Asia* 9(2):59–65.

Raghavendra Rao, S. N., and B. O. Juliano. 1970. Effect of parboiling on some physicochemical properties of rice. *J. Agr. Food Chem.* 18(2):289–294.

Rayarappan, B. 1979. An economic analysis of the contractual padi combine harvesting system in the Muda irrigation project. Undergraduate Project. Universiti Pertanian Malaysia, Serdang, Selangor, Malaysia. 84 pp.

Samson, B. T., and B. Duff. 1973. The pattern and magnitude of field grain losses in paddy production. Paper presented at a Saturday seminar, 7 July 1973, International Rice Research Institute, Los Baños, Philippines. 30 pp. (unpubl. mimeo.)

Seetanun, W., and S. K. De Datta. 1973. Grain yield, milling quality, and seed viability of rice as influenced by time of nitrogen application and time of harvest. *Agron. J.* 65:390–394.

Toquero, Z. F., and B. Duff. 1974. Survey of post-production practices among rice farmers in Central Luzon. Paper presented at a Saturday seminar, 7 September 1974, International Rice Research Institute, Los Baños, Philippines. 44 pp. (unpubl. mimeo.)

Toquero, Z., C. Maranan, L. Ebron, and B. Duff. 1977. *Assessing quantitative and qualitative losses in rice post-production systems.* Agricultural Engineering Department, International Rice Research Institute. Pap. 77-01. Los Baños, Philippines. 25 pp.

Willson, J. H., ed. 1979. *Rice in California.* Butte County Rice Growers Association, Richvale, California. 254 pp.

Wimberley, J. 1972. *Review of storage and processing of rice in Asia.* Agricultural Engineering Deparment, International Rice Research Institute. Pap. 72-01. Los Baños, Philippines. 20 pp.

Wimberley, J. E. 1980. *Technical handbook for the paddy-rice post harvest industry in developing countries.* International Rice Research Institute, Philippines. (in press)

14

Modern Rice Technology, Constraints, and World Food Supply

Populations in most rice-consuming nations are among the highest in the world and in many of them the rate of population increase is alarming (Table 14.1). Demands for rice are expected to increase in the next decade approximately in line with population increases, and countries that have traditionally exported rice may face difficulty in producing food enough for their own population.

DEMAND FOR RICE

To determine the demand for rice, planners generally consider four factors:

- Population growth rates.
- Current levels of income and expected increase in real income.
- The income elasticity of demand for rice, which provides the relationship between changing incomes and changing purchases of rice.
- Anticipated changes in rice prices when compared to substitute foods (for example, maize and wheat).

Of these factors, population growth rate is generally the most important determinant of the increase in the demand for rice. With a world population of more than 4.5 billion in 1980, the world's population will increase by about 90 million people in the coming year. At least 45 million of those will depend on rice as their major source of food. An annual production increase of 9 million tons of rice will be required to maintain the present level of food consumption. Thus, the best estimates are that the needed growth rates in cereals will be in the 2.2–3.5% range. Of this increase, 90% will be due to population effect, and about 10% to an income effect. For output to grow within this range and for rice to retain its current share of food supplied, rice supplies must grow at the same rate they have for the past 15 years (Colombo et al. 1977, IFPRI 1977, FAO 1980).

Table 14.1 Present and Projected Population in Selected Rice-Growing
Countries in Asia (Adapted from FAO 1978)

Country	Population		
	Present (million)	Rate of Increase (%)	Forecast 2000 A.D. (million)
Bangladesh	83.3	2.8	154.9
Burma	31.8	2.2	53.3
People's Republic of China	900.0	1.6	1126.0
India	622.7	2.5	1023.7
Indonesia	136.9	2.6	226.9
Japan	114.2	1.0	133.4
Pakistan	74.5	2.9	145.5
Philippines	44.3	3.4	83.7
Sri Lanka	14.1	1.4	20.7

The world food situation in different areas, however, could cause increase in demand.

- The world's major rice-producing region, which is South and Southeast Asia, appears to be no nearer to solving the food problem than it was a decade ago.
- Rice, which has tended to be a prestige food, is becoming an increasingly important part of the diet in Latin America because of population growth. The increasing world population should also result in a potential export market for Latin American rice (Kirpick 1977).
- In Africa, rice is important to economies of many countries, particularly in West Africa. These countries are among the least self-sufficient in rice production, importing about 30% of their total requirements. There is increasing dependence on imported rice because of urbanization and higher wage rates. Rice is also progressively replacing sorghum and millet, the traditional cereals, in the savanna and Sahelian areas, and the root and tuber crops in the forest zones (Aw 1978).

In environments favorable to rice growth, modern technology has increased rice yields and productivity. Development of modern technology alone, however, may not increase yields and production of rice in a given country. The technology must be tested and adopted on farms and any constraints at the farmer level determined. In some instances, the technology will need to be modified based on the results from the farm-level testing.

The issues of development of modern rice technology and understanding of constraints must be addressed together to generate enough food for the world's growing needs.

TECHNOLOGICAL CHANGES IN RICE PRODUCTION

Perhaps the most significant development in agriculture in South and Southeast Asia during the past decade has been the shift from traditional agriculture to modern agriculture using science-based technologies. This shift was made possible mainly by the adoption of new, fertilizer-responsive varieties of rice and wheat. These high-yielding varieties occupy about 28% of the world's rice-growing area and 44% of the wheat-growing area (Table 14.2).

The rate of adoption of the new cereal varieties has varied greatly between countries and between areas within countries. The differences in the adoption

Table 14.2   Areas of the High-Yielding Varieties Expressed as a Proportion of Total Area of Wheat and Rice, 1976–1977 (Adapted from Dalrymple 1978)

Region	Wheat (%)	Rice (%)
Asia (South and East)	72.4	30.4
Near East including West Asia and North Africa[a]	17.0	3.6
Africa, excluding N. Africa[a]	22.5	2.7
Latin America	41.0[a]	13.0
Total	44.2	27.5

[a]Particularly rough estimate of area.

Table 14.3   Crop Area Planted to Modern Rice in Some South and Southeast Asian Countries (Adapted from Dalrymple 1978)

Country	Area of Modern Rice Varieties Planted or Harvested (ha)	
	1966–1967	1976–1977
Bangladesh	200	1,550,800[c]
Burma	8	349,000
India	888,400	15,000,000[c]
Indonesia	198,000[a]	3,428,900
Pakistan	80	677,900
Philippines	82,600	2,416,700
Sri Lanka	7,000[a]	331,000[d]
Thailand	3,000[b]	960,000

[a]1968–1969.
[b]1969–1970.
[c]1977–1978.
[d]1975–1976.

rate are largely explained by differences in environment. Table 14.3 shows the trend in adoption rate of modern rice varieties in selected countries in South and Southeast Asia.

In many countries, however, yield per hectare has not increased at a rate corresponding to the rate of adoption of modern varieties. For example, in Pakistan where climate and adequate irrigation favored the new rice technology, annual growth in yield has exceeded 6%. On the other hand, in Thailand and Burma, two traditional net exporters of rice, yields have been relatively stagnant and the annual production growth has been less than 1%. This low growth rate is due largely to uncontrolled crop water, which has prevented greater adoption of modern rice varieties. In countries such as Indonesia, Malaysia, Philippines, and Sri Lanka, a rapid growth in output of rice has been achieved by increasing the area irrigated and good adoption of modern technology such as improved seed and increased fertilizer rates.

DEVELOPMENT, TESTING, AND ADOPTION OF MODERN RICE TECHNOLOGY

The concept of a good plant type for high fertilizer response and high rice yields was developed and has been perfected in Japan. High-yielding varieties are grown almost entirely under complete irrigation. As a result, Japan's rice yields have steadily increased over the years.

Despite abundant fertilizers and other agricultural chemicals, and a price support of rice to promote increased rice production in Japan, there still remains a gap between the potential and actual yield of rice in that country. Average rice yields of Japanese farmers are 30–40% of the 15–17 t/ha potential yield for temperate regions suggested by plant scientists (IRRI 1978a).

In tropical Asia, rice yields have been low for many centuries. For India, Indonesia, and the Philippines, rice yields average 12–15% of the potential of 13–15 t/ha per crop (IRRI 1978a). Reasons for those low yields are warm day and night temperatures, high humidity, low light intensity, and the poor rice varieties.

The tall, vigorous-growing, profuse-tillering, late-maturing, photoperiod-sensitive indica varieties have been grown in most of South and Southeast Asia for centuries. Rice yields have been low but stable. The tall types of rice tolerate relatively deep water, compete well with weeds, and grow on soils with low fertility.

Another reason for the low rice yields in the tropics is that pest management and fertilizer practices are not adequate at the farm level to achieve the potential yields in farmers' fields, even in irrigated fields. In rainfed areas, hardly any modern technology has been available that could increase grain yield of rice. In 1962, the rice breeders at IRRI put together a set of objectives for development of high-yielding varieties with a high nitrogen responsiveness that was not found in commercial varieties grown in the tropics (see chapter 6).

The Modern Rice Technology

The new plant type for rice that received worldwide attention in the mid-1960s is represented by the development of IR8. It is characterized by short stature and stiff straw, and differs markedly from the tall, weak-stem plant type of the traditional indica rice varieties. In favorable environments, the new rice varieties respond to increased applications of chemical fertilizer with high yields. Thus, the modern technology associated with the high-yielding varieties is often called the "seed-fertilizer" technology. Associated technology was developed but is less known, primarily because of the overriding impact of the IR8 plant type. Thus, modern rice technology has been associated primarily with the development of modern rices.

Testing of Modern Rice Technology

All rices in the early IRRI rice breeding program were evaluated in fields with good control of irrigation water and the best crop production technology available at that time. Breeding lines were tested under continuous flooding, and the fertilizer rates applied to the tests far exceeded those practiced by the best farmers of the region. Agronomic testing was limited to the advanced breeding lines and was structured to cater to the varying soil fertility conditions. But even the poorest Philippine soils (see Chapter 3) on which the early breeding lines were tested are rich soils compared with many rice-growing soils in the region.

Adoption and Spread of New Rice Technology

It became apparent in the early 1970s that the original new rice varieties are environment-specific and perform best where there is an adequate supply of water and high solar energy. Because the often-used term "high-yielding varieties," or HYV, gave the misleading impression that such rices give high yields under all conditions, the terminology for the new rices has been changed to "modern varieties," or MV.

The subsequent development of modern rice varieties with more resistance to insects and diseases and with improved quality (taste and appearance), and expansion of irrigated areas helped to extend the area grown to modern varieties. Newer modern varieties more suited to local conditions helped increase the spread of the improved plant type. But, despite these technological advances, the modern varieties had spread to only about 30% of the rice-growing area in Asia by 1976–1977 (Table 14.2).

The most rapid adoption of modern rices has taken place in India, Indonesia, and the Philippines. There has been adoption of the new varieties, but at a slower rate, in Burma, Thailand, and Pakistan.

Two reasons are cited for slow adoption of modern rices in Thailand:

- Poor grain quality.
- Short-statured rices have not been suitable to the poor water control conditions, which is also true for Burma.

In Pakistan's Punjab region the spread of the modern varieties has been slow because premium prices have been paid for locally produced Basmati rice for export.

Studies by agricultural economists in six countries in South and Southeast Asia determined that wide variability exists in the level of adoption of modern varieties across the rice-growing areas of Asia, including the irrigated areas covered by the study. Factors contributing to the variability are seasonal suitability of the rice variety, the degree of water control, and profitability (IRRI 1978b).

The first modern varieties were short-statured and, under the ideal conditions of good irrigation and high level of cultural practices, responded well to high levels of fertilizer, particularly during the sunny dry season. Although medium-statured varieties appear to have a lower yield potential than the short-statured rice, the medium-statured rices have fared well because of greater tolerance for unfavorable soil-water environments and they are still popular in many areas of South and Southeast Asia. In some study areas, local varieties have remained popular because these areas have poor drainage and are unsuitable for the modern rices.

Impact of New Rice Varieties on Cultural Practices

One very significant change that has taken place with the introduction of modern varieties is the increased use of fertilizer, particularly nitrogen. In a six-country economic study, it was found that the average village yield positively correlates with the amount of nitrogen fertilizer applied (Fig. 14.1). It was also found that a high level of nitrogen (80 kg N/ha) is associated with a high proportion of the land planted to modern varieties. When village average yields are calculated separately for modern varieties and local varieties, most of the average yields exceeding 4 t/ha have been obtained with modern varieties using 90 kg N/ha or more (Fig. 14.2). The range in fertilizer input has been greater for modern varieties than for local varieties. With the modern varieties, wherever the rate of fertilizer application has been low it has been due either to high fertilizer price or low price of rice. Sometimes both factors have deterred use of high or optimum rates of fertilizer.

Year-to-year and site-to-site variability in fertilizer response has been another factor that often discourages use of the recommended rate of fertilizer for modern varieties (Fig. 14.3).

Weed control is another cultural practice that has been significantly altered with the introduction of modern varieties. For years, rice farmers in tropical Asia have depended on tall, vigorous-tillering plants to provide competition to weeds.

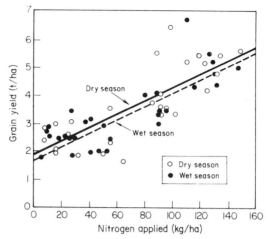

Figure 14.1   Relationship between nitrogen level and average farm yield per hectare by village, wet and dry seasons, selected villages in Asia, 1971–1972. (From Barker 1978)

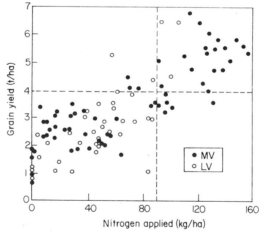

Figure 14.2   Relationship between nitrogen applied and average farm yield of rice per hectare, 36 villages in Asia, 1971–1972 wet and dry seasons. MV = modern varieties; LV = local varieties. (From Barker 1978)

With the introduction of the modern semidwarf rices, weed problems have become more serious. This is because the modern rices provide less of a crop canopy, which allows more light to reach the weeds. Furthermore, the use of high fertilizer rates also increases weed problems with modern cultivars. Nevertheless, the IR8 plant type, with its high tillering capacity, is fairly competitive with weeds. With the introduction of modern varieties, timing of weeding has had to be altered. It is critical to weed early-maturing rices earlier than late-maturing

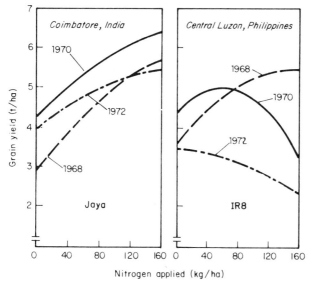

Figure 14.3 Yield response to nitrogen of Jaya and IR8 in two sites, Tamil Nadu Paddy Breeding Station, Coimbatore, India, and Maligaya Rice Research and Training Center, Central Luzon, Philippines, respectively, 1968–1972 wet seasons. (Adapted from Barker 1978)

rices because fast growing barnyard grass-type (*Echinochloa* spp.) weeds damage early-maturing rice more than late-maturing rice.

Manipulation of plant population is another cultural practice that has had to be altered with the introduction of modern rices. Generally, with modern rices, close spacing is essential to minimize weed infestation and obtain high yields (see Chapter 12).

CONSTRAINTS TO HIGH RICE YIELDS IN FARMERS' FIELDS

In studies in 30 Asian villages, 25 social scientists in six countries cooperated to determine socioeconomic constraints (restraints) to high yields of rice where modern rice varieties have been well accepted (IRRI 1978b). It was found that farmers grow modern varieties more widely in the dry than in the wet seasons. The farmers consider lack of good water control, insects, and diseases the most serious constraints to high yields for modern varieties.

In some countries, government policies discourage acceptance of the new technology. In others, particularly those where deep flooding is common, suitable modern varieties have not been developed. And, in villages where substantial land is planted to crops other than rice, incomes are significantly higher than in villages where only lowland rice is grown (IRRI 1978b).

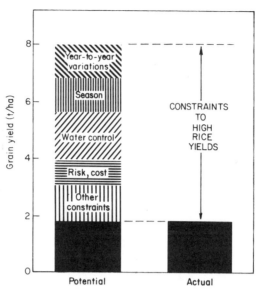

Figure 14.4 Lack of water appears to be the single most important constraint to high yields in a preliminary aggregate study on a national basis in the Philippines. (From Herdt and Wickham 1975)

In an aggregate study of the factors that restrict yields on a national basis in the Philippines, it was reported that lack of water control is the single biggest yield constraint and is responsible for about 35% of the difference between potential and actual yields (Fig. 14.4). Seasonal factors, such as available solar radiation, account for another 20% of the yield difference, and economic factors, including risk, account for 15% of the difference.

Many of these constraints can be reduced or manipulated by appropriate financial investments, policy changes, research, and selective breeding. For example, the construction of irrigation and drainage systems, and modification of their management, can alleviate poor water control. Policy measures to insure available credit and favorable prices can ease economic constraints. Fertilizer inputs might be reduced if varieties or farming systems can be developed to increase efficiency of soil or fertilizer nutrients. Year-to-year variability due to weather conditions can be eased by developing rices that tolerate drought and submergence or that yield well under the low solar radiation of the wet season. Insect and disease damages can be reduced by incorporating a higher level of pest resistance into modern varieties. Research in national and international programs has focused on these issues since the early 1970s.

An interdisciplinary study of the technical and socioeconomic constraints to increased rice production was undertaken by scientists at IRRI and in national rice research programs in several Asian countries. They have developed a methodology to investigate the constraints to high rice yields in farmers' fields combining an experimental and a survey approach.

Development of Methodology for Studying Farm Yield Constraints

Rice yields on farmers' fields in tropical Asia are lower than those commonly obtained in experimental plots. The gap between experiment station yields and actual farm yields is referred to as the yield gap. The factors responsible for the gap are referred to as yield constraints. Reasons for the gap are:

- The so-called "improved technology," that is, varieties and associated crop management, has not been adopted by farmers.
- Expectations for the technology on-farm have been unrealistically high.
- Potential yields of the technology are not fully expressed in the poor environments often found on farmers' fields.
- Farmers' objectives are different from researchers' and farmers do not attempt to maximize yields.
- The supply of inputs, markets, and infrastructure prevents the farmer from realizing the higher yields he wishes (Herdt and Wickham 1975).

The General Approach and Basic Concept

The approach to determining yield constraints involves the conduct of controlled agronomic experiments on farmers' fields and farm surveys. A research team is usually composed of agronomists, agricultural economists, and at times, statisticians.

The concept on which the approach is based is illustrated in Fig. 14.5. The model divides the yield gap into two parts by introducing an intermediate yield level, representing the potential farm yield or yield obtained in farmers' fields using the modern technology. The first part, yield gap I, is the difference between experiment station yield and potential farm yield. It exists mainly because of environmental differences between experiment stations and the actual rice farms. The technology that gives high yields on experiment stations may not give nearly as high yields in the less favorable environments that exist in much of the rice-growing areas of Asia. There may also be some components of the technology that are not transferable from the experiment station to the farmers' fields.

The primary objective of the constraints research at the farm level is not to examine yield gap I, although its size can be assessed, but to focus on yield gap II—the difference between the potential farm yield and the actual farm yield. This gap exists because farmers use inputs or cultural practices that result in lower yields than those possible on their farms. It is possible to examine the gap in two parts. One is to identify what biological or physical inputs or cultural practices account for the gap. The other is to identify why farmers are not using the inputs or cultural practices that would result in higher yields on their own farms.

The biological explanation of yield gap II indicates that farmers' yields would be higher if they would use the highest-yielding variety, apply maximum-yield

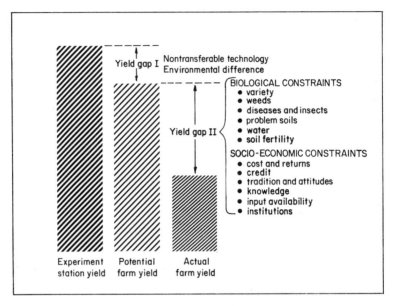

Figure 14.5 The concept of yield gaps between an experiment station rice yield, the potential farm yield, and the actual farm yield. (From Gomez 1977)

levels of fertilizer and insecticide, correct existing soil problems, and use the best cultural practices. The most critical factors differ from one region to another, but before any remedy can be taken, such as recommending a package of improved practices, the biological nature of the gap must be understood. That makes experiments on farmers' fields essential.

The socioeconomic constraints explain why farmers do not use the practices and inputs necessary to obtain maximum yields. The reasons may include economic calculations of costs and returns, lack of knowledge of how to use the technology, lack of credit, poorly operated irrigation systems, nonavailability of inputs, or traditional beliefs. The importance of these factors will differ from area to area, but understanding them will help in designing programs to provide the missing biological components to overcome the yield gap. Farm surveys provide the main research inputs for this aspect (Gomez et al. 1979).

IRRI-initiated constraints research that led to formalization of a network called International Rice Agro-Economic Network (IRAEN) focused attention on yield gap II.

Objectives of the Methodology

The general objective of the methodology for studying farm-yield constraints is to identify the factors that explain the difference between actual and potential rice yields in selected farm environments. The specific objectives are:

- To measure the gap between the actual yield of farmers and the potential yield or the yield with production techniques or input levels of maximum yields in their physical environments.
- To determine the contribution of each tested input or cultural practice to the gap between actual and potential yield.
- To determine the extent to which use of each tested technical factor can be profitably increased.
- To determine what social and institutional factors prevent farmers from using technology that gives higher, more profitable yields.
- To determine whether it is possible to change the physical and socioeconomic factors constraining yields, and the difficulty, if any, of making such a change.

Methodology for Quantifying Biophysical Constraints

Standard research methodologies have been developed to study biophysical constraints to high rice yields (De Datta et al. 1978). Three fundamental characteristics of the constraints research methodology were considered before embarking on the task:

- The first objective of the research approach is to understand constraints from the viewpoint of the farmer.
- Second, the focus of the research is on farmers' fields with an attempt to simulate reality to ensure that the results would be indicative of what farmers could achieve.
- Third, the project is collaborative between agronomists, economists, and statisticians.

Components of the Methodology

A new field plot technique has been developed to identify and quantify the constraints to high yields in farmers' fields. Several field experiments have been conducted in the Philippines and in five other South and Southeast Asian countries to accurately estimate the potential farm yield, the actual farm yield, and the intermediate levels representing varying combinations of input use. Two sets of treatments, the factorial component and the management package component, have been tested in the experiments.

FACTORIAL COMPONENT The concept and design of the factorial component were developed by Gomez et al. (1973) and modified later by Gomez (1977). The treatment consists of either complete or incomplete factorial combinations of n factors (or production inputs) each at two levels. The two levels of each factor are the farmer's practice, and the improved practice.

The farmer's practice refers to what the farmer is actually doing in the current crop season, and hence, varies from one farm to another. The improved practice is the one recommended for maximum yield, and should be fixed for all farms in a given location.

Data collected on different test farms will estimate the:

- Yield gap (gap II of Fig. 14.5).
- The individual contributions of the n test factors to the yield gap.

MANAGEMENT PACKAGE COMPONENT   The management package concept was developed by De Datta et al. (1976) to test intermediate levels of inputs or technology between the farmer's set of practices and the improved, or recommended, set of practices. The incremental steps between treatments usually involve a simultaneous change in more than one input.

The management package component tests the different input combinations selected to represent different yield levels and production costs. Study of this component allows a meaningful look at the question of costs and returns for each management package. Furthermore, one or more of these intermediates should be good candidates for an immediate recommendation to farmers.

Socioeconomic Studies for Determining Farm Yield Constraints

The socioeconomic studies include economic analysis of the experimental data and farm surveys.

ECONOMIC ANALYSIS OF EXPERIMENTAL RESULTS   To meet the third objective of the methodology (see Objectives of the Methodology), economic analysis on cost and return of all experimental data is carried out by country and by season.

FARM SURVEYS   To meet the third and fourth objectives, farm surveys are carried out in the same region of a country where the experiments are laid out. However, surveys could cover more areas than those used for the experimental sites.

One preliminary survey is done prior to the establishment of the experiments. It obtains information on farm size, tenure status of the farmers, farming practices, productivity level, technology awareness, irrigation facilities, credit, labor availability, and prices.

A detailed survey is carried out as a follow-up study at the end of the crop season in which experiments are carried out. The follow-up surveys include information on:

- Variety grown.
- Farmer's management and cultural practices.

- Yield level.
- Farmer's perception of factors that limit the yield and the remedial measures.

Constraints Research Process

Figure 14.6 diagrams the constraints research process and the relationship of the methodology to the goal of agricultural development—increased production, profit, and rural employment. Attaining that goal requires that new technology appropriate to the environment be extended to farmers. An opportunity for profitable sales of farm output must also exist. If production does not increase, or does not increase at an appropriate rate, the reasons are sought.

- When the new technology is used in farmers' fields, are yields increased over those with use of the farmers' technology? If not, what factors restrict yields? Answers here provide feedback to researchers and enable them to design more appropriate technology.
- If yes, is the new technology more profitable than the technology now used by farmers? If not, it is not reasonable to expect it to be adopted.
- Is the level of profitability restricted by government policies? If so, can the policies be changed, or can the technology be redesigned? Answers here provide feedback to policy makers and researchers.
- Do farmers have a level of knowledge adequate for effective use of the technology? Are the necessary inputs and cash or credit required to use the technology available to farmers? If not, how can government programs ensure the availability of such inputs? Answers here provide feedback to government programs and to the extension system.

Actual farm yields are easy to define. They are what farmers get. For purposes of the project, potential yield was defined as the highest yield obtained in farmers' fields with improved technology (De Datta and Herdt 1979).

The constraints project did not attempt to explain or identify the constraints to rice production imposed by circumstances outside the control of farmers—policies of government, weather, soil conditions, water available in the irrigation canals, and similar factors operating at a level beyond the control of farmers working on their farms. Finally, the constraints project did not attempt to explain the existing level of land use intensity or explore how that could be overcome through engineering or technological innovations; it did not identify constraints to intensification (Herdt 1979).

A detailed methodology is described in a handbook developed at IRRI (De Datta et al. 1978).

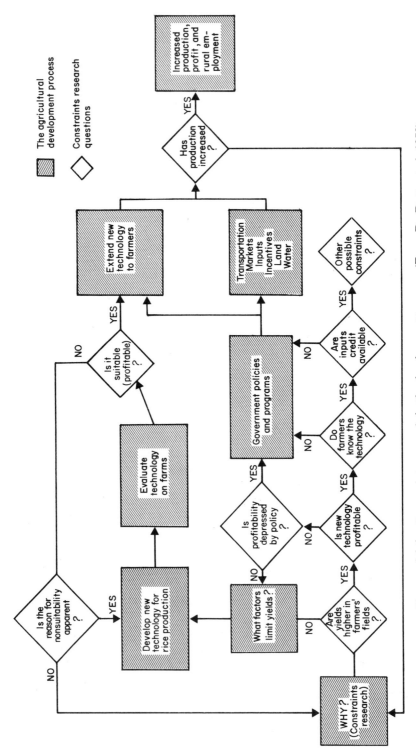

Figure 14.6 Constraints research in the development process. (From De Datta et al. 1978)

The agricultural
development process

Constraints research
questions

Results of Yield Constraints Research

Experiments were initiated during the 1974 wet season in the Philippines and subsequently carried out in five other countries in Asia—Bangladesh, Indonesia, Sri Lanka, Taiwan, and Thailand. Results summarized here cover data collected during 1974–1977.

The Yield Gap

Four hundred ten trials were conducted in the wet season resulting in an average yield of 3.6 t/ha using farmers' inputs and 4.5 t/ha using high inputs, thereby demonstrating an average yield gap of 0.9 t/ha. Table 14.4 shows the average results from each area. At Yogyakarta, Indonesia and Taiwan, farmers' wet-

Table 14.4 Summary of Rice Yields Obtained in Farmers' Fields by Season in Asia, 1974–1977 (Adapted from IRRI 1979)

Site	Yield (t/ha)		
	Farmers'	High	Gap
Wet season			
Subang, Indonesia	1.4	2.0	0.6
Dry zone,[a] Sri Lanka	2.9	3.4	0.5
Camarines Sur, Philippines	3.9	4.5	0.6
Yogyakarta, Indonesia	4.7	5.1	0.4
Taichung, Taiwan	5.0	5.6	0.6
Joydebpur, Bangladesh	2.9	3.8	0.9
Nueva Ecija, Philippines	3.4	4.5	1.1
Laguna, Philippines	3.6	5.4	1.8
Iloilo, Philippines	3.8	5.0	1.2
Central Plain,[b] Thailand	4.0	4.9	0.9
All	3.6	4.5	0.9
Dry season			
Dry zone,[a] Sri Lanka	2.5	2.9	0.4
Subang, Indonesia	3.8	4.4	0.6
Taichung, Taiwan	6.6	7.3	0.7
Joydebpur, Bangladesh	3.6	4.6	1.0
Iloilo, Philippines	3.9	5.3	1.4
Yogyakarta, Indonesia	4.8	6.1	1.3
Camarines Sur, Philippines	4.0	5.8	1.8
Central Plain,[b] Thailand	4.2	6.1	1.9
Laguna, Philippines	4.5	6.5	2.0
Nueva Ecija, Philippines	4.6	6.8	2.2
All	4.3	5.6	1.3

[a]Polonnaruwa and Kurunegala Districts.
[b]Suphan Buri, Sam Chook, Pho Phaya, and Chanasutr irrigation projects.

season yields were quite high (about 5 t/ha) and the high levels of inputs failed to increase yields. Farmers in those two areas were already using high inputs and getting high yields.

Dry-season constraints experiments were on 366 farmers' fields at the 10 sites between 1974 and 1977. Farmers' yields averaged 4.3 t/ha whereas yields with the high inputs were 5.6 t/ha, for an average yield gap of 1.3 t/ha (Table 14.4).

At three sites the yield gap was only about 0.5 t/ha. In the dry zone of Sri Lanka and Subang, Indonesia, both farmers' yields and the yield gap were low, indicating that the technology applied could not overcome the constraints. Taiwan also showed a small gap, but in that case, it was because the farmers were already close to the potential yield.

Contribution of Biophysical Constraints to Yield Gaps

The constraints experiments were designed to enable the analyst to separate the yield gap in the components attributable to each variable factor tested in the experiments. At each site, the researchers identified practices that farmers were not considering or inputs that were being applied at inadequate levels. Those factors became the test factors in the experiments. Two to four factors were generally chosen because a larger number made the experiment too large to conveniently conduct in farmers' fields.

Fertilizer and insect control were the most frequently used test factors. Others included weed control, variety, plant spacing, land preparation, organic manure, and various separate fertilizer elements.

Based on the averaged data for all six collaborating countries, fertilizer application rate and timing have been the most limiting factors for high yields in the dry season. In the wet season, insect control and fertilizer management have been about equal in importance in contributing to high rice yields (Fig. 14.7).

Detailed results on all of these studies from six countries were published by IRRI (IRRI 1979).

FARM-TO-FARM VARIABILITY IN YIELDS It is unclear why, in a given climatic zone, variability in rice yields exists not only among farmers but also among researchers managing farmers' fields. To explain the farm-to-farm variability, experiments were conducted during 1974–1977 in one Philippine province and the results were classified according to farmers' yields: the 25% of sites with the highest farmers' yields and the 25% of sites with the lowest farmers' yields were identified for the wet and dry seasons (Herdt and Mandac 1979). For convenience, these were called top-yield and bottom-yield farms and grouped as shown in Fig. 14.8.

Herdt and Mandac (1980) and IRRI (1980) identified the factors contributing to the yield differences in the wet and dry seasons (Fig. 14.9). In the wet season, the managed inputs—fertilizer, insect control, weed control, and seedling age—contributed little to explain the difference between the low- and the high-

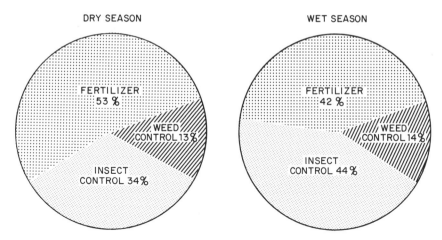

Figure 14.7 Difference in yields attributed to inputs during dry and wet seasons. Data are averages of 1974–1977 in six countries in Asia (Bangladesh, Indonesia, Philippines, Thailand, Taiwan, and Sri Lanka).

yield crops. Environmental parameters and a combination of weather-related factors and insects and diseases, were the major factors causing the yield differences among farms and accounted for some 80% of the yield differences. The dry-season results were in contrast to the wet season: nonmanaged factors accounted for some 40–50% of the yield differences; the remainder was due to the levels of managed inputs used and their interaction with environmental variables.

These analyses indicate that the yield potential of the existing modern rice technology can be realized only with favorable environmental conditions.

Socioeconomic Constraints

Economic evaluation of new technology must be an integral part of constraints research. Economic evaluation, however, is complex. One major problem is that when farmers adopt technology, they utilize it at a completely different level than the researchers. As a result, yields and rates of return are quite different. An economic analysis of the experimental data from six South and Southeast Asian countries was made in 1978.

In experiments subjected to economic analysis of cost and return on the farmers' yields, the average yield was 3.5 t/ha and the yield gap about 1 t/ha. The value of increased rice output was calculated at prevailing local prices and local currencies were converted to dollars at the rates prevailing in 1975–1977. The high inputs increased yields by 0.4 t/ha and value of output by $50/ha or more in all wet-season cases (Table 14.5). The cost of the high inputs averaged 40% more than farmers were spending on the same inputs. In some cases, the difference was much greater.

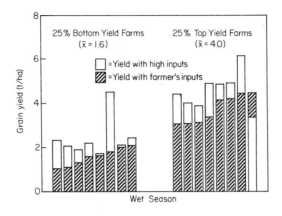

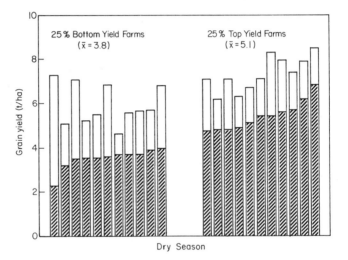

Figure 14.8   High- and low-yield farms by season, Nueva Ecija, 1974–1977. Figures in parentheses are mean yields with farmers' inputs. (From Herdt and Mandac 1979)

In the dry season, the high level of inputs increased net benefits in 9 of 10 sites (Table 14.6). At four sites, the benefit-cost ratio exceeded 2.1 (Herdt 1979).

Figure 14.10 summarizes the results obtained. Where yield with high inputs is plotted against the yield with farmers' inputs, the vertical distance above the line shows the yield gap, which is positive in every case. In the summary of the economics of the gap, points below the line are sites where the farmers' practices were more profitable than the "high" practices.

Table 14.7 shows costs and returns for fertilizer and insect control. The high level of insect control usually entailed a much greater increased cost over the farmers' level than did the high level of fertilizer. The pattern was similar for both

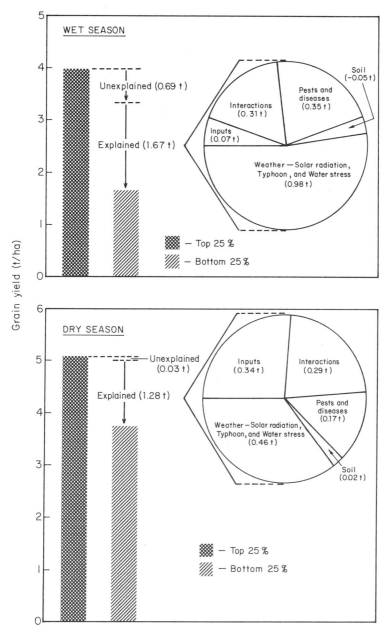

Figure 14.9 Accounting for differences between low- and high-yield farms, Nueva Ecija, Philippines. (Adapted from Hert and Mandac 1980, IRRI 1980)

Table 14.5 Economic Comparisons[a] of Wet Season Yields in Constraints Experiments in Farmers' Fields in Six Asian Countries, 1974–1977 (Adapted from Herdt 1979)

Site	Trial (no.)	Yield (t/ha) Farmers'	Yield (t/ha) Gap	Increased Output Value[b] ($/ha)	Input Cost ($/ha) Farmers'	Input Cost ($/ha) Increase to High	Increased Net Benefits from High ($/ha)	Benefit-Cost Ratio of Increased Inputs
Joydebpur, Bangladesh	52	2.85	0.91	143	35	39	104	3.67
Yogyakarta, Indonesia	14	5.36	0.39	56	92	88	−32	0.64
Subang, Indonesia	8	3.05	1.00	122	132	66	56	1.85
Dry zone,[c] Sri Lanka	17	2.75	0.89	168	80	140	28	1.20
Taichung, Taiwan	9	4.90	0.40	130	834[d]	161	−31	0.81
Central Plain,[e] Thailand	18	3.66	1.02	134	26	160	−25	0.84
Laguna, Philippines	41	3.75	1.77	209	51	183	26	1.14
Nueva Ecija, Philippines	29	3.51	1.10	143	73	104	39	1.38
Camarines Sur, Philippines	27	3.38	0.51	61	52	136	−75	0.45
Iloilo, Philippines	23	3.51	1.30	156	51	151	5	1.03
All	239	3.48	1.03	141	85	118	23	1.19

Data source: IRRI, 1979.

[a] Exchange rates used: Bangladesh TK14 = $1.00; Indonesia Rp145 = $1.00; Sri Lanka Rs8 = $1.00; Taiwan NT$38 = $1.00; Thailand ฿20 = $1.00; Philippines ₱7.3 = $1.00

[b] In Philippine sites, harvesting costs are taken account of but not in other locations.

[c] Polonnaruwa and Kurunegala Districts.

[d] Includes labor cost of farmers' practices.

[e] Suphan Buri, Sam Chook, Pho Phaya, and Chanasutr irrigation projects.

Table 14.6 Economic Comparisons[a] of Dry Season Yields in Constraints Experiments in Farmers' Fields in Six Asian Countries, 1974–1977 (Adapted from Herdt 1979)

Site	Trial (no.)	Yield (t/ha)		Increased Output Value ($/ha)	Input Cost ($/ha)		Increased Net Benefits from High ($/ha)	Benefit-Cost Ratio of Increased Inputs
		Farmers'	Gap		Farmers'	Increase to High		
Joydebpur, Bangladesh	29	3.62	0.94	145	77	47	98	3.09
Yogyakarta, Indonesia	12	4.22	1.63	236	90	84	152	2.81
Subang Indonesia	44	3.99	0.69	115	14	19	96	6.05
Dry zone,[b] Sri Lanka	17	2.99	0.58	108	63	201	-93	0.54
Taichung, Taiwan	8	5.94	0.64	186	755	124	62	1.50
Central Plain,[c] Thailand	17	3.94	2.11	225	41	171	54	1.32
Laguna, Philippines	28	4.43	2.01	274	81	188	86	1.46
Nueva Ecija, Philippines	19	4.56	2.28	294	111	134	160	2.19
Camarines Sur, Philippines	20	3.51	2.07	242	57	207	34	1.17
Iloilo, Philippines	11	3.84	1.60	187	58	182	4	1.03
All	205	4.00	1.40	192	87	119	73	1.61

Data source: IRRI, 1979.
[a]See footnotes for Table 14.5
[b]Polonnaruwa and Kurunegala Districts.
[c]Suphan Buri, Sam Chook, Pho Phaya, and Chanasutr irrigation projects.

Table 14.7 Comparison of Tested High Levels of Fertilizer and Insect Control in Constraints Experiments in Farmers' Fields in Six Asian Countries, 1974–1977 (From Herdt and Mandac 1979)

	Average Increases ($/ha) of High Level Compared to Farmers'							
	Fertilizer				Insect Control			
	Wet Season		Dry Season		Wet Season		Dry Season	
Site	Cost	Net Return	Cost	Net Return	Cost	Net Return	Cost	Net Return
Dry zone,[a] Sri Lanka	9	15	50	-24	53	-27	95	-77
Subang, Indonesia	19	-10	36	8	27	93	4	-4
Taichung, Taiwan	23	49	24	68	n.t.[b]		n.t.[b]	
Yogyakarta, Indonesia	35	-12	63	121	43	-42	22	3
Camarines Sur, Philippines	52	-52	82	68	76	-43	116	-34
Joydebpur, Bangladesh	23	68	31	42	7	24	5	10
Central Plain,[c] Thailand	63	23	84	67	135	-103	151	-121
Iloilo, Philippines	45	17	66	62	95	52	109	-79
Nueva Ecija, Philippines	35	7	38	124	81	-10	87	34
Laguna, Philippines	15	59	18	113	160	-34	170	-74
All	30	20	45	71	86	-16	65	12

[a]Polonnaruwa and Kurunegala Districts.
[b]Factor was not tested.
[c]Suphan Buri, Sam Chook, Pho Phaya, and Chanasutr irrigation projects.

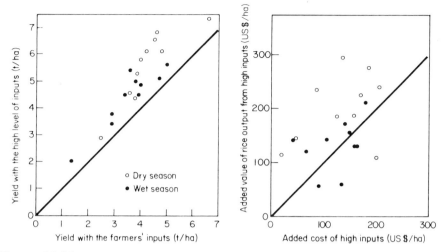

Figure 14.10   Yield gap and economics of the yield gap in wet and dry seasons (each point represents the average result from one of 10 sites in six Asian countries, 1974–1977). (From Herdt and Mandac 1979)

seasons, but the added yield contributed by high inputs was greater in the dry season.

This breakdown by input shows that the application of higher levels of fertilizer in the dry season appears to give a sufficiently high return above added cost to generate a strong incentive for its use. Generally, however, high insect control is not sufficiently attractive for farmers to use it.

SEPARATING TECHNICAL AND ECONOMIC INEFFICIENCY   When the rice constraints experiments were analyzed at IRRI, it became apparent that researchers were recording higher yields than farmers even when they both used about the same levels of inputs—in this case, fertilizer (Fig. 14.11). This observation has raised the issue of how much of the yield gap is a result of farmers apparently being less technically efficient than researchers, and how much because farmers simply use lower levels of inputs.

Figure 14.12 shows how the efficiency issue can be conceptualized by demonstrating how the yield gap can be partitioned into three components. The first segment of the gap, that due to profit-seeking behavior, reflects the difference in input level resulting in maximum profit versus maximum yield. The second segment, referred to as price or allocative inefficiency, reflects the farmers' failure to use inputs to the points of maximum profit. The third gap, technical inefficiency, is defined as the failure to produce on the most efficient production function.

If one or more of the inputs has been systematically varied in the management packages, the researcher is then provided with the data (De Datta et al. 1979) to estimate the response function depicted in Fig. 14.12.

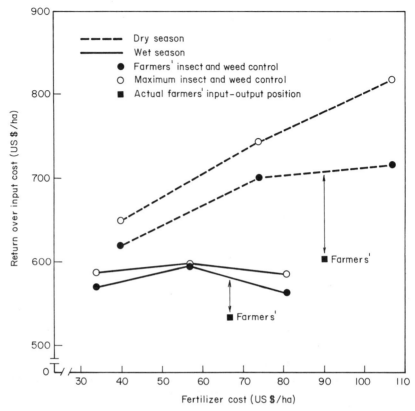

Figure 14.11   Return over cost of variable inputs for four levels of fertilizer inputs and two levels of weed and insect control inputs, average of nine experiments in farmers' fields, Nueva Ecija, 1977. (From Mandac and Herdt 1979)

Institutional Constraints

Considerable attention has been given in the literature to the potential impact of tenure, credit limitation, and relative prices as factors depressing farmer incentives. Generally, prices of rough rice and urea fertilizer represent the relative incentive to use modern technology and the ratio of these prices varied greatly among the sites in the IRAEN studies. The different price ratios had a direct effect on the farmer's use of inputs and therefore affected the yield gap.

Figure 14.13 illustrates the effect. In the dry season, in those areas where the price of urea relative to rice was high, the yield gap attributable to fertilizer was also high. In the areas where it took less than 0.8 kg of rice to buy 1 kg of urea, the yield gap attributed to fertilizer was 0.5 t/ha or less. Where it took 1.8 kg of rice to buy 1 kg of urea, the yield gap exceeded 1 t/ha.

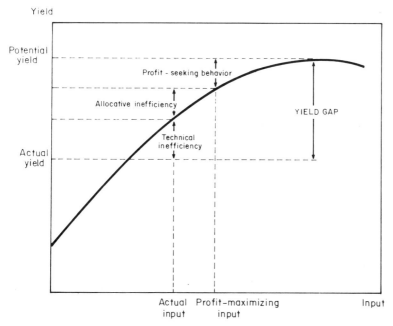

Figure 14.12 Three economic components of the yield gap. (From Barker 1979)

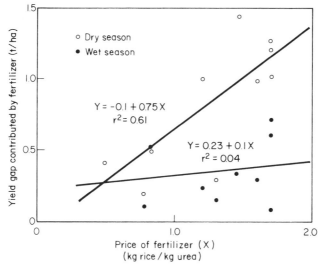

Figure 14.13 Relationship of the real price of fertilizer to the yield gap contributed by fertilizer. (From Herdt 1979)

Source of credit is another institutional factor that determines the level of inputs used by the farmer. Farmers using institutional credit that is available in most countries at an annual interest rate of 9–12% have slightly higher effective fertilizer prices. However, any combination of high interest rate, private credit, or disincentive share tenure results in reduced incentives to use fertilizer (Herdt 1979).

FACTORS LIMITING INCREASED RICE PRODUCTION

In the mid-1960s and early 1970s, there was optimism that technological changes and modern inputs would revolutionize rice yields of tropical Asia. By the mid-1970s there was increased appreciation for the environmental constraints limiting rice yields and production.

To determine the potential of technology it is essential to determine the potential of the crop in the farmers' environments. This was attempted in the constraints study just described. Also to be considered is the farmers' ability and willingness to achieve the yield potential on their farm.

Two contrasting ideas are suggested to explain low yields and production of rice in many tropical countries in Asia. Agricultural scientists feel that both the farmers and the institutional system are not taking full advantage of technology. Social scientists believe that the technology developed in the experiment station in many instances is not appropriate for the farmer's environment.

Recent evidence from the Asian Development Bank's Second Asian Agricultural Survey is that institutional constraints, not lack of technology, are the major reason for a farmer's failure to realize the full potential of modern technology. These constraints are slowing the rate of agricultural and economic development in Asia (Asian Development Bank 1977).

Variability in Environmental Factors

Recent studies (IRRI 1977, 1978b, 1979) suggest that environmental factors are of critical importance in explaining the gap between yield potential of the modern rices and their average farm yields. This is particularly true if the definition of environment includes quality of irrigation.

Diversity and variability of environment in different rice-growing countries and within a country's different rice-growing regions would make development of appropriate technology to match particular environmental factors extremely difficult (Herdt and Barker 1977). Nevertheless, attempts are underway to develop a somewhat more rigorous definition of the rice environment and include characteristics such as water depth, soils, rainfall, and possibly temperature.

The issue of environmental effects on yield, the yield gap I in Fig. 14.5, focuses on the differences between farmers' environments and experiment station

Table 14.8 Factors Perceived to Limit Rice Production Versus Frequency of Breeding for Tolerance to Those Limiting Factors; Responses of 35 Asian Plant Breeders, 1975 (Adapted from Hargrove and Cabanilla 1979)

Perceived Rice Production Limiting Factor	Relative Frequency[a] of Perception of Limiting Factor	Relative Frequency of Breeding for Tolerance to Limiting Factor
Diseases and insects	100	100
Drought	85	7
Excessive monsoon cloudiness	48	0
Injurious soils	60	13
Low temperature	30	18
Deep water	12	15
Floods	36	6
High temperature	9	0
Waterlogged soils	3	3
Other stresses	6	0

[a]The most frequently mentioned problem and objective are scored 100; all other problems and objectives are scored as a proportion of the one most frequently mentioned.

environments. In a recent study of 35 rice breeders throughout Asia, Hargrove and Cabanilla (1979) found that all breeders perceived the important limiting factors to rice production but few of those factors were included in the breeding objectives. For example, the relative frequency of drought as a perceived problem was 85% that of diseases and insects but drought's relative frequency as a breeding objective was only 7% that of pests (Table 14.8). The relative frequency of the breeders' perceptions of injurious soils as a limiting factor was 60%, but the relative frequency of tolerance for injurious soils as a breeding objective was 13%.

Classification of the Variability of Environmental Factors Affecting Rice Production

It is nearly impossible to list and classify all the factors that limit rice production. However, Herdt and Barker (1977) reported a list of environmental factors that vary across sites, seasons, and years (Table 14.9). Among that list of factors, some, such as water depth, can be overwhelmingly more important than all other factors.

With all the available information on soils, climate, and so on, information is still grossly lacking on the quantitative dimensions of the rice-growing environment. Quantification of variability and classification of the environment that

Table 14.9   A Classification of the Variability of Environmental Factors Influencing Rice Production (Adapted from Herdt and Barker 1977)

		Variable		
Factor	Subfactor	Across Locations	Across Seasons	Across Years
Soils	Physical properties	X		
	Chemical properties	X		
	Biotic properties	X	X	
	Origin	X		
Geographic	Topography	X		
	Position	X		
	Photoperiod	X	X	
Water	Depth	X	X	X
	Rate of increase, decrease	X	X	X
	Temperature	X	X	
Climatic	Solar radiation	X	X	X
	Precipitation	X	X	X
	Air temperature	X	X	X
	Wind	X	X	X
Biotic	Competitive plants	X	X	X
	Dependent plants	X	X	X
	Insect predators	X	X	X
	Microbes	X	X	X

relate to rice production are a major challenge to the development of technology appropriate to specific environmental regime.

MINIMIZING TECHNICAL CONSTRAINTS

The past decade heralded remarkable scientific progress in rice production. Modern varieties were adopted on millions of hectares by farms of all sizes. Large sums of money were invested in irrigation and other agricultural development projects. Nevertheless, the green revolution bypassed millions of small rice farmers.

Rice research emphasis shifted early in the 1970s to serve those bypassed farmers. Impetus for the shift came from IRRI, which turned from a discipline-oriented approach to a multidisciplinary team approach to serving the needs of rice farmers. The IRRI Genetic Evaluation and Utilization (GEU) program is detailed in Chapter 6.

Today, teams of GEU scientists from most of Asia's rice-growing countries collaborate closely to identify and test genetic materials resistant to pests and environmental stresses, ranging from fungal and viral diseases to devastating insects, drought, toxic soils, and deep water. The GEU objective is to develop a technology that will provide farmers in disadvantaged cropping environments with the benefits of modern rice technology enjoyed by farmers living in the more attractive environments.

As national research organizations have gained strength and staff capability improved, IRRI has served more as a catalyst for international cooperation and less as a model for a national research organization. Instead of releasing varieties, IRRI has greatly stepped up the supply of genetic materials for evaluation and testing in national programs. Early generation breeding materials (F_2) are now sent in bulk directly to national cooperators without being tested at IRRI. That arrangement assures evaluation under a range of environments and permits local scientists to select cultivars based on best performance in the local environment. Through the International Rice Testing Program (IRTP), rices from national programs are evaluated along with IRRI genetic materials. Many technical constraints to high yields that appeared nonsolvable under limited testing now appear partially solvable through internationalization of testing and use of genetic materials. There is now a chance to succeed in developing an appropriate technology whereas only partial success was achieved with the modern technology that was developed under limited testing.

SCIENCE AND TECHNOLOGY RELEVANT TO SMALL RICE FARMERS

Among those that the green revolution bypassed are farmers who depend solely on the unpredictable monsoonal rains to water their crops. Some grow upland rice and manage it like wheat, others bund their fields to hold water on the land they puddle. But the rains often fail. With the modern rices often even less tolerant of dry weather than the traditional varieties, drought makes it risky to spend money on the fertilizers and other chemicals needed to make modern rices pay best. So the farmers continue to struggle with the same rice technology that their ancestors used and often produce barely a subsistence crop.

At the other extreme are millions of farmers on whose fields floodwaters annually rise to depths of 1–6 m (floating rice-growing areas) marking the modern short-statured rices for failure (Barker and Pal 1979).

Mainly as a result of adverse environmental conditions in a substantial portion of the world's rice lands—estimates range from half to three-fourths—farmers have no improved varieties or technology available that are suited to their areas. To overcome this problem, three steps are suggested:

- Researchers must work closely with farmers to determine their constraints. Researchers must find out why farmers have not adopted the new technology or if they have adopted it, why their yields are so low.
- Research managers must organize research and training programs to encourage interdisciplinary approach in solving farmers' problems in the field.

- Researchers must collaborate their research and development programs with those of cooperating scientists and educators in the rice-growing countries, for theirs is the ultimate task of developing and adopting the technology needed to produce more rice for the developing world.

UNRESOLVED CHALLENGES AND RESEARCH STRATEGIES

Recent progress in the application of science and technology for the optimum use of available soil, water, air, sunlight, and biological resources in agriculture has raised mankind's hope for the future. However, considerable work remains in the development of agricultural balance sheets based on the understanding of the production assets and liabilities of each area and in developing technology to suit specific agroclimatic and management conditions.

Whenever possible, steps for increasing the area under assured irrigation should receive the highest priority to increase production of rice and other food crops (Chandler 1979). This is particularly critical in the tropics and subtropics where rainfall distribution is often skewed. In Asia there is a strong association between the proportion of the area under irrigation and average rice yield, which explains why much of the rice technology now available gives the best results where there is controlled irrigation. However, irrigation is becoming increasingly expensive. To increase rice production further, greater productivity must also be achieved in rainfed areas.

In a world perspective, the urgency for constant expansion of total production remains strong and continues to demand attention. It is true that world food production as a whole, and rice production in particular, has expanded notably since the introduction of modern technology in the mid-1960s. But the need for more food to improve nutritional requirements in large regions, and, more urgently, to ward off severe deficits in selected countries continues to dominate the international food picture.

This situation, compared with the energy crisis and related problems, poses great challenges to all rice researchers and research managers and policy makers in different countries to develop and adopt suitable technology for rice farmers.

Energy Requirements and Mechanization

During the past two decades, agriculture in many developing countries has undergone dramatic change, much of it centered in the small farmer sector. Most of the farms in South and Southeast Asia average less than 2 ha in size and in the past a substantial share of the energy to operate them has come from on-farm sources such as human labor and animal power. Recent breakthroughs in agriculture have stimulated introduction of modern seed and chemical inputs such as fertilizers, insecticides, and herbicides. Recently, the mechanical technologies such as power pumps, tractors, and mechanical equipment have been introduced in some areas.

Recently, Kuether and Duff (1979) studied the energy issues in small farm production systems (Table 14.10). A small amount of Asian rice land is still tilled completely by human power. On the other hand, there are a few large, highly mechanized rice farms that employ the same machinery and cultural techniques that are used in the developed nations.

Studies show that overall energy consumption increases 26% when mechanical land preparation and mechanical threshing and drying replace human and labor power. Fuel is the major contribution to the increase, although a decrease in human and animal energy offsets some of the increase. Use of a four-wheel tractor for primary tillage in the traditional system increases total energy requirements by only 5% and tillage energy by 20%, yet shortens the time required for land tillage from 105 to 43 hours. Many farmers pay for the added energy input to obtain the benefits of timely land preparation and reduced labor requirements. In fact, mechanization of land preparation in some form (mostly hand tractors and some contractual four-wheel tractors) is spreading gradually in the Philippines and in some other countries in South and Southeast Asia. With the availability of the shorter-duration rice varieties, a distinct possibility exists for growing two crops where one grew before, or three crops where two grew before. Therefore, shortening the land preparation period by some degree of mechanization would help maximize the potential of new rice technology.

Reduced tillage or adoption of a minimum tillage practice could also contribute to energy reduction but weed-control problems increase (see Chapter 8), so better herbicides or more efficient cultural methods that do not require a high amount of chemical energy are required before net energy savings can be realized. In tropical Asia, human effort contributes a relatively small portion of the total energy consumed in each production system (Fig. 14.14), never exceeding 13% of the total, although 735 labor hours are required to produce that share of energy (Fig. 14.15). These studies suggest that human energy contributes a small portion of the total energy input into most traditional Asian rice farming systems. As food production systems modernize, they become less energy efficient and more dependent on commercial energy (Kuether and Duff 1979).

Based on input data from the Philippines, energy return per hour of human labor ranged from 12,000 to 15,000 kcal/hour (Table 14.11).

In the United States, where rice production is highly mechanized, only 20–30 labor-hours/ha are used compared with 800 or more labor-hours/ha used in developing countries in Asia (Rutger and Grant 1980). Energy return per hour of human labor in the United States ranges from 400,000 to 800,000 kcal/hour. Table 14.12 shows the energy inputs per hectare of rice in Sacramento Valley in California.

Fossil energy inputs for rice production are much lower in tropical Asia than in the United States because rice is grown in conditions ranging from no mechanization, irrigation, or fertilization to conditions as intensive as those in the United States. However, it is becoming increasingly clear that as higher production per unit of land is required to feed an increasing world population, use of greater cropping intensity, through substitution of mechanical for human and animal power, higher chemical rates, and increased irrigation, will be neces-

Table 14.10 Components of Alternative Rice Production Systems (From Kuether and Duff 1979)

Production Stage	Traditional	Mechanical	Transitional
Land preparation Primary tillage	Water buffalo and plow	5–7 HP tiller and plow	50-HP tractor and rototiller
Secondary tillage	Water buffalo and harrow	5–7 HP tiller and harrow	Water buffalo and harrow
Levee repair	Manual	Manual	Manual
Planting Seedbed	Wetbed system	Wetbed system	Wetbed system
Transplanting	Manual	Manual	Manual
Crop care	Manual	Manual	Manual
Harvesting	Hand sickle	Hand sickle	Hand sickle
Threshing and drying	Manual threshing Sun-drying	IRRI-type power thresher Mechanical batch dryer	Manual threshing Sun-drying

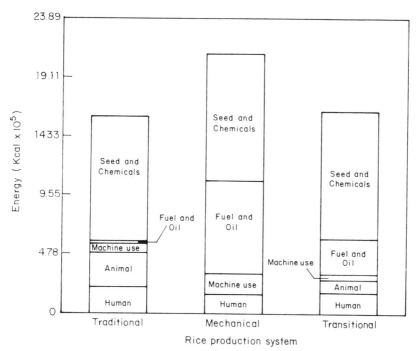

Figure 14.14 Total energy inputs by source in kilocalories per hectare. (Adapted from Kuether and Duff 1979)

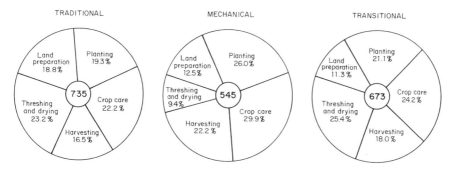

Figure 14.15 Labor distribution (%) by major tasks. The figure at the center of each system is total hours. (From Kuether and Duff 1979)

sary to supplement high-yielding varieties and improved management. These inputs are high energy consumers and will probably reduce the energy efficiency of the rice production system (Kuether and Duff 1979). Some degrees of mechanization and dependence on chemical energy must take place if rainfed rice production has to be increased. Even in a labor surplus area, some degree of mechanization must take place to speed up an operation such as planting.

Table 14.11   Energy Inputs per Hectare of Rice, Wet and Dry Seasons, Philippines, 1972–1973 (Adapted from Rutger and Grant 1980)

Input	Quantity/ha[a]		kcal/kg	
	Wet Season	Dry Season	Wet Season	Dry Season
Labor	814.4 hr	814.4 hr	—	—
Machinery	4.5 kg[b]	4.5 kg[b]	81,000	81,000
Gasoline	131.3 l[c]	131.3 l[c]	1,327,312	1,327,312
Nitrogen	33.0 kg	88.0 kg	485,100	1,293,600
Herbicide	0.7 kg	0.3 kg	69,937	26,073
Insecticide	3.2 kg	0.7 kg	255,664	69,937
Seed	88.0 kg	103.9 kg	352,000	415,600
Irrigation	15.0 cm[d]	30.0 cm[d]	227,090	454,180
Total			2,798,103	3,667,702
Rice yield	3.2 t/ha	4.2 t/ha	9,540,864	12,324,600
kcal output/kcal input			3.41	3.36
Protein yield	186 kg	240 kg		
kcal output/hour of labor			11,715	15,133

[a]For wet season, input data developed from Pecadizo et al. (1973); for dry season, input data developed from Roman et al. (1973).
[b]Assumes 18,000 kcal/kg.
[c]A total of 69.4 hours of small tractor use per hectare.
[d]Well irrigation was used on about 25% of the farms. For wet season: Assume 0.152 m/ha (0.61 m/ha × 25% of area) delivered to the field. Assumes that 0.190 m/ha were lifted 9.1 m (30 ft), with 0.038 m/ha loss (20% of total pumped) in conveyance and application. For dry season: Assume 0.305 m/ha (1.22 m/ha × 25% of area) delivered to the field. Assumes that 0.380 m/ha were lifted 9.1 m (30 ft), with 0.076 m/ha loss (20% of total pumped) in conveyance and application. Energy use calculated by formula adapted from Knutson et al. (1977): kcal/ha = 131,342 × L × A, where L = lift in meters, and A = amount of water pumped in m/ha.

Unresolved Challenges

In developing an agricultural balance sheet for self-sufficiency in food production, various forms of energy, collectively called cultural energy, have been introduced to enable crop plants to give higher and stable yields (Fig. 14.16).

Minimizing Energy Requirement

Sunlight provides the energy for the biochemical processes that reduce carbon dioxide in the air to carbohydrate in the crop. But only a small portion of the

Table 14.12   Energy Inputs per Hectare of Rice, Sacramento Valley, California, 1977 (Adapted from Rutger and Grant 1980)

Input	Quantity/ha[a]		kcal/ha
Labor	23.6	hr	—
Machinery	37.7	kg	742,460
Gasoline	55.2	l	558,017
Diesel	225.4	l	2,572,716
Electricity	29.7	kwh	85,031
Nitrogen (units N)	132.3	kg	1,944,810
Phosphorus (units P_2O_5)	56.0	kg	168,000
Zinc	9.8	kg	49,000
Carbofuran	0.5	kg	42,650
Parathion	0.085	kg	7,387
Molinate	3.4	kg	294,440
MCPA	0.6	kg	59,946
Copper sulfate	11.2	kg	56,000
Seed	180.5	kg	722,000
Irrigation	250.0	cm	2,138,886
Drying	6,969.0	kg	1,393,800
Transportation	451.3	kg	115,984
Total			10,951,127
Yield	6.5	t/ha	19,226,376
kcal output/kcal input			1.76
Protein yield	373.8	kg	
kcal output/hour of labor			814,677

[a]All input quantities except machinery and irrigation are from FEDS (1977). FEDS inputs are 104% of actual to account for approximately 4% reseeding. Energy values per unit are from the Advisory Board, except as noted.

energy in available sunlight is conserved in photosynthesis; most of it escapes as heat.

In addition to sunlight, the production of food, feed, and fiber is supplemented by cultural energy, which is that derived from human and animal labor, fossil fuels burned by tractors, harvesters, and threshers, and the energy used in transportation and in processing of rice. A small portion of the cultural energy is conserved when the crop utilizes plant nutrients from fertilizers, but most of it is ultimately dissipated into the environment as heat and not transformed into harvestable energy (Heichel 1973). The relative contributions of different forms of cultural energy in agricultural production have varied over time and

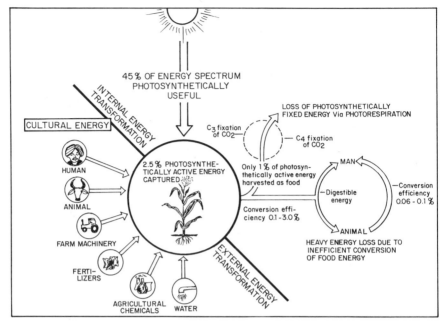

Figure 14.16 Solar and cultural energy input and output cycle in plants. (From Swaminathan 1979)

geographic regions. Modern agriculture derives practically all of its cultural energy from fossil fuels or other energy sources that replace labor. On the other hand, much of the energy in the traditional agricultural system is derived from on-farm sources. Human labor and animal power are notable examples. But as agriculture modernizes, more resources are purchased outside the traditional sector.

The study by Kuether and Duff (1979) suggests that in most traditional, small-farm, food production systems, human energy represents a relatively small portion of total input energy. For example, rice production in the Philippines is highly labor-intensive but is considered the most energy-efficient agricultural system.

As food production systems modernize, energy consumption increases and the system becomes less energy efficient and increasingly more dependent on commercial forms of energy. Therefore, development of methods and machines to more efficiently use high-energy inputs offers potential for reducing the total input energy of the food production systems. For example, a framework should be considered to develop, evaluate, and adopt improved nitrogen sources and management practices for rice. Unless high losses and low efficiency of nitrogen fertilizers are corrected, most of the potential benefits from increased use of fertilizer nitrogen may not be realized. It is critical to develop inexpensive techniques for making large urea granules for point placement in the reduced layer in lowland soils. Such techniques should be developed for factory- and village-level production. In addition, simple machines are needed to apply the urea particles beneath the surface of flooded soils and increase fertilizer efficiency from the current level of 30–35% to 60–65%.

It is imperative to develop integrated systems of nitrogen management, which utilize compost and crop residues, biological nitrogen fixation, and improved fertilizer practices for rice. Finally, varieties and cultural practices should be sought that will be less dependent on chemical energy in the forms of pesticides and fertilizers and still able to produce high and stable yields.

Developing Rainfed Rice Technology

Much of today's modern rice technology performs well where there is controlled irrigation. To increase rice production, greater productivity must be achieved in rainfed areas. Constraints on high yields in rainfed farms are innumerable.

Description of Environment

A better description is needed of the various environmental regimes. It is important to know the condition in which farmers grow rice and then test the technology under those environments. Planning of research and international testing must include newer strategies in developing new rice technology appropriate in diverse environmental conditions. In other words, the goals of developing new varieties should be in terms of environments instead of in terms of plant characteristics. A better description is needed of the various environments and also of the institutional structures of research as they relate to those environments.

Tailoring New Varieties

Varieties need to be tailored not only for the physical environment but also for cropping systems and financial resources of farmers. Increased cropping intensity has been a major source of increased rice production since the introduction of the early-maturing modern rices. These early-maturing rices may offer greater profits by increasing cropping intensity and productivity per hectare per day, than by attempting to increase yield per hectare of a single crop of rice. However, technology for increased cropping intensity should cater to financial resources of small farmers.

Options for Increasing Rice Supplies

The most important options for increasing world rice supplies are through:

- An expansion in the area grown to rice.
- An increase in the productivity of existing rice lands.

Table 14.13 Total Land Area and Arable Land by Continents in Millions of Hectares (Adapted from Burnigh 1977)

Continent	Total Land Area	Cultivated Land	% of Land Area Cultivated	Potential Arable Land	Cultivated to Potential Arable Land (%)
Africa	3,010	158	5.2	734	22
Asia	2,740	519	18.9	627	83
Australia and New Zealand	820	32	3.9	153	21
Europe	480	154	32.1	174	88
North America	2,110	239	11.3	465	51
South America	1,750	77	4.4	681	11
USSR	2,240	227	10.6	356	64
Total	13,150	1,406	10.6	3,190	44

East Asian countries, where the land constraint was encountered in the early part of the century, have long been exploiting the potential for better farm productivity through irrigation development and the application of science-based technologies. However, by the early 1960s there was little land available in the rest of Asia where intensive rice farming could be developed at reasonable cost. Needed production increases could be maintained only through the diffusion of yield-increasing technology and an increase in cropping intensity. Compared to the rest of the world, Asia has little new land to bring into cultivation (Table 14.13). Some expansion in land is possible in the outer islands of Indonesia, some parts of Malaysia, Burma, Thailand, and the Philippines. However, the new cultivated lands in Asia are generally remote, and often are not farmed because of their low productivity. Thus, to develop these lands in general—and for rice in particular—requires considerable investment in research, capital, infrastructure, and human energy.

In essence, the possibilities for increasing the area under rice in Asia are limited. Thus, a substantial part of the required increase in production must come from higher yields per crop on existing rice lands and from increasing cropping intensity on both irrigated and nonirrigated rice lands.

Even then, if supplies of rice are to keep pace with projected demand there must be an increase in investments for rice research and production. Technologies must be designed and pricing policies set to make it attractive for the farmer to adopt, on a continuing basis, the production-increasing technology (Flinn and Gascon 1980).

Irrigation Development

Analyses by the Asian Development Bank (1977), FAO (1978), and the Trilateral Commission (Colombo et al. 1977) have indicated that irrigation is a critical component of the infrastructure for agricultural development in rice-growing areas of Asia. Irrigation enables higher yields per hectare (Table 14.14) and

Table 14.14   Example of Rice Yields in the Monsoonal Season for Irrigated and Upland Rice (Adapted from Palacpac 1979)

Country	Rice Yield (t/ha)	
	Irrigated	Upland
Bangladesh	1.9	1.2
Indonesia	3.0	1.3
Japan	5.9	2.2
Republic of Korea	6.8	2.6
Philippines	2.9	1.1
Taiwan	4.5	2.3

stabilizes yields for a given technology. Water management also facilitates the production of modern varieties of rice, and makes it profitable for farmers to use a higher level of associated inputs such as fertilizers and increase number of rice cropping. Thus, it is widely considered that good water management is the single most important factor necessary for sustained increase in rice production in Asia. The opportunities for increasing the productivity of rice in an irrigated environment are predicted to be higher than in alternative rice-growing environments (Hsieh et al. 1980).

There appears to be a close relationship between irrigation rate and mean national average yield of a given country (Fig. 14.17):

$$\text{irrigation rate} = \frac{\text{total irrigated rice area (harvested)}}{\text{total rice area (harvested)}}$$

By using this concept the countries in Asia can be divided into three groups. The first group contains only Japan, with an irrigation rate of 98% and rice yields of

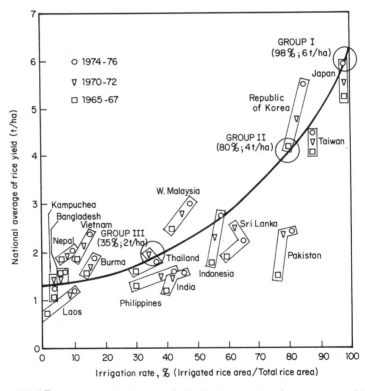

Figure 14.17 Relationship between irrigation rate and national average yield in rice-growing countries in Asia. (Adapted from Colombo et al. 1977) Note: 1979 estimate for rice yields for Republic of Korea is 6.2 t/ha, and for Taiwan, 4.5 t/ha.

about 6 t/ha. The second group includes the Republic of Korea and Taiwan with irrigation rates of about 80% and average yields of about 4 t/ha. The third group includes the remaining 13 countries with a mean irrigation rate of 35% and yields of about 2 t/ha in 1974–1976 (Colombo et al. 1977). However, current statistics show that the Republic of Korea has increased its irrigation rate further (86% in 1978) and, with increased rate of inputs—particularly chemical fertilizers, the national average for 1979 was 6.2 t/ha (see Chapter 6).

On the other hand, the People's Republic of China, with almost the entire rice-growing area grown to irrigated rice, had a 3.5 t/ha national average in 1976–1978. This indicates that complete water control with irrigation and drainage is a major factor but not the only factor in increasing national average yield of rice in a given country. Nevertheless, the overriding benefits of improvement of existing irrigation facilities are not only increased yield per hectare of a given crop but also increased cropping intensity, and hence, productivity and stability of rice and other crop production in the rural areas. However, because of a high irrigation cost of $3000 or more per hectare the increase in irrigated area will be slow in many countries in South and Southeast Asia.

Research Priorities

Research priorities for rice should focus on potential benefits and costs of alternative methods of increasing rice production by minimizing or eliminating constraints (Barker et al. 1975). In 1977, an attempt was made to develop plans for appropriate research based on potential benefits. Estimates were made on the probable sources of increased rice production in major rice-growing systems in Asia (Table 14.15). A team of scientists with diverse disciplines serving on IRRI's Long Range Planning Committee concluded that about 55% of the future food production in Asia would come from increased yields and 45% from increased cropping intensity (IRRI 1978a).

Irrigated areas in South and Southeast Asia will likely provide about 46% of the increased production and rainfed lowland and deepwater areas about 34%. While the anticipated increased production from upland and deepwater areas is much less, the fact that these production areas have largely been ignored in the past justifies increased research attention. A large, and perhaps the poorest, sector of population depends on their basic food need from the upland, deepwater, and floating rice-growing areas.

Global interest in upland rice, which is the major system of rice culture in West Africa and Latin America, dictates continuing, if not increased, attention to research on that crop.

Thus, in reevaluating research priorities, relative emphasis and consideration must be given not only to potential production increases but also to the welfare need of the farmers, the landless laborers, and the rice consumers.

Boyce and Evenson (1975) compiled international data on investment in research and extension directed toward the improvement of agricultural crops.

Table 14.15   Rice Production Benefits Expected from Research under Different Environmental Complexes as Estimated by an IRRI Long-Range Planning Committee during 1977 (Adapted from IRRI 1978a)

Environmental Complex	Rice Production Benefits Expected (%)		
	Yield Increase	Crop Intensification	Total
Irrigated	22	24	46
Shallow rainfed lowland	15	11	26
Medium-deep rainfed lowland and deepwater	6	2	8
Floating	2	1	3
Upland	3	1	4
Arid, high temperature	6	5	11
Long day, low temperature	1	<1	1
Total	55	45	100

In their judgment the expectations created by the rice varieties developed by IRRI and other national programs have been only partially met. Using a statistical model, they estimated the production increase associated with high-yielding varieties. Their results suggest that the first generation investment on research yielded extraordinarily high income streams, which could not be maintained in the second generation investment on research. The authors finally concluded that investment policy on research has been far from optimal, particularly in the low-income countries.

According to Wortman and Cummings (1978), payoff from investment in agricultural research can be very high if research is organized to achieve ambitious development goals. Research should be an important catalyst of rapid agricultural development. Past results clearly demonstrate that the opportunities for technology payoff from research still remain high in irrigated rice-growing areas. Furthermore, considerable research is needed to maintain the high yield levels already achieved under irrigation. Nevertheless, rainfed lowland, deepwater, and upland areas must receive higher priorities than have been given them in the past.

Recently, Herdt and Barker (1977) suggested a system of technology development processes that should be considered by scientists and institutions involved in rice research. Figure 14.18 shows four types of research activity between basic biological research and farm production. Type I research, which deals with development of new technologies that might raise production potentials, can take place in a narrow range of environments such as in experiment station fields, laboratories, greenhouses, or phytotrons. Because rice production takes place under a wide range of environmental conditions represented by the vertical axis of Fig. 14.18, the intermediate steps of

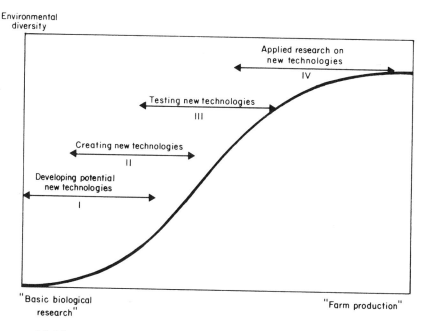

Figure 14.18   The relationship between components of an agricultural research system and the degree of environmental diversity in which each component should be operated. (From Herdt and Barker 1977)

developing, testing, and applying new technologies should take place across a wide range of environments. Naturally, there would be some overlapping of these activities because there is a continuum from basic biological research to farm production.

To increase rice production 3.5% per year, which is assumed as the need to equal population growth, considerable efforts and commitments are needed to examine old strategies of rice research and to develop new ones. However, research and technology development alone may not be adequate to increase rice production in the coming decade. Government policy to provide farmers with incentive for adopting yield-increasing technology and with infrastructure, including increased irrigation facilities, are seemingly the most crucial factors in the use of technology by small farmers in Asia.

REFERENCES

Asian Development Bank. 1977. *Asian agricultural survey, 1976.* Manila, Philippines. 490 pp.

Aw, D. 1978. Rice development strategies in Africa. Pages 69–74 *in* Academic Press. *Rice in Africa.* London.

Barker, R. 1978. Yield and fertilizer input. Pages 35–66 *in* International Rice Research Institute. *Interpretive analysis of selected papers in changes in rice farming in selected areas of Asia.* Los Baños, Philippines.

Barker, R. 1979. Adoption and production impact of new rice technology–the yield constraints problem. Pages 1–26 *in* International Rice Research Institute. *Farm-level constraints to high rice yields in Asia: 1974–1977.* Los Baños, Philippines.

Barker, R., and T. K. Pal. 1979. *Barriers to increased rice production in Eastern India.* IRRI Res. Pap. Ser. 25. 23 pp.

Barker, R., H. E. Kauffman, and R. W. Herdt. 1975. Production constraints and priorities for research. Paper presented at the International Rice Research Conference, 21–25 April 1975, International Rice Research Institute, Los Baños, Philippines. Paper 75-8.

Boyce, J. K., and R. E. Evenson. 1975. *Agricultural research and extension programs.* Agricultural Development Council, Inc., New York. 229 pp.

Burnigh, P. 1977. Food production potential of the world. *World Dev.* 5:477–485.

Chandler, R. F., Jr. 1979. *Rice in the tropics: a guide to the development of national programs.* Westview Press, Boulder, Colorado. 256 pp.

Colombo, U., D. G. Johnson, and T. Shishido. 1977. *Reducing malnutrition in developing countries: increasing rice production in South and Southeast Asia.* The Trilateral Commission. The Triangle Pap. Ser. 16. 55 pp.

Dalrymple, D. G. 1978. *Development and spread of high-yielding varieties of wheat and rice in the less developed nations.* USDA Foreign Agric. Econ. Rep. 95. 134 pp.

De Datta, S. K., and R. W. Herdt. 1979. Relating development, testing and spread of modern rice technology to constraints to high yields in tropical Asia. Paper presented at the workshop on Pre-release Testing of Agricultural Technology, 19–21 March 1979, Centro Internacional de Agricultura Tropical (CIAT), Colombia, South America. 62 pp. (unpubl. mimeo.)

De Datta, S. K., K. A. Gomez, R. W. Herdt, and R. Barker. 1978. *A handbook on the methodology for an integrated experiment-survey in rice yield constraints.* International Rice Research Institute, Los Baños, Philippines. 60 pp.

De Datta, S. K., W. N. Obcemea, W. P. Abilay, M. T. Villa, B. S. Cia, and A. K. Chatterjee. 1976. Identifying farm yield constraints in tropical rice using a management package concept. Paper presented at the 7th annual meeting of the Crop Science Society of the Philippines, 10–12 May 1976, Davao City. (unpubl. mimeo.)

De Datta, S. K., F. V. Garcia, A. K. Chatterjee, W. P. Abilay, Jr., J. M. Alcantara, B. S. Cia, and H. C. Jereza. 1979. *Biological constraints to farmers' rice yields in three Philippine provinces.* IRRI Res. Pap. Ser. 30. 69 pp.

FAO (Food and Agriculture Organization). 1978. Information notes on water for agriculture. Issued at FAO Regional Office for Asia and the Far East. No. 11. July 1978. 10 pp.

FAO (Food and Agriculture Organization). 1980. Increasing rice production in Asia and the Pacific, including rainfed rice. Fifteenth FAO Regional Conference for Asia and the Pacific, New Delhi, India, 5–13 March.

FEDS. 1977. (Farm Enterprise Data System for 1977). CED, ESCS, U.S. Department of Agriculture, Washington D.C.

Flinn, J. C., and F. Gascon. 1980. Projecting the demand and options for increasing rice supplies in Asia. Notes for the Rice Production and Training class, International Rice Research Institute, Los Baños, Philippines. 43 pp. (unpubl. mimeo.)

Gomez, K. A. 1977. On-farm assessment of yield constraints: methodological problems. Pages 1–16 in International Rice Research Institute. *Constraints to high yields on Asian rice farms: an interim report*. Los Baños, Philippines.

Gomez, K. A., D. Torres, and E. Go. 1973. Quantification of factors limiting rice yield in farmer's field. Paper presented at a Saturday seminar, 24 November 1973, International Rice Research Institute, Los Baños, Philippines..

Gomez, K. A., R. W. Herdt, R. Barker, and S. K. De Datta. 1979. A methodology for identifying constraints to high rice yields on farmers' fields. Pages 27–47 in International Rice Research Institute. *Farm-level constraints to high rice yields in Asia: 1974–1977*. Los Baños, Philippines.

Hargrove, T. R., and V. L. Cabanilla. 1979. The impact of semidwarf varieties on Asian rice breeding programs. *BioScience* 29:731–735.

Heichel, G. H. 1973. *Comparative efficiency of energy use in crop production*. Connecticut Agric. Exp. Stn. Bull. 739. 26 pp. illus.

Herdt, R. W. 1979. An overview of the constraints project results. Pages 395–411 in International Rice Research Institute. *Farm-level constraints to high rice yields in Asia: 1974–1977*. Los Baños, Philippines.

Herdt, R. W., and R. Barker. 1977. *Multi-site tests environments and breeding strategies for new rice technology*. IRRI Res. Pap. Ser. 7. 32 pp.

Herdt, R. W., and A. M. Mandac. 1979. Overview, findings and implications of constraints research: 1975–1978. Paper presented at the workshop on rice yield constraints, 30 April–3 May 1979, Kandy, Sri Lanka. Agricultural Economics Department, International Rice Research Institute. Pap. 79-04. (unpubl. mimeo.)

Herdt, R. W., and A. M. Mandac. 1980. *Modern technology and economic efficiency of Philippine rice farmers. Economic Development and Cultural Change*. International Rice Research Institute, Los Baños, Philippines.

Herdt, R. W., and T. H. Wickham. 1975. Exploring the gap between potential and actual rice yield in the Philippines. *Food Res. Inst. Stud.* 14:163–181.

Hsieh, S. C., J. C. Flinn, and N. Amerasinghe. 1980. The role of rice in meeting future needs. *In* International Rice Research Institute. *Rice research strategies for the future*. Los Baños, Philippines. (in press)

IFPRI (International Food Policy Research Institute). 1977. *Food needs of developing countries: projections of production and consumption to 1990*. Res. Rep. 3. Washington, D.C.

IRRI (International Rice Research Institute). 1977. *Constraints to high yields on Asian rice farms: an interim report*. Los Baños, Philippines. 235 pp.

IRRI (International Rice Research Institute). 1978. *Research highlights for 1977*. Los Baños, Philippines. 122 pp.

IRRI (International Rice Research Institute). 1978b. *Interpretive analysis of selected papers from changes in rice farming in selected areas of Asia*. Los Baños, Philippines. 166 pp.

IRRI (International Rice Research Institute). 1979. *Farm-level constraints to high yields in Asia: 1974–1977*. Los Baños, Philippines. 411 pp.

IRRI (International Rice Research Institute). 1980. *Research highlights for 1979.* Los Baños, Philippines. 133 pp.

Kirpick, P. Z. 1977. Development of lowland tropical floodplains in Latin America. Paper presented at the United Nations World Water Conference, March 1972, Buenos Aires. 32 pp.

Knutson, G. D., R. G. Curly, E. B. Roberts, R. M. Hagan, and V. Cervinka. 1977. *Pumping energy requirements for irrigation in California.* Division of Agricultural Sciences, University of California, Spec. Publ. 3215.

Kuether, D. O., and B. Duff. 1979. Energy requirements for alternative rice production systems in the tropics. Paper presented at the annual meeting of the Society of Automotive Engineers, September 1979, Milwaukee, Wisconsin. Pap. AE 79-04.

Mandac, A. M., and R. W. Herdt. 1979. *Environmental and management constraints to high rice yields in Nueva Ecija, Philippines.* IRRI Agric. Econ. Pap. 79-03. 25 pp.

Palacpac, A. C. 1979. World rice statistics. International Rice Research Institute, Los Baños, Philippines. 131 pp. (unpubl. mimeo.)

Pecadizo, L. M., E. R. Roman, N. M. Fortuna, and E. P. Abarientos. 1973. *Cost of producing palay in Laguna.* University of the Philippines at Los Baños. Ser. 3. 31 pp.

Roman, E. R., N. M. Fortuna, and E. P. Abarientos. 1973. *Cost of producing palay in Laguna.* University of the Philippines at Los Baños. Ser. 4. 34 pp.

Rutger, J. N., and W. R. Grant. 1980. Energy use in rice. *In* D. Pimentel, ed. CRC Press. *Handbook of energy utilization in agriculture.* Boca Raton, Florida.

Swaminathan, M. S. 1979. Global aspects of food production. World Climate Conference, 12–23 February 1979, Geneva. World Meteorological Organization. Overview Pap. 14. 37 pp.

Wortman, S., and R. W. Cummings, Jr. 1978. *To feed this world, the challenge and strategy.* The Johns Hopkins Press, Baltimore, Maryland. 440 pp.

Author Index

Subject Index